Developments in Antenna Analysis and Design

Related titles on electromagnetic waves:

Dielectric Resonators, 2nd Edition Kajfez and Guillon
Electronic Applications of the Smith Chart Smith
Fiber Optic Technology Jha
Filtering in the Time and Frequency Domains Blinchikoff and Zverev
HF Filter Design and Computer Simulation Rhea
HF Radio Systems and Circuits Sabin
Microwave Field-Effect Transistors: Theory, design and application, 3rd Edition Pengelly
Microwave Semiconductor Engineering White
Microwave Transmission Line Impedance Data Gunston
Optical Fibers and RF: A natural combination Romeiser
Oscillator Design and Computer Simulation Rhea
Radio-Electronic Transmission Fundamentals, 2nd Edition Griffith, Jr
RF and Microwave Modeling and Measurement Techniques for Field Effect Transistors Jianjun Gao
RF Power Amplifiers Albulet
Small Signal Microwave Amplifier Design Grosch
Small Signal Microwave Amplifier Design: Solutions Grosch
2008+ Solved Problems in Electromagnetics Nasar
Antennas: Fundamentals, design, measurement, 3rd Edition Blake and Long
Designing Electronic Systems for EMC Duff
Electromagnetic Measurements in the Near Field, 2nd Edition Bienkowski and Trzaska
Fundamentals of Electromagnetics with MATLAB®, 2nd Edition Lonngren *et al.*
Fundamentals of Wave Phenomena, 2nd Edition Hirose and Lonngren
Integral Equation Methods for Electromagnetics Volakis and Sertel
Introduction to Adaptive Arrays, 2nd Edition Monzingo *et al.*
Microstrip and Printed Antenna Design, 2nd Edition Bancroft
Numerical Methods for Engineering: An introduction using MATLAB® and computational electromagnetics Warnick
Return of the Ether Deutsch
The Finite Difference Time Domain Method for Electromagnetics: With MATLAB® simulations Elsherbeni and Demir
Theory of Edge Diffraction in Electromagnetics Ufimtsev
Scattering of Wedges and Cones with Impedance Boundary Conditions Lyalinov and Zhu
Circuit Modeling for Electromagnetic Compatibility Darney
The Wiener–Hopf Method in Electromagnetics Daniele and Zich
Microwave and RF Design: A systems approach, 2nd Edition Steer
Spectrum and Network Measurements, 2nd Edition Witte
EMI Troubleshooting Cookbook for Product Designers Andre and Wyatt
Transmission Line Transformers Raymond Mack and Jerry Sevick
Electromagnetic Field Standards and Exposure Systems Grudzinski and Trzaska
Practical Communication Theory, 2nd Edition Adamy
Complex Space Source Theory of Spatially Localized Electromagnetic Waves Seshadri
Electromagnetic Compatibility Pocket Guide: Key EMC facts, equations and data Wyatt and Jost
Antenna Analysis and Design Using FEKO Electromagnetic Simulation Software Elsherbeni, Nayeri and Reddy
Scattering of Electromagnetic Waves by Obstacles Kristensson
Adjoint Sensitivity Analysis of High Frequency Structures with MATLAB® Bakr, Elsherbeni and Demir

Developments in Antenna Analysis and Design

Volume 1

Edited by
Raj Mittra

The Institution of Engineering and Technology

Published by SciTech Publishing, an imprint of The Institution of Engineering and Technology, London, United Kingdom

The Institution of Engineering and Technology is registered as a Charity in England & Wales (no. 211014) and Scotland (no. SC038698).

First published 2018

The Institution of Engineering and Technology
Michael Faraday House
Six Hills Way, Stevenage
Herts, SG1 2AY, United Kingdom

www.theiet.org

British Library Cataloguing in Publication Data
A catalogue record for this product is available from the British Library

ISBN 978-1-78561-888-8 (Hardback Volume 1)
ISBN 978-1-78561-889-5 (PDF Volume 1)
ISBN 978-1-78561-890-1 (Hardback Volume 2)
ISBN 978-1-78561-891-8 (PDF Volume 2)
ISBN 978-1-78561-892-5 (Hardback Volumes 1 and 2)

Typeset in India by MPS Limited

Contents

Preface **xi**

1 Applications of the characteristic mode theory to antenna design **1**
Ting-Yen Shih and Nader Behdad

1.1 Introduction 1
1.1.1 Background 1
1.1.2 Characteristic mode theory 1
1.2 Antenna design examples using the characteristic mode theory 6
1.2.1 Circularly polarized antennas 7
1.2.2 Wideband antennas 11
1.2.3 Chassis-based MIMO antennas 12
1.2.4 Bandwidth enhancement of platform-based antennas 15
1.3 Summary 31
References 32

2 Design of antennas mounted on complex platforms using the characteristic mode (CM) and characteristic basis (CB) function methods **35**
Raj Mittra, Ashwani Kumar, and Chao Li

2.1 Introduction 35
2.2 TCM approach to designing antennas for mobile phone platforms 37
2.3 Characteristic basis method for locating antennas on mobile phone platforms 42
2.4 Placement of multiple antennas on a complex platform 45
2.4.1 TCM-based approach 45
2.4.2 CB-based approach 50
2.5 Illustrative examples 52
2.5.1 Four microstrip patch antennas on an FR4 substrate 52
2.5.2 Topside of a ship excited by monopoles 55
2.5.3 Four PIFA antennas on FR4 substrate 57
2.5.4 Chassis excited by six dipoles 58
2.6 Conclusion 62
Acknowledgment 63
Appendix 64
Appendix A1 Characteristic modes and bases 64
A1.1 Generation of characteristic modes (CMs) 64
A1.2 Generation of CBs 64

Appendix A2 73
A2.1 TCM analysis of mobile phone antenna and antenna-plus-platform 73
References 74

3 Wideband L-probe patch antenna **77**
Hau Wah Lai and Kwai Man Luk

3.1 Introduction 77
3.2 Basic characteristics 78
3.2.1 L-probe feeding mechanism 78
3.2.2 M-probe feeding mechanism 83
3.3 Parametric studies 91
3.3.1 Performance with different P_h 91
3.3.2 Performance with different aspect ratio 93
3.4 Development of L-probe and M-probe fed patch antenna 97
3.4.1 Circular polarization 97
3.4.2 Dual polarization 98
3.4.3 Dual band 99
3.4.4 Conformal ground plane 99
3.4.5 Printed circuit board 100
3.4.6 Fusion of the L-probe and M-probe in antenna design 100
3.5 Conclusion 103
References 103

4 Advancements in MIMO antenna systems **109**
Mohammad S. Sharawi

4.1 Introduction 109
4.2 MIMO antenna system performance metrics 110
4.3 Major MIMO antenna system design challenges 115
4.4 MIMO antenna system examples 116
4.4.1 Mobile phones and handheld devices 116
4.4.2 Cognitive radio front-ends 119
4.4.3 USB dongle MIMO implementations 120
4.4.4 Wireless access point MIMO implementations 121
4.5 MIMO antenna solutions for 5G-enabled systems 122
4.5.1 Mobile terminal 5G solutions 122
4.5.2 Base station 5G solutions 122
4.6 Conclusions 123
References 124

5 Reconfigurable leaky-wave antennas **129**
Yingjie Jay Guo, Debabrata K. Karmokar, and Trevor S. Bird

5.1 Introduction 129
5.2 History of LWAs 130
5.2.1 Basic operating principle 130

5.2.2 Classification of LWAs 131
5.3 Passive frequency-scanning LWA structures 132
5.3.1 One-dimensional (1-D) Fabry–Pérot LWA 132
5.3.2 Composite right/left-handed transmission line and LWA 135
5.3.3 Half-width microstrip LWA 137
5.4 Reconfigurable LWAs 138
5.4.1 1-D FP-reconfigurable LWAs 138
5.4.2 Two-dimensional (2-D) FP-reconfigurable LWA 146
5.5 Experimental results 149
5.5.1 CRLH-based reconfigurable LWA 152
5.5.2 Reconfigurable half-width microstrip LWA 156
5.6 Conclusion 167
References 168

6 Reconfigurable high-gain antennas for wireless communications 171
Yingjie Jay Guo, Pei-Yuan Qin, and Raj Mittra

6.1 Introduction 171
6.2 Reconfigurable array antennas 172
6.3 Reconfigurable PRS antennas 182
6.3.1 Frequency-reconfigurable PRS antenna 182
6.3.2 Pattern-reconfigurable PRS antenna 183
6.3.3 Polarization-reconfigurable PRS antenna 189
6.4 Conclusions 197
References 197

7 Microfluidically reconfigurable antennas 203
Gokhan Mumcu

7.1 Introduction 203
7.2 Fabrication and actuation techniques 205
7.3 Flexible and stretchable liquid metal antennas 210
7.4 Frequency-reconfigurable liquid metal antennas 213
7.5 Reconfigurable antennas using dielectric liquids 220
7.6 Beam-steerable liquid metal antennas 224
7.7 Reconfigurable antennas using microfluidically repositionable metallized plates 227
7.8 Concluding remarks 236
References 237

8 Flexible and wearable antennas 243
Muhammad M. Tahseen and Ahmed A. Kishk

8.1 Introduction 243
8.2 Wearable antennas for biomedical applications 246
8.3 AMC-based flexible wearable antennas 247
8.4 Inkjet-printed wearable antennas 249

8.5 Textile-based wearable antennas 250
8.5.1 Single- and multi-layer multi-Bowtie conformal antennas 250
8.5.2 Dielectric resonator antennas for wearable application 253
8.5.3 Wearable artistic antennas for WLAN-band 255
References 271

9 Wearable technology and mobile platform for wearable antennas for human health monitoring 279
Vijay K. Varadan, Pratyush Rai, Se Chang Oh, Prashanth Shyam Kumar, Mouli Ramasamy, and Robert E. Harbaugh

9.1 Introduction 280
9.2 Smart textile for health monitoring 282
9.3 Electrical signals from the brain and heart 286
9.4 Cardiovascular anatomy and electrophysiology 291
9.4.1 The dipole theory for ECG 294
9.4.2 Derivation of ECG from dipole vector 296
9.5 Monitoring and diagnosis: neurological signal measurements 298
9.6 Monitoring and diagnosis: cardiological signal measurements of diagnostic value 302
9.7 Monitoring systems 307
9.8 Neurological disorder monitoring by wearable wireless nano-bio-textile sensors 311
9.9 Cardiovascular health monitoring 327
9.9.1 Hardware system 328
9.9.2 ECG signal acquisition 329
9.10 Biofeedback system for therapeutics 337
9.11 Conclusion 340
References 341

10 Meta-atoms and artificially engineered materials for antenna applications 351
Ravi Kumar Arya, Shiyu Zhang, Shaileshachandra Pandey, Ashwani Kumar, Yiannis Vardaxoglou, William Whittow, and Raj Mittra

10.1 Introduction 351
10.2 Lens designs using MTMs 353
10.3 Lens design using RO 355
10.4 3D-Printing technique 356
10.5 Design of artificially engineered materials 357
10.5.1 Designing higher-permittivity materials from low-permittivity COTS material: method-1 358
10.5.2 Designing higher-permittivity materials from low-permittivity COTS material: method-2 359

10.5.3 Designing lower-permittivity materials from high-permittivity COTS material 360
10.5.4 Designing lower-permittivity materials from high-permittivity 3D-printing material 360
10.6 Different lens designs 362
10.6.1 PLA Lens design 362
10.6.2 DaD lens design 371
10.6.3 ABS lens design 377
10.6.4 Comparison of DaD and ABS lenses 381
10.7 Summary 384
10.8 Metal-only reflectarray antenna designs using metasurfaces 386
10.9 Performance enhancement of antenna and array antennas using metasurface superstrates 394
10.9.1 Example-1 394
10.9.2 Example-2 398
10.9.3 Summary 402
References 402

11 Microwave antennas based on metamaterials and metasurfaces 407
Wen Xuan Tang and Tie Jun Cui

11.1 GRIN MTM lens antennas 408
11.1.1 MTM flat lens antenna 408
11.1.2 MTM curved lens antennas 419
11.2 MTM antennas using transformation optics 425
11.2.1 MTM flattened reflectors 427
11.2.2 MTM flattened convex and hyperbolic lenses 431
11.2.3 MTM Luneburg lens with flattened focal surface 434
11.3 Metasurface antennas 436
11.3.1 Holographic metasurfaces for beam scanning 438
11.3.2 Spoof SPP radiations 438
11.3.3 Coding metasurfaces 439
References 440

12 Metamaterial-based zero-phase-shift-line loop antennas 445
Zhi Ning Chen, Xianming Qing, Jin Shi, and Yunjia Zeng

12.1 Introduction 445
12.2 State-of-the-art ZPSL loop antennas 446
12.3 Modeling of zero-phase-shift-line structure 447
12.3.1 Dispersion analysis of zero-phase-shift-line structure 447
12.3.2 Design guidelines 452
12.4 Design and applications 456
12.4.1 Electrically large zero-phase-shift-line loop antennas for UHF near-field RFID readers 456

12.4.2 Horizontally polarized omnidirectional antenna for WLAN access points 470
12.4.3 CP omnidirectional antenna for UHF far-field RFID readers 472
12.5 Summary 477
References 478

Index **483**

Preface

Antenna design is a mature field, with a long history dating back to Hertz, Marconi, and Bose, among others. Of course, a wide variety of antennas have been developed over the years since the simple wire antenna configurations were introduced by the pioneers in the field, some examples being microstrip patch antennas (MPAs), reflectors, phased arrays, etc., to name just a few. Not unexpectedly, antenna design methodologies have also evolved over the years to meet the design specifications, which appear to become more and more challenging with time, since the devices utilizing these antennas have progressively increased the number of functionalities they offer, while still placing significant constraints on the real estate available for the platforms upon which these antennas are placed. Additionally, new materials have been developed and these developments continue unabated. The developments have even picked up the pace noticeably with the advent of metamaterials (MTMs) and recently discovered applications of graphene-based antennas. In light of this background, if one wonders whether there is anything new that is left to do for the antenna design engineers to keep themselves gainfully occupied, the short answer is: “Definitely, yes.” In support of this assertion, we point out that the number of research publications on antennas have been rising progressively in recent years though their focus has shifted to emerging areas.

Let us turn now to the problem of antenna analysis, which plays a pivotal role in modern approaches to designing antennas. Before the advent of the computer, the antenna designers relied solely on theoretical approaches for analysis, which were limited in their application to simple configurations that were tractable by using analytical or quasi-analytical techniques. However, all that changed once the computers became sufficiently powerful, and sophisticated computational electromagnetic (CEM) techniques were developed. These techniques included the method of moments (MoM), the finite element method (FEM), and the finite difference time domain (FDTD) method, and commercial codes based on these methods became widely available and affordable to the antenna designers. Hence, one might again ask the question: “Is there any point in looking for new or more advanced techniques for antenna analysis, given the fact that the commercial codes are already well capable of handling complex antenna configurations comprised of both perfectly conducting, or PEC, as well as dielectric and/or magnetic materials?” Once again the short answer is “yes.” One of the reasons is that the time and memory requirements of the commercial codes can be exorbitant for complex antennas mounted on large platforms, and the designers are always looking for EM simulation codes, whose solve times are reasonable so that they can help improve the time for the design cycle, and thereby reduce the “time to market,” which the manufactures consider to be a very important factor when strategizing the introduction of new antennas in their next-generation products. This is also true when the antennas are integrated with packages and active devices.

Given this backdrop, the concept of this book, which is titled, *Developments in Antenna Analysis and Design*, was developed to present recent developments in antenna design and modeling techniques for a wide variety of applications. The application areas were chosen because they are contemporary in nature, have been receiving considerable attention in recent years, and also because they are crucial for future developments. The book includes topics such as body-worn antennas that play an important role as sensors for Internet of Things (IoT), and millimeter wave antennas that are vitally important for 5G devices. It also covers a wide frequency range that includes terahertz and optical frequencies. Additionally, it discusses topics such as theoretical bounds of antennas and aspects of statistical analysis that are not readily found in the existing literature. To ensure that the book covers a wide scope, which includes as many aspects of modern antenna design and analysis as possible, it is comprised of 23 chapters, written by authors who are leading experts in their fields. Since we made an editorial judgment to be liberal with the page budget so as not to compromise the quality, readability, and usefulness of the presentations, the manuscript grew to become considerably larger than it was originally planned. Hence, a second round of editorial judgment was made, and the decision was to split the book into two parts. The first volume includes chapters covering the topics of Theory of Characteristic Modes (TCM) and Characteristic bases; Wideband antenna element designs; MIMO antennas; Antennas for Wireless Communication; Reconfigurable antennas employing microfluidics; Flexible and Body-worn antennas; and Antennas using Meta-atoms and Artificially Engineered Materials, or Metamaterials (MTMs).

The second volume of the book covers the topics of: Graphene-based antennas; Millimeter-wave Antennas; Terahertz Antennas; Optical Antennas; Fundamental Bounds of Antennas; Fast and Numerically Efficient techniques for analyzing antennas; Statistical analysis of antennas; Ultra-wideband arrays; Reflectarrays; and antennas for small satellites, viz., CubeSats.

We believe that together the two volumes of the book represent a unique combination of topics pertaining to antenna design and analysis, not found elsewhere. It is hoped that the antenna community including designers, students, researchers, faculty engaged in teaching and research of antennas, and the users as well as decision-makers would find the book useful and timely, and their feedback and comments are most welcome. They can post these comments on the website for the e-journal FERMAT (www.e-fermat.org).

The editor takes this opportunity to thank all the contributors who invested a considerable amount of their precious time to participate in this gargantuan task of preparing their contributions for the book and ensuring the quality of their contributions that helps to make the book so unique. Finally, the editor thankfully acknowledges the help of Ravi Arya, who served as the editorial assistant throughout the entire period of the manuscript preparation, and did such an excellent job of handling all the details that go with the task of putting together such a monumental piece of work. Before closing, I would like to mention that the book often uses the terms S11 and return loss interchangeably. It has recently been pointed out by the IEEE Standards Committee that the two are not the same and that there is a sign difference between the two. So, we want the reader to be aware of this when going through the book so that there is no confusion.

Chapter 1
Applications of the characteristic mode theory to antenna design

Ting-Yen Shih[1] *and Nader Behdad*[2]

1.1 Introduction

1.1.1 Background

Antennas are used in a wide range of applications including terminal devices (e.g., cell phones), base stations, and advanced communication systems on large platforms (e.g., aircrafts, ships, vehicles). The rapid growth of wireless communications has been an important driving force for the advancement of antenna engineering. Modern antennas are often expected to have small size, light weight, low cost, wideband, or multiband operation, and even reconfigurable capabilities. All of these requirements issue strong demands for new innovations in antenna design. To achieve promising antenna performance, various techniques have been developed to approach the physical limitation in antenna engineering. The characteristic mode theory is one of such methods that has drawn much attention in the recent years.

The characteristic mode theory can be used to examine the eigen-functions of an antenna (e.g., eigen-currents, eigen-fields, etc.) and provide physical insights on antenna design. With the results generated from characteristic mode analysis, antenna designers can build systematic design methods with clear physical understandings to reach the desired goals (e.g., bandwidth, polarization, etc.) and the design iteration to achieve optimal antenna performance are reduced. This is a major advantage over other techniques that rely heavily on the personal experience of the antenna designers.

1.1.2 Characteristic mode theory

The characteristic mode theory was first described by Garbacz *et al.* [1,2] and the expansion method for the computation of characteristic modes was reformulated and refined by Harrington *et al.* in 1971 [3,4]. The characteristic mode computations implemented in recent years are primarily based on Harrington's approach.

[1]Department of Electrical and Computer Engineering, University of Idaho, USA
[2]Department of Electrical and Computer Engineering, University of Wisconsin, USA

Characteristic modes are eigen-modes obtained numerically for an arbitrarily shaped conducting object, and they form a basis set in which to expand the currents and fields scattered or radiated by the conducting object. The characteristic modes of an conducting object are determined by the shape and the size of the object. The radiation pattern of any object is a linear combination of modal patterns that are orthogonal to each other. The current distribution on the object can be decomposed into an infinite number of eigen-currents (or characteristic currents), which are also orthogonal to each other.

The eigen-currents J_n can be determined from the following generalized eigen-value equation [3]:

$$[X]J_n = \lambda_n[R]J_n \tag{1.1}$$

where R and X are, respectively, the real and imaginary parts of the impedance matrix of the electric field integral equation. λ_n is the nth eigen-value associated with the nth eigen-current J_n.

The recent advancement in computer-aided design tools has made antenna design easier by allowing antenna researchers to examine the physical characteristics of antennas using characteristic mode analysis in commercially available software such as Altair FEKO (https://altairhyperworks.com/product/FEKO#) and CST Microwave Studio (https://www.cst.com/) and has brought much excitement to the research community. To demonstrate important parameters of characteristic modes, a rectangular patch of width $W = 3$ mm and length $L = 5$ mm in free space[1] will be used as an example.

Figure 1.1 illustrates the eigen-current distribution at 28 GHz for the first six characteristic modes of this rectangular patch. The simulations were carried out using Altair FEKO[2]. All currents were normalized to their maximum value in order to carry out fair comparison. Modes 1 and 2 are characterized by vertical ($\hat{x}$ direction) and horizontal ($\hat{y}$ direction) currents, respectively. Hence, the first two modes act as half-wavelength dipole modes. The normalized eigen-current distribution of mode 3 resemble those of a full-wavelength loop mode. The normalized eigen-current distribution of mode 4 resemble the super position of two diagonal full-wavelength dipole modes. Mode 5 is the high-order mode of mode 1. Mode 6 is composed of two half-wavelength loop modes. The electric field (E_n), and the magnetic field (H_n) produced by an eigen-current (J_n) on the surface of the conducting body are the eigen-fields or characteristic fields in correspondence with J_n [3]. The normalized 3-D E-field patterns of the first six characteristic modes of the rectangular patch at 28 GHz are shown in Figure 1.2. Note that the 3-D field patterns of characteristic modes are orthogonal to each other. In Figure 1.2, mode 1 possesses an omni-directional radiation pattern in the $x - z$ plane, while the radiation patterns of modes 2 and 3 are omni-directional in the $y - z$ and $x - y$ planes, respectively.

[1]If the patch is placed above a ground plane, the presence of the ground plane does not significantly change the eigen-current distribution, but it may impact the resonance frequency and the radiation bandwidth.
[2]FEKO numbers the characteristic modes according to their modal significance values at the lowest-examined frequency.

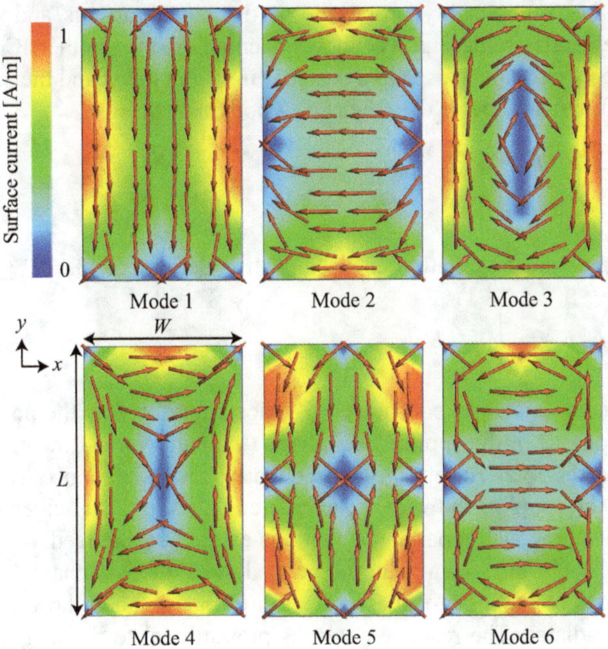

Figure 1.1 *The normalized eigen-current distribution of the first six characteristic modes of the rectangular patch with W = 3 mm and length L = 5 mm in free space at 28 GHz*

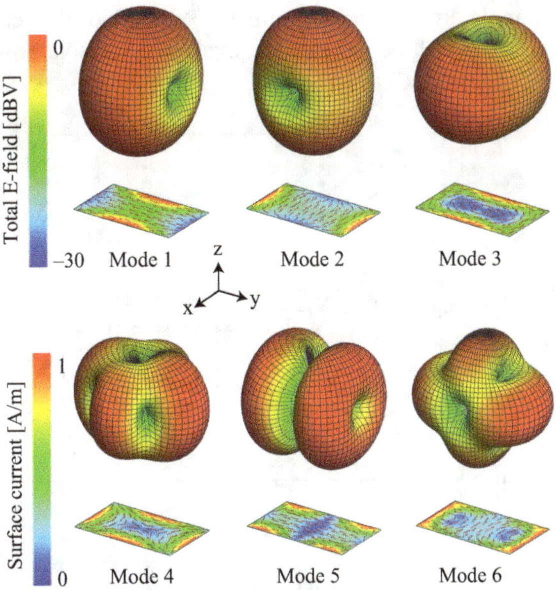

Figure 1.2 *The normalized E-field patterns of the first six characteristic modes of the rectangular patch*

Due to the orthogonality properties of the eigen-currents and eigen-fields, characteristic modes radiate power independently of one another. Consequently, characteristic modes can be used as a basis set in which to expand the arbitrary total current (J) on the surface of the conducting body as [3]:

$$J = \sum_n \frac{V_n^i J_n}{1 + j\lambda_n} \tag{1.2}$$

where V_n^i is the modal excitation coefficient. It is defined as [3]:

$$V_n^i = \langle J_n, E^i \rangle = \oint_S J_n \cdot E^i ds \tag{1.3}$$

where the modal excitation coefficient (V_n^i) accounts for how the positions, magnitudes, and phases of the applied excitations (the feeds) affect the contribution of each mode to the total current, J. Consequently, the excitation coefficient V_n^i in (1.2) reflects the coupling between the excitation (E^i) and the nth eigen-current (J_n), and determines how well a particular mode is excited by the feed.

In (1.2), λ_n is the eigen-value associated with the nth characteristic mode. The magnitudes of the eigen-values provide information about how well the associated mode radiates. The reactive power is proportional to the magnitudes of the eigen-values [3]. A characteristic mode is considered resonant when $\lambda_n = 0$. In other words, the smaller the magnitude of the eigen-value is, the more efficiently the mode radiates. The sign of the eigen-value determines whether the mode contributes to the stored magnetic energy ($\lambda_n > 0$) or the stored electric energy ($\lambda_n < 0$).

The eigen-values associated with the first six characteristic modes of the rectangular patch example are shown in Figure 1.3. Mode 1 of the rectangular patch radiates effectively, while the other modes have more reactive power and do not

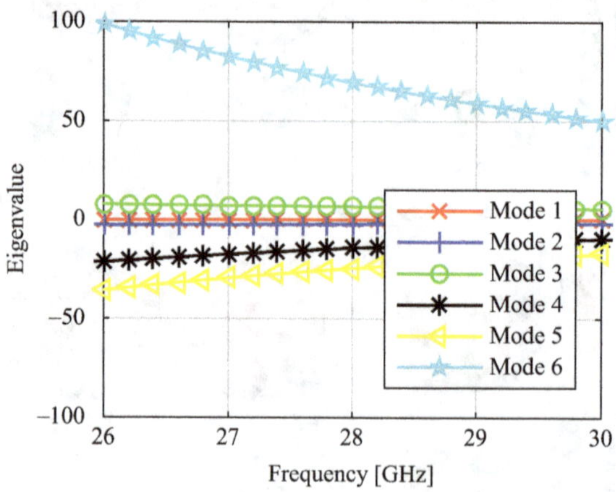

Figure 1.3 The eigen-values associated with the first six characteristic modes of the rectangular patch

radiate as effectively. Modes 3 and 6 mainly contribute to the stored magnetic energy, and modes 2, 4,and 5 mainly contribute to the stored electric energy.

An alternative parameter to measure the potential radiation contribution of each characteristic mode is the modal significance (MS) [5]. With the assumption that each characteristic mode is 100% excited by an ideal external source, the MS is defined as

$$\mathrm{MS} = \left| \frac{1}{1 + j\lambda_n} \right| \tag{1.4}$$

MS represents the normalized magnitude of the eigen-values of the characteristic modes. A mode is considered resonant when the MS of that mode attains a value of 1 [6], and a mode is considered significant when the MS is greater than or equal to 0.707 [7]. The closer the value of MS is to 1 (the maximum value), the more effectively the associated mode contributes to radiation.

The MSs of the first six characteristic modes of the rectangular patch are shown in Figure 1.4. As can be observed, in the examined frequency range of 26 GHz 30 GHz, mode 1 is the dominant mode and it is significant (>0.707). MS and J_n are determined by the shape and size of the radiating object, whereas V_n^i is influenced by the number, type, and position of the feed(s) or coupling element(s). Note that the products of the excitation coefficient (V_n^i) and the MS are defined as weighting coefficient in Altair FEKO.

Yet another parameter to evaluate characteristic modes is the characteristic angle, α. Characteristic angles are defined in [8] as

$$\alpha = 180^\circ - \tan^{-1}(\lambda_n). \tag{1.5}$$

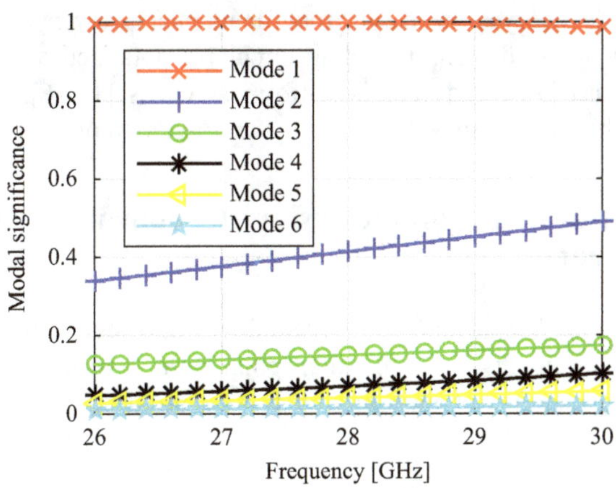

Figure 1.4 Simulated modal significances of the first six characteristic modes of the rectangular patch

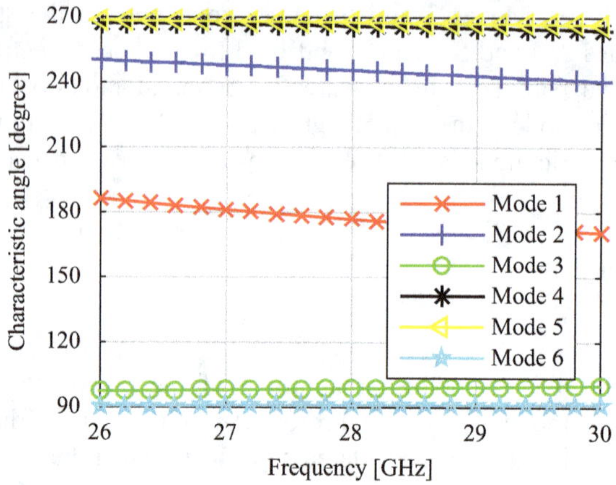

Figure 1.5 The characteristic angles of the first six characteristic modes of the rectangular patch

Modes with characteristic angles close to 180° are effective radiators, while those with characteristic angles close to 90° or 270° are ineffective radiators. The characteristic angle models the phase difference between an eigen-current (J_n) and the associated eigen-field (E_n). Characteristic angles also provide information on the phases of characteristic modes. Hence, this parameter is very useful when designing reflect arrays or circularly polarized antennas, as will be discussed later in this chapter. The characteristic angles of the first six characteristic modes of the rectangular patch are shown in Figure 1.5.

Characteristic modes provide a physical interpretation of the radiation phenomena taking place on the antenna and provide valuable information for antenna design. The characteristic mode theory has been applied to antenna shape synthesis [9,10] and to the control of radar scattering by reactive loading [11]. The great potential of the characteristic mode theory as an analytic tool for antenna researchers can also be seen in a variety of antenna designs including circularly polarized antennas [6,12,13], broadband antennas [14], and chassis/platform antennas [15,16]. Antenna design examples using the characteristic mode theory are discussed in Section 1.2.

1.2 Antenna design examples using the characteristic mode theory

The characteristic mode analysis has been employed to design radiators that have to meet certain requirements (e.g., circular polarization, wide bandwidth, omnidirectional patterns, etc.). Antennas in such systems are designed to play one of the two major roles:

- Antennas act as major radiators
- Antennas act mainly as coupling elements and a platform acts as the major radiator (in the vicinity of the antennas).

This section discusses four design examples, including the application of the characteristic mode theory to the design of circularly polarized antennas, wideband antennas, chassis antennas, and platform antennas. The antennas act as major radiators in the circularly polarized antenna and the wideband antenna design examples [6,12,13,14], while the antennas primarily act as coupling elements, and a chassis or a platform acts as the major radiator in the latter two design examples [7,15–20].

1.2.1 Circularly polarized antennas

One of the methods to achieve circularly polarized radiation patterns is to combine two orthogonal and linearly polarized modes, with the same current amplitude and in phase quadrature. The orthogonality of characteristic modes makes the synthesis of circular polarization possible in an easy and intuitive way.

Take a square patch antenna with dimensions of $W=L=4.8$ mm as an example. The square patch is placed above an infinite ground plane, as shown in Figure 1.6. The normalized eigen-current distributions of the first four characteristic modes are shown in Figure 1.7. The resonant frequencies of the modes are determined by the size and the shape of the patch. Since characteristic modes are all orthogonal to one another, any two characteristic modes with 90° phase difference could be used to generate a combined circularly polarized pattern. However, modes with MSs closer to 1 or larger than 0.707 are preferred since it is easier to excite them. The MSs of the first four characteristic modes of the square patch antenna are shown in Figure 1.8. In this case, modes 1 and 2 are chosen[3]. The next step is to fulfill the phase quadrature (90°) requirement. By examining the characteristic angles of the two modes in Figure 1.9, we can see that there is no phase difference between them. This is because modes 1 and 2 are degenerate modes. In order to achieve a 90° phase difference, one common method is to use multiple feeds such as dual feeds (1∠0° and 1∠90°) or quadruple feeds (1∠0°, 1∠90°, 1∠180°, and 1∠270°, as shown in Figure 1.6).

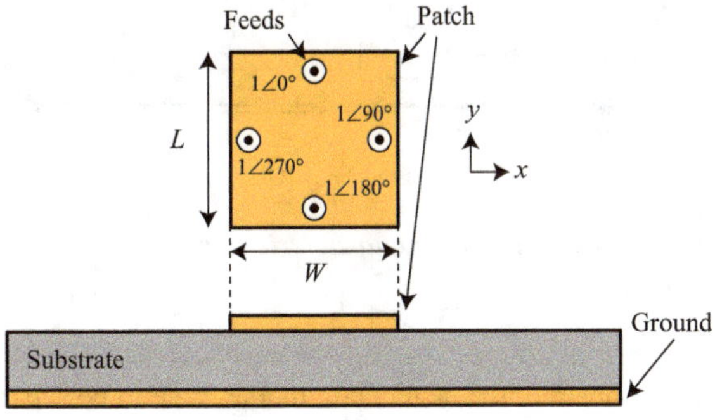

Figure 1.6 The structure of the square patch antenna placed above an infinite ground plane

[3]Since FEKO numbers the characteristic modes according to their modal significance values at the lowest examined frequency, the modes numbered with smaller numbers are generally in favor.

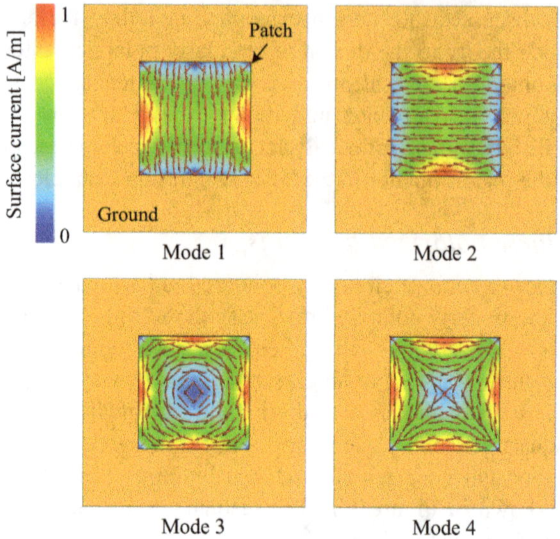

Figure 1.7 The normalized eigen-current distributions of modes 1–4

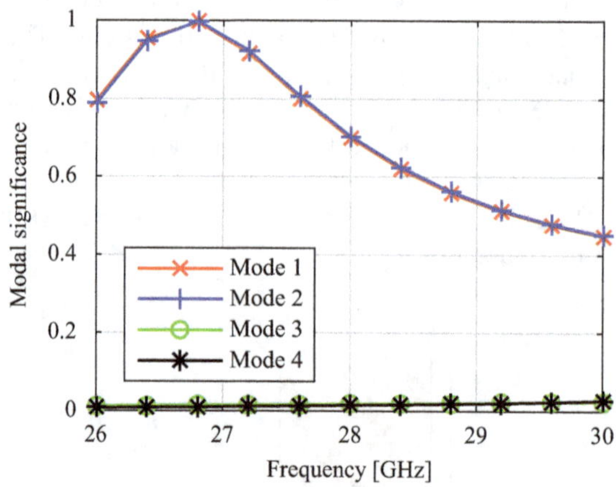

Figure 1.8 The MSs of the first four characteristic modes of the square patch antenna

If a single feed is desired, the degenerate modes have to be separated. This can be done by modifying the geometry of the antenna. As an example, let us transform the square patch antenna into an isosceles trapezoid patch antenna, as shown in Figure 1.10(a). Adjusting W_1 in Figure 1.10(a) alters the characteristic angles of mode 2. When $W_2 = L = 4.8$ m and $W_1 = 3.84$ m, modes 1 and 2 present a 90°

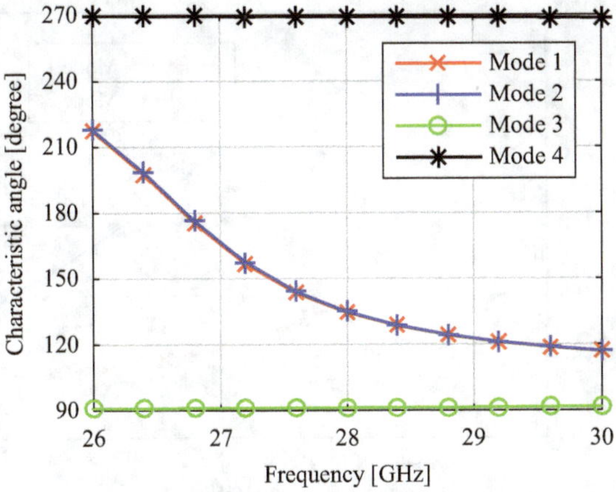

Figure 1.9 The characteristic angles of the first four characteristic modes of the square patch antenna

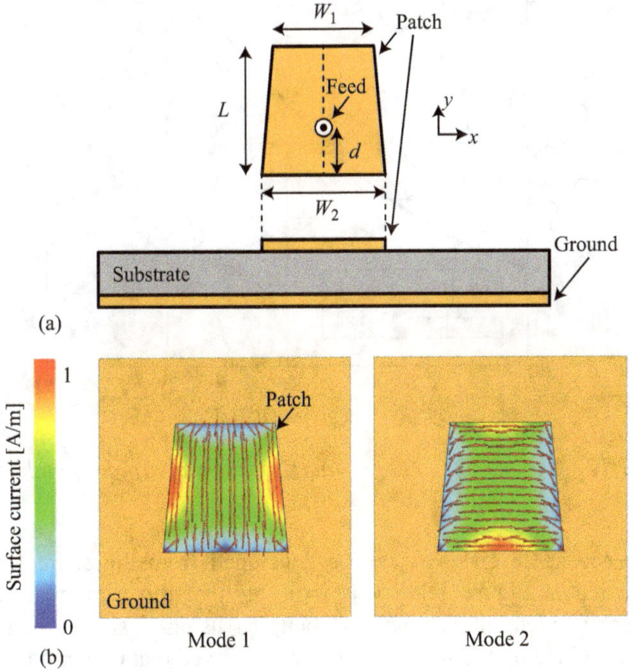

Figure 1.10 (a) The structure of the isosceles trapezoidal patch antenna placed above an infinite ground plane and (b) the normalized eigen-current distribution of modes 1 and 2 of the trapezoidal patch antenna

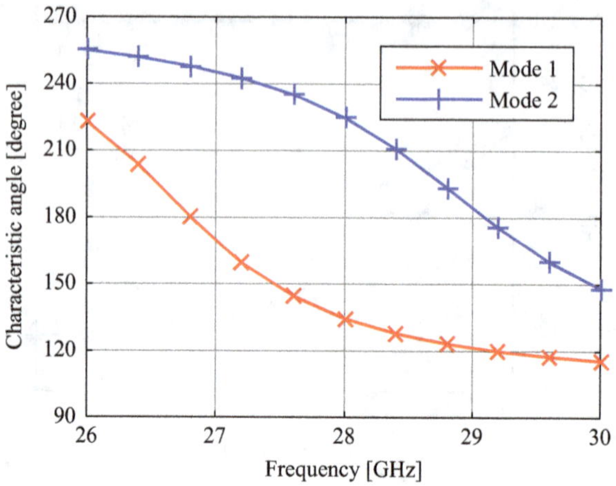

Figure 1.11 The characteristic angles of modes 1 and 2 of the trapezoidal patch antenna

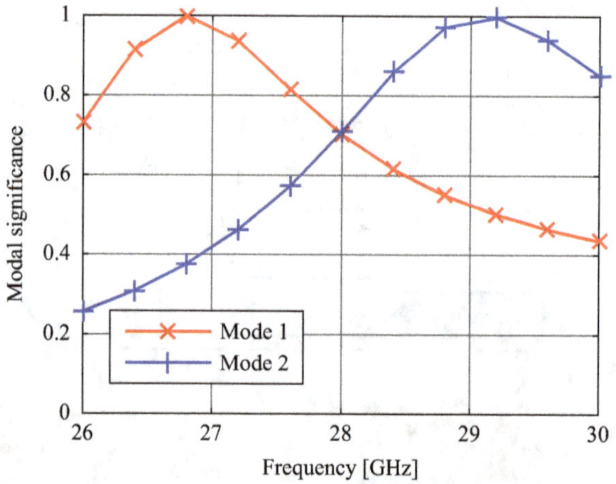

Figure 1.12 The MSs of modes 1 and 2 of the trapezoidal patch antenna

phase difference at 28 GHz, as can be concluded from the characteristic angle curves presented in Figure 1.11. This indicates that this pair of modes can be excited to achieve circularly-polarized patterns using a single feed. From the information provided by Figure 1.12, it can be observed that both modes present the same magnitude of MSs at 28 GHz. The MSs of the two modes also indicate that they could radiate efficiently at this frequency. As can be observed in Figure 1.10 (b), the eigen-current distributions of the first two characteristic modes are slightly different from those of the first two modes of the square patch antenna (Figure 1.7) but this does not change the fact that they are orthogonal to each other. Hence, if these two modes were properly excited and combined using a single feed, they will yield circular polarization at 28 GHz. Figure 1.10(a) shows the feed position where the two modes present exactly the same current amplitude. Design examples of single feed circularly polarized antennas using the characteristic mode theory can be found in [6,12].

1.2.2 Wideband antennas

The characteristic mode theory can also be applied to wideband antenna design [14]. The potential impedance bandwidth and radiation patterns of an antenna can be calculated using the properties of its characteristic modes. For wideband antenna designs, it is important to examine the MSs of the characteristic modes since exciting a broadband characteristic mode allows for achieving wideband impedance bandwidth.

As an example, let us examine the MSs of the first four modes of a planar square monopole of $W = L = 40$ mm placed above an infinite ground plane ($h = 4.5$ mm), as shown in Figure 1.13. As can be observed, mode 1 of this antenna is significant throughout the entire examined frequency band (3–11 GHz) and mode 2 remains significant from 3 to 8.2 GHz. This indicates that, with a proper feeding configuration to excite these modes, this planar square monopole could provide ultra-wide impedance bandwidth (e.g., 3:1 bandwidth). The radiation characteristics of modes 1 and 2 of the planar square monopole[4] are shown in Figure 1.14. If we were to design an ultra-wideband antenna, for example, omni-directional radiation patterns are required, and mode 1 falls off the list. Mode 2, on the other hand,

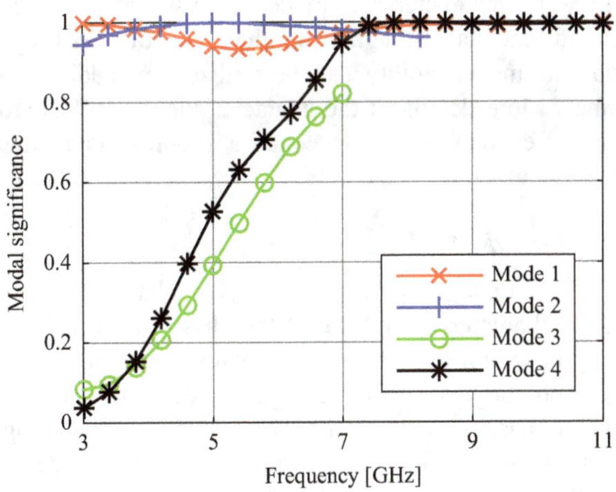

Figure 1.13 The MSs of the first four characteristic modes of the planar square monopole antenna

[4]Note that the planar square monopole has less pattern degradation over the impedance bandwidth than that of a circular planar monopole antenna.

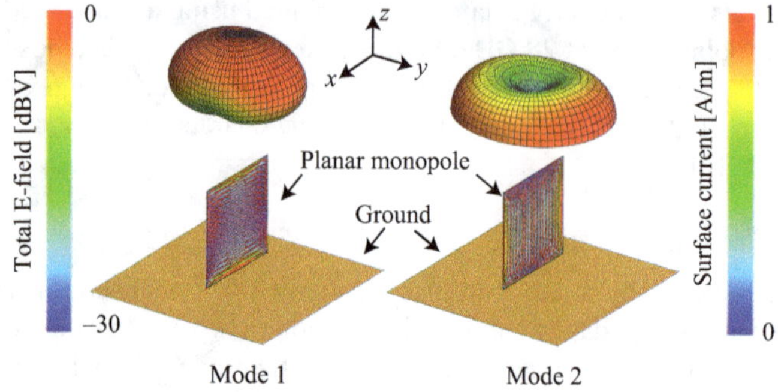

Figure 1.14 The normalized eigen-current distributions and radiation patterns (E-field) of the first two characteristic modes of the planar square antenna

meets this requirement, and can be used for designing such an antenna. The analysis of the normalized eigen-current distribution of mode 2 (Figure 1.14) also provides physical interpretation of ways to excite this mode efficiently. Namely, the mode would be excited more effectively with multiple edge-feeds compared to a single edge-feed since less horizontal currents will be induced and fewer undesired modes will be excited. This explains the impedance bandwidth observed by [21]. Note that mode 2 of this planar square monopole is only dominant in the lower band, and mode 4, which is the higher order mode of mode 2, can be excited and used for the upper frequency band to achieve ultra-wide bandwidth. The characteristic mode analysis also provides clues for designing wideband antennas with filtering characteristics. In [14], it was shown that a band-stop filtering capability can be realized by adding a slot in planar monopole antennas. More details of the characteristic-mode-based design of such antennas and the effects of the presence of a ground plane, the slots, or the miniaturization of the antenna on the antenna characteristics can be found in [14].

1.2.3 Chassis-based MIMO antennas

Let us look at an example of how the theory of characteristic modes can be employed to excite characteristic modes of the chassis of a mobile handset. This example is mainly based on the work in [15], where two capacitive coupling elements which are each composed of a rectangular patch and a feed strip are placed on the edges of the chassis, as shown in Figure 1.15. The coupling elements are placed at the locations where the electric current densities of mode 2 and mode 3 of the chassis are the weakest (Figure 1.16).

The normalized eigen-current distribution and E-field radiation characteristics of the first four modes of the chassis with the coupling elements are simulated in Altair FEKO, as shown in Figure 1.17. The coupling elements are included in the analysis since their size is not negligible compared with the dimension of the chassis, and they change the properties of the characteristic modes or even create new modes

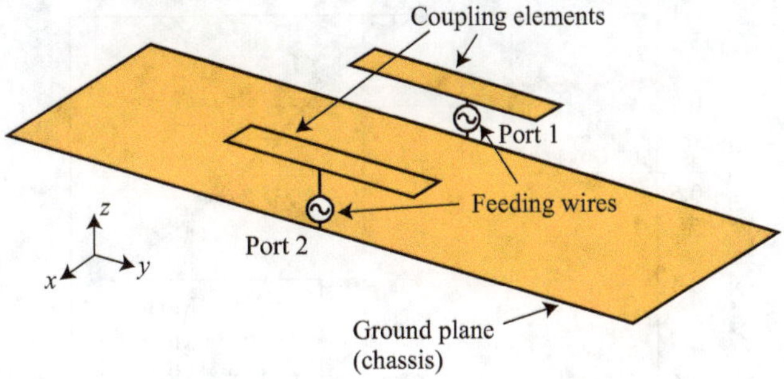

Figure 1.15 Schematic of the chassis antenna with the coupling elements

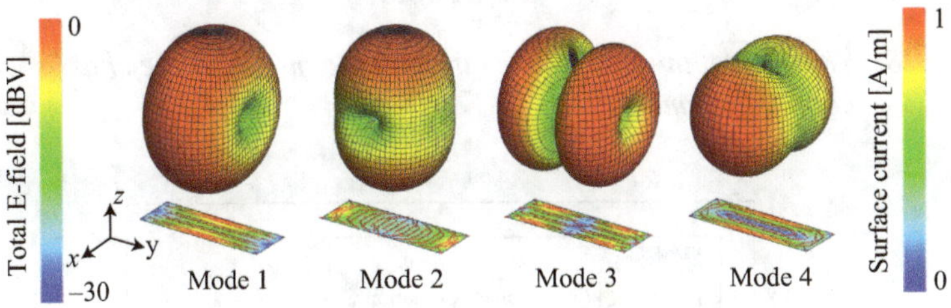

Figure 1.16 The normalized current distribution and E-field patterns of the first four characteristic modes of the chassis

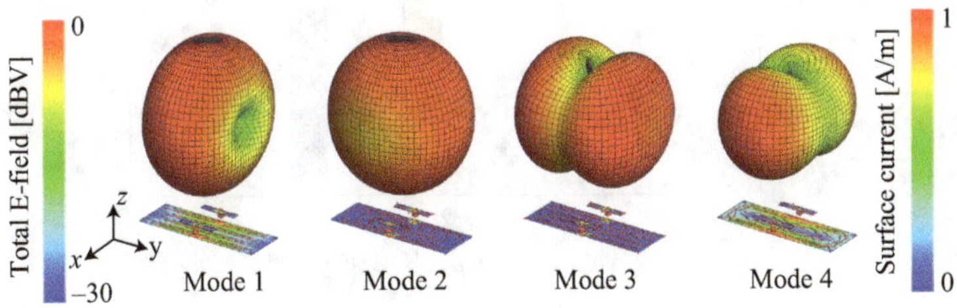

Figure 1.17 The normalized current distribution and E-field patterns of the first four characteristic modes of the chassis with the coupling elements

(compare Figures 1.16 and 1.17). The current distributions in Figure 1.17 show that modes 2 and 3 act like an odd mode and an even mode, respectively. Consequently, modes 2 and 3 of this chassis can be excited using the same set of coupling elements, but they have to be fed out-of-phase and in-phase, respectively. The MSs of the first four characteristic modes of the chassis with the coupling elements are shown in Figure 1.18. In this case, the operating frequency is 2.06 GHz since characteristic modes 2 and 3 of the chassis have similar significant MS at this frequency. Note that the properties of the characteristic modes can be controlled by the size, shape, and

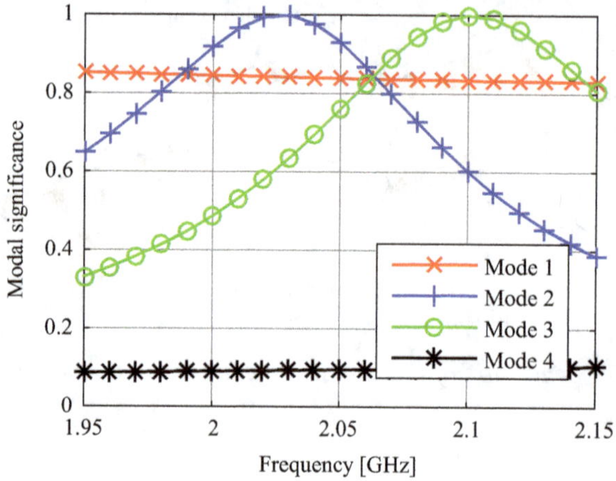

Figure 1.18 The MSs of the first four characteristic modes of the chassis with the coupling elements

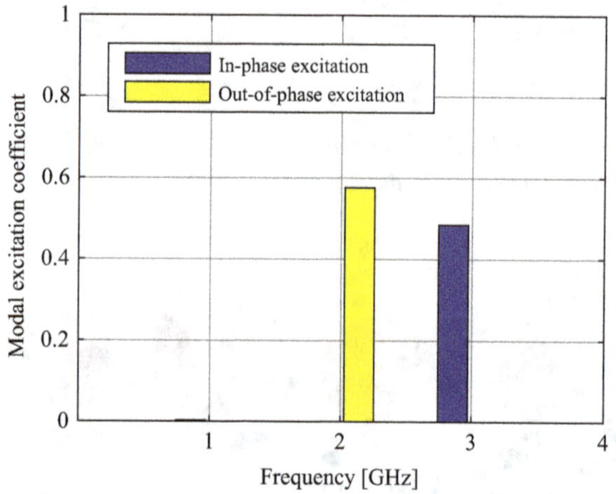

Figure 1.19 The modal excitation coefficients of the modes excited using the capacitive coupling elements

position of the coupling elements as long as their size is not negligible compared to the chassis. Hence, the operating frequency can be tuned by changing these parameters. The modal excitation coefficients of the modes excited using the out-of-phase and in-phase feeding setups are shown in Figure 1.19. As can be observed, with out-of-phase feeding, mode 2 is excited; whereas with in-phase excitation, mode 3 is excited. This multiple-mode capability and the orthogonality between the excited modes make this chassis-base antenna a good candidate for multiple-input and multiple-output (MIMO) applications. More details of the chassis-mode MIMO antenna can be found in [15].

1.2.4 Bandwidth enhancement of platform-based antennas

In addition to having broad bandwidth and satisfying certain radiation properties, many antennas are required to have small physical size. However, there exists a tradeoff between the size and bandwidth of antennas. The relationship between antenna size, bandwidth, and operating frequency has been described by Chu in 1948, and the upper physical bound for the bandwidth of an antenna enclosed in a Chu sphere can be derived [22]. Antennas that have significantly smaller dimensions than the wavelength at which they operate (usually in the lower frequencies) suffer from narrow bandwidths. This is a limiting factor for certain applications.

In many civilian and military applications, antennas are mounted on metallic platforms that are physically larger than the antennas themselves (e.g., mobile phones, ships, military vehicles, and airplanes, etc.). To address the challenge of narrow bandwidth, antenna designers have developed characteristic-mode-based bandwidth enhancement techniques that exploit the presence of the platform in the vicinity of the antenna. In this subsection, we examine how a platform-mounted electrically-small antenna can be used to excite the characteristic modes of the metallic platform to increase the overall bandwidth of the system. In this case, the platform will act as the main radiator and the mounted antennas act primarily as the coupling mechanism between the antenna and the external circuit. The theory of characteristic modes is used to identify the appropriate platform modes and to determine the practical means of exciting them. In these situations, since the platform is generally larger than the antenna mounted on it, if the platform can be used as part of the antenna, the maximum linear dimension of the antenna can be increased significantly, resulting in an enhanced bandwidth. To accomplish this, the platform and the antenna must be designed together and the platform must be considered to be a major part of the radiating structure from the beginning. A systematic method for designing an antenna system that takes advantage of the presence of the platform to achieve significantly enhanced bandwidth compared to a stand-alone antenna is presented in [16]. This approach is employed to enhance the bandwidth of a horizontally polarized loop antenna system by as much as 10 times compared to a stand-alone loop antenna. We will borrow the examples used in [16] to demonstrate this method.

The platform used for this example is a simple metallic platform in the shape of a rectangular box with dimensions of 3.7 m × 10.7 m × 3.3 m. The design process starts by examining the resonant behavior of the platform and the associated eigen-current distributions within the frequency band of interest, which is the lower high-frequency band (3–15 MHz) in this case.[5] The MSs of the platform examined are shown in Figure 1.20. As can be observed, the dominant mode for this platform is mode 1, and this mode is significant above approximately 9 MHz. Figure 1.21 shows the normalized eigen-current distributions and the radiation patterns of the

[5]The commercial full-wave software Altair FEKO was used to carry out the analysis of characteristic modes in this subsection.

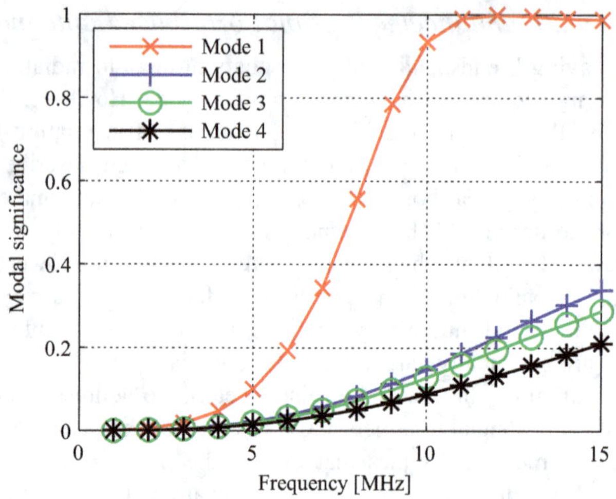

Figure 1.20 Simulated MSs of the first four modes of the platform

first four characteristic modes of the platform. The radiation patterns of modes 1 and 2 indicate that their maximum radiations are toward zenith, which makes them suitable for near vertical incidence skywave, or NVIS, applications. Modes 3 and 4 generate radiated fields that are useful for ground-wave and skywave communications. Depending on the desired radiation characteristics of the application, we can choose which mode to excite. In this example, we focus on exciting mode 1 because it is the dominant mode of the platform within the examined frequency band. However, the concepts are equally applicable to the excitation of other modes.

To examine the upper limit of the bandwidth that can be achieved by exciting the platform's characteristic modes, the lower bound for the quality factor (Q) of the desired platform mode is calculated. Given the platform dimensions and considering a frequency of operation of 10 MHz, the platform Q based on the Chu-McLean limit [22,23] is calculated to be $Q_{\mathrm{Chu}} = 1.3391$:

$$Q = \frac{1}{ka} + \frac{1}{(ka)^3} \tag{1.6}$$

where Q is the quality factor, k is the wave number ($2\pi/\lambda$), a is the radius of the minimum size sphere that encloses the antenna. The 10 dB return loss fractional bandwidth [23,24] associated with this Q is approximately 50%:

$$B_V = \frac{1}{\eta}\frac{1}{Q}\frac{\mathrm{VSWR} - 1}{\sqrt{\mathrm{VSWR}}} \tag{1.7}$$

where B_V is the upper bound of the matched VSWR fractional bandwidth, η is the radiation efficiency, and VSWR is the voltage standing wave ratio. This bandwidth, however, is too optimistic a value because the platform shape does not utilize the

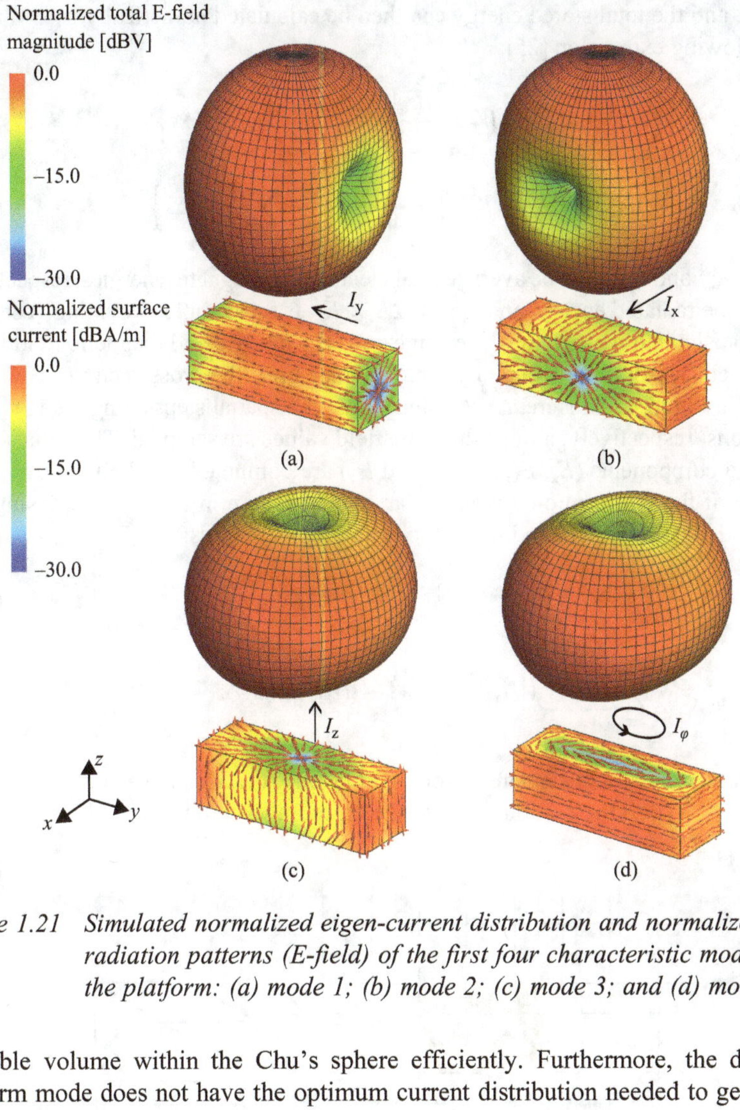

Figure 1.21 Simulated normalized eigen-current distribution and normalized radiation patterns (E-field) of the first four characteristic modes of the platform: (a) mode 1; (b) mode 2; (c) mode 3; and (d) mode 4

available volume within the Chu's sphere efficiently. Furthermore, the desired platform mode does not have the optimum current distribution needed to generate the lowest possible Q. A more realistic calculation of the desired platform mode's Q can be performed by examining the actual stored energy and radiated power associated with the current distribution of that specific mode. The radiation Q can be specified mathematically as:

$$Q = \omega \frac{2\text{Max}(W_e, W_m)}{P_{rad}} = 4\pi \frac{\text{Max}(W_e, W_m)}{W_{rad}} \tag{1.8}$$

where W_e and W_m are the total stored electric energy and magnetic energy, respectively, P_{rad} is the radiated power, and W_{rad} is the radiated energy. The electric and magnetic fields associated with the desired platform mode can be obtained from Altair

FEKO, and the total stored energy can then be calculated numerically by evaluating the following expression [25]:

$$W_e = \frac{1}{2}\left[W_e^{tot} + W_m^{tot} - W_{rad}\right] - \frac{1}{4\omega}\mathrm{Im}\left\{\sum\sum \vec{S}\cdot\vec{n}\Delta s\right\} \tag{1.9}$$

$$W_m = \frac{1}{2}\left[W_e^{tot} + W_m^{tot} - W_{rad}\right] + \frac{1}{4\omega}\mathrm{Im}\left\{\sum\sum \vec{S}\cdot\vec{n}\Delta s\right\} \tag{1.10}$$

where W_e^{tot} and W_m^{tot} are the average total electric and magnetic energies, respectively, W_{rad} is the radiated energy, and $\vec{S} = 1/2\vec{E}\times\vec{H}^*$ is the complex Poynting vector.

To calculate these values, we can export the electric and magnetic field values calculated by Altair FEKO along a rectangular grid, the cross section of which is shown in Figure 1.22. Here, Δx, Δy, and Δz are the spatial steps along the $\hat{x}$, $\hat{y}$, and $\hat{z}$ directions, respectively, along which the field values are sampled. The values of the six field components (E_x, E_y, E_z, H_y, and H_z) are computed by FEKO at each point outside of the platform on a rectangular grid as shown in Figure 1.22. Using this, W_e^{tot} and W_m^{tot} are calculated using the following equations:

$$W_e^{tot} = \frac{\varepsilon}{4}\sum_x\sum_y\sum_z(|E_x|^2 + |E_y|^2 + |E_z|^2)\Delta x\Delta y\Delta z \tag{1.11}$$

$$W_m^{tot} = \frac{\mu}{4}\sum_x\sum_y\sum_z(|H_x|^2 + |H_y|^2 + |H_z|^2)\Delta x\Delta y\Delta z \tag{1.12}$$

The expression calculated from (1.11) and (1.12), however, includes the total energy (i.e., the combination of the stored energy and the radiated energy).

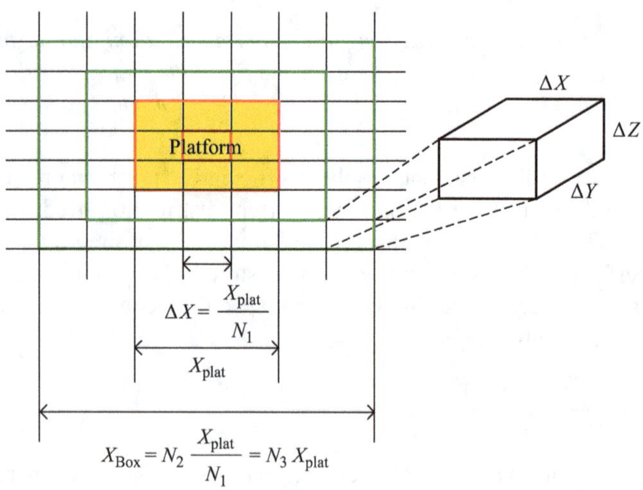

Figure 1.22 The scheme of the radiation Q calculation

To obtain the total stored energy, the contribution of the radiated energy must be subtracted from the total stored energy. The radiated energy is obtained using:

$$\begin{aligned} W^{\mathrm{rad}} = \frac{1}{c}\mathrm{Re}\Big\{ & 2\Delta x \sum\sum \vec{S}_x \cdot \hat{x}\Delta y \Delta z \\ & +2\Delta y \sum\sum \vec{S}_y \cdot \hat{y}\Delta x \Delta z \\ & +2\Delta z \sum\sum \vec{S}_z \cdot \hat{z}\Delta x \Delta y \Big\} \end{aligned} \tag{1.13}$$

where $\vec{S} = 1/2\vec{E} \times \vec{H}^*$ is the complex Poynting vector and c is the speed of light. The numerical calculations expressed by (1.11)–(1.13) must be carried out over a sufficiently large volume in the vicinity of the antenna to ensure that all of the stored energy in the vicinity of the antenna is taken into account. To determine the size of the domain over which these calculations must be performed, one can increase the volume successively and monitor for the convergence of the results obtained. The mesh scheme for the Q calculations is shown in Figure 1.22. The mesh size and the size of the computational domain are determined by N_1 and N_3, respectively. N_2 is the number of grids along each direction ($\hat{x}$, $\hat{y}$, and $\hat{z}$). In the calculations for this example, $N_1 = 9$, $N_2 = 171$, and $N_3 = 19$ result in a computational domain large enough to allow for accurate calculation of the Q, and hence, these values are used to perform the Q calculations for the platform. Following this procedure, the Q of mode 1 is calculated to be $Q_{\mathrm{Mode}_1} = 8.6655$. This corresponds to a matched 10 dB return loss fractional bandwidth of 7.69% at 10 MHz. These results are summarized in Table 1.1.

Examination of the current distribution of mode 1 of the platform reveals that the entire antenna acts as a fat dipole for this mode, as shown in Figure 1.21(a). Such a dipole antenna can be fed by a voltage gap placed at the center of the dipole, as shown in Figure 1.23(a). However, this requires dividing the platform in two, which is often not practical. Alternatively, capacitive or inductive coupling techniques can be used to excite this mode, as shown in Figure 1.23(b) and (c). Figure 1.23 (b) shows an example of using capacitive coupling elements to excite mode 1. In this case, two monopole antennas, fed with the same magnitudes and 180° phase

Table 1.1 Comparison of simulation results of the different antennas

Scenario[a]	f_0 (MHz)	BW	η_{tot}
Chu limit	10.00	49.79%	
Mode 1	10.00	7.69%	
One full loop alone	11.99	0.33%	97.1%
One half-loop (Center)	12.32	0.64%	99.1%
One half-loop (Edge)	11.48	1.31%	99.5%
Two half-loops (Edge)	10.76	2.20%	99.7%
Four half-loops (Edge)	10.16	3.27%	99.7%

[a]The one full loop antenna is stand-alone. Everything else is on the platform.

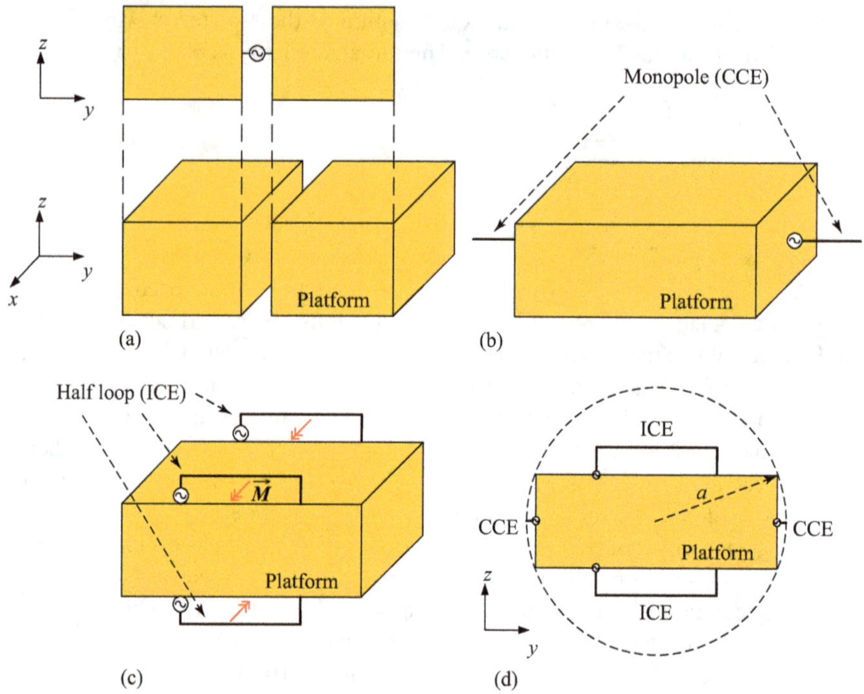

Figure 1.23 Mode 1 of the platform is excited by (a) a voltage source; (b) monopole antennas (functioning as capacitive coupling elements, CCEs); (c) half-loop antennas (functioning as inductive coupling elements, ICEs); and (d) comparison of the two coupling methods with a given maximum linear dimension of the system, 2a

difference, are used to excite mode 1. In Figure 1.23(c), multiple half-loops, all fed with the same magnitudes and phases, are placed on the periphery of the platform. The loops act as inductive coupling elements and couple with the magnetic field of mode 1 to excite this mode. Both of these coupling methods can be used to efficiently excite mode 1 without cutting the platform. In this example, we employ the inductive coupling element excitation method shown in Figure 1.23(c) since we assume that the maximum linear dimension of the antenna/platform is to be maintained (i.e., the coupling antennas should not increase the radius of the Chu sphere, *a*, circumscribing the entire structure, as seen in Figure 1.23(d)). With this constraint, Figure 1.23(d) shows that within the Chu sphere, very limited space is available to place the capacitive coupling elements whereas ample space is available to accommodate the inductive coupling elements. The presence of the loops or monopoles on the platform creates new modes. However, in this example, these modes are associated with the resonances of the coupling elements themselves and become significant at higher frequencies only. Specifically, characteristic mode simulations in Altair FEKO show that the presence of the coupling elements does not significantly impact the properties of mode 1 and its current distribution.

Therefore, in the analysis conducted in the remainder of this example, we focus on examining the characteristic modes of the platform alone.

Figure 1.24(a) shows an example of using inductive coupling elements to excite mode 1. Here, only a single-loop antenna is used to excite the desired platform mode. To magnetically excite mode 1, where the current direction on the platform is in the $\hat{y}$ direction and the H-plane of the fat dipole is in the $x-z$ plane,

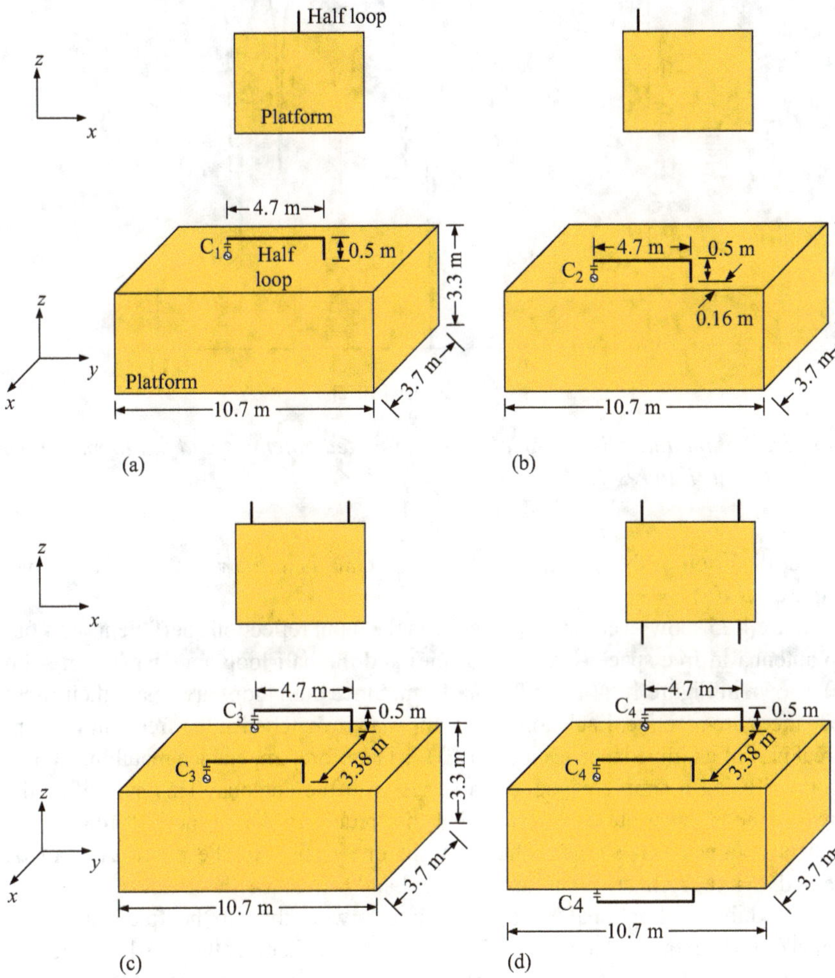

Figure 1.24 (a) One half-loop antenna on the top of the platform (center); (b) one half-loop antenna on the top of the platform (edge); (c) two half-loop antennas on the top of the platform (edge); and (d) four half-loop antennas on the top and bottom of the platform (edge). $C_1 = 12$ pF, $C_2 = 18$ pF, $C_3 = 25$ pF, and $C_4 = 32$ pF. The wire diameter of the loops is 81.92 mm

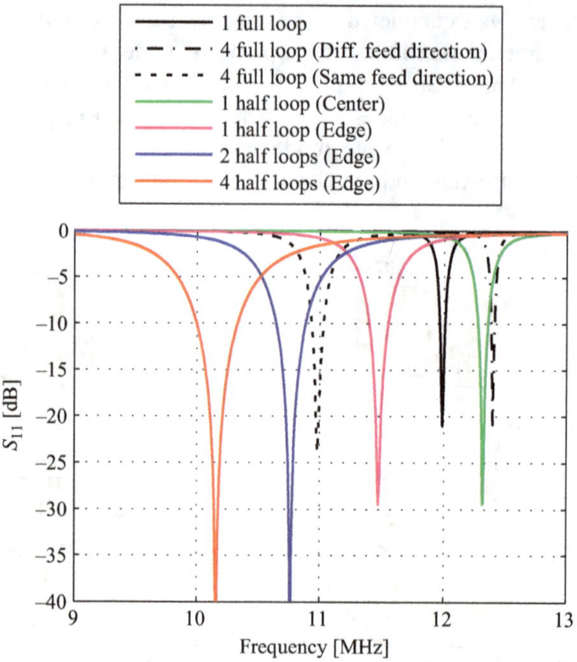

Figure 1.25 Simulated S_{11} of the platform-mounted antennas and full loops shown in Figures 1.24 and 1.27

the ideal direction of the matched half-loop antenna is in the $y-z$ plane, as shown in Figure 1.24(a).

Figure 1.25 shows a comparison between the input reflection coefficients of a full loop antenna in free space (black solid line) and the half-loop antenna (green solid line), shown in Figure 1.24(a), on the platform. Since both loops are small, their input impedances are reactive. Each antenna is impedance matched at the frequency where the real part of its impedance is close to 50 Ω. Therefore, impedance matching can be performed using a simple capacitor in series with the antenna. This simplifies the design of the antenna but also means that the frequency where the best impedance match is obtained can slightly change from one design to the other. As can be observed, while both antennas are narrow-band antennas, the platform mounted antenna exhibits a fractional bandwidth almost twice that of the free-space loop antenna (0.64% compared to 0.33%). This is due to the fact that the half-loop antenna excites currents on the platform and thus has an effectively larger electrical dimension. The current distribution and the radiation pattern of the antenna shown in Figure 1.24 (a) are also calculated using full-wave EM simulations in Altair FEKO and the results are shown in Figure 1.26(a). As can be observed, the current distribution in this case is different from that shown in Figure 1.21(a). This is to be expected since a single loop does not effectively synthesize the required magnetic current distribution on the periphery of the platform.

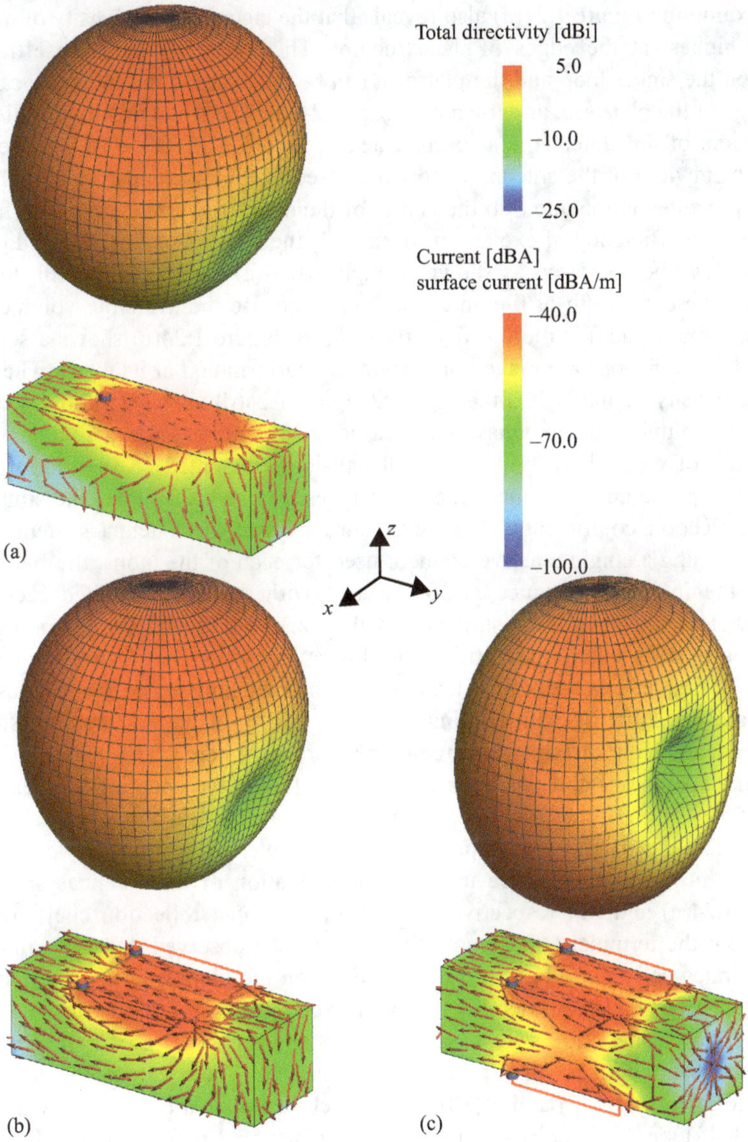

Figure 1.26 Simulated normalized current distributions and radiation patterns of the platform-mounted antennas: (a) one half-loop antenna, placed on the top surface of the platform, is used to excite currents on the surface of the platform; (b) two half-loop antennas, placed at the edges of the top surface, are used to excite electric currents on the platform; and (c) four half-loop antennas, two placed at the edges of the top surface and the other two placed at the edges of the bottom surface, are used to excite electric currents on the platform

Examining Figure 1.21(a) also reveals that the eigen-current density of mode 1 is the highest at the edges of the structure. Therefore, the coupling efficiency between the single loop and the platform can be enhanced if the loop is located at the edge of the platform, as shown in Figure 1.24(b). The simulated input reflection coefficient of this antenna is also presented in Figure 1.25. Observe that the fractional bandwidth of the antenna is now increased to 1.31% compared to 0.64% for the single antenna mounted on the center of the platform's top surface. To further increase the efficiency of excitation of mode 1, the number of loops placed on the platform can be increased as shown in Figure 1.24(c) and (d). This will, in turn, increase the bandwidth of the antenna system because the available volume (i.e., the platform surface) is more efficiently utilized. Figure 1.24(c) shows a scenario where two half-loops are placed on top of the platform and at its edges where the current density of mode 1 is the highest. Figure 1.24(d) shows a similar scenario where the number of antennas is increased to four to provide a more uniform excitation of mode 1. In both cases, all half-loops mounted on the platform are excited in phase and with the same magnitudes. Figure 1.25 shows the simulated input reflection coefficients of these antennas. Since the antenna systems were matched using a single reactive element used for each of the loops, the frequency where the internal resistance is 50 Ω varies with the different feed scenarios. Nevertheless, the frequency shift is less than 2.16 MHz, and the operating frequencies of all scenarios are still in the lower high-frequency band. If identical operating frequencies are required, a two-element matching network can be used to eliminate the shift. Notwithstanding this, the validity of the concept can still be demonstrated using the simple matching network used here.

The dimensions and positions of the loops (inductive coupling elements), as well as the values of the commercially available matching capacitors, are shown in Figure 1.24 and its caption, respectively. S-parameter files of ideal two-way and four-way power splitters were used in the simulation of the antennas shown in Figure 1.24(c) and (d), respectively, to obtain the input reflection coefficient as seen from the terminals of a single feed. As can be observed, by using multiple loops strategically placed at different locations on the platform, the bandwidth of the antenna system is considerably increased. Specifically, for the case of two loops the fractional bandwidth is increased to 2.20% and for the case of four loops, the fractional bandwidth is further increased to 3.27%. These enhancements correspond to factors of 3 (7) and 5 (10), respectively, compared to the bandwidth of a single half-loop (full loop) mounted on the platform (in free space). The total efficiency (η_{tot}) of the platform-mounted antennas shown in Figure 1.24 are greater than 99% (Table 1.1), which is expected because the only source of loss considered in the simulations is the ohmic losses in the conductors. This was done deliberately because the feed network loss increases with the number of coupling loops, primarily due to the loss of the power splitters. Therefore, considering a more realistic loss scenario in the simulations makes it difficult to distinguish the main source of enhancement of the bandwidth. However, as can be observed from the ideal simulations performed here, the bandwidth of the antenna increases significantly as the number of coupling loops is increased. This confirms that the increase in

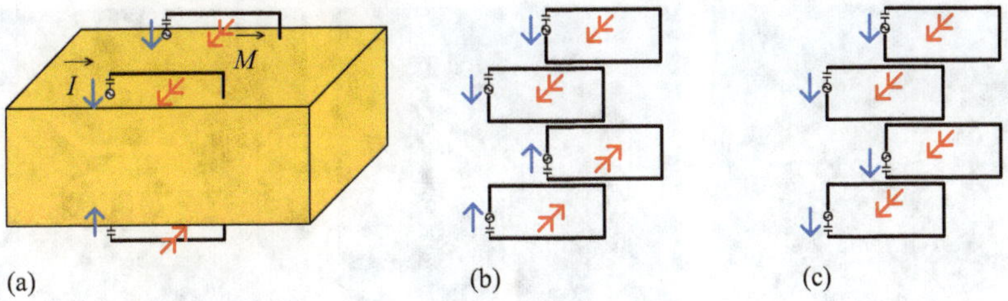

Figure 1.27 (a) Four half-loops on the platform; (b) four full loops in free space. The feed directions of the bottom two loops are 180° different from the top two loops (as in (a)); (c) four full loops in free space (feed directions are the same)

bandwidth is not due to the losses of the system because all systems considered in these simulations have efficiencies close to 100%.

To confirm that the bandwidth enhancement is achieved by exciting the electric currents of the platform, as opposed to increasing the volume over which the four loops are placed, let us examine the performance of a four-element antenna array composed of four full loops in free space, as shown in Figure 1.27. Each loop has dimensions of 4.7 m × 1 m, and the relative positions of the loops in Figures 1.27(b) and (c) are the same as those shown in Figures 1.24(d) and 1.27(a) in the absence of the platform. In Figure 1.27(b), all four elements of this array are fed and impedance matched similar to what was described earlier and the same ideal 4-1 power splitter is used to feed them. Using this approach, the 10 dB return loss fractional bandwidth of this array is simulated to be 0.22%, as shown in Figure 1.25, which is even less than that of the one full loop in free space. This is because the direction of the feed ports of the bottom two loops is 180° different from the upper two loops as is the case in the structure shown in Figure 1.24(d). When the feed directions are the same (Figure 1.27(c)), the fractional bandwidth becomes 0.87%, as shown in Figure 1.25. Either way, the fractional bandwidth of the four full loops in free space is significantly smaller than the 3.27% bandwidth obtained from the structure shown in Figure 1.24(d). This verifies that the presence of the platform and efficient excitation of the platform modes are indeed critical in achieving the enhanced bandwidths reported in Figure 1.25. Additionally, the comparison of the bandwidths of the structures shown in Figure 1.24 (a) and (b) reveals that efficient coupling between the coupling loops and the platform is also important in enhancing the bandwidth of the antenna system. Figures 1.26(b) and (c) show the current distribution and the radiation pattern of the antennas for the cases where two and four coupling loops are used to excite the desired platform mode. Observe that by increasing the number of coupling loops, the radiation pattern and the current distribution of the structure become closer to that shown in Figure 1.21(a). This distribution of the radiating current over a larger volume is in part responsible for the bandwidth improvement observed.

Scaled versions of the structures used in this example were fabricated in [16]. The platforms were fabricated by 3D printing and the 3D printed structures were covered with copper tape. The feed networks for the scaled versions are shown in

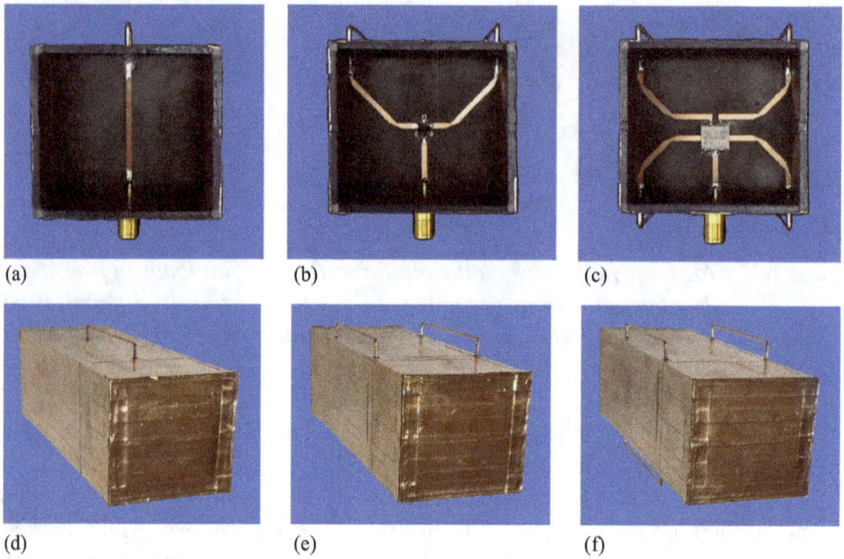

Figure 1.28 Photograph of the scaled platform-mounted antennas. One half-loop scenario (center): (a) feed network (d) prototype; two half-loop scenario (edge): (b) feed network (e) prototype; four half-loop scenario (edge): (c) feed network (f) prototype. The dimensions of the scaled platforms are 46.25 mm × 133.75 mm × 41.25 mm. The dimensions of the scaled half-loops are 58.75 mm × 6.25 mm. The values of the capacitors for the one, two and four half-loop scenarios are 0.25 pF, 0.4 pF, and 0.4 pF, respectively

Figure 1.28(a)–(c).To excite the multiple half-loop antennas on the same platform with the same magnitudes and phases, a two-way power splitter (Mini-Circuits SP-2C1+) and a four-way power splitter (Mini-Circuits SCA-4-10+) were used[6]. Figure 1.28(d)–(f) shows photographs of the fabricated prototypes. The dimensions of the fabricated scaled models are provided in the caption of Figure 1.28. The radiation parameters of the fabricated prototypes were measured using a multi-probe spherical near field system. The measured maximum directivity of the scaled versions of the one, two, and four half-loop antennas are 4.2, 3.5, and 3.9 dBi, respectively. Figure 1.29 shows the measured S_{11} of the fabricated prototypes. The measured 10 dB return loss fractional bandwidth of the one, two, and four half-loop antennas on the platform are 0.91%, 1.78%, and 5.12%, respectively, as shown in Figure 1.29 and Table 1.2. A relatively good agreement between the measured bandwidths of the scaled antennas and the simulated bandwidths of the full-scale antennas can be observed (see Tables 1.1 and 1.2). The normalized radiation patterns of the antennas are also measured and shown in Figure 1.30. From Figures 1.21(a) and 1.30(c), it can be observed that the radiation patterns of the four

[6]The realistic 2-1 and 4-1 power splitters have typical insertion losses of 0.4 and 1.2 dB, respectively.

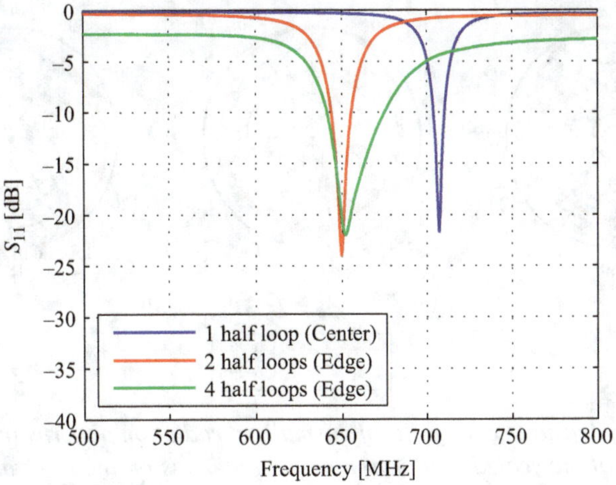

Figure 1.29 Measured S_{11} of the scaled platform-mounted antennas for one half-loop antenna on the platform (center), two half-loop antennas on the platform (edge), and four half-loop antennas on the platform (edge), as shown in Figure 1.28

Table 1.2 Comparison of measurement results of the different scaled antennas

Scenario	f_0 **(MHz)**	**BW**	**Max. D (dBi)**	η_{tot}
One half-loop (Center)	707.55	0.91%	4.2	26%
Two half-loops (Edge)	650.00	1.78%	3.5	37%
Three half-loops (Edge and Center)	634.15	1.97%	3.8	33%
Four half-loops (Edge and Center)	639.60	2.84%	2.8	30%
Four half-loops (Center)	672.65	3.75%	3.1	26%
Four half-loops (Edge)	652.60	5.12%	3.9	32%

half-loop case are very much similar to those of the mode 1 of the platform: omni-directional in the $x-z$ plane, and figure-eight shaped in the $y-z$ plane. The results presented in Figure 1.29 and Table 1.2 demonstrate the validity of using this characteristic-mode-based technique in designing platform-mounted antennas with enhanced bandwidth.

As can be observed from the results, the bandwidth of the platform-based antenna increases when the number of our coupling loops increases. The bandwidth enhancement factor, however, will eventually saturate due to the increasing mutual coupling between the loops. Therefore, an optimum number of coupling loops should exist that, when arranged properly, can provide the bandwidth closest to the maximum available bandwidth.

In addition to the scaled versions of the scenarios shown in Figure 1.24, [16] also examined several alternative structures that use multiple coupling loops where three

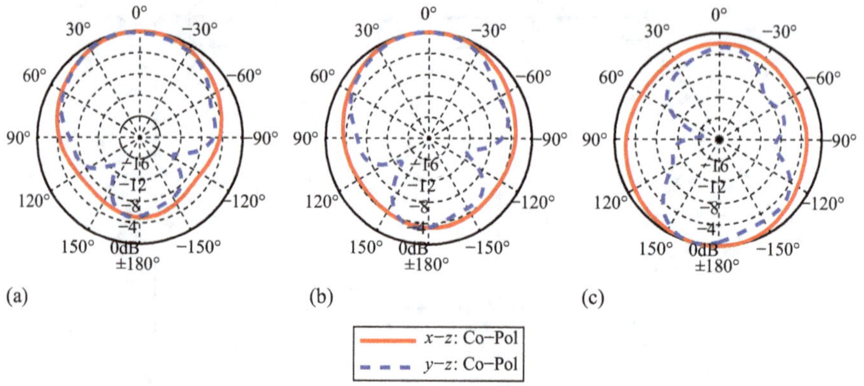

Figure 1.30 Measurement results of normalized radiation patterns (realized gain) of the scaled platform-mounted antennas on the x − z and y − z planes for (a) one half-loop antenna on the platform (center); (b) two half-loop antennas on the platform (edge); and (c) four half-loop antennas on the platform (edge), as shown in Figure 1.28

half-loops are placed on the top of the platform, two half-loops are placed on the top and one half-loop is placed on each side of the platform, or one half-loop is placed on the centers of the top, the bottom, and the two sides as shown in Figure 1.31. The feed networks used to feed each prototype are shown in Figure 1.31(a)–(c). The prototypes of these structures were also fabricated and characterized as discussed in the previous paragraph and are shown in Figure 1.31(d)–(f). The feed networks use three-way (Mini-Circuits SCA-3-11+) and four-way (Mini-Circuits SCA-4-10+) power splitters to excite the loops with the same magnitudes and phases.[7] The measured fractional bandwidth of the three half-loops (Edge and Center) shown in Figure 1.31(a), four half-loops (Edge and Center) shown in Figure 1.31(b), and four half-loops (Center) shown in Figure 1.31(c) are 1.97%, 2.84%, and 3.75%, respectively. Comparison of the results shown in Table 1.2 demonstrates that the bandwidth of these antenna systems can be enhanced by distributing the coupling loops over the entire surface of the platform. Moreover, for the cases that a similar number of coupling loops are used, placing the loops at locations where the current density of the desired mode is strongest improves the bandwidth of the system. This is clearly seen by examining the performances of the antennas shown in Figures 1.28(f) and 1.31(f). In both cases, four coupling loops and the same type of power divider are used. However, the structure shown in Figure 1.28(f) has a fractional bandwidth of 5.12% while the one shown in Figure 1.31(f) has a fractional bandwidth of 3.75%. In all cases, the radiation patterns are horizontally polarized and the direction of maximum radiation is orthogonal to the *y*-axis of the platform. The maximum directivity values of the antennas are respectively 3.8 dBi, 2.8 dBi, and 3.1 dBi for the structures shown in Figures 1.31(d)–(f).

[7]The realistic 3-1 and 4-1 power splitters have a typical insertion loss of 0.7 and 1.2 dB, respectively.

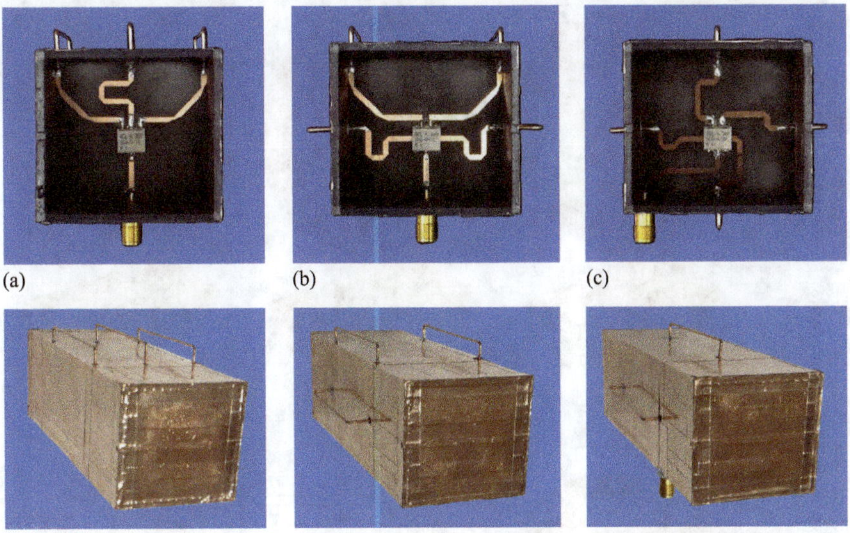

(a) (b) (c) (d) (e) (f)

Figure 1.31 Photograph of the scaled platform-mounted antennas. Three half-loop scenario (edge and center): (a) feed network (d) prototype; Four half-loop scenario (edge and center): (b) feed network (e) prototype; Four half-loop scenario (center): (c) feed network (f) prototype. The dimensions of the scaled platforms are 46.25 mm × 133.75 mm × 41.25 mm. The dimensions of the scaled half-loops in these scenarios are: (d) 58.75 mm × 7.75 mm (center) and 58.75 mm × 6.25 mm (edge); (e) 58.75 mm × 6.25 mm (edge) and 58.75 mm × 8.55 mm (center); (f) 58.75 mm × 6.4 mm (center, top, and bottom) and 58.75 mm × 6.25 mm (center, sides). The values of the capacitors in these scenarios are: (d) 0.3 pF (center) and 0.35 pF (edge); (e) 0.4 pF (edge) and 0.35 pF (center); and (f) 0.25 pF (center)

To demonstrate the applicability of the proposed approach to a more realistic platform, the same design concepts were applied to the U.S. Marine Corps Expeditionary Fighting Vehicle (EFV) in [16]. To accomplish this, a simplified model of the EFV was used in conjunction with full-wave EM simulations in Altair FEKO. Figure 1.32 shows the eigen-current distributions associated with the first few modes of this platform as well as their MSs. It was observed that the first four modes of this realistic platform are similar to those of the simplified platform shown in Figure 1.21. Specifically, in the EFV model, mode 1 of the platform is the dominant mode and it is significant above approximately 11 MHz. When excited in this mode, the entire platform acts as a horizontally polarized dipole antenna with an electric current oriented along the $\hat{y}$ direction. In this case, it was assumed that the coupling loops are placed only on the top and the two side surfaces of the platform. The EFV model was examined when one, two, three, and four half-loops

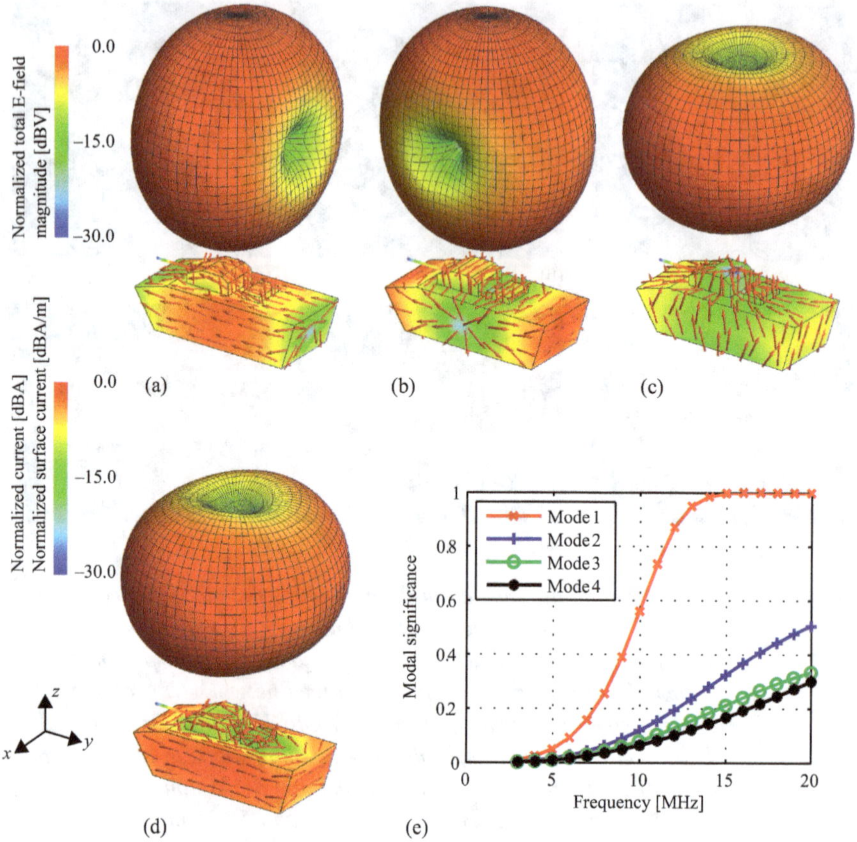

Figure 1.32 Simulated normalized eigen-current distribution and radiation patterns of the first four characteristic modes of the simplified EFV platform: (a) mode 1; (b) mode 2; (c) mode 3; (d) mode 4; and (e) simulated MS of the simplified EFV platform

were placed on the top surface of this platform. Another topology, where two half-loops are placed on the top and one half-loop on each side of the platform was also considered. The particular topologies were chosen for practical reasons. Table 1.3 presents the simulated bandwidth[8] of the different topologies. The bandwidth increases as the number of coupling loops increases, except the four half-loops scenario where all four half-loops are placed on top of the EFV. The fractional bandwidth of this scenario is less than that of the one where three half-loops are placed on the top of the EFV because the strong mutual coupling between the four half-loops increases the input impedance, which decreases the bandwidth. In the two arrangements where four feed loops are used, the fractional

[8]The S_{11} of the antennas are obtained by FEKO simulations that include the feed loops, the Expeditionary Fighting Vehicle, and the related ideal power splitters.

Table 1.3 Comparison of simulation results of the different antennas on the simplified EFV platform

Scenario	f_0 (MHz)	BW
One half-loop (Top)	13.53	0.75%
Two half-loops (Top)	11.94	3.42%
Three half-loops (Top)	11.48	3.52%
Four half-loops (Top)	11.15	3.36%
Four half-loops (Top and Side)	11.12	3.85%

bandwidth of the antenna system where two half-loops are placed on the top and one half-loop on each side of the EFV is wider (3.85% vs. 3.36%). This is due to the fact that the distribution of the coupling loops along the periphery of the EFV more efficiently excites mode 1 of the platform compared to the other case. In all cases examined in [16], the radiation patterns of the EFV antenna systems show a direction of maximum radiation toward zenith and are horizontally polarized. The results obtained using this more realistic platform exhibit the same trends as the ones demonstrated for a simplified platform through simulations and scaled-model measurements. Hence, the technique discussed in this subsection can be applied for designing realistic and more complicated platform-based antennas.

1.3 Summary

A summary of the characteristic mode applications discussed in this chapter can be found in Table 1.4. To design circularly polarized antennas (Section 1.2.1), the characteristic mode analysis was done to identify and select two eigen-modes of the antenna that are significant (MS > 0.707) and to obtain information on the phase difference of these modes. With this knowledge, antennas with circularly polarized patterns can be designed by combining the two orthogonal characteristic modes with a 90° phase difference. In the wideband antenna case (Section 1.2.2), the MSs of the antenna's characteristic modes were examined to find modes that are significant over a wide bandwidth. Subsequently, the eigen-currents and the eigen-fields of the wideband modes were evaluated to find modes that possess omni-directional radiation patterns, which is an essential property of ultra-wideband antennas. In the example of chassis antenna design (Section 1.2.3), two significant chassis modes were excited using the same set of capacitive coupling elements. This was accomplished by understanding the MSs and the eigen-current distributions of the two desired modes, and by placing the capacitive coupling elements at the locations where the magnitudes of the eigen-currents are the weakest for these modes. In Section 1.2.4, loop antennas were mainly used as inductive coupling elements to excite a desired mode of a metallic platform, and a systematic approach to enhance the bandwidth of the platform-based antenna was demonstrated. This approach employed the characteristic mode theory to examine the MSs and the eigen-current distributions of the platform modes, and to

Table 1.4 Summary of the characteristic mode applications discussed in this chapter

Section	Goal	Major radiator	# of modes	Key Parameters[b]
1.2.1	Circular-polarization	Antenna	2	MS, J, α
1.2.2	Wide bandwidth	Antenna	1[a]	MS, J, E
1.2.3	Dual-mode MIMO	Chassis	2	MS, J, MEC
1.2.4	Bandwidth enhancement	Platform	1[a]	MS, J, Q_{rad}

[a]Significant mode with wide bandwidth.
[b]MS: Modal significance; J: Eigen-current distribution; α: Characteristic angle; E: E-field pattern; MEC: Modal excitation coefficient; Q_{rad}: Radiation quality factor.

calculate the maximum available bandwidth that the desired characteristic mode could offer. The characteristic mode theory can be used as a tool to aid antenna design by allowing engineers to interpret the physical characteristics of the radiator(s), and its applications are not limited to the examples discussed in this chapter.

References

[1] Garbacz RJ. Modal expansions for resonance scattering phenomena. Proc IEEE. 1965 Aug;53(8):856–864.
[2] Garbacz RJ, and Turpin R. A generalized expansion for radiated and scattered fields. IEEE Trans Antennas Propag. 1971 May;19(3):348–358.
[3] Harrington RF, and Mautz JR. Theory of characteristic modes for conducting bodies. IEEE Trans Antennas Propag. 1971 Sep;19(5):622–628.
[4] Harrington RF, and Mautz JR. Computation of characteristic modes for conducting bodies. IEEE Trans Antennas Propag. 1971 Sep;19(5):629–639.
[5] Austin BA, and Murray KP. The application of characteristic-mode techniques to vehicle-mounted NVIS antennas. IEEE Antennas Propag Mag. 1998 Feb;40(1):7–21, 30.
[6] Cabedo-Fabres M, Antonino-Daviu E, Valero-Nogueira A, *et al.* The theory of characteristic modes revisited: A contribution to the design of antennas for modern applications. IEEE Antennas Propag Mag. 2007 Oct;49(5):52–68.
[7] Chen Y, and Wang CF. HF band shipboard antenna design using characteristic modes. IEEE Trans Antennas Propag. 2015 Mar;63(3):1004–1013.
[8] Newman EH. Small antenna location synthesis using characteristic modes. IEEE Trans Antennas Propag. 1979 Jul;27(4):530–531.
[9] Garbacz RJ, and Pozar DM. Antenna shape synthesis using characteristic modes. IEEE Trans Antennas Propag. 1982 May;30(3):340–350.
[10] Liu D, Garbacz RJ, and Pozar DM. Antenna synthesis and optimization using generalized characteristic modes. IEEE Trans Antennas Propag. 1990 Jun;38(6):862–868.

[11] Harrington R, and Mautz J. Control of radar scattering by reactive loading. IEEE Trans Antennas Propag. 1972 Jul;20(4):446–454.

[12] Chen Y, and Wang CF. Characteristic-mode-based improvement of circularly polarized U-slot and E-shaped patch antennas. IEEE Antennas Wireless Propag Lett. 2012;11:1474–1477.

[13] Ciafardini JP, Daviu EA, Fabrés MC, *et al.* Analysis of crossed dipole to obtain circular polarization applying characteristic modes techniques. In: 2016 IEEE Biennial Congress of Argentina (ARGENCON); 2016. p. 1–5.

[14] Wu W, and Zhang YP. Analysis of ultra-wideband printed planar quasi-monopole antennas using the theory of characteristic modes. IEEE Antennas Propag Mag. 2010 Dec;52(6):67–77.

[15] Kishor KK, and Hum SV. A two-port chassis-mode MIMO antenna. IEEE Antennas Wireless Propag Lett. 2013;12:690–693.

[16] Shih TY, and Behdad N. Bandwidth enhancement of platform-mounted HF antennas using the characteristic mode theory. IEEE Trans Antennas Propag. 2016 Jul;64(7):2648–2659.

[17] Li H, Tan Y, Lau BK, et al. Characteristic mode based tradeoff analysis of antenna-chassis interactions for multiple antenna terminals. IEEE Trans Antennas Propag. 2012 Feb;60(2):490–502.

[18] Miers Z, Li H, and Lau BK. Design of bandwidth-enhanced and multiband MIMO antennas using characteristic modes. IEEE Antennas Wireless Propag Lett. 2013;12:1696–1699.

[19] Kishor KK, and Hum SV. A pattern reconfigurable chassis-mode MIMO antenna. IEEE Trans Antennas Propag. 2014 Jun;62(6):3290–3298.

[20] Chen Y, and Wang CF. Electrically small UAV antenna design using characteristic modes. IEEE Trans Antennas Propag. 2014 Feb;62(2):535–545.

[21] Wong KL, Wu CH, and Su SW. Ultrawide-band square planar metal-plate monopole antenna with a trident-shaped feeding strip. IEEE Transactions on Antennas and Propagation. 2005 Apr;53(4):1262–1269.

[22] Chu LJ. Physical limitations of omni-directional antennas. J Applied Physics. 1948 Dec;19(12):1163–1175.

[23] McLean JS. A re-examination of the fundamental limits on the radiation Q of electrically small antennas. IEEE Trans Antennas Propag. 1996 May;44 (5):672–676.

[24] Yaghjian A, and Best SR. Impedance, bandwidth, and Q of antennas. IEEE Trans Antennas Propag. 2005 Apr;53(4):1298–1324.

[25] Geyi W. A method for the evaluation of small antenna Q. IEEE Trans Antennas Propag. 2003 Aug;51(8):2124–2129.

Chapter 2

Design of antennas mounted on complex platforms using the characteristic mode (CM) and characteristic basis (CB) function methods

Raj Mittra[1], Ashwani Kumar[2], and Chao Li[3]

2.1 Introduction

While the EM simulation techniques for predicting the radiation characteristics of antennas mounted on complex platforms are very well developed, the same cannot be said about deigning antennas to realize specific performance goals, for example, achieving multiband coverage, desired radiation pattern, efficiency level, return-loss characteristic, etc., which meet the user specifications that can often be very challenging to realize, especially because of the size limitations and/or complexities of the platforms.

In this work, we will discuss two types of antenna design strategies, tailored for two different categories of antenna design problems. First of these will address the well-known problem of designing a multiband antenna, to be mounted on a mobile phone platform, to simultaneously cover the typical communication bands, for example, LTE, GSM, Bluetooth, WiMax, etc., with only a *single* antenna tailored for the platform, as is usually the requirement, except for MIMO applications.

Next, we will turn to the problem of designing a plurality of antennas to be located on a complex platform to achieve certain radiation characteristics deemed desirable by the user, in order to receive the desired signals while suppressing the interfering ones. While several methods for radiation pattern synthesis of antennas exist in literature, as for instance [1–4], they are only applicable to specific antenna geometries and they typically do not handle complex structures. The pattern synthesis problem is of great interest in many practical applications, for example, global positioning system (GPS) antenna design, and locating multiple antennas mounted on the rooftop of a large vehicle, or on the topside of a ship.

Recently, there has been considerable interest [5–13] in applying the theory of characteristic modes (TCM) for the purpose of designing antennas to operate

[1]EMC Lab, University of Central Florida, USA
[2]Department of Electronics, Delhi University, India
[3]School of Physics and Technology, University of Jinan, China

on complex platforms. Since this theory is well covered in Chapter 1, we suggest that the readers familiarize themselves thoroughly with the basics of TCM presented in the above chapter before proceeding further with the present chapter. The readers are also encouraged to refer to the extensive body of literature on the subject that provides a variety of examples to which TCM has been applied, for example, for the design of mobile terminals [9,10], antennas on military vehicles [11,12], and topside of a ship [26]. However, for the sake of completeness and convenience of the readers, we do include a brief summary of TCM in Appendix A1 of this chapter, along with some definitions that would be useful for our discussion of TCM in the context of the mobile phone antenna problem, which we deal with below in the Section 2.2.

We begin this chapter by presenting a step-by-step procedure that we would need to follow, in the context of TCM, to design a mobile phone antenna on a platform, whose physical dimensions, geometrical details and material parameters have presumably been provided to us by the user. Next, we identify some of the difficulties we are likely to encounter when attempting to apply the TCM to the mobile phone problem. Our primary concerns—as we will soon see—are: (i) finding the modes that satisfy the modal significance (MS) criterion (see the definition in Appendix A1) for the multiple frequency bands that the phone needs to cover; and (ii) figuring out how to design antennas that can excite these modes on the given platform in a systematic manner —even if we could identify the particular set of modes we would need to realize the desired performance. We will further elaborate on these points in Section 2.2 and will include examples to validate our arguments given above regarding the problems with the TCM approach, when applied to the mobile phone problem at hand.

Next, we will present an alternative method which addresses the same antenna design problem for mobile phone platforms, but approaches it in a way such that the challenging issues with the TCM, which we have just identified above, are obviated. The alternative approach we propose is based on the concept of the characteristic basis function method (CBFM) [14,15], which has been introduced in the past for the solution of radar cross section computation, as well as for numerical modeling of antenna configurations. We will describe how we tailor the CBFM for the problem of antenna placement on complex platforms, by confronting the problem of *excitation* right up front, in a systematic manner, to determine the location of the antenna on the platform, within the constraints that have been specified by the user in terms of the real estate available for the antenna on the phone platform, which is usually quite limited since we have to accommodate the battery, electronic circuits, camera, etc., sharing the same platform.

The principal difference between the characteristic modes (CMs) and the characteristic bases (CBs) is that the former are source-independent, while the latter are derived from sources (antennas) that are chosen either by the user, or the designer, right at the beginning of the design process. The difficulty arising in the CM approach stems from the fact that the "source-free" nature of the CMs neither connects them to any particular set of feed excitations nor to the locations of these sources on a given platform. Consequently, even if we determine a set of CM distributions that would produce the desired pattern when a given platform supports these CM distributions, there is no clue provided by the TCM as to what type of antennas we should use and where we should place them to realize the desired pattern. This is precisely the question for which the CBFM provides a systematic answer, as we will demonstrate below.

Before closing this section, we mention that Newman has suggested in [13] that the optimum approach to launching a particular CM on the platform is to excite the structure at a location corresponding to the maximum value of the modal distribution. While this is true for trivial cases of wire structures, we will show, by some examples, that simply placing an excitation source (antenna)—which is typically specified by the user to satisfy the requirements of frequency coverage and bandwidth—at the location of the "maximum" of the modal current distribution on the platform corresponding to the CMs of interest, often fails to produce the desired current distribution and, hence, the pattern. Furthermore, the issues pertaining to frequency coverage, return-loss bandwidth and antenna efficiency, etc., which can be even more important than the radiation pattern, are not addressed in the TCM-based methods. Moreover, the user typically stipulates a "forbidden zone" where we are not allowed to place these sources, in order to refrain from contesting the real estate occupied by the auxiliary components, such as the battery or the camera, on the platform of a mobile phone. It is not immediately obvious how to effectively launch and to control the various CMs to synthesize the desired radiation pattern while complying with the stipulated constraints at the same time. Perhaps even more important is the fact the user specifications typically place priorities on the frequency coverage and efficiency, rather than on the radiation pattern of the antenna on the platform, and it is difficult to relate these performance metrics to the CMs.

2.2 TCM approach to designing antennas for mobile phone platforms

In this section, we outline the steps necessary to design a mobile phone antenna for a specified platform designated by the user. The simple geometry of the rectangular plate (see Figure 2.1), which we will designate in this work as Platform-1, is often

Figure 2.1 Mobile phone platform-1

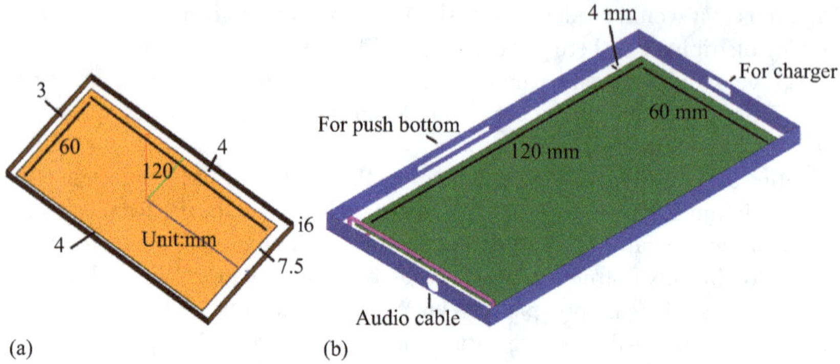

Figure 2.2 Mobile phone platform-2

Table 2.1 Desired frequency coverage bands

Bands	Coverage	Frequency range
Band-1	LTE and GSM	760–1,001 MHz
Band-2	LTE and GSM	1,400–2,100 MHz
Band-3	Bluetooth and WiFi	2,400–2,450 MHz
Band-4	WiMax and WLAN	3,000–3,700 MHz
Band-5	Fixed mobile	4,300–4,700 MHz

chosen [16–20] to illustrate the application of the TCM to the problem at hand. However, we will choose a slightly different version of the platform geometry, namely one with a frame (see Figure 2.2), since this is the geometry which is commonly used in the mobile phone designs today by most manufacturers, and we will refer to this geometry as Platform-2. The discussion of the method we present in this section is equally relevant for both platform geometries, however, as well as for any other platform geometry.

Let us assume that the user has specified, as is typically the case, the performance requirements of the device in terms of frequency coverage as follows, and our task is to explore a systematic application of the TCM for this antenna-on-platform problem.

Our first step is to derive the various CMs for the platform, and the associated eigenvalues as well as "MS" of these modes, as we vary the frequency from 760 MHz– 5 GHz to cover the range of interest indicated in Table 2.1. Associated eigenvalues as well as "MS" of these modes are shown in Figure 2.3.

The CM distributions for Platform-2 and their associated patterns are presented in Figures 2.4 and 2.5. It is evident from these figures that each mode has multiple "hot-spots," that these spots vary with different modes, and that the choice of the preferred hot-spot for the purpose of exciting these modes [21–23], where the current concentration is high, is not at all obvious. What exacerbates the problem even further is that almost all the antennas that are mounted on these platforms, for example, PIFAs

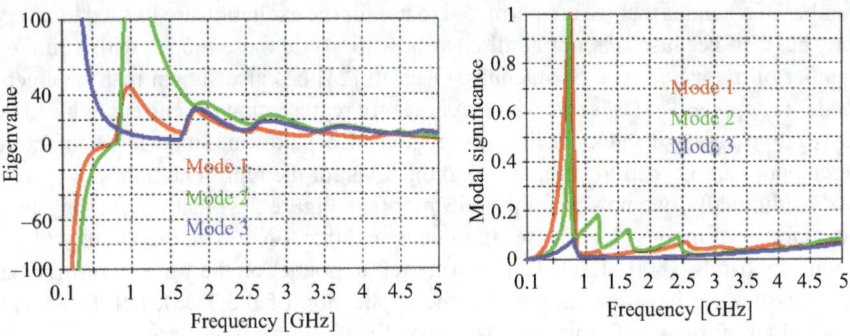

(a) Eigenvalues (b) Modal Significance

Figure 2.3 Eigenvalue and modal significance plots

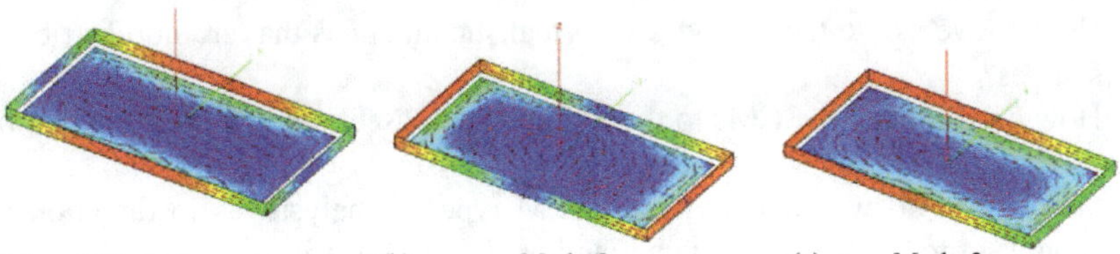

(a) Mode-1 (b) Mode-2 (c) Mode-3

Figure 2.4 CM current distributions for platform-2 for different modes at 800 MHz

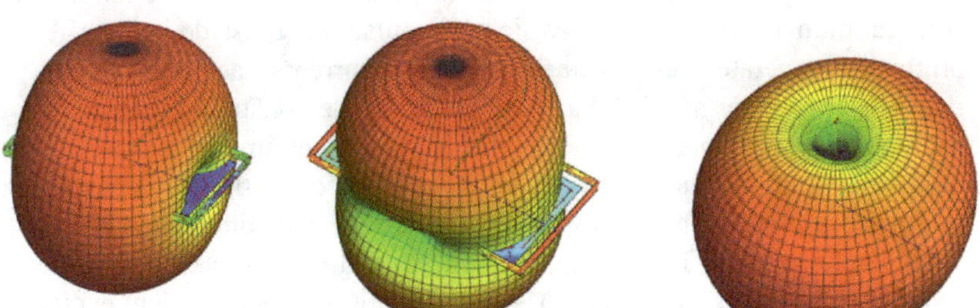

(a) Mode-1 (b) Mode-2 (c) Mode-3

Figure 2.5 CM radiation pattern for platform-2 for different modes at 800 MHz

and their variants, excite current distributions with high imaginary-part contents, especially in the vicinity of the exciting antenna, while the CM distributions are entirely real, as is evident from (Appendix A1), the defining equation for the CMs. In fact, it has recently been shown by Oueslati [24] that one needs upward of hundreds of CMs to accurately represent the current distribution generated by a typical source placed on a platform, which is obviously not practical to do, not only because the number of antennas that we are permitted to use for the excitation are limited to a very few, but also because it is not at all clear how to excite these modes with a limited number of *allowable* type of antennas. Oueslati [24] has also shown that the eigenmodes (as opposed to CMs) are a better choice for representing the current induced by an exciting source located in the near region of a platform, since fewer of these eigenmodes are needed in comparison to the CMs for the representation.

Additionally, we note from the MS plot (see Figure 2.3) that only a very limited number of modes (1 or 2) are significant, and that too only in the vicinity of the frequency band—(800 MHz). Thus, we are left to ponder on the ways to resolve the major issues we have encountered in the application of the TCM, before we can proceed any further with this exercise. Specifically, these issues are:

1. How do we satisfy our requirements for multiple frequency coverage by using the CMs that we have generated for the given platform?
2. How many modes would we need to meet the bandwidth requirements for the return-loss specified by the user?
3. How do we excite these modes with realistic antennas that are not distributed sources?
4. How do we relate the CMs to the efficiency, to ascertain that we have met the user specifications?
5. What clues do we have from the TCM type of analysis, as to direction we should go to modify the antenna and/or the platform in order to achieve our performance goals?
6. How do we handle MIMO problems, and place two or more antennas on the same platform with low levels of coupling between them, which is needed for satisfactory MIMO operation?
7. How will we deal with the situation that the addition of an excitation source configuration to the platform, which of course we must do, will affect the originally generated eigenvalues and eigencurrents, as well as the MS response? Ikram *et al.* [25] have convincingly demonstrated that adding the excitation source on the platform has a significant impact on eigenvalues, modal distributions and the MS. They have pointed out that any change in the original platform geometry will alter all of these quantities and, hence, they must be reevaluated. The CB approach described in Section 2.3 addresses this issue up front since, unlike the CMs, the CBs are source-dependent excitations, and we only need to calculate them just once.

Unfortunately, there is no simple and/or clear-cut answers to any of the questions we have posed above, with the exception of #3. Appendix A2 examines this question, provides a theoretically rigorous solution for the Electric Field distribution that would be needed to excite a given CM, and demonstrates that a *distributed* source must be used to excite them, as shown in Figures 2.6 and 2.7, for instance. A slightly different approach to answering #3 is also provided in Appendix A1, which appeals to Equation (A.1.1) in Appendix A1 to determine what type of excitation will excite a single CM. As a corollary, it also shows why such a distributed source would be very difficult to realize in practice.

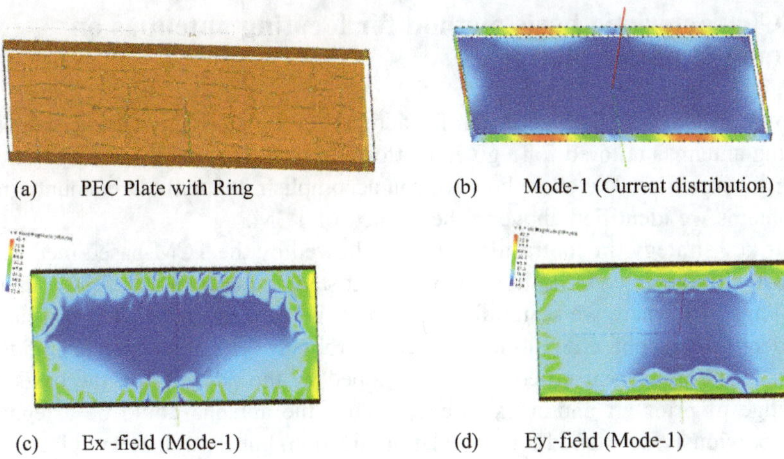

Figure 2.6 *(a) Mobile phone chassis (plate with a metal ring) and (b) mode-1 distribution. The Ex- and Ey-fields needed to excite the mode are shown in (c) and (d), respectively*

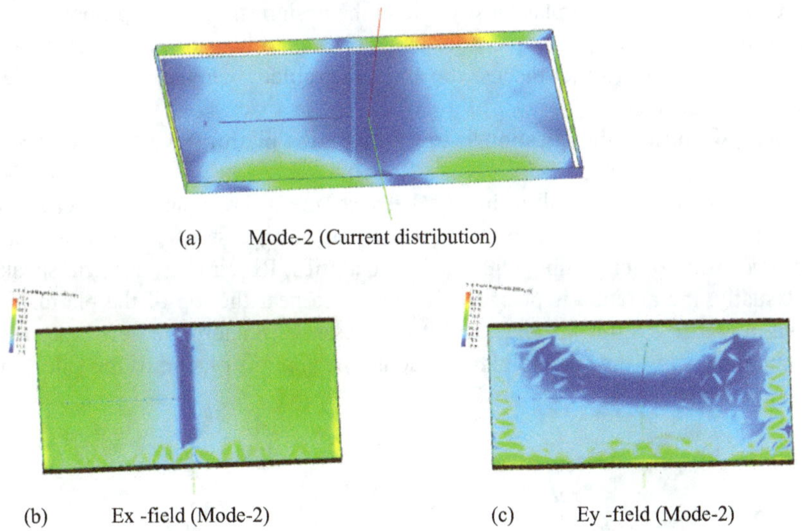

Figure 2.7 *(a) Mode-2 current distribution for mobile phone chassis (plate with a metal ring), shown in Figure 2.6(a); (b) Ex-field; and (c) Ey-field*

Next, below we discuss an alternative approach, based on the CBFM, for addressing the same problem as we have discussed in this section, namely deigning and placing an antenna in an appropriate location on Platform-2, but without the burden of the problems we encountered when attempting to use the TCM for the same purpose.

2.3 Characteristic basis method for locating antennas on mobile phone platforms

We now turn to the CBFM approach for addressing the problem at hand, namely designing antennas tailored for a given platform to meet the desired user-specified performance metrics. We show how we can accomplish this without encountering the problems we identified above in the context of TCM.

Our key strategy, in contrast to the one followed by the TCM-based methods, is to start with the antenna itself as the very first step, rather than as a follow-up of the CM analysis where we search for ways to excite the "desired" CMs that have been identified as potential solutions of our problem. The initial design of the antenna could be user-specified, or be developed by the designer on the basis of knowledge of prior art and/or experience; hence the antenna could be a legacy device, previously developed for a similar application, but for a different platform. However, it should have the important feature that it can be tuned for various frequency bands by adjusting certain parameters. Additionally, it would be highly desirable for the antenna to have the feature that varying one of the parameters of the antenna, for example, the length of one of the arms, primarily controls its response at one frequency, with only a little influence on the others. The availability of such an antenna plays a key role in the design strategy we propose herein, and the modal analysis alone provides us no clue as to how we might design such an antenna. We will further discuss the antenna problem below and will provide an example of the same.

Next, we turn to the question as to where on the platform we should place the antenna.

Our choices are usually limited because the user typically specifies the "allowable" locations at a late stage of the design process of the platform, after accommodating other components, such as batteries, RF circuits, camera, speaker, etc. Usually, the antenna is placed in a small area near the top of the platform, as shown in Figure 2.8, for instance, and it is necessary to determine how best to feed the antenna against the ground plane. Placing the antenna on one of the sides, or at

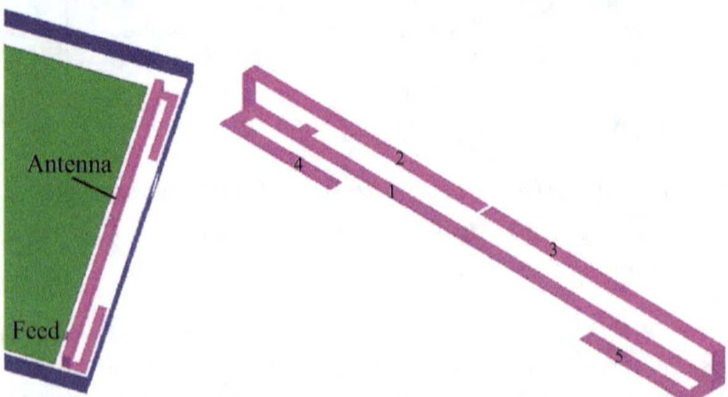

(a) Feed at top edge (b) Antenna

Figure 2.8 Antenna near the top of the platform

the bottom, are also options that can be explored, provided such placements are not entirely excluded *a priori* by the user, of course.

To answer the question of placement of the antenna—assuming that we already have a preliminary design for the same—we now turn to the CBFM, and generate the results for the current distribution, the radiation pattern, the return-loss characteristics and the efficiency, across the entire frequency range of interest, for a number of different allowable locations to determine which one delivers the performance that comes close to the desired one. The distributions themselves are referred to as the CBs that are obviously source-dependent, which distinguishes them from the source-independent CMs; hence we do not need to concern ourselves later with the question as to how we are going to excite them, which posed the difficulty for us when we attempted to follow the TCM-based design strategy, instead of working with the CB-based procedure we are proposing now.

The final step is to co-design the antenna and the platform, by modifying one or both, within the allowable constraints of size, and without violating the mechanical guidelines as well as requirements driven by cosmetic considerations. For instance, we are usually permitted to change the size of the platform only a little and can only modify the metal frame slightly by introducing some gaps to interrupt the current flow in the frame. The current distributions we generate for each location are the CBs, which we work with to see how well they satisfy our needs, and we modify them on as-needed basis. Typically, we zoom in on a single CB that comes closest to fulfilling our requirements, rather than using a plurality of the same, since that would require using more than one antenna on the mobile phone platform, which the mobile form manufactures do not permit us to do, often because of cost considerations. This is in contrast to the TCM-based approach in which we often need to work with multiple CM distributions to meet the various performance specifications, and must concern ourselves with the difficult—if not impossible—task of exciting these CM distributions by using a single launcher, as stipulated by the user.

We now illustrate the CB-based procedure for the mobile phone platform (Platform-2), for which we have presented the frequency coverage specifications in Table 2.1. The antenna design that we initially choose is shown in Figure 2.9, and its return-loss characteristic vs. frequency is presented in Figure 2.10, when the antenna operates either in free-space environment or on another platform. We realize that these characteristics would be modified when the antenna is placed on the phone platform we are working with, which would alter both its original near- and far-field characteristics. Our next step is to place this antenna at various locations on the platform to determine the best choice for the same, and we find that

Figure 2.9 Initial design of the antenna feeding at (a) center; (b) edge; and (c) folded strip

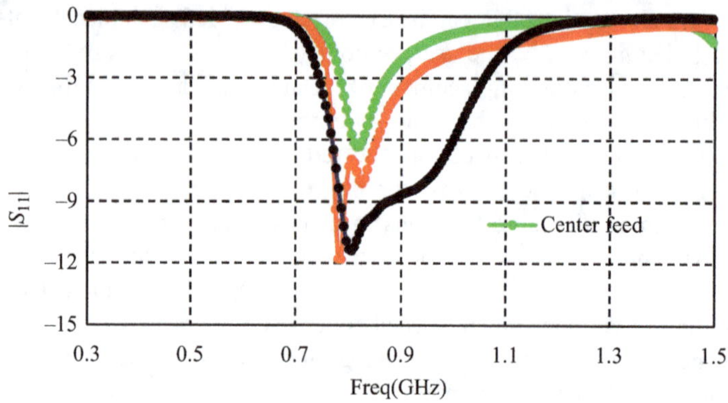

Figure 2.10 Effect on $|S_{11}|$ *of the antenna with feed located at (a) center; (b) edge; and (c) folded strip*

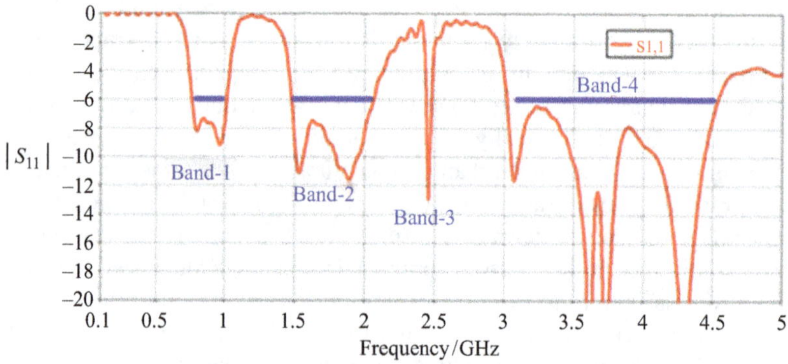

Figure 2.11 $|S_{11}|$ *characteristics of the mobile antenna designed for platform-2*

placing it between the frame and the ground near the top (see Figure 2.9(b) and (c)) is the preferred location. The return-loss characteristic for this choice of antenna placement is shown in Figure 2.10, which obviously does not satisfy our requirements at all five frequency bands. We then modify the antenna, by adding additional arms, and also adjust both the feed location as well as the lengths of the arms to realize the desired performance, as shown below.

Figure 2.11 displays the performance characteristics of the antenna mounted on the mobile phone platform-2 as shown in Figure 2.12. It is evident that the co-designed antenna-plus-platform indeed meets our design goal of multiband coverage as specified in Table 2.1.

Before closing this section, it is worthwhile to ask the following question: "Did the CM analysis provide us any clue as to how we might have modified the platform to generate "modally significant" CMs which cover the user-specified frequency range?" And the short answer is "No." Next, we might ask a related

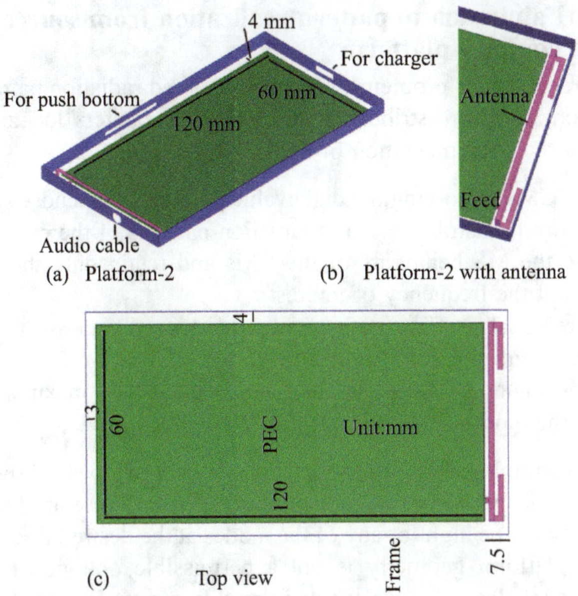

Figure 2.12 (a) Mobile antenna platform-2; and (b) antenna mounted on platform-2

question: "Does the antenna we designed, on our own (without help from TCM), generate the CMs which do cover some of the higher frequency bands that were missed by the CMs associated with the platform alone?" The answer to this question is affirmative, and, as shown in Figure A2.1 of Appendix A2, a few additional CMs that cover the higher frequencies can be generated by certain antenna configurations that are designed to be multiband. Unfortunately, as shown in Figure A2.2, the combination of antenna plus platform loses some of the CMs that covered the high frequency bands, once the antenna is mounted on the phone platform. For this reason, we recommend following the CB approach, and design the antenna to cover multiple frequency bands, as a first step, and then determine its best location on the platform.

2.4 Placement of multiple antennas on a complex platform

2.4.1 TCM-based approach

We now turn to the problem belonging to the second category, namely the realization of the desired radiation pattern from a *plurality* of antennas placed on a given platform, at desired frequencies specified by the user. We will briefly review, once again, the TCM-based method for addressing this problem, identify the difficulties encountered in the process, and then go on to apply the CB-based method to a few illustrative examples to circumvent the aforementioned difficulties.

2.4.1.1 TCM approach to pattern realization from sources located on a complex platform

The TCM approach to the problem of realizing desired radiation pattern at a given frequency f from current distributions generated by exciters located on a given platform entails the implementation of the following steps:

(a) Derive the CMs by solving the eigenvalue problem (Appendix A1), as before.
(b) Find the current distributions and radiation patterns of the modes.
(c) Determine the MS behaviors of the CMs and retain only the ones that are significant at the frequency of interest f.
(d) Determine modal coefficients by fitting the desired patterns, using optimization (or least square), for instance.
(e) Place the excitations (User specified antennas) at the "maxima" of the CMs to launch the modes.

Although seemingly straightforward, the procedure given above encounters some practical issues when we attempt to implement it. These are listed below:

The MS may not be high for any of the modes at the desired frequency f. Then, if altering the platform geometry is not a permissible option, which is almost always true, it is not clear what we can do to modify our strategy so that we can still work with the CMs. An example of this scenario was provided earlier, in the last section for a mobile phone platform. There the MS issue was found to arise at frequencies above 1 GHz, for instance, as may be seen from Figure 2.3.

There are some scenarios, especially those that involve electrically large structures with a large number of degrees of freedom (DoFs) associated with them, for example, the topside of a ship, or an airplane, where the number of modes can be unmanageably large, even upward of millions (see Figure 2.1 in [26], for instance). As an illustrative example, we consider a large plate whose size is 10 m × 30 m, and we derive a relatively large number of modes (say 100—the actual number of modes is much larger) by using CMA in FEKO. Solving for the modes of large matrices can be very time-consuming, and the run-time for extracting the eigenvalues and eigenvectors of these matrices can be hours. Hence, it is pertinent to ask whether going through the exercise of extracting the modes is worthwhile investment of our time. To answer this question, we plot the MS of the computed modes in Figure 2.13, which shows that only two modes are significant. Next, in Figure 2.14, we show the CM current distributions for some of the modes on the large plate, derived from FEKO, and we plot the radiation pattern of these modes in Figure 2.15.

As mentioned earlier, generating a large number of CMs associated with large matrices can be very time-consuming. What is even more significant is the fact that there is no clear-cut strategy spelled out by the TCM for handling this situation in a systematic and efficient manner. (We will show later in this section how the CB-based approach deals with this problem of large structures relatively easily, systematically, and efficiently, without having to solve the eigenvalue problem associated with large matrices, which can be unwieldy.) Figure 2.16 displays the

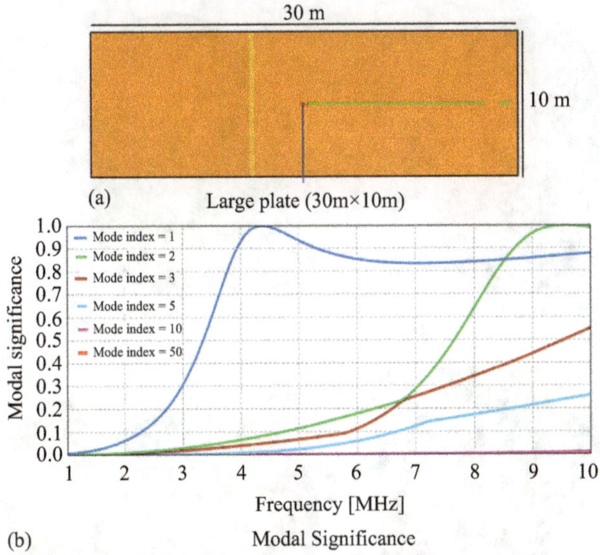

Figure 2.13 Modal significance of large plate (10 m × 30 m)

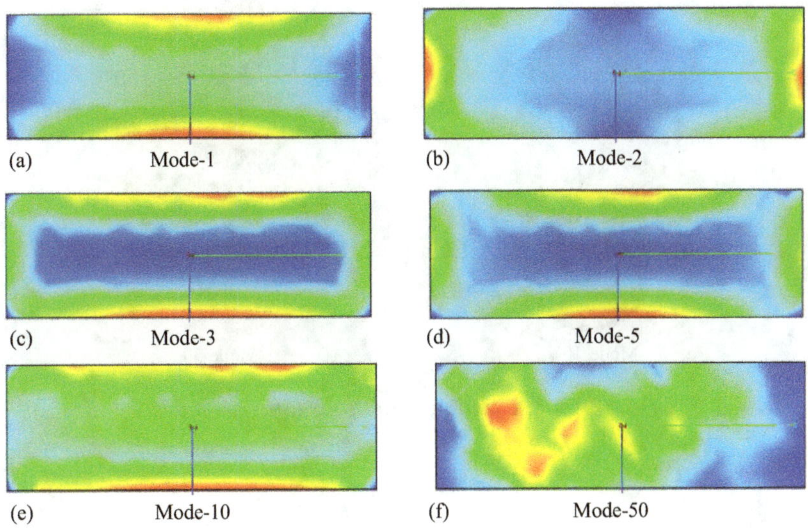

Figure 2.14 Modal current distribution for different modes

placement of the monopole on a large plate, and Figure 2.17 shows the current distribution and radiation pattern for these monopoles. Note that neither the current generated by the monopole nor its pattern matches the CM current distribution and/or its associated radiation pattern.

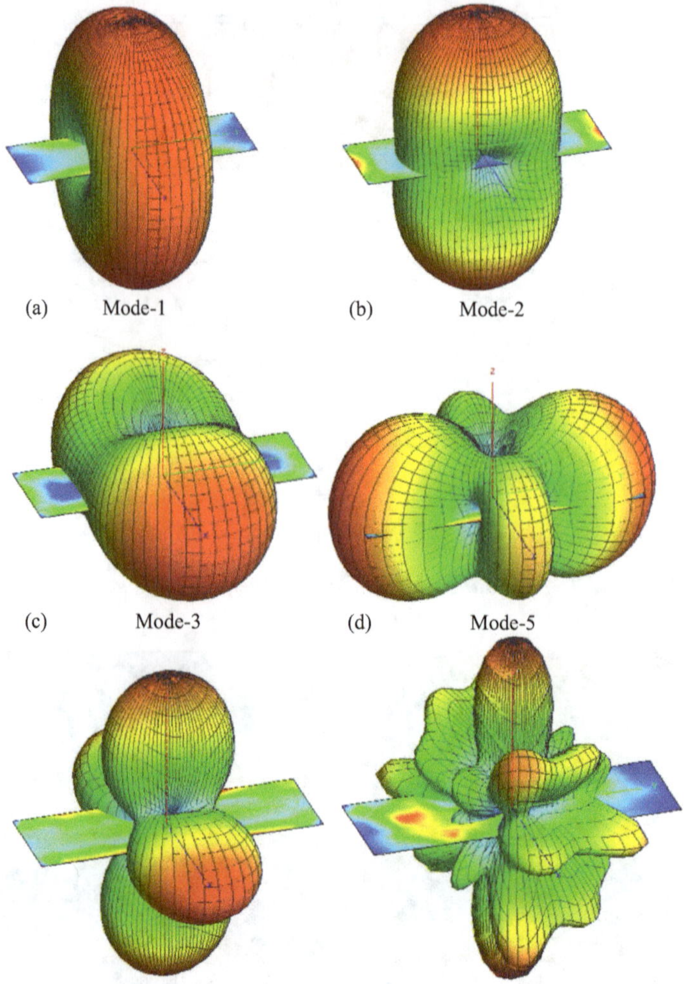

(a) Mode-1 (b) Mode-2

(c) Mode-3 (d) Mode-5

(e) Mode-10 (f) Mode-50

Figure 2.15 Radiation pattern for different modes

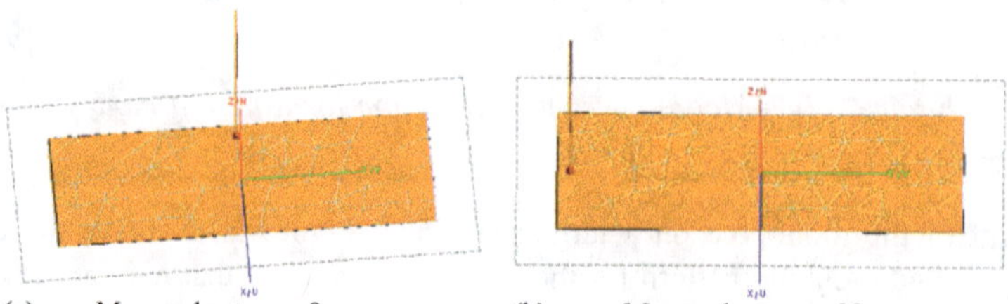

(a) Monopole at $x = -9$ m (b) Monopole at $y = -29$ m

Figure 2.16 Monopole exciter at two positions on a large plate (10 m × 30 m)

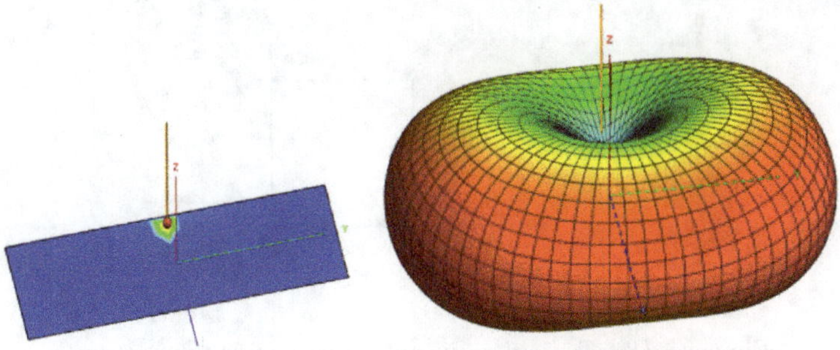

(a) Current distribution (monopole at $x = -9$ m); (b) Radiation pattern

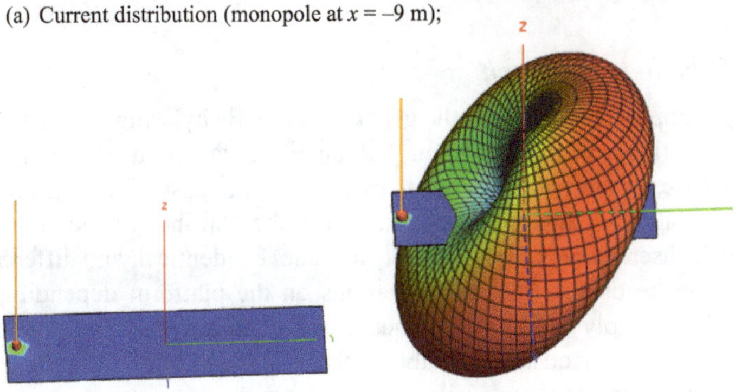

(c) Current distribution (monopole at $y = -29$ m); (d) Radiation pattern

Figure 2.17 Current distribution and radiation pattern of a monopole on a large plate (10 m × 30 m)

(i) Even if we were able to somehow resolve the issue of handling a large number of modes associated with a structure whose method of moments (MoM) matrix is large, the issue of exciting these modes would still have to be confronted with, and this problem is expected to be exacerbated manifold as compared to the single-mode case we have discussed in the last section.

(ii) An important issue related to the excitation problem is the determination of the modal coefficients, and we need to follow the procedure outlined in [21–23] to represent the desired radiation pattern in terms of the modal patterns. However, to use this method we would need the user to specify the pattern over the entire angular (θ, ϕ) range, which is not realistic. Instead, often the user simply designates the preferred angular region in which the radiation should be strong, and the complementary angular range where it would be relatively weak. In other words, typically only the amplitude envelope of the pattern is available, as shown in Figure 2.18, for instance, and it is not a trivial task to extract the magnitude and phase of the radiated field from this type of pattern specification, needed to determine the weight coefficients of the CMs. We will show below how we handle this practical issue in the context of the CB approach in a systematic and efficient manner.

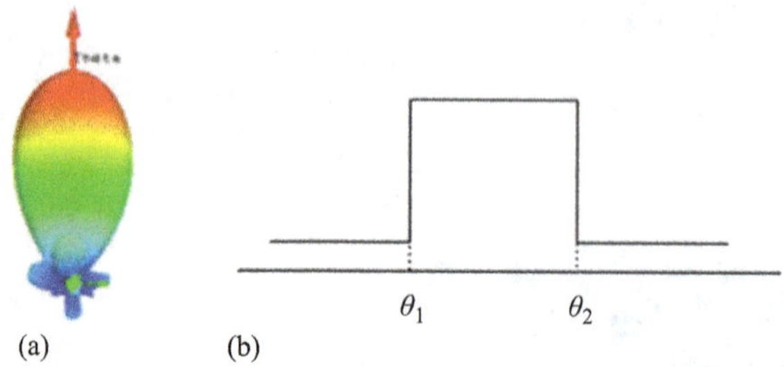

Figure 2.18 Typical amplitude envelope of the pattern

2.4.2 CB-based approach

The CB-based approach starts with the generation of CBs by using antenna elements at different locations on the platform. The antennas can be user-specified, as is typically the case, or they may be chosen by the designer based on the promise they show as potential candidates for the task, within the confines of the acceptability limits set by the user. The antenna elements need not be identical, and different types of antennas can be used at different locations on the platform depending on the application. To comply with the geometrical constraints specified by the user, the antennas are only placed at "allowable" locations, also specified by the user. The CBs are generated for the complex structure under analysis by using a suitable full-wave solver, which could either be a MoM- or FEM-type, home-grown or commercial. A singular value decomposition (SVD) can be employed to remove the redundancy of the CBs, if necessary. As pointed out earlier, the CBs are fundamentally different from the CMs—the latter being "source-free" solutions—and they (CBs) not only enable us to approximate the desired pattern in a systematic manner but also to meet the requirements on the return-loss, bandwidth, multiple frequency coverage, efficiency, and other performance metrics that are typically specified by the user, but are not easily related to the source-free CM distributions. An additional advantage of the proposed CB-based approach is that it does not require dealing with the problem of extracting the eigenvalues and eigenvectors of matrices associated with the complex platform, and to keep track of the order of the characteristic modes, which may be difficult to do unless the number of CMs we need for the problem at hand is small. Note also that the coupling between different antennas and the effect of material parameters, for example, those of single or multilayer substrates, lossy or lossless, can be taken into account rigorously in the process of deriving the CBs, while the same is not true in the TCM approach, which is neither too well suited for handling the excitation issue involving multiple antennas (since this issue is skirted when we generate the CMs) nor for application to platforms with non-PEC type of material characteristics, as shown in [24; Chapter 3].

2.4.2.1 Pattern control using CBs

The concept of characteristic base pattern (CBP) was first introduced in [27] for the design of the MeerKAT/SKA radio telescope. By measuring only a few calibration sources, the expansion coefficients for CBPs can be derived using a least square approach. In this work, we follow an approach that is different from that of described in [28] and [27], and use the CBPs as basis functions for synthesizing the pattern instead. Furthermore, we develop a novel strategy for the synthesis problem that transforms the pattern design process into an eigenvalue problem, which is considerably more numerically efficient to solve than applying a nonlinear optimization procedure. The proposed approach finds the best-fit solution that maximizes the energy radiated in the desired angular range, while minimizing the undesired radiation outside of this range. Since the CBs are generated by placing the antennas only at "allowable" locations, the corresponding CBPs are well suited for assessing the realizability of the specified pattern, which is important from a designer's point of view. Rather than using a nonlinear optimization procedure to realize the desired pattern, we transform the pattern design process into a small-size eigenvalue problem—which is considerably more efficient to solve numerically—to find the best-fit solution that maximizes the energy radiated in the desired angular range, while minimizing the undesired radiation outside of this range. The rank of this eigenvalue equation is very small, by several orders of magnitude, as compared to the rank of the matrix in [7], which is used to derive the CMs, and this is true especially for large problems with many DoFs. The details of this pattern-synthesis algorithm will now be provided below.

The step-by-step CB-based pattern control approach is summarized below:

1. Choose suitable user-designed feed configurations for the given platform and place them at locations designated as "allowable" by the user.
2. To derive the CBs rigorously, use either Commercial 3D EM software (such as HFSS, FEKO) or an in-house numerical CEM code.
3. Perform an SVD on the pre-SVD (original) CBs set to remove the redundancies. From the SVD analysis. We can readily show that the post-SVD CBs (coefficient vectors) are orthogonal, and that this step is only necessary when the number of CBs is large.
4. Next, use the CB currents to compute their corresponding radiation patterns, called the CBPs. The CBPs take both the design constraints and the EM structure properties into consideration; hence, they are well suited for assessing the realizability of the specified pattern.
5. Find the best-fit solution to the specified pattern by transforming the problem into an eigenvalue problem, as explained below.

We now explain the transformation method mentioned in step 5 above. Suppose the desired pattern we wish to realize is $\mathbf{E}^{\text{des}}(\Omega)$. We begin by defining $\mathbf{E}^{\text{mod}}$ as

$$\mathbf{E}^{\text{mod}}(\Omega) = \sum_{n=1}^{N} C_n \mathbf{E}_n(\Omega) \tag{2.1}$$

where $\mathbf{E}_n$ is the nth CBP. Our goal is to determine the weight coefficients C_n such that the $\mathbf{E}^{\text{mod}}$ closely mimics the specified radiation pattern $\mathbf{E}^{\text{des}}$, which is realized by maximizing the coefficient μ, defined below:

$$\mu = \frac{\int_{\theta_1}^{\theta_2} \mathbf{E}^{\mathrm{mod}*}(\Omega) \cdot \mathbf{E}^{\mathrm{mod}}(\Omega) d\Omega}{\int_{0}^{\pi} \mathbf{E}^{\mathrm{mod}*}(\Omega) \cdot \mathbf{E}^{\mathrm{mod}}(\Omega) d\Omega} \tag{2.2}$$

Our next step is to transform (2.2) into a generalized eigenvalue problem, which reads:

$$\overline{\overline{\mathrm{M}}}_1 \overline{\mathrm{I}}_C^T = \mu \overline{\overline{\mathrm{M}}}_2 \overline{\mathrm{I}}_C^T \tag{2.3}$$

where

$$\overline{\mathbf{I}}_C^T = [C_1, C_2, \cdots, C_N]^T \tag{2.4}$$

$$\overline{\overline{\mathrm{M}}}_1(i,j) = \int_{\theta_1}^{\theta_2} \mathbf{E}_i^*(\Omega) \cdot \mathbf{E}_j(\Omega) d\Omega \tag{2.5}$$

$$\overline{\overline{\mathrm{M}}}_2(i,j) = \int_{0}^{\pi} \mathbf{E}_i^*(\Omega) \cdot \mathbf{E}_j(\Omega) d\Omega \tag{2.6}$$

It is evident that the matrices $\overline{\overline{\mathrm{M}}}_1$ and $\overline{\overline{\mathrm{M}}}_2$ are both Hermitian. The original problem is therefore reduced to the determination of the weight excitation coefficient vector $\overline{\mathbf{I}}_C^T$ so that μ is maximized. It has been already proven in [29] that the eigenvector corresponding to the largest eigenvalue in (2.2) is the solution we are seeking. Solving for the eigenvalue and eigenvectors is particularly straightforward and simple. The interested reader may refer to [29] for more details. It is also important to realize that the dimensions of the matrices in (2.5) are typically only 4 or 5, as opposed to hundreds and thousands, if not millions, for large complex platforms needed to derive the CMs by solving the eigenvalue equation (2.1).

2.5 Illustrative examples

We now present in this section a number of examples to illustrate the CB-based procedure for placement of a plurality of antennas on a specified platform.

2.5.1 *Four microstrip patch antennas on an FR4 substrate*

The first example is that of four microstrip patch antennas placed on an FR4 substrate, as shown in Figure 2.19. The patch shape is chosen to be rectangular, with dimensions of 12 mm × 13 mm. The reflection coefficient (return-loss) characteristic of the antenna is shown in Figure 2.20. All antennas resonate around 5.75 GHz, which is chosen to be the operating frequency.

Next, the CBPs are derived by exciting one port at a time while terminating the others using 50 Ω resistance, and the patterns obtained are shown in Figure 2.21. It is worthwhile to point out, once again, that the CBPs incorporate *a priori* information

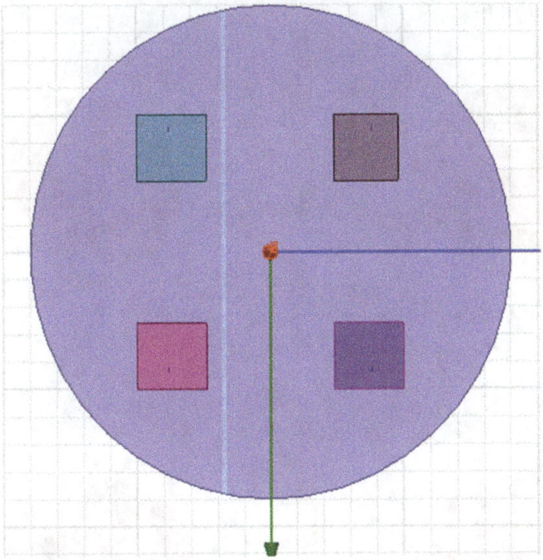

Figure 2.19 Four patch antennas placed on substrate

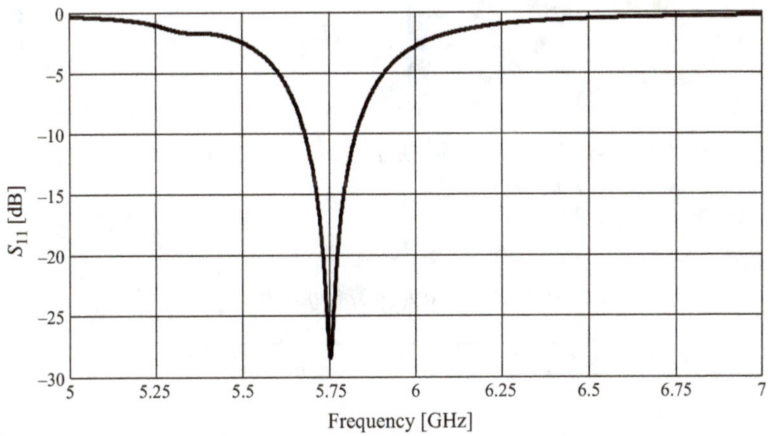

Figure 2.20 Return-loss characteristic of the microstrip patch antennas

about the platform and are tailored for the same; therefore, they can help assess the realizability of the specified pattern. The CBPs, shown in Figure 2.21, are seen to all have main beams around zenith. Hence, it would be reasonable to argue that a realistic desired pattern should have a maximum directivity around $\theta = 0°$, and such a broadside pattern is assumed to be specified by the user (see Table 2.2). Our next step is to transform the problem into an eigenvalue approach, as discussed in Section 2.4.2, and to determine the weight excitation coefficients presented in Table 2.3.

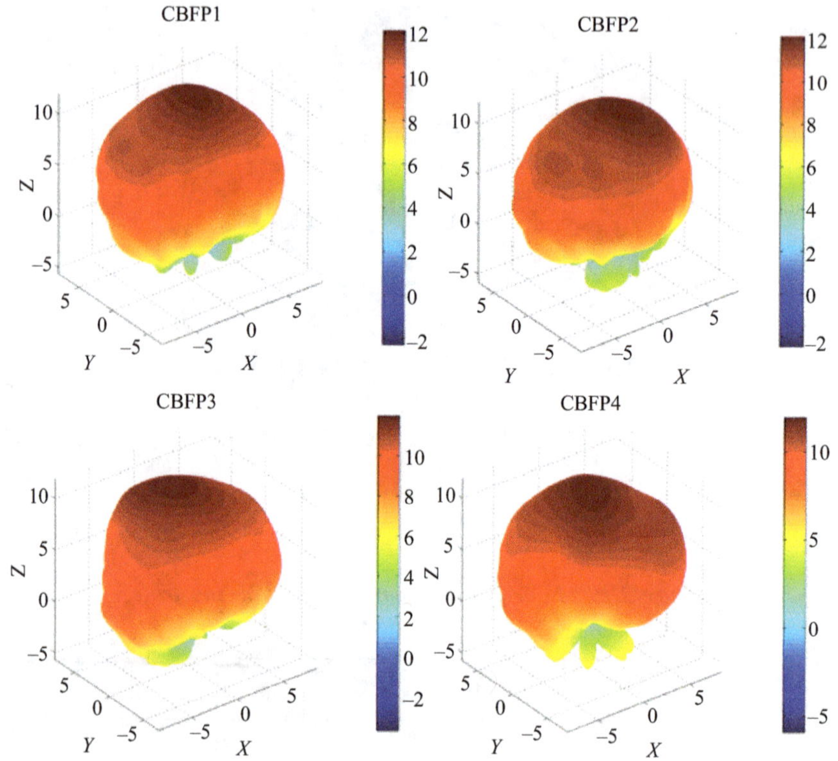

Figure 2.21 CBPs for microstrip patch antennas on the substrate. All the color scales are in decibel

Table 2.2 Pattern specification

	Angular range
θ	$0° \le \theta \le 30°$
ϕ	$0° \le \phi \le 360°$

Table 2.3 Weight coefficients of excitations

Excitation Index	**Amplitude (V)**	**Phase (degree)**
1	7	−167
2	6.7	−170
3	7.1	5.11
4	7.1	0 (Reference)

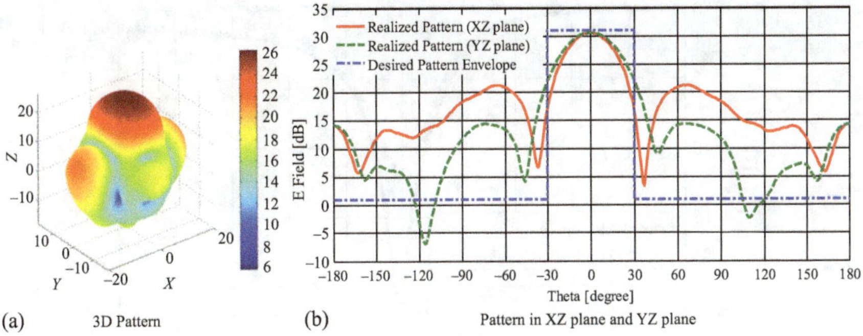

Figure 2.22 The realized pattern

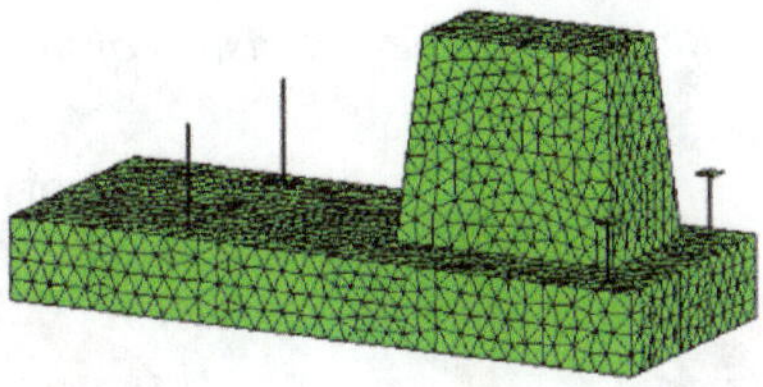

Figure 2.23 Topside of a ship

The realized pattern, shown in Figure 2.22, meets the required specifications very well. Its main lobe has a 38° beamwidth and the antenna has a 17 dB front-to-back ratio.

2.5.2 Topside of a ship excited by monopoles

Next, we consider a simplified version of the topside of a ship shown in Figure 2.23. The total length of the ship is 30 m and its width is 10 m. The main superstructure is a frustum of a pyramid with the height of 10 m. Figure 2.17 also shows the positions where the monopoles and top-loaded monopoles were "allowed" to be placed. As shown in Figure 2.24, the two different monopoles both resonate (after tuning) around 14 MHz, which is chosen to be the operating frequency. Clearly, choosing antennas with broader bandwidths would enable us to cover a wider frequency range, for example, 2–30 MHz (HF).

Next, we generate the CBPs by using an MoM code and display them in Figure 2.25. It is observed that the ship's structure can have a considerable impact on the pattern of its on-board antennas. Consequently, the CBPs are obviously different from the omnidirectional radiation pattern of a vertical whip monopole over a perfect ground. However, CBPs have a good coverage in the elevation planes. Taking into both the characteristics of the CBPs and the application requirement into consideration, we then specify the desired pattern in Table 2.4 as an ideal monopole-like pattern.

To measure the pattern uniformity, we borrow the concept from [30], and specify gain variation on the horizontal plane with respect to the peak value, to be

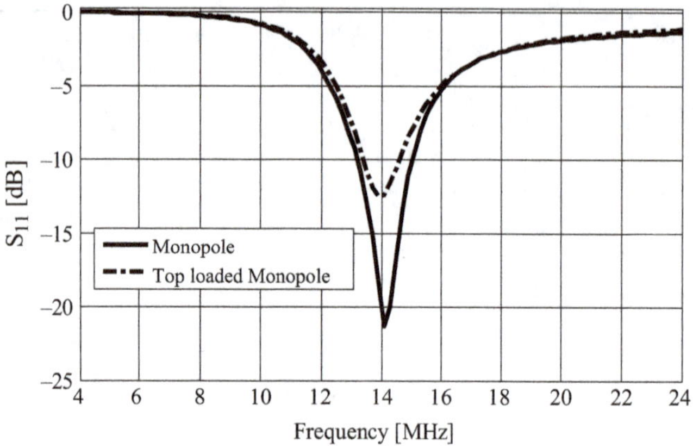

Figure 2.24 Return-loss characteristics of the monopoles

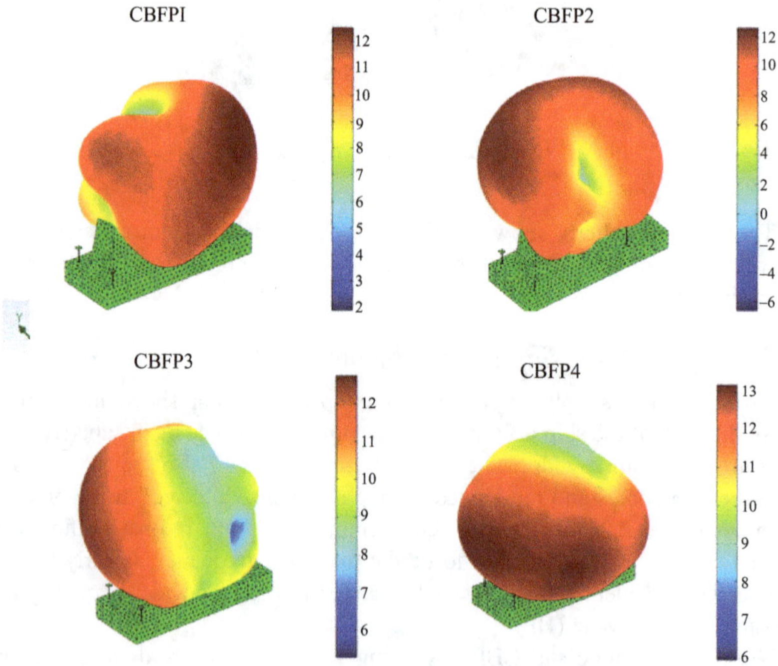

Figure 2.25 The CBPs

Table 2.4 Pattern specifications

	Angular range
θ	$20^\circ \leq \theta \leq 160^\circ$
ϕ	$0^\circ \leq \phi \leq 360^\circ$

Table 2.5 Weight coefficients of excitations

Excitation index	Amplitude (V)	Phase (degree)
1	2.5	−71.5379
2	3.7	−83.8951
3	0.1	−140.9263
4	1.4	180 (Reference)

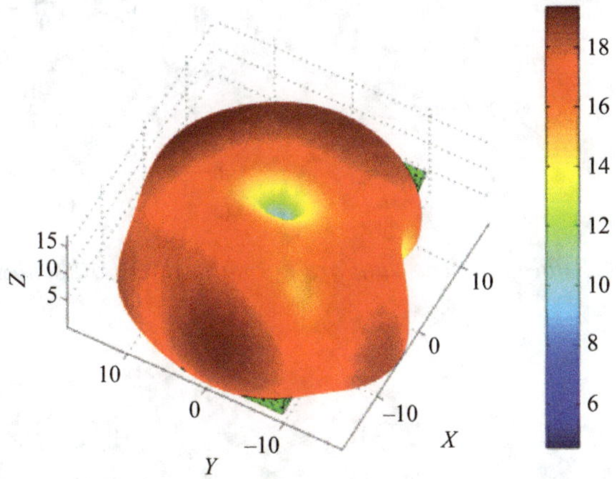

Figure 2.26 Realized pattern in dB

less than 10 dB in the 80% of the whole azimuthal angle, is the desired performance. A function U is then defined to indicate the percentage of the azimuthal angular range, where the gain requirement is met.

The coefficients produced by the eigenvalue approach are presented in Table 2.5, the corresponding realized pattern is displayed in Figure 2.26, while Figure 2.27 shows the normalized total gain. It is demonstrated that the realized pattern has a good coverage in the elevation plane, while it retains the pattern uniformity up to 89%.

2.5.3 Four PIFA antennas on FR4 substrate

This example has four PIFA antennas placed on an FR4 substrate, as shown in Figure 2.28. The operating frequency of the four antennas is 2 GHz. The reflection coefficient characteristic of the antenna is shown in Figure 2.29. All antennas resonate around 2 GHz, which is chosen to be the operating frequency.

The corresponding CBPs are derived by exciting one port at a time while terminating the others using 50 Ω resistance, and the patterns obtained are shown in Figure 2.30. The desired pattern with specification is shown in Table 2.6 and the excitation information is shown in Table 2.7. The realized pattern is shown in Figure 2.31.

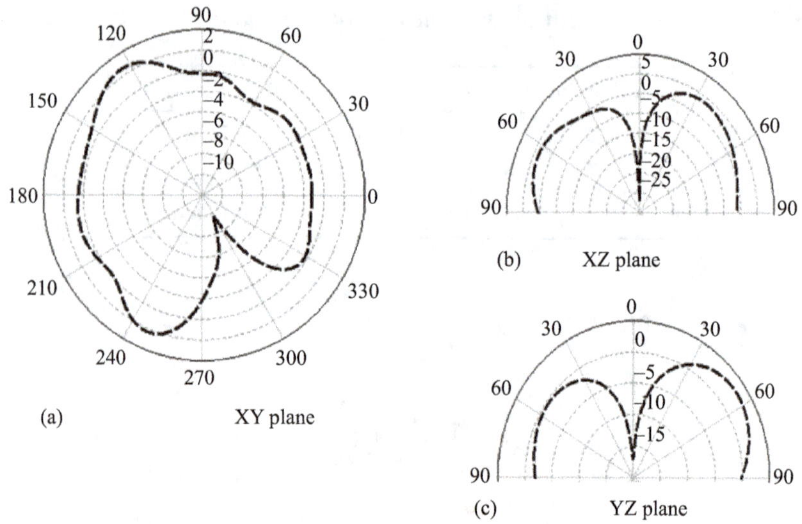

Figure 2.27 Total gain. The data are normalized and all scales are in dBi

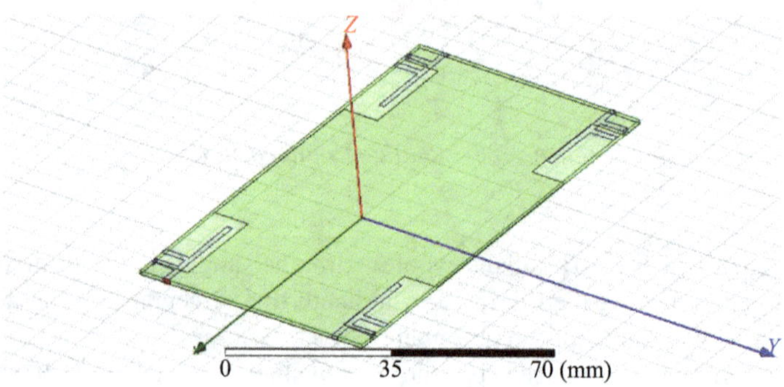

Figure 2.28 Four PIFA antennas

2.5.4 Chassis excited by six dipoles

For the next example, we consider six dipoles placed on a PEC rectangular plate, as shown in Figure 2.32. The six antennas are used to generate the desired radiation pattern. Next, the CBPs are derived by exciting one dipole at a time while terminating the others using 50 Ω resistance, and the current and patterns obtained are shown in Figures 2.33 and 2.34. The spacing of the dipoles are shown in Table 2.8 and the weight excitation coefficients are presented in Table 2.9. Figure 2.35 shows the desired radiation pattern.

In summary, this section has demonstrated that, in contrast to the CM approach, the problem of antenna pattern synthesis can be systematically handled by using the CB method, which addresses the excitation problem up front. An eigenvalue problem,

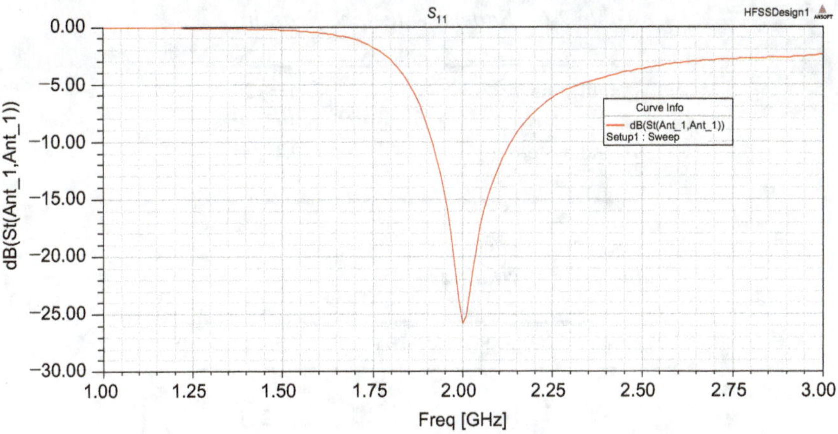

Figure 2.29 Return-loss characteristics of the PIFA antenna

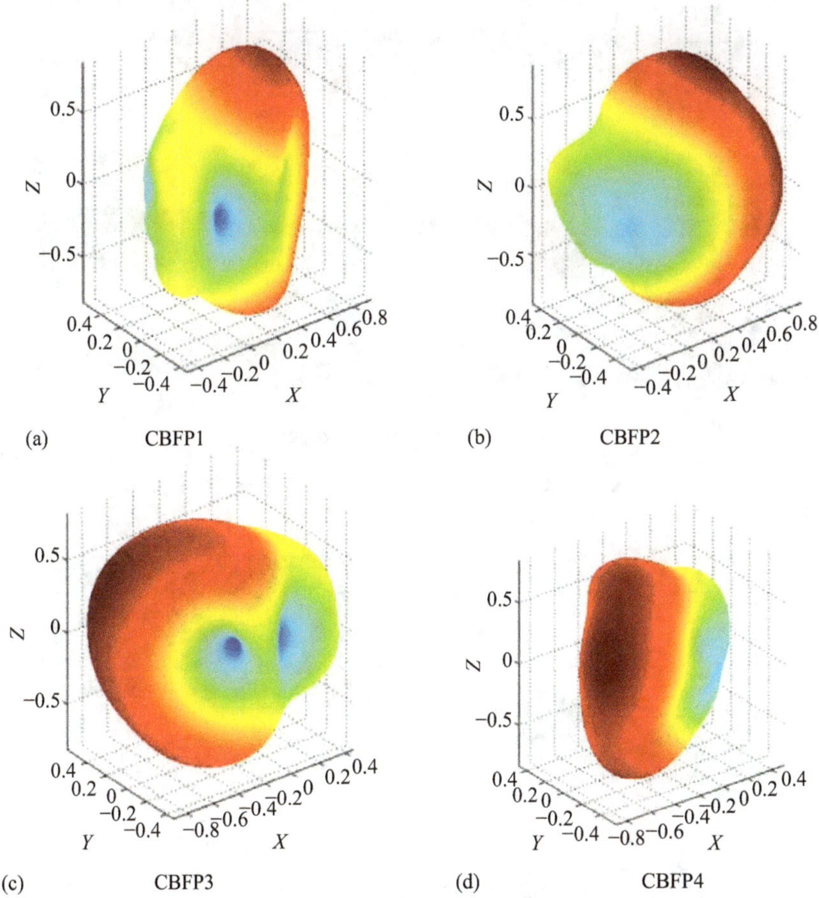

Figure 2.30 The CBPs

Table 2.6 Pattern with specification

Angular range	$0^\circ \leq \phi \leq 20^\circ$, $80^\circ \leq \theta \leq 100^\circ$
Angular step	$\Delta\phi = 0.1^\circ$, $\Delta\theta = 0.1^\circ$

Table 2.7 Excitations

Port	Amplitude/(mV)	Phase/(rad)
1	3.4	3.0433
2	2.1	1.221
3	1.6	−2.7727
4	3.0	1.9176

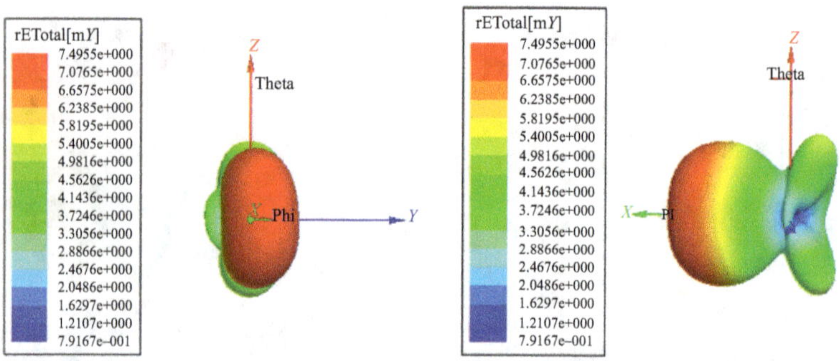

(a) *YZ* plane (b) *XZ* plane

Figure 2.31 The realized pattern

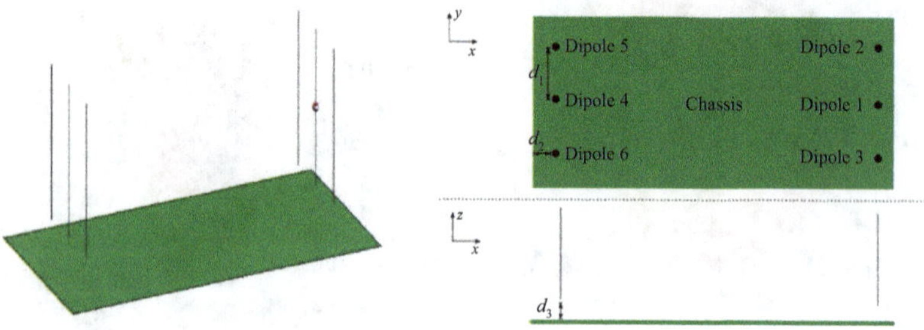

(a) 3D view (b) *XY* and *XZ* plane view

Figure 2.32 Chassis excited by six dipoles

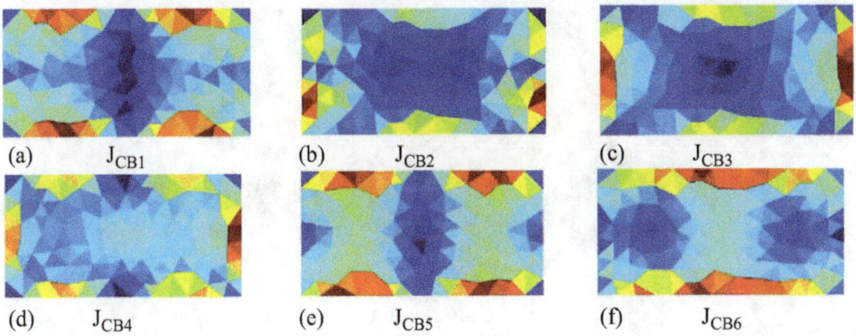

Figure 2.33 Current distribution for different character basis

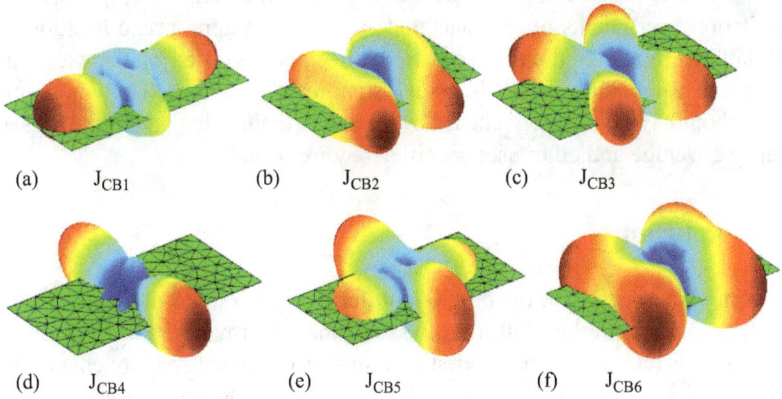

Figure 2.34 Radiation pattern for different characteristic basis

Table 2.8 Dimensions

d_1	d_2	d_3
$\lambda/10$	$\lambda/8$	$\lambda/10$

Table 2.9 Excitations

CB index	Abs(v0)/(V)	Phase(v0)/(degree)
1	4.978720	0.0000(ref.)
2	2.948916	−166.3712
3	3.251143	−177.2109
4	4.908803	−181.7048
5	2.983191	12.2425
6	3.148979	181.0971

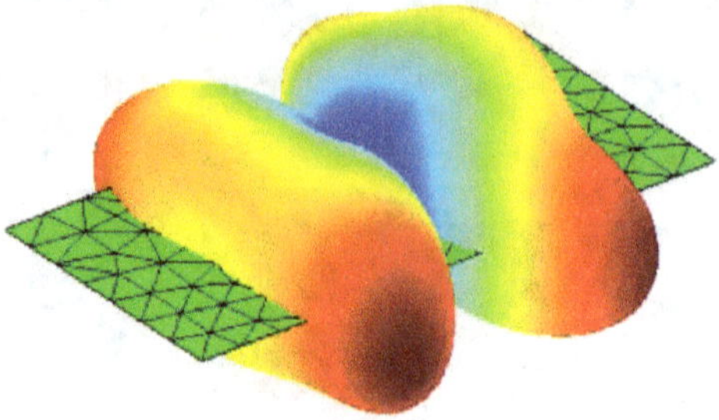

Figure 2.35 Desired radiation pattern

whose size is much smaller than that needed in the CM method, is solved to determine the excitation coefficients of antennas that are placed at appropriate locations on a given platform to realize the desired pattern envelope specified by the user. This is done without calculating the CMs, their "significance" and perhaps, most importantly, without having to worry about how to excite them using antennas that satisfy frequency coverage and other user-specified requirements.

2.6 Conclusion

This chapter has focused on the problem of designing of one or more antennas, and their placement on complex platforms, to satisfy the performance metrics specified by the user, such as return-loss characteristics, radiation efficiency, pattern envelope, etc. It began with a review of the TCM, which has been extensively used recently to address this type of problem. Next, we have identified certain issues that arise during the process of applying the TCM to the problem at hand, for example designing and placing an antenna on a realistic mobile phone platform. Following this we have introduced an alternative approach, based on a modification of the classical CBFM—which has been used in the past to solve scattering problems involving complex structures—to address the problem at hand, namely realizing the desired performance characteristics of antenna systems mounted on a given platform, and operating under the constraints of location as well as the use of permissible type of excitation sources. It has been pointed out that since the CBs are closely related to the excitation sources, they are well suited for the type of problem at hand, and that they can work directly with the available choices for the type of excitation sources and their locations.

The chapter has discussed two different antenna design and placement scenarios. The first of these involved a single antenna placed on a platform and addressed the problem of co-designing the same. The second group of problems discussed in this chapter has dealt with multiple antennas and has presented a systematic approach to realizing the desired radiation patterns for the combination of specified geometries of the platforms and types of acceptable excitation sources, as well as their possible locations, at least in the sense of "best-fit". Numerous examples of the latter class of problems have been illustrated for the application of the CB-based approach, which circumvents the difficulties encountered in TCM-based method for the type of problems discussed herein. The presented approach can be readily extended to platforms with more complex geometries, with large sizes, and with inhomogeneous material properties, by using the concepts already developed in the context of CBFM that have been described in numerous recent publications [14,15,–31] on the Characteristic Basis Function Method.

Acknowledgment

This chapter is derived, in part, from an article published in the *Journal of Electromagnetic Waves and Applications* on June 29, 2016, available online: http://dx.doi.org/10.1080/09205071.2016.1201435.

Appendix

Appendix A1 Characteristic modes and bases

A1.1 Generation of characteristic modes (CMs)

The CM theory was first introduced in the pioneering works of Garbacz [5], and Harrington used the method of moments (MoM) [6,7] to numerically generate the CMs and their properties of an object, for example, a platform. Following the strategy in [6], the modal currents $\mathbf{J}_\mathrm{n}$ can be derived from the solution of the following generalized eigenvalue equation:

$$\mathbf{X}\cdot\mathbf{J}_\mathrm{n} = \lambda_\mathrm{n}\mathbf{R}\cdot\mathbf{J}_\mathrm{n} \tag{A1-1}$$

$$\mathbf{ZI} = \mathbf{V} \tag{A1-2}$$

where **I** is the induced current on the object and **V** is derived from the incident field. The eigenmode equation corresponding to (A1-2) is expressed as:

$$\mathbf{Z}\,\mathbf{I}_\mathbf{n} = \lambda_\mathbf{n}\mathbf{I}_\mathbf{n} \tag{A1-3}$$

It is obvious that (A1-3) is complex, whereas (A1-1) is entirely real. Hence, (A1-1) has been the preferred choice over (A1-3) for modal analysis. Nonetheless, there are some advantages to using the eigenmodes over CMs for certain type of scenarios, as has been pointed out by Oueslati [24]. In this work we propose to use (A1-2) instead, which gives rise to solutions referred to as the characteristic bases or CBs (see below) that are *source-dependent,* owing to the presence of **V** in the RHS of (A1-2), as opposed to modal solutions of (A1-1) or (A1-3) that are obviously *source-independent.*

Before leaving this discussion, we just want to mention that we can definitively and unambiguously answer the question that we raised briefly in Section 2.2, namely what type of source excitation is needed to launch one of the CMs, say J_m. The short answer is Simply substitute J_m in (A1-2) and generate its RHS to obtain the incident field (tested), which is obviously a distributed source in general (see, for example, Figures 2.6 and 2.7). This unequivocally challenges the claim often made in the literature that one could simply use a confined source and locate it at the maximum of a modal current distribution CM to excite it.

A1.2 Generation of CBs

The solutions for the induced currents on the platform, derived for the number of sources (N_e) we have chosen for the excitations, form the "initial" set of macro basis functions, which we refer to herein as CBs. They are derived by solving (A1-2) above for a given set of excitations, for instance using antennas at different locations on a platform which we are analyzing. A singular value decomposition (SVD) procedure can be implemented to remove the redundancy from the set of original CBs in the process. Toward this end, we arrange the coefficients of each of the CBs in columns to form a matrix JCBF and apply the SVD, to derive the following factorized representation

$$\mathbf{J}^{\mathrm{CBFs}} = \mathbf{U}\boldsymbol{\Sigma}\mathbf{V} \tag{A1.4}$$

where **U** and **V** are orthogonal matrices with dimensions of $N_i \times N_i$ and $N_e \times N_e$, respectively, N_i is the number of unknowns. The matrix Σ is a diagonal matrix containing the singular values that are normalized relative to a suitable threshold value, typically 10E-3. Only the K_i columns in the matrix **U** above the threshold are retained to generate the "final" post-SVD CBs.

A1.2.1 Comparison between CMs and CBs

It is instructive to present a comparative review of the CMs and CBs, introduced above, and we will do that now. The CMs and CBs are similar in several ways—they both depend on the characteristics (geometry, material properties, etc.) of the object, and they can provide physical insight into the current distributions that need to be excited on the structure under consideration to realize a given pattern.

However, the CMs and CBs have fundamental differences. The CMs are "source-free" solutions, whereas the CBs are closely related to the excitations that we plan to use later to realize the desired radiation pattern. We note also that different sets of CBs can be generated by using different excitation methods. Hence, the CBs are tailored to both the excitations and platform characteristics.

As an illustrative example, we consider a thin wire structure with length and radius of 0.36 m and 0.3 mm, respectively (see Figure A1-1(a)). The operating frequency is chosen to be 1 GHz. We use two different excitations to generate the CBs, namely the plane waves (PWs) and delta-gap voltages (DVs), as shown in Figure A1-1(b) and (c), respectively.

Figure A1-2 illustrates the CMs and CBs for the thin wire geometry shown in Figure A1-1. It is evident that the post-SVD CBs have distributions very similar to

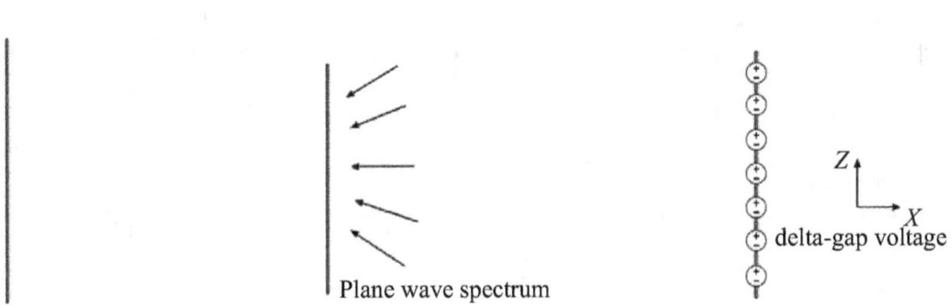

(a) Wire structure (b) Wire with plane waves (c) Wire with delta-gap voltages

Figure A1-1 Wire structure and excitations

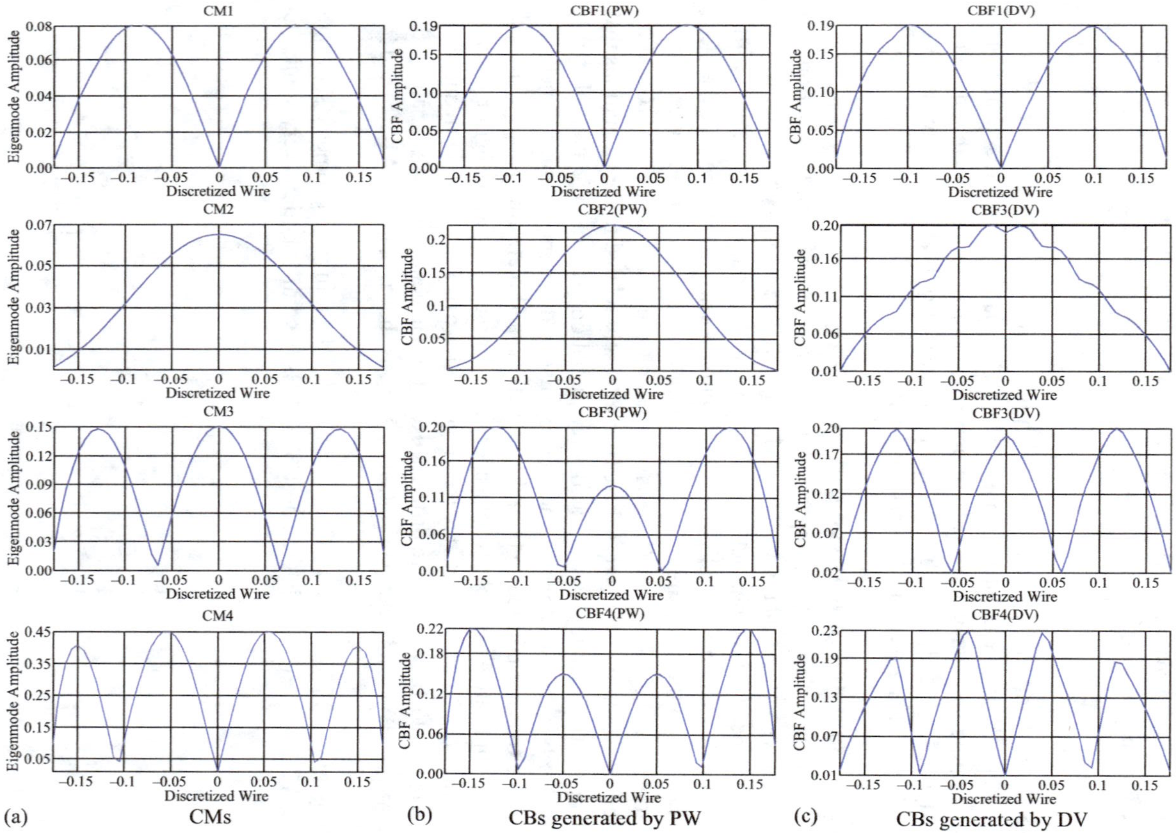

Figure A1-2 Comparison between CMs and CBs

those of the CMs, as may be seen by comparing Figure A1-2(a) with Figure A1-2 (b) and (c). Though the levels are different (this is a normalization issue), the spatial distributions are quite similar for this simple example of *thin-wire* geometry when a ***PW** excitation* is used to generate the CBs. It is worthwhile mentioning here that the same is *not* true in general, either for sources close to the platform where the near fields dominate, or for platforms with complex geometries, as pointed out earlier.

As mentioned earlier, the CBs are dependent on the excitations, and we can observe this to be the case by comparing the CBs shown in Figure A1-2(b), which correspond to the PW excitation case, to those shown in Figure A1-2(c) for delta-gap excitation. This is due primarily to the fact that the Fourier spectrum of the delta-source is comprised of fields belonging to both visible and invisible regions, whereas we have limited the spectral range of the PW excitation to cover only the visible range.

We should point out that the CMs and CBs can be very different for surface or volume type of structures, as opposed to thin wires. Several case examples which demonstrate this fact have been provided in Section 2.5, and assuming that the simple wire example above is representative of practical antenna geometries is erroneous and should be avoided.

It is interesting to consider another example, shown in Figure A1-3 through A1-5 below, which is comprised of two dipoles, both with a radius of 5 mm. Let us suppose that we are tasked to find the lengths of the two dipoles (unequal in general) and the separation distance between them that would provide the widest return-loss bandwidth when the center frequency is 2.4 GHz, together with a directive pattern in the azimuth plane. As shown below, the TCM provides us little help in addressing this problem, but the CB approach is well suited for the task at hand. We aim to show, via this simple example, the advantage of using the CB approach, over its CM counterpart, when designing practical antenna configurations, be they for sensing or for communication.

The CMs for the two dipoles of several different lengths and separation distances, their associated MS characteristics, as well as their radiation patterns are shown in Figures A1-3 through Figure A1-5. They do not provide us any clues as to what combination of the lengths and separation distance would help us achieve an optimal performance in terms of the S_{11} bandwidth; which mode combination should be excited; and how (i.e., which one of the two dipoles should be excited and where). In contrast to this, the CB approach directly answers these questions and addresses the excitation issue up front. Figure A1-3 shows the geometry of the two-dipole configuration, and our goal is to determine the length as well as the separation distance between the driven and the parasitic dipole to achieve the best return-loss bandwidth, as well as the desired pattern.

Figure A1-3 illustrates the geometry of the two dipoles of equal length (=60 mm), and the modal significance (MS) of the combination of the dipoles in the frequency range of 1.5–3.0 GHz, with their separation distance varying from 10 to 40 mm. Figure A1-3(a) shows the geometry of the two dipoles used for the Characteristic Mode (CM) analysis, neither of which is excited, since we are solving for

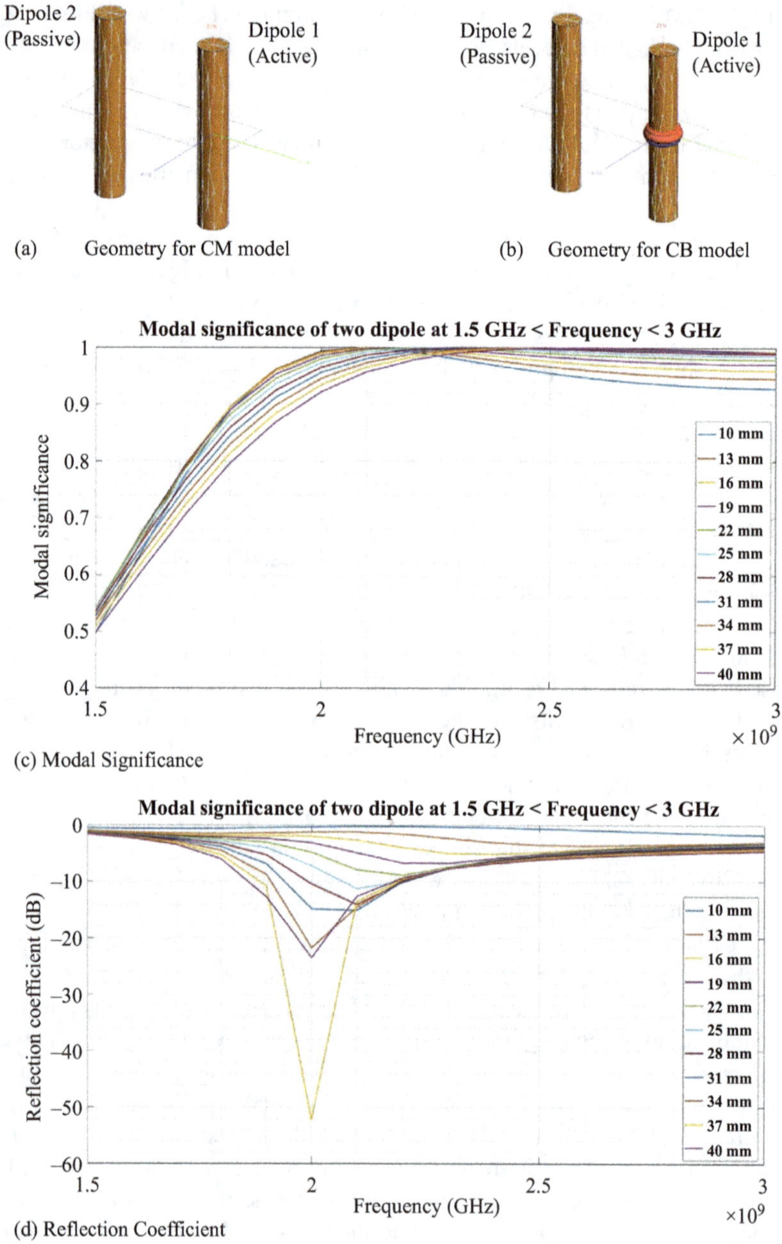

Figure A1-3 Comparison of CM and CB and their modal significance and reflection coefficient

Table A1-1 Reflection coefficient of two dipoles with separation distance = 10 mm–40 mm

First dipole Length (mm)	Second dipole Length (mm)	Distance (mm)	−10 dB Bandwidth (GHz)
60			0.4009
60	60	10	–
60	60	13	–
60	60	16	–
60	60	19	–
60	60	22	–
60	60	25	0.1192
60	60	28	0.2099
60	60	31	0.2585
60	60	34	0.2834
60	60	37	0.2997
60	60	40	0.3180

the modes. Figure A1-3(b) shows the geometry used to derive the CBs, and it consists of an active dipole placed in front of a passive dipole, with the latter located behind the driven dipole. Figure A1-3(c) displays the CM results, and it is obvious that their MS plots do not provide us a clue as to which combination of lengths and separation distance between the dipole would provide us the widest return-loss bandwidth when one of the dipoles is excited while the other is passive.

Next, we turn to Figure A1-3(d), which presents the CB results. The figure shows that the −10 dB bandwidth of the reflection coefficient increases as the distance between the dipoles is increased from 10 to 40 mm. Table A1-1 shows the result for the −10 dB bandwidth, achieved for various separation distances. We observe that while the results for the −10 dB bandwidths improve with an increase in the distance, these bandwidths are always smaller than that of the isolated dipole. Hence, we need to modify the design to improve the performance and achieve our goal of a bandwidth which is wider than that of the single dipole. We note that the CM analysis does not provide us a clue as to how we might modify the design of the combination of the dipoles to enhance the performance of the two-dipole system.

Figure A1-4 illustrates the geometry of two dipoles, with a separation distance of 19 mm between them. (The distance was chosen based on a previous parametric study of the two-dipole system.) We vary the length of the second dipole from 30 to 90 mm, and study the performance of the two dipole-system in the frequency range of 1.5–3.0 GHz. Figure A1-4(a) shows the two-dipole model we use, as before, for the CM study—neither of them driven—as per requirement of a modal analysis of the system. Next, Figure A1-4(b) shows the model we employ for the CB function method. It consists of one *active* dipole, placed in front of a *passive* dipole to form the two-dipole system, whose performance we wish to optimize.

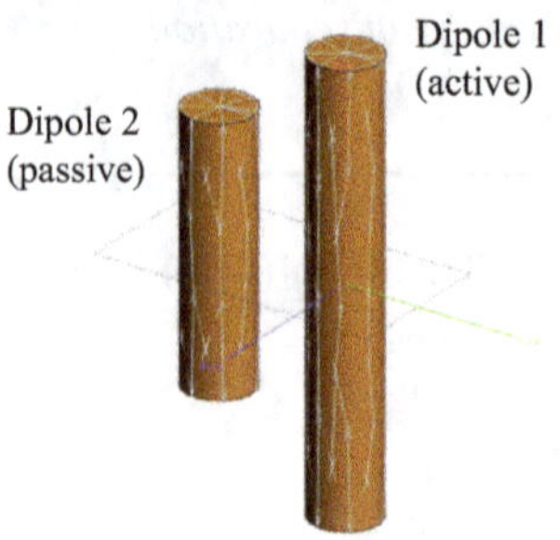

(a) Geometry for CM model

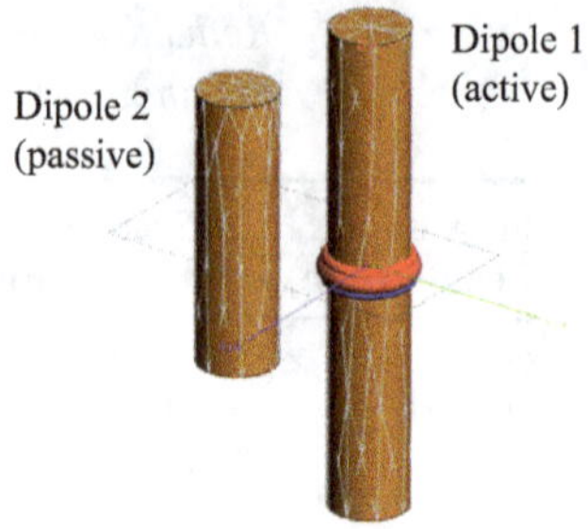

(b) Geometry for CB model

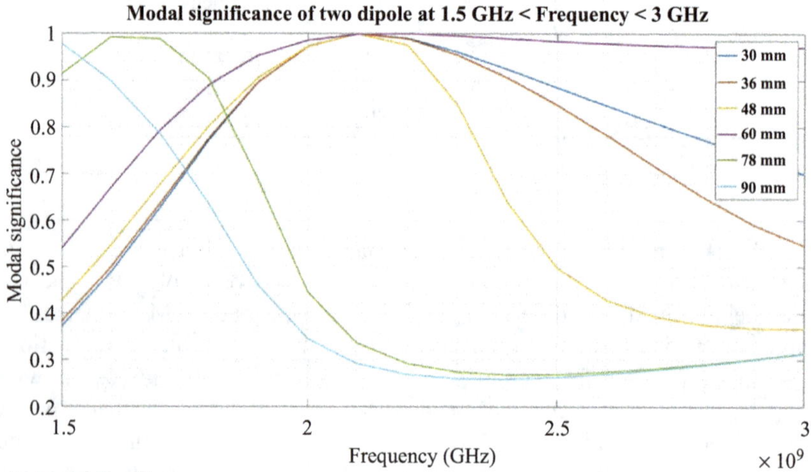

(c) Modal Significance

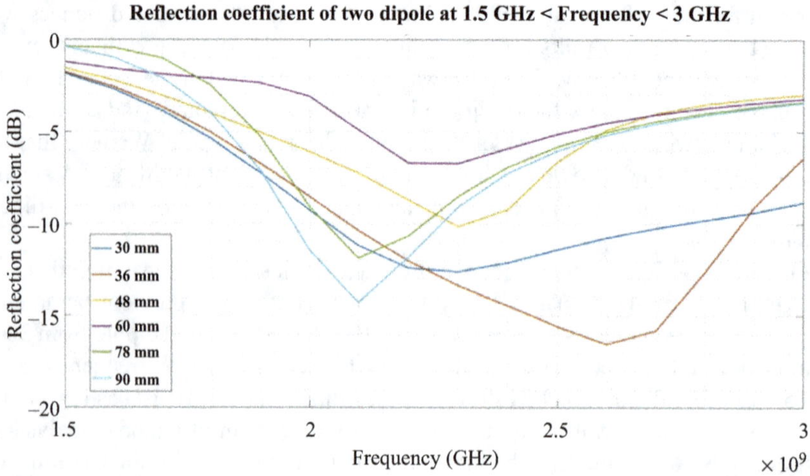

(d) Reflection Coefficient

Figure A1-4 Comparison of CM and CB, and their modal significance and reflection coefficient characteristics

Table A1-2 Reflection coefficient of two dipoles with a second dipole's length = 30 mm–90 mm, and separation distance = 19 mm

First dipole Length (mm)	Second dipole Length (mm)	Distance (mm)	−10 dB Bandwidth (GHz)
60			0.4009
60	30	19	0.7126
60	31	19	0.8463
60	32	19	0.8883
60	33	19	0.8839
60	34	19	0.8615
60	36	19	0.7932
60	48	19	0.0247
60	60	19	–
60	78	19	0.1968
60	90	19	0.2983

Figure A1-4(c) displays the CM results for MS. We note that it is difficult to identify a pattern in the MS plots that would tell us how to choose the length of the passive dipole which will provide us the widest return-loss bandwidth, since the modes are excitation-free solutions; furthermore, the bandwidths of the MS plots do not correlate with those of S_{11}, as we will soon see below. So, once again, we turn to the CB analysis to generate the S_{11} results presented in Figure A1-4(d). The figure shows that the results for the reflection coefficient bandwidth improves from 0.7126 GHz, when the length of the passive dipole is 30 mm, to its maximum value of 0.888 GHz—which is 2.2 times that of the isolated dipole—when the dipole length is increased to 32 mm, and steadily declines thereafter when the length is further increased. Table A1-2 summarizes the results for the −10 dB bandwidth as we vary the length of the passive dipole.

Figure A1-5 presents the results obtained by using the Characteristic Basis (CB) Function Method for the optimal choice of the two parameters—length of the passive dipole and the separation distance between the dipoles. Figure A1-5(a) shows that the −10 dB reflection coefficient bandwidth ranges from 2.053 to 2.94 GHz for this choice of the dipole parameters. Figure A1-5(b) displays the surface currents on the two dipoles and shows that the surface current on the driven dipole is maximum at the middle, and is similar to that of CM Mode-1. No evidence of Mode-2 excitation was found for this case, even though CM analysis had predicted that Mode-2 would be significant over a frequency range around 2 GHz, for some choices of the dipole parameters. Finally, the radiation pattern of the system, displayed in A1-5 (c), shows that it is nearly omnidirectional at 1.5 GHz, which is below the operating frequency range, that it becomes become and more directional with an increase in the frequency, and that it has the desired character at the operating frequency. We close with the observation that neither of these characteristics could have been predicted from the CM analysis of the same problem geometry, without excitation of course.

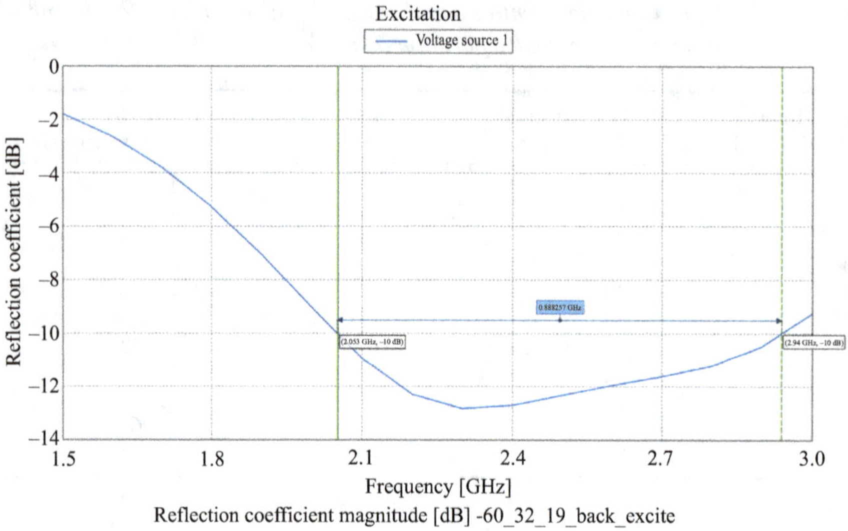

(a) Reflection Coefficient

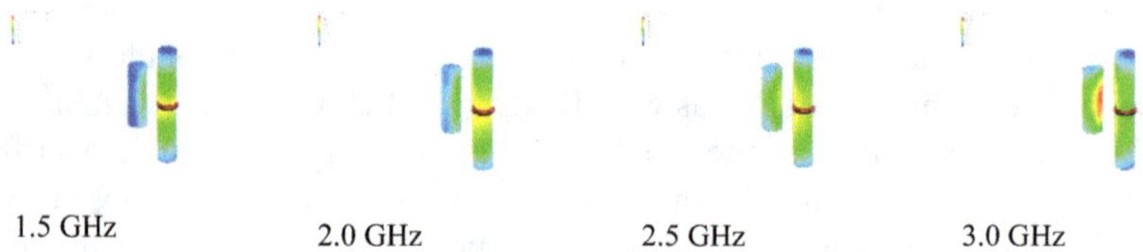

(b) Surface Current distributions induced on the driven and passive dipoles

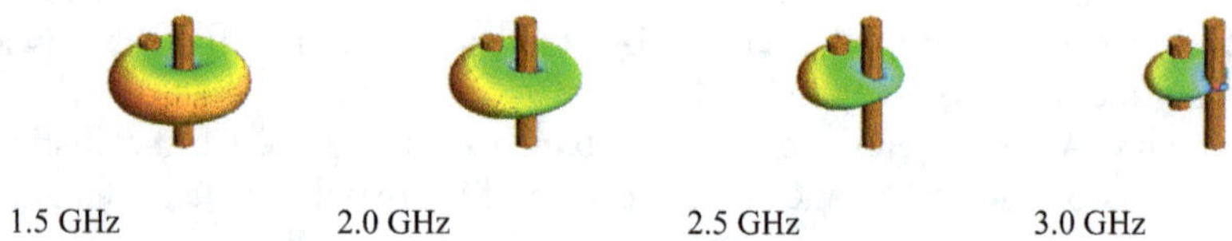

(c) Radiation Pattern of the two-dipole system providing optimum bandwidth performance

Figure A1-5 *CB results for two dipoles separated by a distance of 19 mm. The length of the driven dipole is 60 mm while that of the passive dipole length is 32 mm, with the combination providing the widest* S_{11} *bandwidth. The directional pattern of the system is evident from the radiation patterns*

Appendix A2

A2.1 TCM analysis of mobile phone antenna and antenna-plus-platform

We have carried out a TCM analysis of the mobile phone frame (Platform-2); the antenna used to excite the frame; and a combination of mobile phone frame with the antenna mounted on it. The mobile phone frame is shown in Figure 2.4, and its eigenvalues, MSs, and its modal distribution are presented in Figures 2.3–2.5. The antenna and the mobile phone frame with antenna (but without the excitation) are shown in Figure A2-1. Both the frame and frame-plus-antenna combination have two modes that are significant in the range of 850–950 MHz, as may be seen from Figures 2.3 and A2-2, while the isolated antenna structure supports four additional significant modes close to the frequencies 1,450, 2,450, 3,000, and 4,000 MHz, as shown in Figure A2-1.

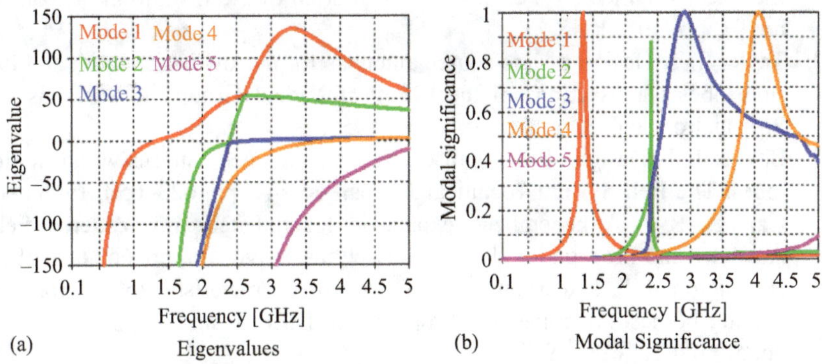

Figure A2-1 Eigenvalues and modal significances of the isolated antenna

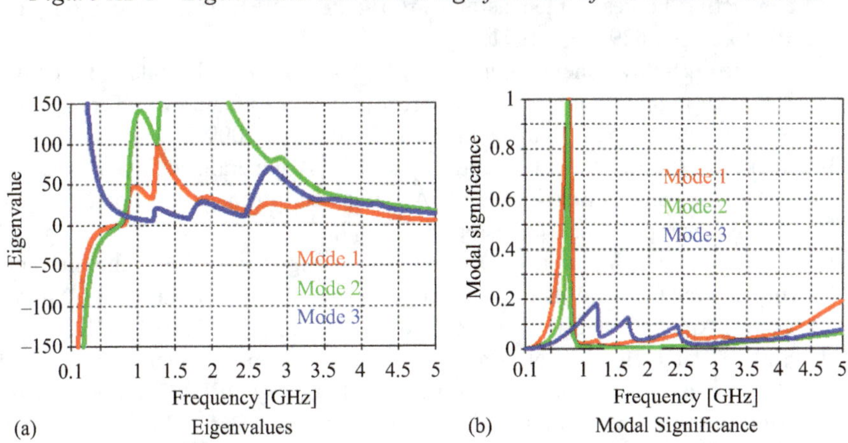

Figure A2-2 Eigenvalues and modal significances of antenna-plus-platform-2

We can conclude, on the basis of the above study, that the TCM and the associated "MS" analyses fail to provide us a clue as to whether the modes we might be interested in will actually be excited significantly, what their relationship with the return-loss characteristic of the antenna-plus-platform will be, and what might be our best choice for the excitation location of the platform. These are obviously the critical design questions that we need answers to, and we can find them through a CB analysis as we have demonstrated earlier in Section 2.5.

References

[1] Dolph C L. A current distribution for broadside arrays which optimizes the relationship between beam width and side-lobe level. Proceedings of the IRE. vol. 34, no. 6, pp. 335, 348, 1946.

[2] Woodward P M, and Lawson J D. The theoretical precision with which an arbitrary radiation pattern may be obtained from a source of finite size. Electrical Engineers – Part I: General, Journal of the Institution of. vol. 95, no. 93, pp. 405, 1948.

[3] Taylor T T. Design of line source antennas for narrow beam width and low side lobes. Transactions of the IRE Professional Group on Antennas and Propagation. vol. 3, no. 1, pp. 16–28, 1955.

[4] Bayliss E T. Design of monopulse antenna difference patterns with low side lobes. The Bell System Technical Journal. vol. 47, pp. 623–650, 1968.

[5] Garbacz R J. A generalized expansion for radiated and scattered field. Antennas and Propagation, IEEE Transactions on. vol. 19, pp. 662,668, 1971.

[6] Harrington R F, and Mautz J R. Theory of characteristic modes for conducting bodies. Antennas and Propagation, IEEE Transactions on. vol. 19, no. 5, pp. 622, 628, 1971.

[7] Harrington R F, and Mautz J R. Computation of characteristic modes for conducting bodies. Antennas and Propagation, IEEE Transactions on. vol. 19, no. 5, pp. 629, 639, 1971.

[8] Shaker G, Safavi-Naeini S, and Sangary N. Q-Bandwidth relations for the design of coupled multi-element antennas. IEEE Antennas and Propagation Society International Symposium, Charleston, SC, 2009.

[9] Shaker G, Safavi-Naeini S, and Sangary N. A generalized modal analysis method for antenna design. IEEE Antennas and Propagation Society International Symposium, Charleston, SC, 2009.

[10] Martens R, and Manteuffel D. Systematic design method of a mobile multiple antenna system using the theory of characteristic modes. IET Microwaves, Antennas and Propagation. vol. 8, no. 12, pp. 887, 893, 2014.

[11] Shih T, and Behdad N. Bandwidth enhancement of platform-mounted HF antennas using the characteristic mode theory. IEEE Transactions on Antennas and Propagation. vol. 64, no. 7, pp. 2648–2659, 2016.

[12] Shih T Y, and Behdad N. Bandwidth enhancement of HF antennas mounted on military platforms using a characteristic-modes-based design approach.

International Symposium on Antennas and Propagation (ISAP), Hobart, Tasmania, Australia, pp. 1, 3, 2015.

[13] Newman E H. Small antenna location synthesis using characteristic modes. Antennas and Propagation, IEEE Transactions on. vol. 21, no. 4, pp. 530, 531, 1979.

[14] Prakash V, and Mittra R. Characteristic basis function method: A new technique for efficient solution of method of moments matrix equation. Microwave and Optical Technology Letters, vol. 36, no. 2, pp. 95,100, 2003.

[15] Lucente E, Monorchio A, and Mittra R. An iteration-free MoM approach based on excitation independent characteristic basis functions for solving large multiscale electromagnetic scattering problems. Antennas and Propagation, IEEE Transactions on. vol. 56, no. 4, pp. 999, 1007, 2008.

[16] Antonino-Daviu E, Suarez-Fajardo C A, Cabedo-Fabre´s M, and Ferrando-Bataller M. Wideband antenna for mobile terminals based on the handset PCB resonance. Microwave and Optical Technology Letters. vol. 48, no. 7, 2006.

[17] Manteuffel D, and Martens R. Systematic design method of a mobile multiple antenna system using the theory of characteristic modes. IET Microwaves, Antennas and Propagation, vol. 8, no. 12, pp. 887–893, 2014.

[18] Miers Z, Li H, and Lau B K. Design of bandwidth-enhanced and multiband MIMO antennas using characteristic modes," IEEE Antennas Wireless Propagation Letters, vol. 12, pp. 1696–1699, 2013.

[19] Kishor K K, and Hum S V. A two-port chassis-mode MIMO antenna," IEEE Antennas Wireless Propagation Letters, vol. 12, pp. 690–693, 2013.

[20] Li H, Tan Y, Lau B K, Ying Z, and He S, Characteristic mode based tradeoff analysis of antenna-chassis interactions for multiple antenna terminals. IEEE Transactions on Antennas and Propagation, vol. 60, no. 2, pp. 490–502, 2012.

[21] Martens R, Safin E, and Manteuffel D. Inductive and capacitive excitation of the characteristic modes of small terminals," in 2011 Loughborough. Antennas Propagation Conference, IEEE, 2011, pp. 1–4.

[22] Li H, Miers Z T, and Lau B K. Design of orthogonal MIMO handset antennas based on characteristic mode manipulation at frequency bands below 1 GHz. IEEE Transactions on Antennas and Propagation, vol. 62, no. 5, pp. 2756–2766, 2014.

[23] Antonino-Daviu E, Cabedo-Fabres M, Sonkki M, Mohamed MohamedHicho N, and Ferrando-Bataller M. Design guidelines for the excitation of characteristic modes in slotted planar structures. IEEE Transactions on Antennas and Propagation, vol. 64, no. 12, pp. 5020–5029, 2016.

[24] Oueslati D. Modal analysis methods and design of chipless RFID tags with natural fractal shapes," Ph.D thesis, Catholic University of Louvain, 2016.

[25] Ikram M, Hussain R, Ghalib A, and Sharawi M S. Compact 4-element MIMO antenna with isolation enhancement for 4G LTE terminals. IEEE International Symposium on Antennas and Propagation (APSURSI), 2016.

[26] Chen Y and Wang C-F. HF band shipboard antenna design using characteristic modes. Antennas and Propagation, IEEE Transactions on, vol. 63, no. 3, pp. 1004, 1013, 2015.

[27] Maaskant R, Ivashina M V, and Wijnholds S J. Efficient prediction of array element patterns using physics-based expansions and a single far-field measurement. Antennas and Propagation, IEEE Transactions on. vol. 60, no. 8, pp. 3614, 3621, 2012.

[28] Young A, Maaskant R, and Ivashina M V. Accurate beam prediction through characteristic basis function patterns for the MeerKAT/SKA radio telescope antenna. Antennas and Propagation, IEEE Transactions on. vol. 61, no. 5, pp. 2466, 2473, 2013.

[29] Cheng D K. Optimization techniques for antenna arrays. Proceedings of the IEEE. vol. 59, no. 12, pp. 1664, 1674, 1971.

[30] Marrocco G, Mattioni L, and Martorelli V. Naval structural antenna systems for broadband HF communications–Part II: Design methodology for real naval platforms. Antennas and Propagation, IEEE Transactions on. vol. 54, no. 11, pp. 3330, 3337, 2006.

[31] Maaskant R, Mittra R, and Tijhuis A. Fast analysis of large antenna arrays using the characteristic basis function method and the adaptive cross approximation algorithm. Antennas and Propagation, IEEE Transactions on. vol. 56, no. 11, pp. 3440, 3451, 2008.

Chapter 3

Wideband L-probe patch antenna

Hau Wah Lai[1] and Kwai Man Luk[2]

The two related feeding techniques, namely, the L-shaped probe and meandering probe (M-probe), for broadening the bandwidth of patch antenna are reviewed. The L-probe has a simple structure and wide bandwidth whereas the M-probe has low cross polarization and a symmetric radiation pattern within the operating frequencies. Design guidelines of the two kinds of patch antennas for Wi-Fi applications are presented. All the impedance bandwidths of the antennas were optimized to standing wave ratio (SWR) less than 1.5. The performance of the patch antennas with different patch height and aspect ratio is demonstrated. The performance of a wide M-probe fed patch antenna that has an impedance bandwidth of around 44% is presented. Finally, a review of various designs for different applications is discussed.

3.1 Introduction

The microstrip patch antenna has been investigated and developed extensively in the past 40 years. Although disclosed in 1953 [1] and was patented in France in 1955 [2], it did not attract any attention until early 1980s when there was a growing demand for low profile antennas for various commercial, defense, spacecraft and satellite applications [3–9]. The antenna can be fed by either a coaxial probe or a microstrip line. It can be made conformal to the mounting surface, such as a cylindrical post or any curved structure [3,10] and can be integrated with printed circuitries. It can be designed for applications with linearly polarization, circularly polarization and dual polarization. In its basic form, the patch antenna, however, has a very narrow impedance bandwidth of a few percent, which is insufficient for contemporary mobile communication systems.

A strict forward approach to increase the bandwidth of a patch antenna is to introduce additional resonances by adding parasitic elements, on top of the original patch or beside the original patch. This method increases the complexity, size and material costs of the antenna. The achievable bandwidth is around 15% if the antenna is excited by a coaxial probe, a commonly used feeding technique at lower

[1]State Key Laboratory of Millimeter Waves, City University of Hong Kong, Hong Kong
[2]Department of Electronic Engineering, City University of Hong Kong, Hong Kong

microwave frequencies. Alternatively, by etching a U-shaped slot in the patch and increasing the thickness of a single-layer probe-fed patch antenna to about 10% wavelength, a substantial increase in bandwidth (SWR < 2), over 30%, was disclosed [11]. After conducting a detailed parametric study [12] on the characteristics of this U-slot patch antenna, it was revealed that the U-shaped slot can provide a capacitance to counteract the probe inductance of the coaxial feed, resulting in wideband performance. The U-slot antenna, however, can only work with linear polarization radiation which was believed in that time. In 1998, inspired by Nakano's work [13] on the use of an L-shaped probe for improving the matching performance of his curl antenna, the L-shaped probe was found to be highly effective to excite a thick patch antenna with about 28% bandwidth [14]. This versatile feeding technique was then successfully applied for developing wideband patch antennas with linear polarization [14–22], circular polarization [23–25] and dual linear polarization [26–28].

Even though the L-probe feeding technique can enhance the impedance bandwidth of patch antennas, it has inherent unwanted radiation which increases the cross polarization and degrades the co-polar radiation pattern. Therefore, several types of modified feeding probes were proposed to improve the radiation pattern of a wideband patch antenna [29–31]. Among them, the attractive one may be the meandering probe (M-probe) as the patch antenna can achieve the characteristics of wide bandwidth, low cross polarization and symmetric radiation pattern, if the M-probe position is selected properly [32]. Similar to the L-probe feed, the M-probe is also useful for designing antennas with linear polarization [30–34], circular polarization [35–37] and dual polarization [38,39].

Parametric studies and design guidelines for both L-probe and M-probe feeding techniques are presented in this chapter. For comparison, all the impedance bandwidths of the antennas were optimized to SWR < 1.5. In addition, antenna designs in the literature with various features are also discussed, which are circular polarization and dual linear polarizations, conformal surface, dual-band and printed circuit board structure. This chapter is organized as follows. Section 3.2 provides parametric studies and design guidelines of patch antennas by feeding with L-probe and M-probe at 2.4 GHz for possible Wi-Fi applications. Section 3.3 compares the performance of the two feeding techniques with different aspect ratio and patch height, respectively. Section 3.4 introduces various designs by L-probe and M-probe for different applications. Finally, a brief conclusion is given in Section 3.5.

3.2 Basic characteristics

3.2.1 L-probe feeding mechanism

Figure 3.1 shows the geometry of the L-probe fed patch antenna. It can be seen that the antenna consists of a square ground plane, a patch and an L-probe. The ground plane has a length G_L and its center is the original in the HFSS simulation. The patch has a length P_L and a width P_w, and is located above the ground plane with a height of P_h. The edges of the patch and the ground plane are in parallel and their

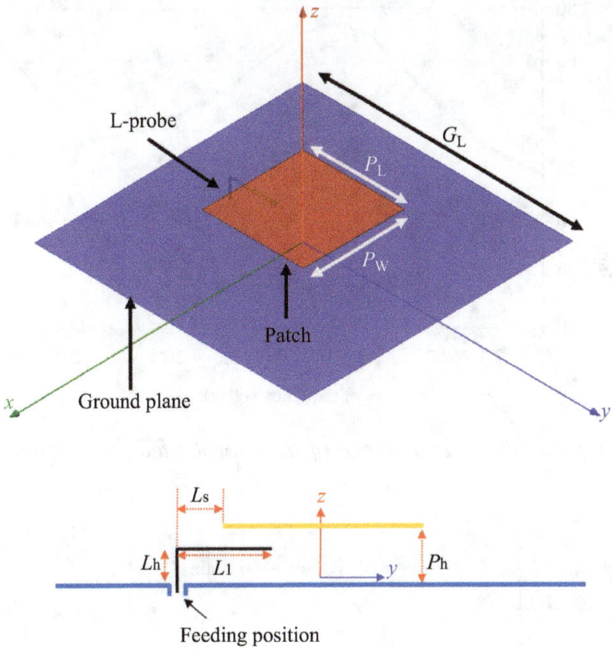

Figure 3.1 Geometry of the L-probe fed antenna

centers are aligned together. The L-probe is bended from a strip with wide L_w and it has a vertical portion L_h and a horizontal portion L_l. The horizontal portion of the L-probe is sandwiched between the patch and the ground plane. Unlike the conventional coaxial feed in which the inner conductor is directly connected to the patch, the L-probe is proximity coupled to the patch.

In this part, an L-probe fed patch antenna for Wi-Fi application was designed and the corresponding performances were simulated by HFSS. To simplify the design procedure, a square patch with length of P_L and a square ground plane with one wavelength were applied. There are several steps to find out the value of P_L. The first step is to calculate the patch height, which is 10% of wavelength at the center frequency and it is 12.5 mm. The second step is to calculate the resonant length of the patch by tuning the values of L_h and L_l to achieve three conditions, which are tuning the peak of the resistance at 2.4 GHz, optimizing the maximum point within the bandwidth at SWR < 1.5 and optimizing the two minimum points having the same values. The corresponding results are, respectively, shown in Figure 3.2 and Figure 3.3. The third step is to tune the position of the L-probe to achieve a symmetric radiation pattern at the E-plane at the center frequency, which is shown in Figure 3.4(b). Table 3.1 shows the optimized values for the L-probe fed patch antenna. It is found that the optimized value of P_L is 45.3 mm. The total length of the L-probe is around $0.25\lambda_0$. This optimized antenna has an impedance bandwidth (voltage standing wave ratio (VSWR) < 1.5) of 24.6%, which is from

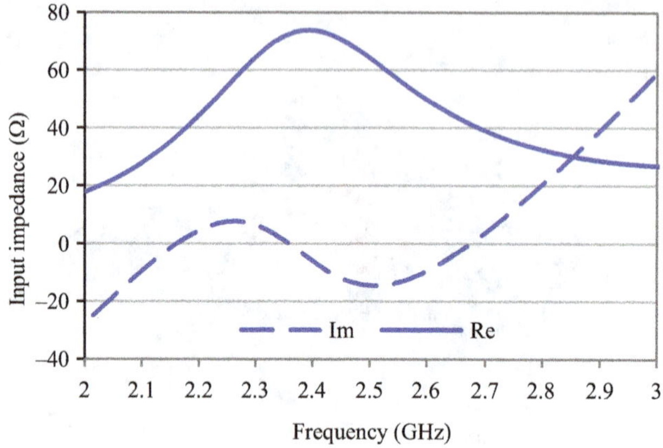

Figure 3.2 Input impedance of the L-probe fed patch antenna

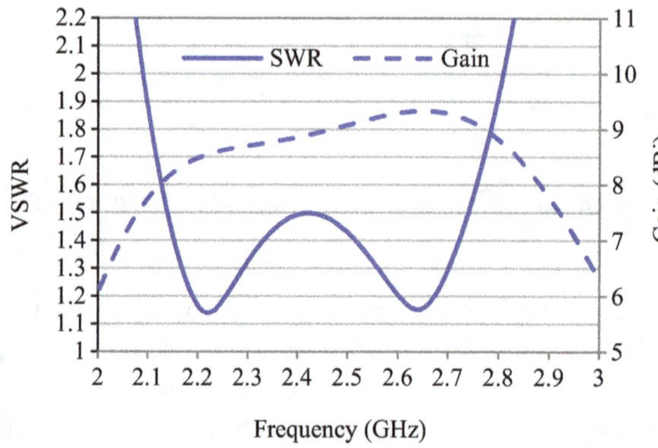

Figure 3.3 VSWR and gain of the L-probe fed patch antenna

Table 3.1 Optimized parameters for the L-probe fed patch antenna

Parameter		P_L	P_h	L_h	L_l	L_w
Value	(mm)	45.3	12.5	8.45	21.3	2
	(λ_0)	0.368	0.102	0.069	0.173	0.016

$^{*}f_c = (f_{start} + f_{stop})/2$ of the bandwidth with VSWR < 1.5.

2.139 to 2.741 GHz. It also has a stable gain within the bandwidth with a peak gain of 9.33 dBi at 2.66 GHz. Figure 3.4 shows the radiation patterns of the antenna at 2.14, 2.4 and 2.74 GHz. As we have optimized the position of the L-probe (which is 10.85 mm), the radiation pattern at the E-plane of the proposed version is much

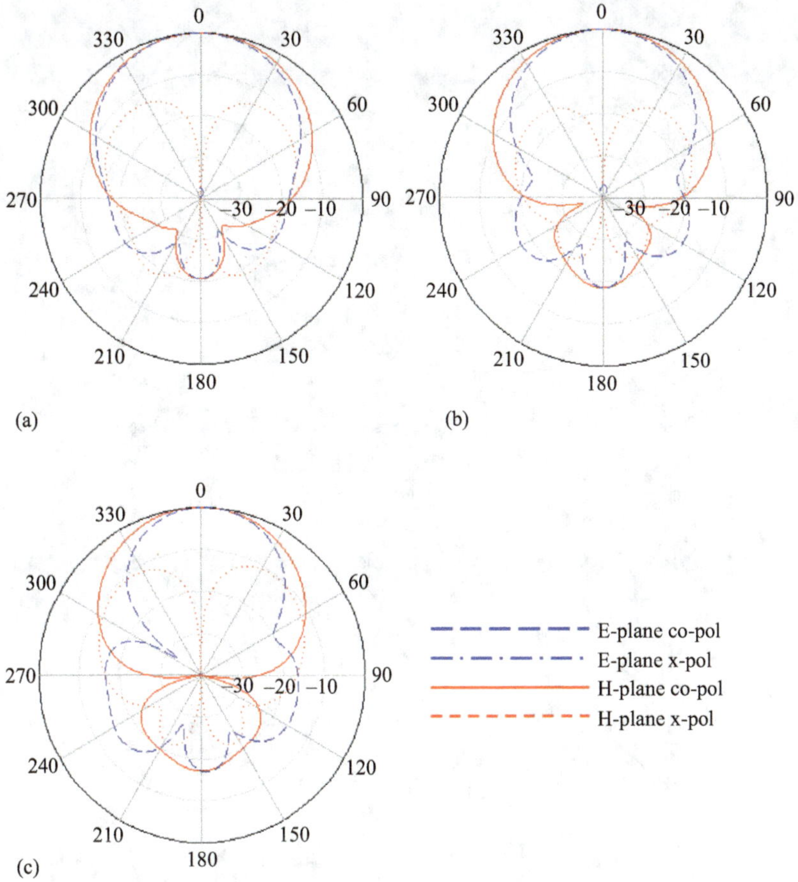

Figure 3.4 Radiation pattern of the L-probe fed patch antenna at (a) 2.14 GHz; (b) 2.4 GHz; and (c) 2.74 GHz

symmetric than the corresponding results in [14–16]. Figure 3.5 shows the radiation pattern of the antenna with different values of L_s. It can be seen that the radiation pattern at the E-plane is sensitive to the change of the value of L_s. The peak in the elevation angle is changed from negative to positive when the value of L_s is increased. Even though a symmetric radiation pattern can be achieved by tuning the L-probe position properly, the cross polarization level is quite high, which is around -15 dB at the center frequency. Figures 3.6 and 3.7 show the impedances of the antenna with different values of L_l and L_h, respectively. It is found that the value of L_l is sensitive with both resistance and reactance, while the value of L_h is much sensitive with resistance. Figure 3.8 shows the optimized bandwidth of the L-probe fed patch antenna with different values of L_w, with the value of P_h is fixed at 12.5 mm. It is found that the bandwidth is decreased with the increase of the value of L_w. Figure 3.8 also shows that the optimized value of L_w is around 2 mm.

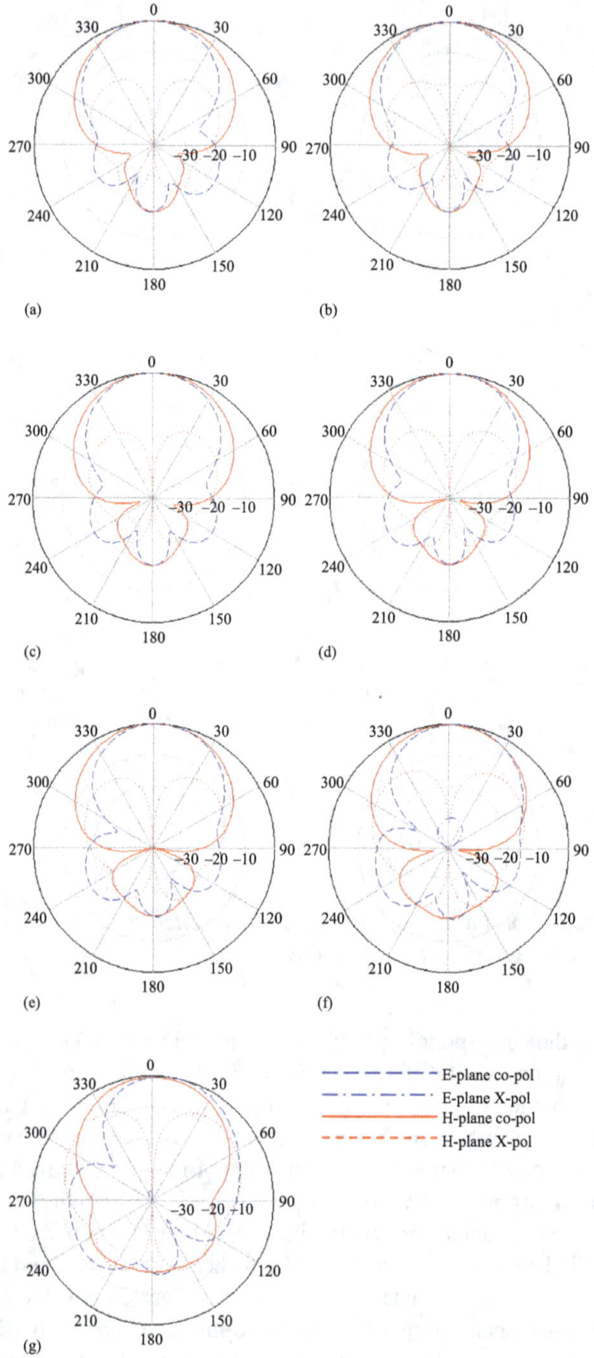

Figure 3.5 Radiation pattern with different values of L_s (a) 1.35 mm; (b) 5.35 mm; (c) 9.35 mm; (d) 13.35 mm; (e) 17.35 mm; (f) 21.35 mm; and (g) 27 .35 mm

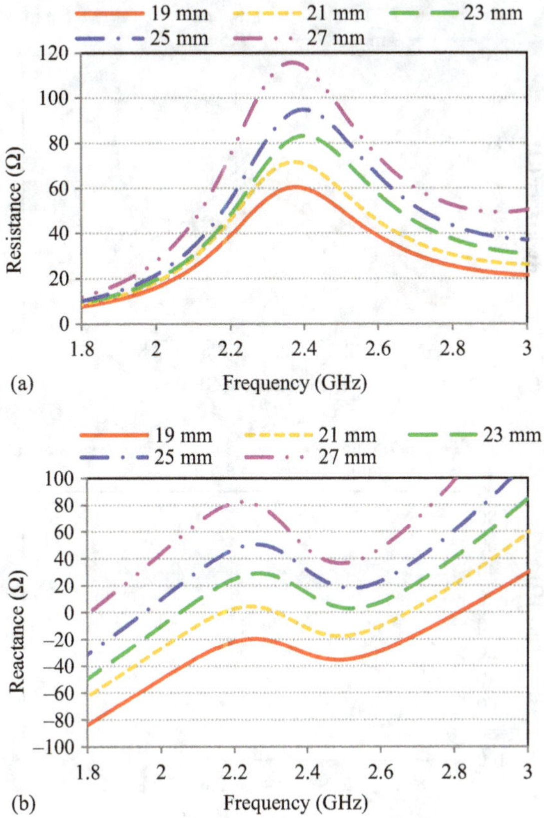

Figure 3.6 Input impedance with different values of L_l

3.2.2 M-probe feeding mechanism

Figure 3.9 shows the geometry of the M-probe fed patch antenna. Basically, the geometry of this antenna is the same as the L-probe fed patch antenna in the last part. The only difference is the L-probe is replaced by the M-probe. The M-probe is made by folding a strip by four times, and the M-probe has three vertical portions and two horizontal portions. The side view in Figure 3.9(b) clearly shows that the two outer vertical portions have lengths M_h and the two horizontal portions have length M_l. The center vertical portion has a length of "$P_h - 2 \times (P_h - M_h)$." The M-probe is sandwiched by the patch and the ground plane, with its center vertical portion lies along the z-axis. It has one end connected to the feeding position and the other end directly connected to the patch.

In this part, an M-probe fed patch antenna for Wi-Fi application was designed and the corresponding simulated results were obtained by HFSS. The patch height and ground plane size are same as the L-probe fed patch antenna which mentioned in the last part. To find out the value of P_L for the patch, same tuning steps are followed as the L-probe case. Table 3.2 lists the optimized values of the M-probe

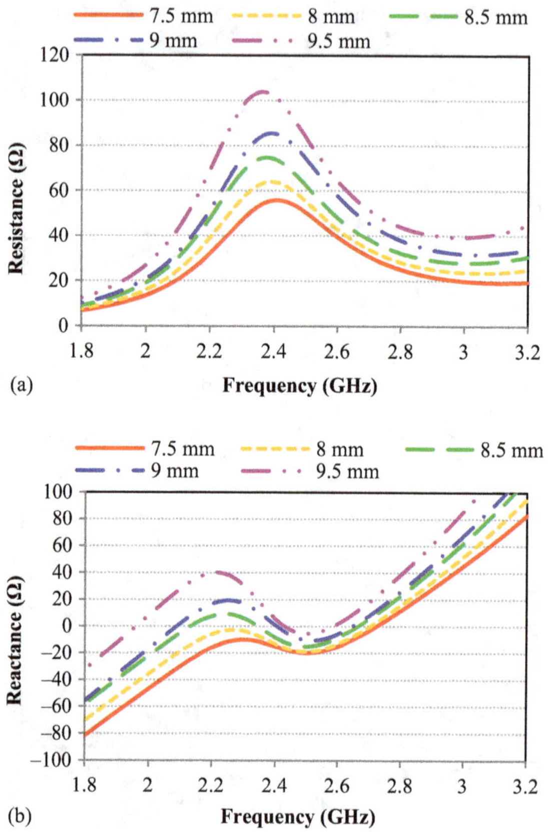

Figure 3.7 Input impedance with different values of L_h

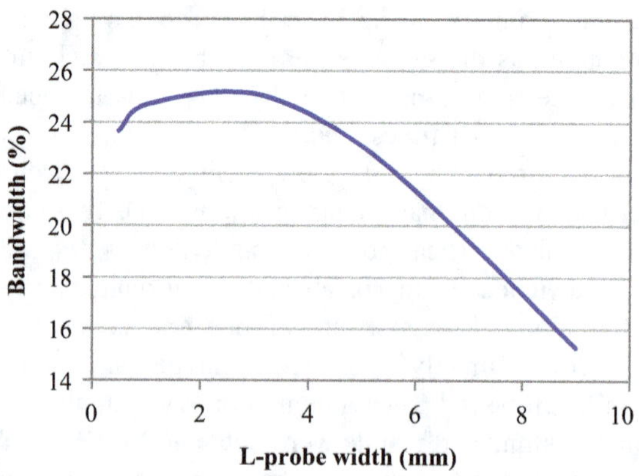

Figure 3.8 Optimized bandwidth of the L-probe fed patch antenna with different values of L_w with P_h fixed at 12.5 mm

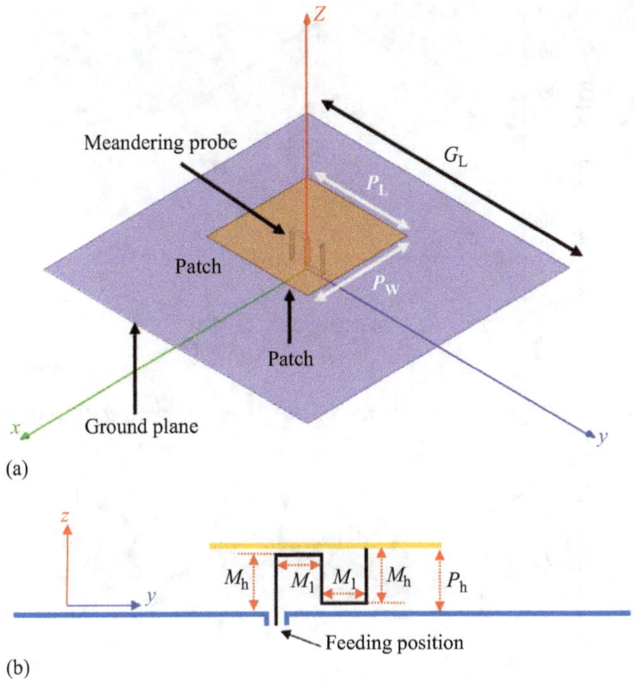

Figure 3.9 Geometry of the meandering probe fed antenna

Table 3.2 Optimized parameters for the M-probe fed patch antenna

Parameter		P_L	P_h	M_h	M_l	M_w
Value	(mm)	46.5	12.5	11.82	6.98	2
	(λ_0)	0.377	0.101	0.096	0.057	0.016

$^*f_c = (f_{start} + f_{stop})/2$ of the bandwidth with VSWR < 1.5.

fed patch antenna. It is found that the optimized value of P_L is 46.5 mm and the total length of the M-probe is $0.395\lambda_0$. Figure 3.10 shows the simulated resistance and reactance of the optimized M-probe fed patch antenna. It is found that the peak resistance is around 75 Ω. Figure 3.11 shows that the antenna has an impedance bandwidth (VSWR < 1.5) of 22.2%, which is from 2.164 to 2.7045 GHz. The antenna has a stable gain across the operating frequency band and has a peak gain of 10 dBi at 2.68 GHz. The above results show that patch antenna feeding by L-probe has wider bandwidth but lower gain than feeding by M-probe. Figure 3.12 shows the radiation patterns of the M-probe fed patch antenna at 2.16, 2.4 and 2.7 GHz. It is noted that the antenna can radiate a very symmetric radiation pattern and low cross polarization level across the operating bandwidth with the use of M-probe feeding mechanism. A co-to-cross level of over 32 dB can be achieved across the

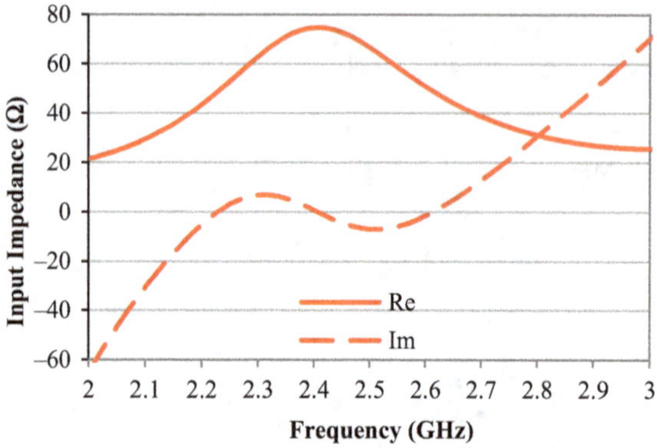

Figure 3.10 Input impedance of the M-probe fed patch antenna

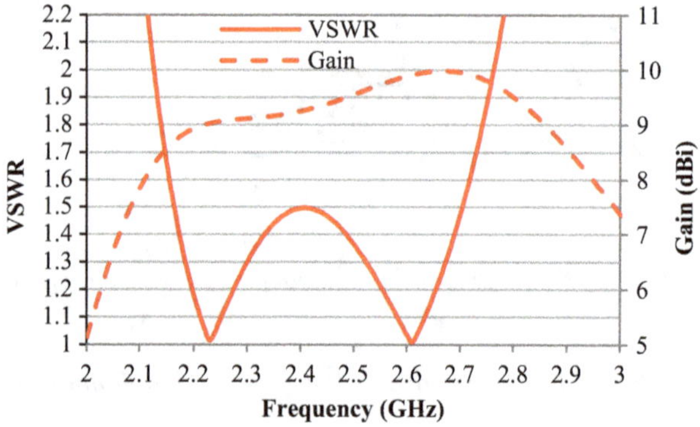

Figure 3.11 VSWR and gain of the M-probe fed patch antenna

operating frequency band. The impedances of the antenna with different values of M_h, M_l and M_w are, respectively, plotted in Figures 3.13, 3.14 and 3.15. It is found that both M_h and M_w are sensitive with the reactance; while M_l is sensitive with both the resistance and reactance. The performances at center frequency of the patch antenna feeding by the L-probe and he M-probe are tabulated in Table 3.3. It clearly shows that the L-probe case has wider bandwidth and the M-probe case has much lower cross polarized radiation and high gain. For the front-to-back ratio and half power beamwidths at the E- and H- planes, there are similar results were obtained for the L-probe case and M-probe case.

The impedance bandwidth of the M-probe fed patch antenna can be further increased by using a wide M-probe with thicker patch height. Table 3.4 listed the optimized values of the antenna. It is noted that the patch height is $0.174\lambda_0$ and the width of the M-probe is $0.21\lambda_0$. Figure 3.16 shows that by increasing the width of the M-probe and the thickness of the antenna, another resonant is appeared at 3.2 GHz. In addition, there are more frequency points with reactance values close to

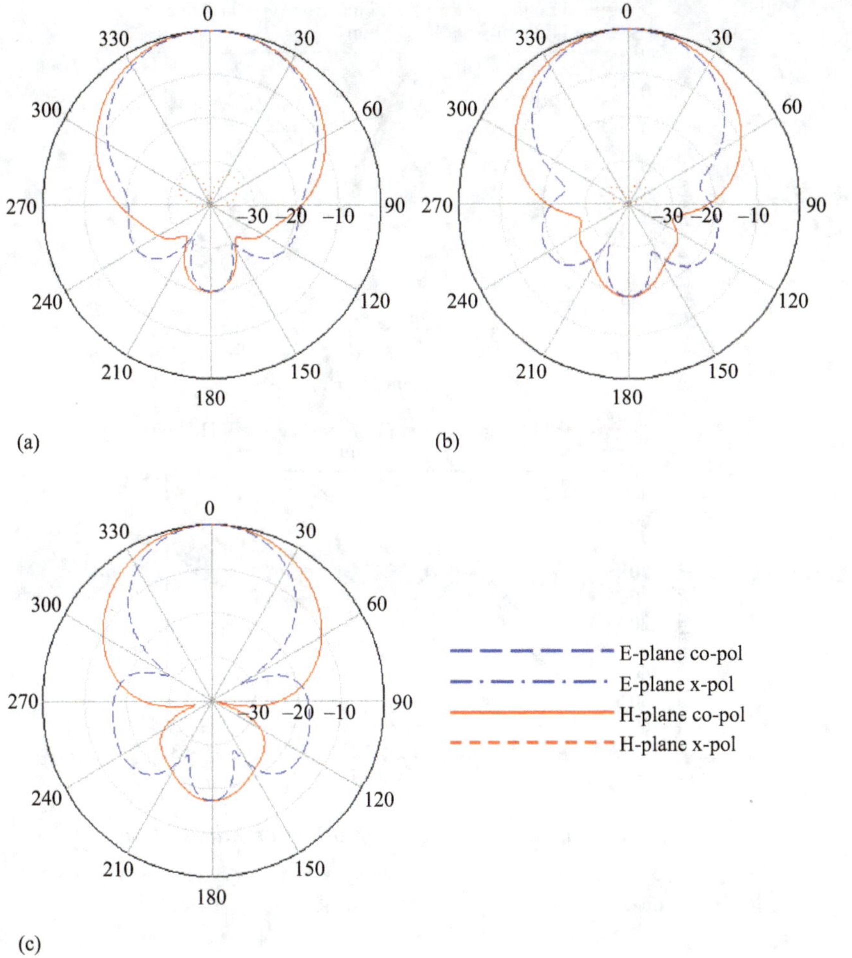

Figure 3.12 Radiation pattern of the M-probe fed patch antenna at (a) 2.16 GHz; (b) 2.4 GHz; and (c) 2.7 GHz

zero. It is believed that the new resonant is generated from the wide M-probe. The wide M-probe has a total length of $0.537\lambda_0$, which is close to the half wavelength and forming a structure of top loaded monopole. Figure 3.17 shows that the wide M-probe fed patch antenna has an impedance bandwidth of 43.4%, which is from 1.9015 to 2.956 GHz. Figure 3.17 also shows that the antenna has a gain of around 9 dBi and a peak gain is 10 dBi at 2.81 GHz. The bandwidth is around two times more than the normal M-probe fed patch antenna. Figure 3.18 shows the radiation patterns of the antenna at 1.9, 2.43 and 2.9 GHz. It is found that the radiation pattern at the E-plane is affected by the monopole mode from the wide M-probe slightly. Fortunately, the cross polarization level is still lower than −18 dB across the operating frequency band.

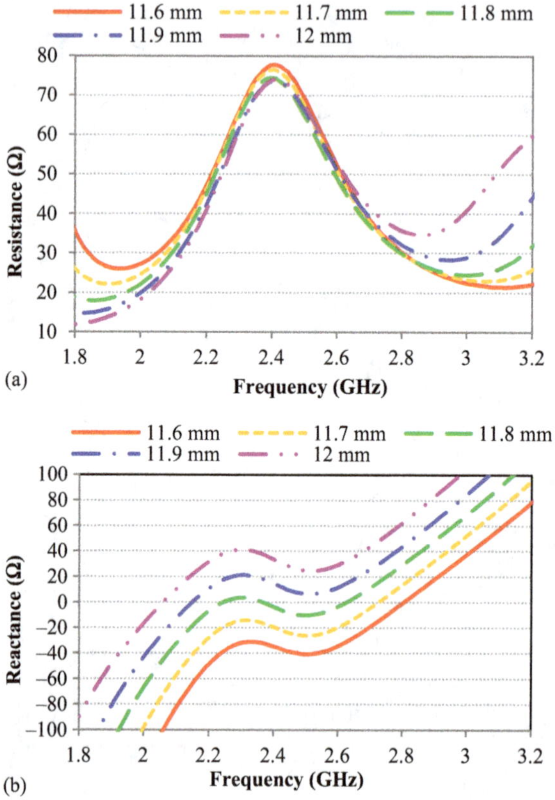

Figure 3.13 Input impedance with different values of M_h

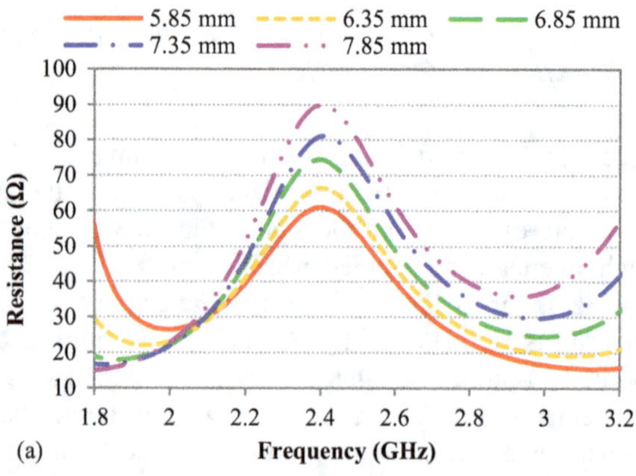

Figure 3.14 Input impedance with different values of M_l

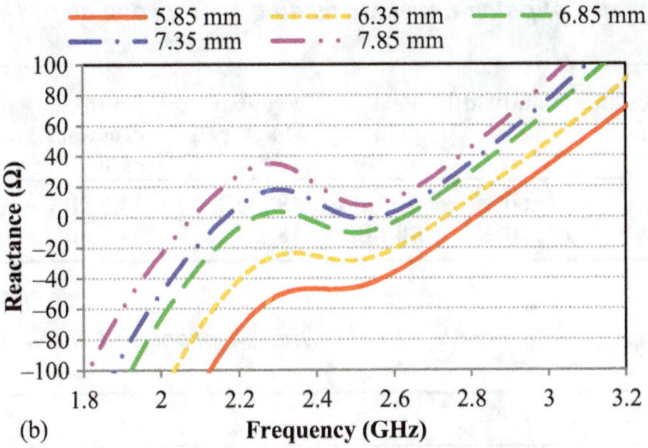

(b)

Figure 3.14 (Continued)

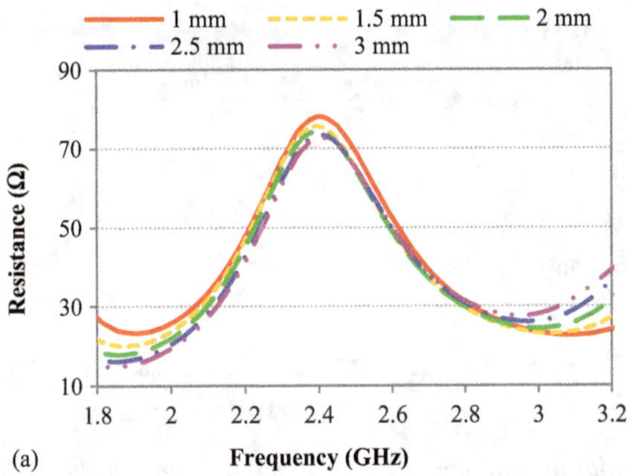

(a)

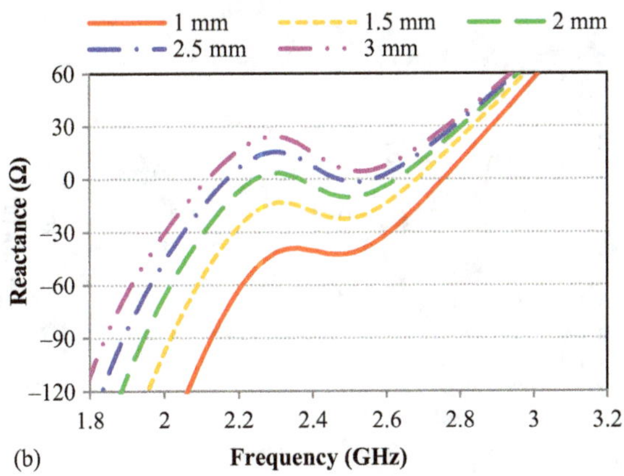

(b)

Figure 3.15 Input impedance with different values of M_w

Table 3.3 Summary of patch antennas feeding by L-probe and M-probe

Feeding method	Connection with the patch	Bandwidth	Peak gain	Front-to -back ratio	Co-to-cross-pol ratio	HPBW	
						E-plane	H-plane
L-probe	Coupling	24.60%	9.33 dBi	18.2 dB	14.5 dB	52.5°	66°
M-probe	Direct	22.20%	10 dBi	18.7	34.3 dB	53°	65°

Table 3.4 Optimized parameters for the wide M-probe fed patch antenna

Parameter		P_L	P_h	M_h	M_l	M_w
Value	(mm)	46.5	21.5	17.95	8	26
	(λ_0)	0.376	0.174	0.145	0.065	0.210

*$f_c = (f_{start} + f_{stop})/2$ of the bandwidth with VSWR < 1.5.

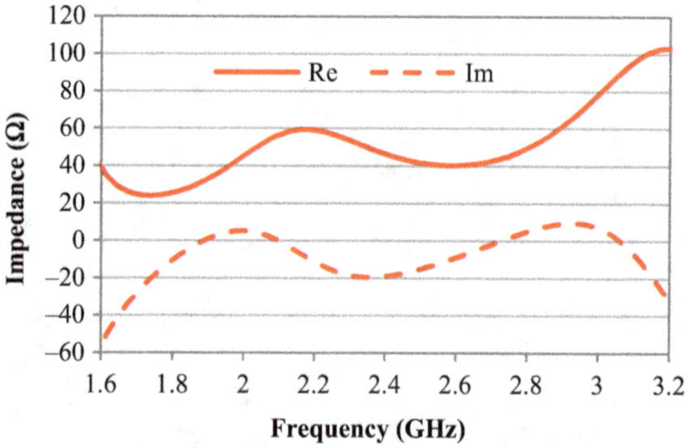

Figure 3.16 Input impedance of the wide meandering probe fed patch antenna

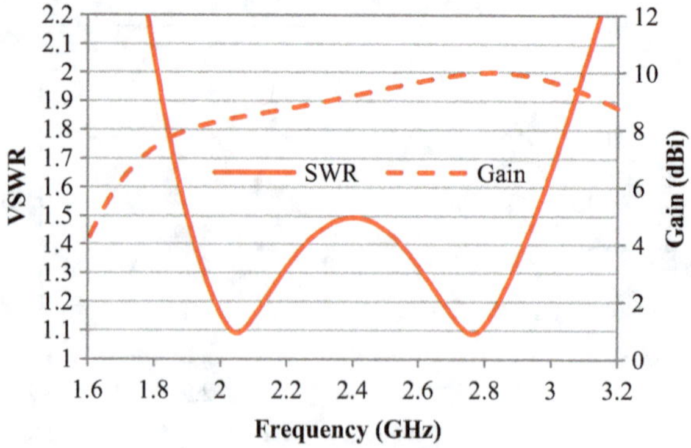

Figure 3.17 VSWR and gain of the wide meandering probe fed patch antenna

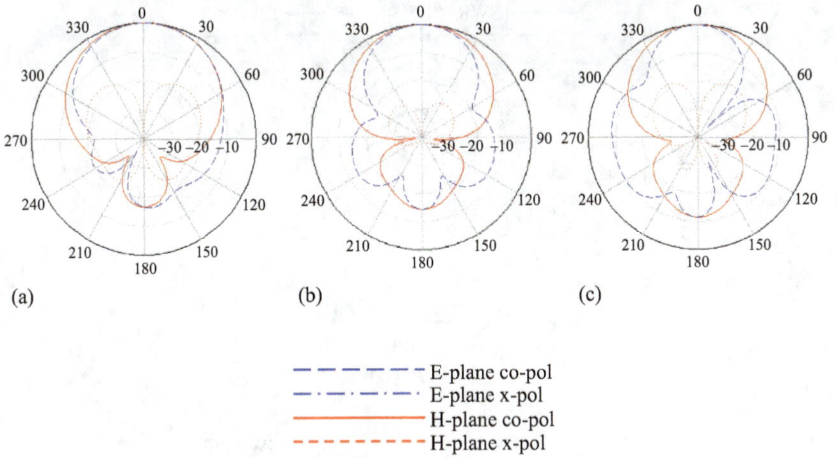

Figure 3.18 Radiation pattern of the wide M-probe fed patch antenna at (a) 1.9 GHz; (b) 2.43 GHz; and (c) 2.9 GHz

3.3 Parametric studies

In this section, parametric studies of patch antennas feeding by L-probe and M-probe are carried out. Their performances with different P_h and aspect ratio at their center frequencies are compared.

3.3.1 Performance with different P_h

In this part, an optimization of impedance matching for the antenna is required when the value of P_h is changed. The length and width of the patch are fixed and the widths of the L-probe and M-probe are fixed to 2 mm. Figure 3.19 shows the impedance bandwidths of the patch antennas with different values of P_h. It is found that the L-probe case can achieve a bandwidth of 30% with a patch height of $0.135\lambda_0$; while the corresponding value for the M-probe case is 22.4% with a patch height of $0.102\lambda_0$. Figures 3.20 and 3.21 show the total lengths of the L-probe and M-probe, and the values of L_l, L_h, M_h and M_l with different values of P_h, respectively. It is found that the total length of the L-probe is very stable and is around $0.24\lambda_0$ for all values of P_h. On the other hand, the total length of the M-probe is unstable and is varying from $0.17\lambda_0$ to $0.55\lambda_0$ when the value of P_h is increased from $0.03\lambda_0$ to $0.216\lambda_0$. Figure 3.22 shows the gains at $(\theta, \phi) = (0°, 0°)$ for the cases of L-probe and M-probe. It is noted that the gain of the L-probe case is dropped from 9.8 dBi to only 4.74 dBi when the value of P_h is increased from $0.03\lambda_0$ to $0.25\lambda_0$; while the gain of the M-probe case is only reduced by 1.2 dB when the value of P_h is increased from $0.03\lambda_0$ to $0.216\lambda_0$.

Figure 3.23 shows the co-to-cross-pol ratio and front-to-back ratio of the two antennas. Results show that the co-to-cross-pol ratio is reducing with the increase

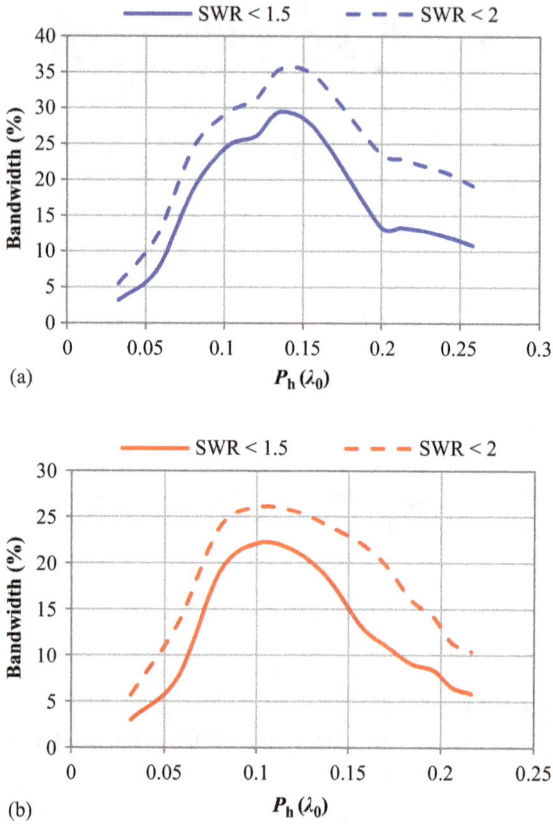

Figure 3.19 Bandwidth of square patch antenna with different patch height feeding by (a) L-probe and (b) M-probe

of patch height for the L-probe case. In addition, a longer L_h is required to match the antenna, and this increases the cross polarized radiation from the L-probe. The co-to-cross-pol ratio is over 26 dB with all values of patch height for the M-probe case. This demonstrated that the M-probe can suppress the cross polarization level effectively. Figure 3.23 also showed that the M-probe case has a higher front-to-back ratio than the L-probe case when the value of P_h is increasing.

Figure 3.24 shows the half power beamwidths at the E- and H- planes for the L-probe case and M-probe case. It is found that the 3 dB beamwidth at the H-plane is around 14 dB wider than the 3 dB beamwidth at the E-plane when the patch height is increasing. However, the values of the 3 dB beamwidths for the two cases are different. Both cases have a stable 3 dB beamwidths of around 65° at the H-plane and around 55° at the E-plane with P_h less than $0.15\lambda_0$. When the values of P_h are increased, the 3 dB beamwidths are reducing for the L-probe case and the 3 dB beamwidths are increasing for the M-probe case. The difference in this variation is that the co-polar radiation from

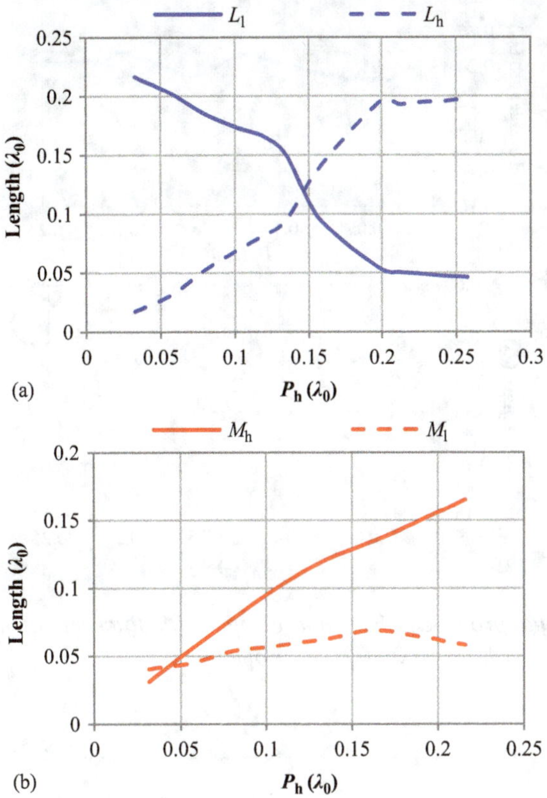

Figure 3.20 Lengths of the feeding probe portions of square patch with different patch height feeding by (a) L-probe and (b) M-probe

the patch mode is distorted by the radiation from the L-probe. While the M-probe can suppress the cross polarized radiation and providing a stable patch mode.

3.3.2 Performance with different aspect ratio

In this part, an optimization of impedance matching for the antenna is required when the aspect ratio is changed. The heights of the patch antennas are fixed and the widths of the L-probe and M-probe are fixed to 2 mm. Figure 3.25 shows the impedance bandwidths of the patch antennas with different values of aspect ratio. It is found that the L-probe case can achieve a stable bandwidth of around 25% with the aspect ratio varying from 1 to 1.77, while the corresponding bandwidth for the M-probe case is 22.5% with the aspect ratio varying from 1 to 2. Figure 3.25 also shows that an impedance bandwidth of 40% can be achieved with the aspect ratio equal to 1.88 if using VSWR less than 2 as reference. This significant bandwidth enhancement is contributed by the resonant mode from the L-probe. Figures 3.26 and 3.27 show the total lengths of the L-probe and M-probe, and the values of L_l,

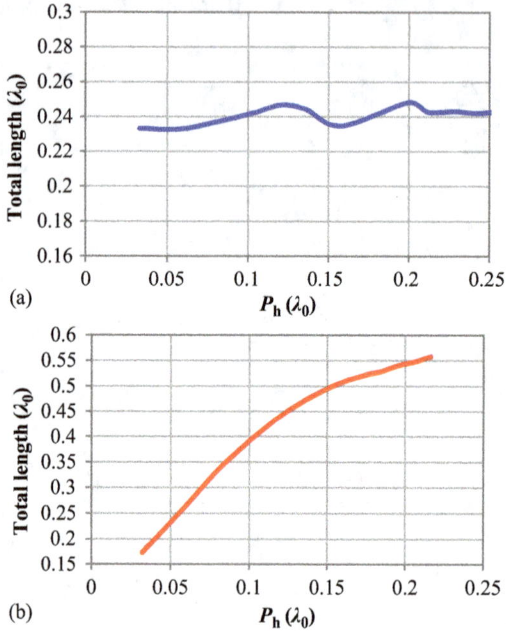

Figure 3.21 Total probe length of square patch with different patch height feeding by (a) L-probe and (b) M-probe

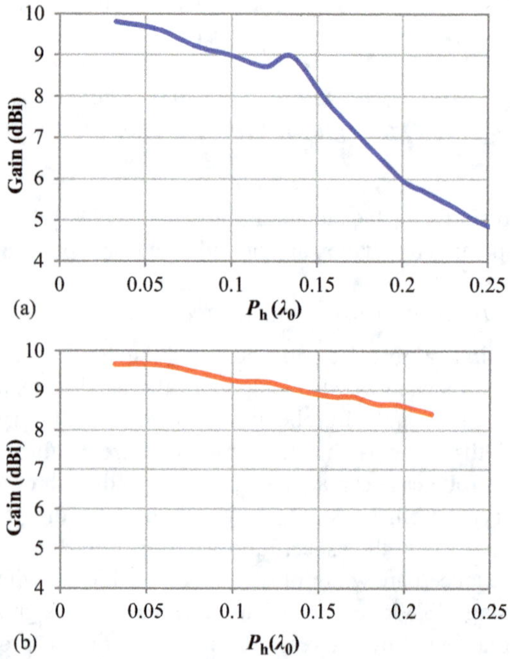

Figure 3.22 Gain of square patch antenna with different patch height feeding by (a) L-probe and (b) M-probe

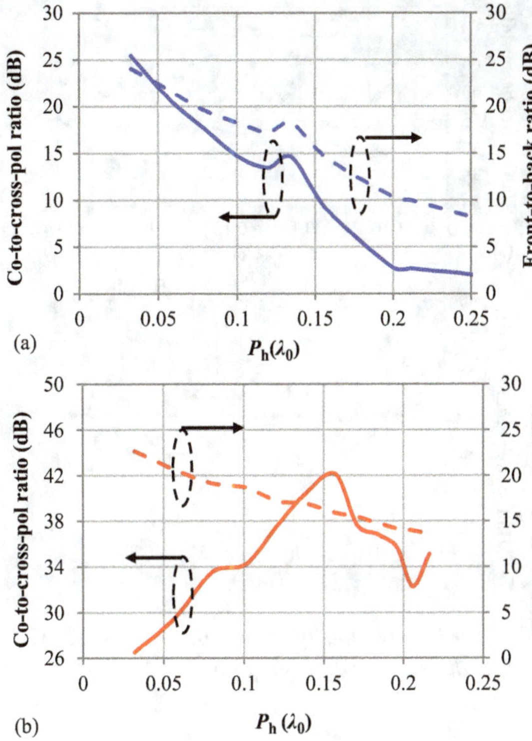

Figure 3.23 Co-to-cross level and front-to-back ratio of square patch antenna with different patch height feeding by (a) L-probe and (b) M-probe

L_h, M_h and M_l with different aspect ratio values, respectively. It is found that the total length of the L-probe case and M-probe case are around $0.23\lambda_0$ and $0.4\lambda_0$, respectively. Figure 3.28 shows the gains at $(\theta, \phi) = (0°, 0°)$ for the two cases. It is noted that the gain of the L-probe case is around 9 dBi with different value of aspect ratio, while the gain of the M-probe case is increasing from 9 dBi to 10.3 dBi when the aspect ratio is increasing from 0.043 to 2.3.

Figure 3.29 shows the co-to-cross-pol ratio and front-to-back ratio of the antennas feeding by L-probe and M-probe. Results show that the co-to-cross-pol ratio is reducing with the increase of aspect ratio value for the L-probe case. This high cross polarized radiation is due to the increase of vertical portion length for optimizing the impedance of the antenna. For the M-probe case, the co-to-cross-pol ratio is over 22 dB with all aspect ratio values. This demonstrated that the M-probe can suppress the cross polarization level effectively with different aspect ratio value. Figure 3.29 also showed that patch antenna feeding by the L-probe and M-probe have front-to-back ratio higher than 18 dB for all aspect ratio values.

Figure 3.30 shows the half power beamwidths at the E- and H-planes for the L-probe case and M-probe case. It is found that the variation of the 3 dB beamwidths at the E- and H-planes for both cases quite similar. In E-plane, both cases have stable 3 dB beamwidths of around 52° for all aspect ratio values. In the H-plane, the 3 dB beamwidths for both cases are reducing from ~76° to ~55° when the value

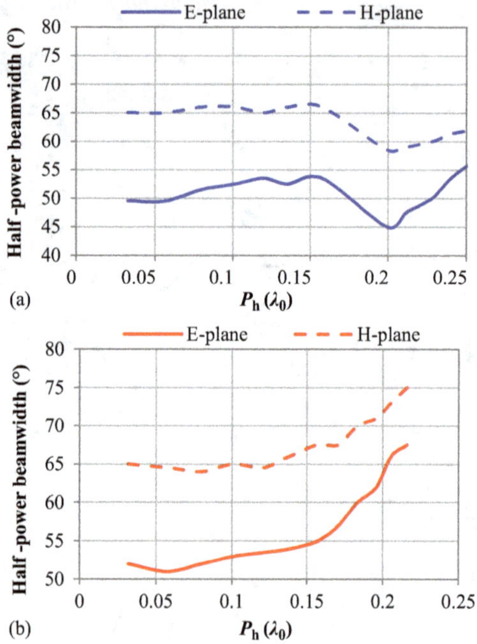

Figure 3.24 Half-power beamwidths of square patch antenna with different patch height feeding by (a) L-probe and (b) M-probe

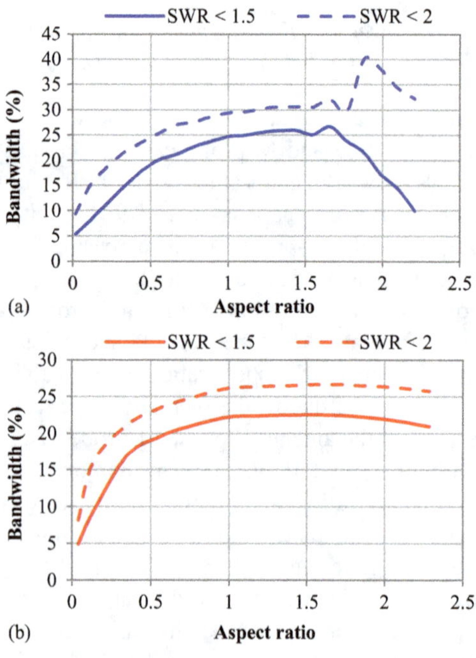

Figure 3.25 Bandwidth of patch antenna in different aspect ratio feeding by (a) L-probe and (b) M-probe

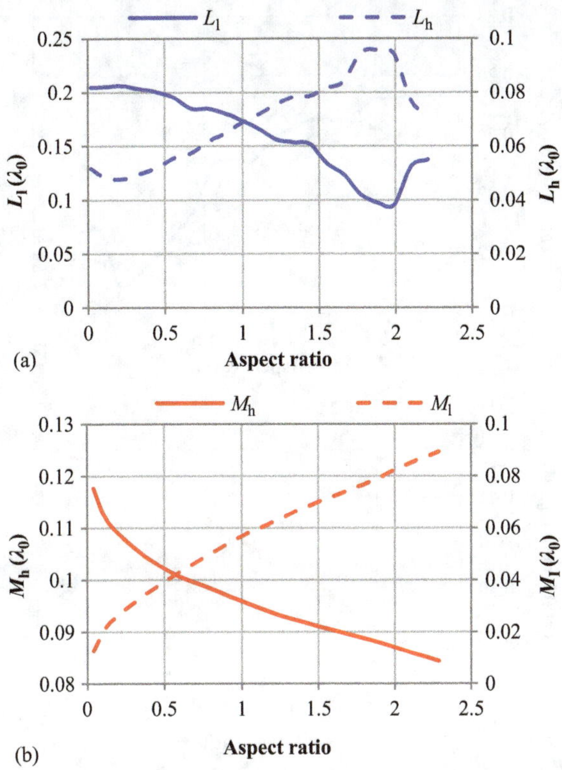

Figure 3.26 Lengths of the feeding probe portions of patch antenna in different aspect ratio feeding by (a) L-probe and (b) M-probe

of aspect ratio is changed from 0.3 to 2. The results showed that patch modes can radiate properly if patch antenna is feeding by either-probe and M-probe.

3.4 Development of L-probe and M-probe fed patch antenna

In this section, some antenna designs with different characteristics for various wireless applications are introduced. Five categories are included in this section, which are circular polarization, dual polarization, dual-band, conformal ground plane and fabricated with printed circuit board (PCB). Finally, a dual polarized patch antenna design feeding by L-probe and M-probe is mentioned in this section.

3.4.1 Circular polarization

Other than linear polarization, circular polarized antennas are be used for wireless communications. It is commonly known that both of the L-probe [25] and the M-probe [35] can be used to feed a truncated corners patch to achieve CP characteristic. To further enhance the axial ratio bandwidth, there are some techniques

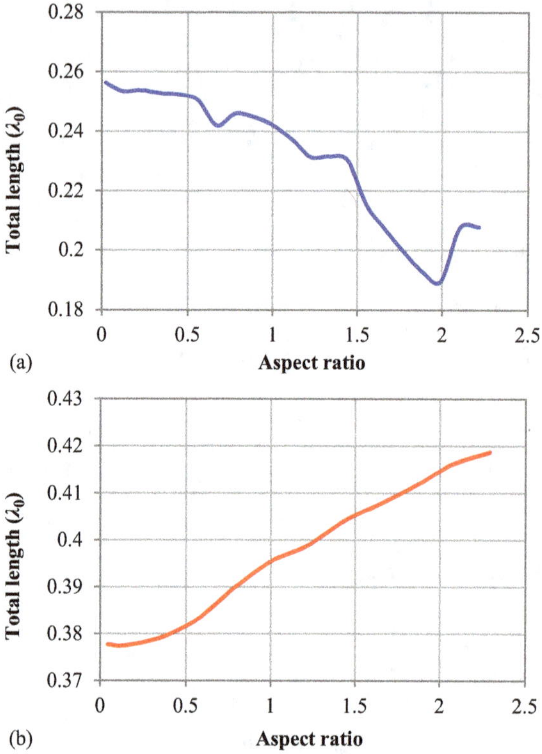

Figure 3.27 Total probe length of patch antenna with different aspect ratio feeding by (a) L-probe and (b) M-probe

that can be applied and are shown in Figure 3.31. By using four L-probes with a very wideband 1 to 4 ports sequential phases power divider to feed a circular patch, an impedance bandwidth (<−10 dB) and axial ratio bandwidth of 71.28% and 81.6% can be achieved [24]. Figure 3.31(b) shows a CP patch antenna fed by a 3D M-probe [36]. With the use of this 3D M-probe, the patch antenna has an impedance bandwidth (SWR < 2) and axial ratio bandwidth of 25.2% and 22.4%, respectively.

3.4.2 Dual polarization

In modern wireless communications, dual polarization is often applied for increasing the channel capacity and polarization diversity. Figure 3.32(a) shows a dual polarized patch antenna feeding by the L-probe [26]. There are four L-probes located at the four corners of a square patch. Two out-of-phase power dividers are, respectively, connected to ports 1 and 3 and ports 2 and 4. This configuration can effectively suppress the cross polarized radiation and enhance the input ports isolation to over 30 dB. For designing dual polarized patch antenna with M-probe feeding mechanism, the two M-probes are arranged in "T-shape" configuration

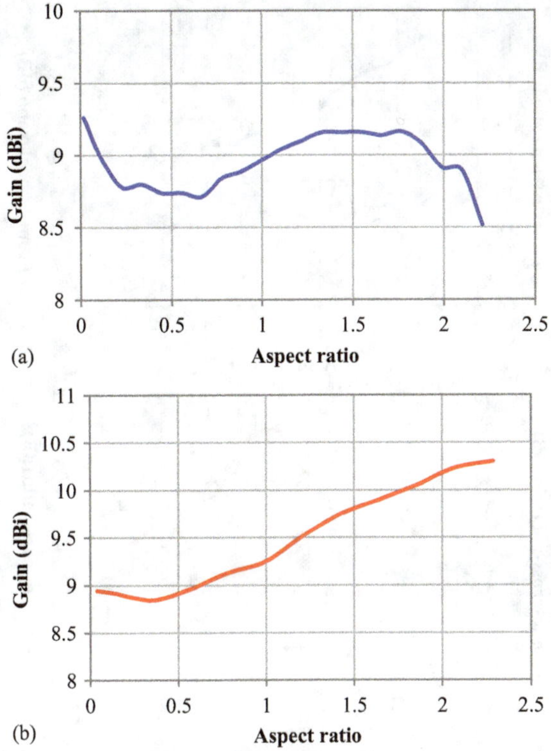

Figure 3.28 Gain of patch antenna with different aspect ratio feeding by (a) L-probe and (b) M-probe

and the antenna geometry is shown in Figure 3.32(b) [38]. To further enhance the isolation, it is necessary to surround the antenna with four vertical walls.

3.4.3 Dual band

There are two types of dual-band antennas in general. One type is an antenna with a single port with dual operating frequency bands [40–43]. The other type is an antenna with two ports and each port has an operating frequency band [44,45]. A dual band L-probe fed patch antenna with two ports [44] and a dual band M-probe fed patch antenna with single port are shown in Figure 3.33 [42] as examples.

3.4.4 Conformal ground plane

One of the advantages of patch antenna is that it can be mounted on a conformal ground plane [3]. The L-probe and M-probe feeding methods are also possible to realize on a cylindrical ground plane [46–48]. Typical designs of patch antennas feeding by L-probe [44] and M-probe [46] on conformal ground planes are shown in Figure 3.34. It is found that the impedance bandwidths on conformal ground planes are wider than on planar ground planes for both cases.

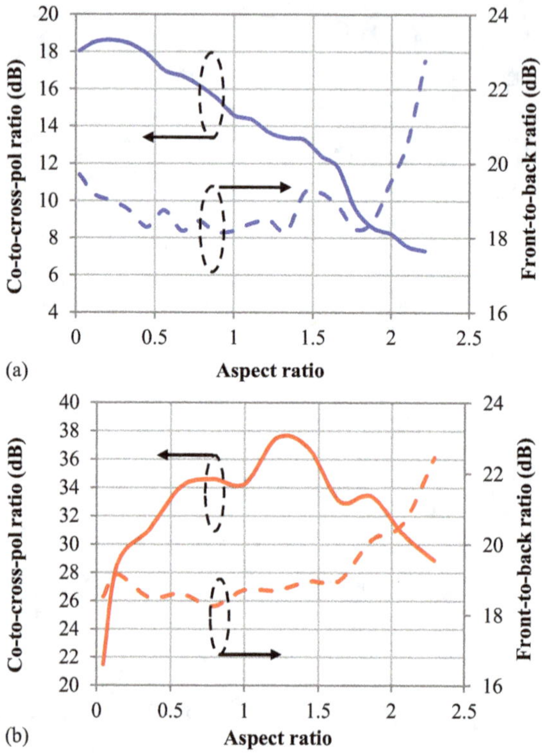

Figure 3.29 Co-to-cross level and front-to-back ratio of patch antenna with different aspect ratio feeding by (a) L-probe and (b) M-probe

3.4.5 Printed circuit board

Using PCB material to design antenna is a common technique for mass production in industry with the advantages of low cost, high reliability and simple in manufactory [49–58]. Figure 3.35 shows the geometries of an L-probe [54] and an M-probe [52] printed on a PCB. It is found that wideband characteristics can still be achieved for both feeding methods even printing the feeding probes on substrate material. It is demonstrated in [52] that a second- or third-order M-probe can be realized. The advantages include further improvement on impedance matching and co-to-cross-pol ratio.

3.4.6 Fusion of the L-probe and M-probe in antenna design

In this part, a dual-polarized patch antenna design feeding by L-probe and M-probe for LTE application [59] and the geometry is shown in Figure 3.36. A pair of twin-L-probes together with a wideband differential power divider underneath the ground are fed the patch for the +45° polarization (port 1); while a wide M-probe is used to feed the patch for the −45° polarization (port 2). By using the two wideband feeding techniques, a dual polarized patch antenna element with the characteristics of high

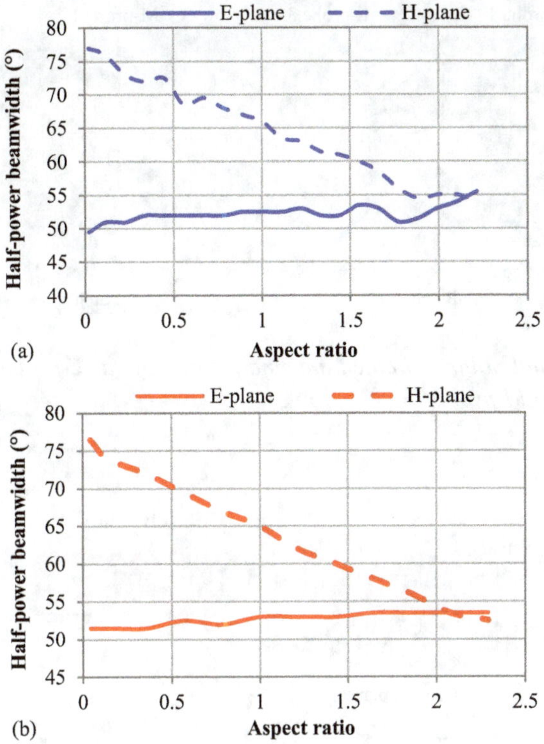

Figure 3.30 Half-power beamwidths of patch antenna with different aspect ratio feeding by (a) L-probe and (b) M-probe

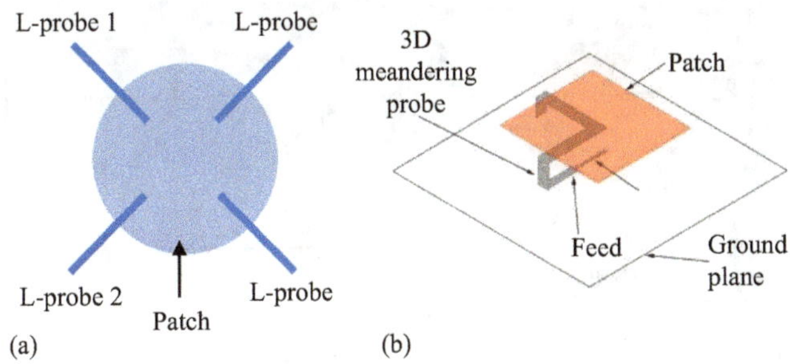

Figure 3.31 Circular polarized patch antenna feeding by (a) L-probe and (b) M-probe

input port isolation and wide impedance bandwidth is designed for LTE application. This antenna has impedance bandwidths (SWR < 2) of over 47% for both input ports. Within the operating frequency band In addition, the antenna has an input port isolation of over 37 dB and a gain of over 8 dBi within the entire operating band.

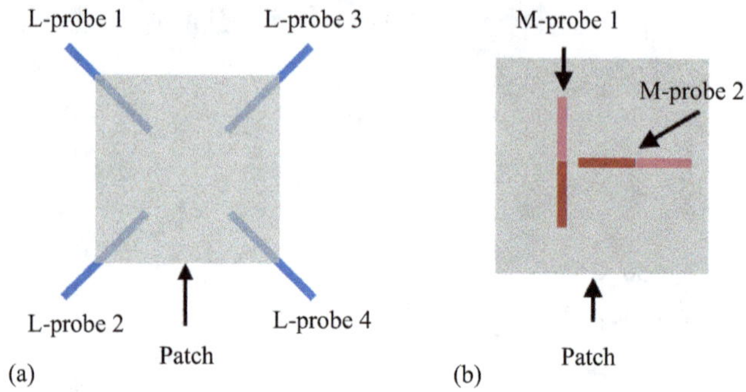

Figure 3.32 Dual polarized patch antenna feeding by (a) L-probe and (b) M-probe

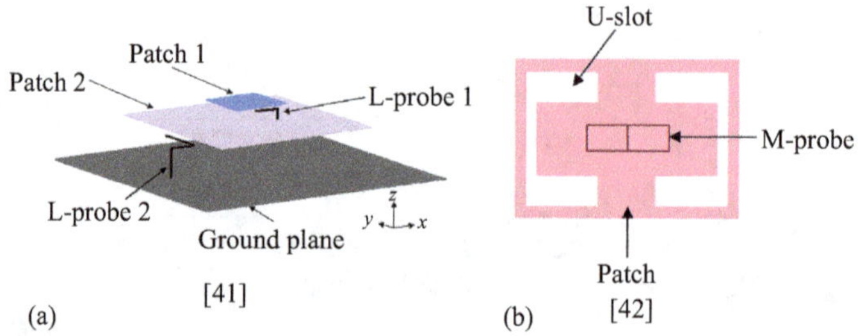

Figure 3.33 Dual-band patch antenna feeding by (a) L-probe and (b) M-probe

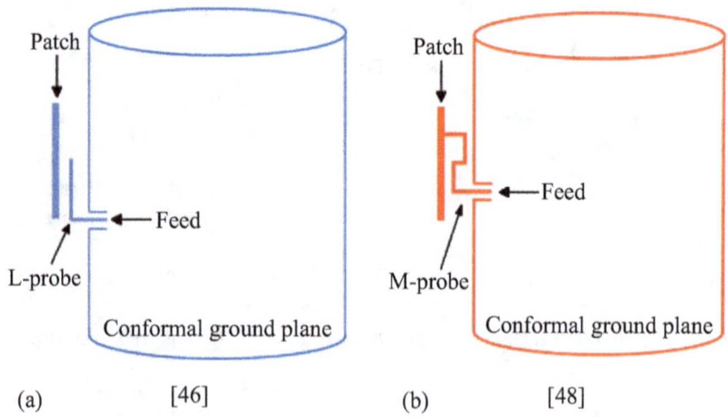

Figure 3.34 Patch antenna with conformal ground plane feeding by (a) L-probe and (b) M-probe

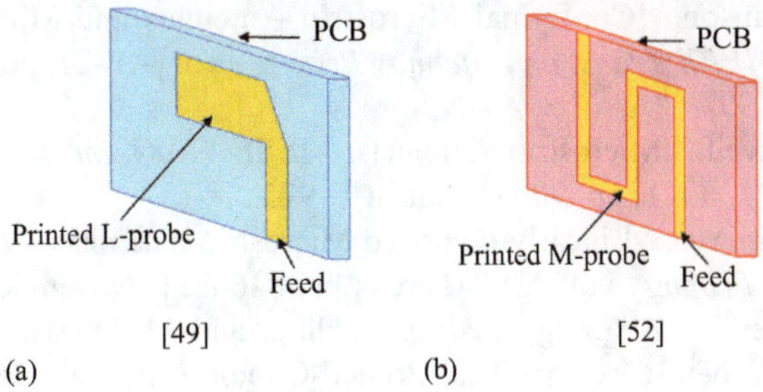

Figure 3.35 Printed circuit board design of (a) L-probe and (b) M-probe

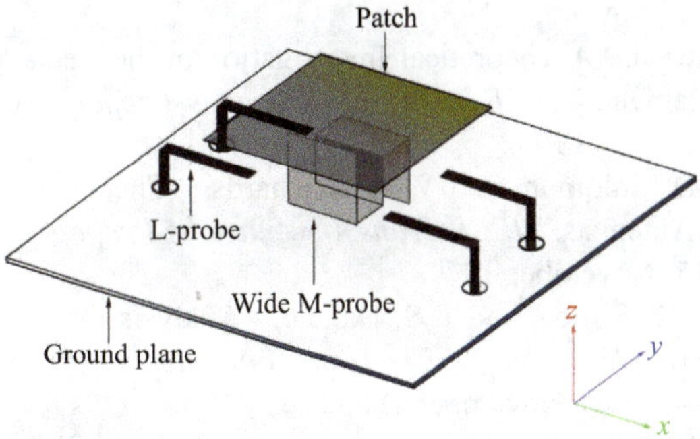

Figure 3.36 Dual polarized patch antenna fed by L-probe and M-probe

3.5 Conclusion

This chapter demonstrates a design guideline of a patch antenna for Wi-Fi application by L-probe and M-probe. Parametric studies of patch antennas feeding by L-probe and M-probe are also included. The performances of patch antennas with different patch height and aspect ratio are demonstrated for comparison. In addition, the performances of a wide M-probe fed patch antenna are also presented. This antenna has an impedance bandwidth (VSWR < 1.5) of around 44% with stable gain and low cross polarization level across the operating frequency bands. Finally, a review of various designs for different applications is introduced.

References

[1] G. A. Deschamps, "Microstrip Microwave Antennas," 3rd USFA Symposium on Antennas, 1953.

[2] H. Gutton and G. Baissinot, "Flat Aerial for Ultra High Frequencies," French Patent No. 703 113, 1955.

[3] R. E. Munson, "Conformal Microstrip Antennas and Microstrip Phased Arrays," *IEEE Trans. on Antennas Propag.,* vol. AP-22, no. 1, pp. 74–78, January 1974.

[4] J. Q. Howell, "Microstrip Antennas," *IEEE Trans. on Antennas Propag.,* vol. AP-23, no. 1, pp. 90–93, January 1975.

[5] A. G. Derneryd, "Linearly Polarized Microstrip Antennas," *IEEE Trans. on Antennas Propag.,* vol. AP-24, no. 6, pp. 846–851, November 1976.

[6] L. C. Shen, S. A. Long, M. R. Allerding, and M. D. Walton, "Resonant Frequency of a Circular Disc, Printed-Circuit Antenna," *IEEE Trans. on Antennas Propag.,* vol. AP-25, no. 4, pp. 595–596, July 1977.

[7] P. K. Agrawal and M.C. Bailey, "An Analysis Technique for Microstrip Antennas," *IEEE Trans. on Antennas Propag.*, vol. AP-25, no. 6, pp. 756–759, November 1977.

[8] A. G. Derneryd, "A Theoretical Investigation of the Rectangular Microstrip Antenna Element," *IEEE Trans. on Antennas Propag.*, vol. AP-26, no. 4, pp. 532–535, July 1978.

[9] Y. T. Lo, D. Solomon, and W. F. Richards, "Theory and Experiment on Microstrip Antennas," *IEEE Trans. on Antennas Propag.,* vol. AP-27, no. 2, pp. 137–145, November 1979.

[10] K. M. Luk, K. F. Lee, and J. S. Dahele, "Analysis of the Cylindrical-Rectangular Patch Antenna," *IEEE Trans. on Antennas Propag.,* vol. AP-37, no. 2, pp. 143–147, November 1989.

[11] T. Huynh, and K. F. Lee, "Single-Layer Single-Patch Wideband Microstrip Antenna," *Electron. Lett.*, vol. 31, no. 16, pp. 1310–1312, August 1995.

[12] K.F. Lee, K. M. Luk, K. F. Tong, S. M. Shum, T. Huynh, and R. Q. Lee, "Experimental and Simulation Studies of Coaxially Fed U-Slot Rectangular Patch Antenna," *IEE Proc. Microw. Antennas Propag.*, vol. 144, pp. 354–358, October 1997.

[13] H. Nakano, M. Yamazaki, and J. Yamauchi "Electromagnetically Coupled Curl Antenna," *Electron. Lett.*, vol. 33, no. 12, pp. 1003–1004, June 1997.

[14] K. M. Luk, C. L. Mak, Y. L. Chow, and K. F. Lee, "Broadband Microstrip Patch Antenna," *Electron. Lett.*, vol. 34, no. 15, pp. 1442–1443, July 1998.

[15] C. L. Mak, K. M. Luk, K. F. Lee, and Y. L. Chow, "Experimental Study of a Microstrip Patch Antenna with an L-shaped Probe," *IEEE Trans. on Antennas Propag.*, vol. AP-48, no. 5, pp. 777–783, May 2000.

[16] Y. X. Guo, C. L. Mak, K. M. Luk, and K. F. Lee, "Analysis and Design of L-Probe Proximity Fed-Patch Antennas," *IEEE Trans. on Antennas Propag.,* vol. AP-49, no. 2, pp. 145–149, February 2001.

[17] C. L. Mak, K. M. Luk, and K. F. Lee, "Microstrip Line-Fed L-Strip Patch Antenna," *IEE Proc. Microw. Antennas Propag.*, vol. 146, no. 4, pp. 282–284, August 1999.

[18] K. M. Luk, L. K. Au Yeung, C. L. Mak, and K. F. Lee, "Circular Patch Antenna with an L-Shaped Probe," *Microw. Opt. Technol. Lett.*, vol. 20, no. 4, pp. 256–257, January 1999.

[19] C. L. Mak, K. M. Luk, K. F. Lee, and Y. L. Chow, "Experimental Study of a Microstrip Antenna with an L-Shaped Probe," *IEEE Trans. Antennas Propag.*, vol. 48, no. 5, pp. 777–783, May 2000.

[20] C. L. Mak, K. M. Luk, and K. F. Lee, "Wideband Triangular Patch Antenna," *IEE Proc. Microw. Antennas Propag.*, vol. 146, no. 2, pp. 167–168, April 1999.

[21] Y. X. Guo, C. L. Mak, K. M. Luk, and K. F. Lee, "Analysis and Design of L-Probe Proximity Fed-Patch Antennas," *IEEE Trans. Antennas Propag.*, vol. 49, no. 2, pp. 145–149, February 2001.

[22] T. P. Wong and K. M. Luk, "A Wideband L-Probe Patch Antenna Array With Wide Beamwidth," *IEEE Trans. Antennas Propag.,* vol. AP-51, no. 10, pp. 3012–3014, October 2003.

[23] K. L. Lau, and K. M. Luk, "A Wideband Circularly Polarized Conical-Beam Patch Antenna," *IEEE Trans. Antennas Propag.*, vol. 54, no. 5, pp. 1591–1594, May 2006.

[24] Y. X. Guo, K. W. Khoo, and L. C. Ong, "Wideband Circularly Polarized Patch Antenna Using Broadband Baluns," *IEEE Trans. Antennas Propag.*, vol. 56, no. 2, pp. 319–326, February 2008.

[25] S. L. S. Yang, and K. M. Luk, "A Wideband L-Probes Fed Circularly-Polarized Reconfigurable Microstrip Patch Antenna," *IEEE Trans. Antennas Propag.*, vol. 56, no. 2, pp. 581–584, February 2008.

[26] H. Wong, K. L. Lau and K. M. Luk, "Design of Dual-Polarized L-Probe Patch Antenna Arrays With High Isolation," *IEEE Trans. Antennas Propag.*, vol. 52, no. 1, pp. 45–52, January 2004.

[27] K. L. Lau and K. M. Luk, "A Wideband Dual-Polarized L-Probe Stacked Patch Antenna Array," *IEEE Antennas Wirel. Propag. Lett.*, vol. 6, pp. 529–532, January 2007.

[28] S. L. S. Yang, K. M. Luk, H. W. Lai, A. A. Kishk, and K. F. Lee, "A Dual-Polarized Antenna with Pattern Diversity," *IEEE Antennas Propag. Mag.*, vol. 50, no. 6, pp. 71–79, December 2008.

[29] Z. N. Chen and M. Y. W. Chia, "Broad-band Suspended Probe-fed Plate Antenna with Low Cross-polarization Levels," *IEEE Trans. Antennas Propag.*, vol. AP-51, no. 2, pp. 345–346, February 2003.

[30] H. W. Lai, P. Li and K. M. Luk, "A Wideband Patch Antenna with Low Cross-polarization," *Proceeding of APMC 2003*, vol. 3, pp. 1718–1721, 2003.

[31] H. W. Lai and K. M. Luk, "Wideband Patch Antenna with Low Cross-Polarisation," *Electron. Lett.*, vol. 40, no. 3, pp. 159–160, February 2004.

[32] H. W. Lai and K. M. Luk, "Design and Study of Wide-band Patch Antenna Fed by a Meandering Shaped Probe," *IEEE Trans. Antennas Propag.*, vol. 54, no. 2, pp. 564–571, February 2006.

[33] H. W. Lai and K. M. Luk, "Wideband Patch Antenna Fed by a Modified L-shaped Probe," *Microw. Opt. Technol. Lett.*, vol. 48, no. 5, pp. 977–979, March 2006.

[34] H. W. Lai and K. M. Luk, "Electrically Long Rectangular Patch Antenna Fed by Meandering Probe with Wideband and High Gain Characteristics," *Microw. Opt. Technol. Lett.*, vol. 50, no. 9, pp. 2289–2292, June 2008.

[35] H. W. Lai and K. M. Luk, "Meandering Probe Fed Patch Antenna with High Gain Characteristic for Circularly Polarized Application," *Microw. Opt. Technol. Lett.*, vol. 49, no. 5, pp. 1095–1098, March 2007.

[36] X. H. Tang, Y. L. Long, H. Wong, and K. L. Lau, "Broadband Circularly-Polarised Patch Antenna with 3D Meandering Strip Feed," *Electron. Lett.*, vol. 47, no. 19, pp. 1060–1062, September 2011.

[37] Q. W. Lin, H. Wong, X. Y. Zhang and H. W. Lai, "Printed Meandering Probe-Fed Circularly-Polarized Patch Antenna with Wide Bandwidth," *IEEE Antennas Wirel. Propag. Lett.*, vol. 13, pp. 654–657, March 2014.

[38] H. W. Lai and K. M. Luk, "Dual Polarized Patch Antenna Fed by Meandering Probes," *IEEE Trans. Antennas Propag*, vol. 55, no. 9, pp. 2625−2627, September 2007.

[39] C. Y. D. Sim, C. C. Chang, and J. S. Row, "Dual-Feed Dual-Polarized Patch Antenna with Low Cross Polarization and High Isolation," *IEEE Trans. Antennas Propag.*, vol. 57, no. 10, pp. 3405–3409, October 2009.

[40] Y. H. Shum, K. M. Luk, and C. H. Chan, "Multi-band Base Station Antenna with Compact Microstrip Resonant Cell Filters," *IEE Proc.-Microw. Antennas Propag.*, vol. 151, no. 6, pp. 545–548, December 2004.

[41] K. F. Lee, K. M. Luk, and K. M. Mak, "Dual and Triple Band Patch Antennas Fed by Meandering Probe," *Microw. Opt. Technol. Lett.*, vol. 52, no. 7, pp. 1498–1504, July 2010.

[42] X. L. Zhao, and Q. W. Lin, "Dual-Band Patch Antenna Fed by Meandering Probe for Low Cross-Polarization," *Int. J. Antennas Propag.*, vol., 2016, pp. 1–7, 2016.

[43] C.-H. Chen and E. K. N. Yung, "A Novel Unidirectional Dual-Band Circularly-Polarized Patch Antenna," *IEEE Trans. Antennas Propag.*, vol. 59, no. 8, pp. 3052–3057, August 2011.

[44] P. Li, K. M. Luk, and K. L. Lau, "A Dual-feed Dual-band L-Probe Patch Antenna", *IEEE Trans. Antennas Propag.*, vol. 53, no. 7, pp. 2321–2323, July 2005.

[45] K. L. Lau, and K. M. Luk, "A Wide-Band Circularly Polarized L-Probe Coupled Patch Antenna for Dual-Band Operation," *IEEE Trans. Antennas Propag.*, vol. 53, no. 8, pp. 2636–2644, August 2005.

[46] P. Li, H. W. Lai, K. M. Luk and K. L. Lau, "A Wideband Patch Antenna with Cross-polarization Suppression," *IEEE Antennas Wirel. Propag. Lett.*, vol. 3, pp. 211– 214, 2004.

[47] H. W. Lai, T. P. Wong, C. L. Mak and K. M. Luk, "Wideband Dual-Polarized Antenna Array on Finite Conducting Cylinder," *Microw. Opt. Technol. Lett.*, vol. 48, no. 5, pp. 862–866, March 2006.

[48] P. Li, K. L. Lau, and K. M. Luk, "A Study of the Wide-Band-Probe Fed Planar Patch Antenna Mounted on a Cylindrical or Conical Surface," *IEEE Trans. Antennas Propag.*, vol. 53, no. 10, pp. 3385–3389, October 2005.

[49] Q. Xue, Y. Liu, and K. M. Shum, "Printed L-Probe Antenna on Multi-layer PCB," *IEEE International Symposium on Antennas Propag.*, vol. 3, pp. 2281–2284, July 2004.

[50] C. M. Su, H. T. Chen, and K. L. Wong, "Printed Dual-band Dipole Antenna with U-slotted Arms for 2.4/5.2 GHz WLAN Operation," *Electron. Lett.*, vol. 38, no. 22, pp. 1308–1309, October 2002.

[51] Z. N. Chen, T. S. P. See, and X. Qing, "Small Printed Ultrawideband Antenna With Reduced Ground Plane Effect," *IEEE Trans. Antennas Propag.*, vol. 55, no. 2, 383–388, February 2007.

[52] H. W. Lai and K. M. Luk, "Wideband Patch Antenna Fed by Printed Meandering Strip," *Microw. Opt. Technol. Lett.*, vol. 50, no. 1, pp. 188–192, November 2008.

[53] K. L. Wong, W. Y. Chen, and T. W. Kang, "On-Board Printed Coupled-Fed Loop Antenna in Close Proximity to the Surrounding Ground Plane for Penta-Band WWAN Mobile Phone," *IEEE Trans. Antennas Propag.*, vol. 59, no. 3, 751–757, March 2011.

[54] J. Lu, Z. Kuai, and X. Zhu, "A Novel Co-planar L-probe Designed to Feed Patch Antenna," *IEEE International Symposium on Antennas Propag.*, pp. 3103–3105, July 2011.

[55] J. Liu, Q, Xue, H. Wong, H. W. Lai, and Y. Long, "Design and Analysis of a Low-Profile and Broadband Microstrip Monopolar Patch Antenna," *IEEE Trans. Antennas Propag.*, vol. 61, no. 1, pp. 11–18, January 2013.

[56] M. Li and K. M. Luk, "A Low-Profile Unidirectional Printed Antenna for Millimeter-Wave Applications," *IEEE Trans. Antennas Propag.*, vol. 62, no. 3, 1232–1237, March 2013.

[57] H. Wong, Q. W. Lin, H. W. Lai, and X. Y. Zhang, "Substrate Integrated Meandering Probe-Fed Patch Antennas for Wideband Wireless Devices," *IEEE Trans. on Components Packaging and Manufacturing Technology*, vol. 5, no. 3, pp. 381–388, March 2015.

[58] H. W. Lai and H. Wong, "Substrate Integrated Magneto-Electric Dipole Antenna for 5G Wi-Fi," *IEEE Trans. Antennas Propag.*, vol. 63, no. 2, pp. 870–874, February 2015.

[59] K. M. Mak, X. Gao, and H. W. Lai, "Low Cost Dual Polarized Base Station Element for Long Term Evolution," *IEEE Trans. Antennas Propag.*, vol. 62, no. 11, pp. 5861–5865, November 2014.

Chapter 4
Advancements in MIMO antenna systems

Mohammad S. Sharawi[1]

Multiple-input-multiple-output (MIMO) technology is currently used in fourth-generation (4G) wireless systems because they can provide higher data rates at fixed power and bandwidth levels when compared to their single-input-single-output conventional counterparts. MIMO-enabled systems require the integration of multiple antenna elements operating at the same band to provide higher data throughput. The design of such antenna solutions is a challenging task especially for small form factor devices where the space is limited and the antennas are placed close to one another. Such close placement of antennas with respect to one another will increase the port coupling between adjacent antennas as well as radiated field coupling, thus degrading the efficiency as well as increasing field correlation. This will affect the overall improvement in the data rates (channel capacity) expected. In this chapter, we will go over the latest advancements in MIMO antenna system designs for 4G and the anticipated 5G wireless standard and will highlight various challenges that can be encountered and provide the solutions to overcome them.

4.1 Introduction

Several new technologies were introduced in the fourth-generation (4G) wireless communication standard to provide users with higher data transfer capabilities. Examples of such technologies include adaptive modulation and coding, the use of new multiplexing techniques (i.e., orthogonal frequency division multiplexing—OFDM) and multiple-input-multiple-output (MIMO). MIMO systems were adopted as a solution to the multipath fading problem that was degrading the performance of legacy wireless communication links. MIMO requires dense multipath environments to provide the advantage anticipated. By sending multiple streams of data, the reliability and data rates of the system can increase dramatically [1]. To have simultaneous multiple transmissions, multiple antenna systems at the transmitter and receiver sides are needed. The design of such MIMO antenna systems to achieve the required increase in reliability and data rates needs careful attention and

[1]Electrical Engineering Department, Polytechnique Montréal, Canada

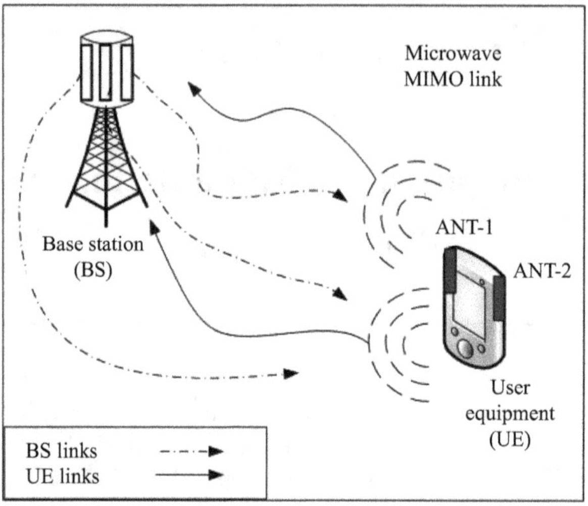

Figure 4.1 A generic diagram showing a microwave MIMO communication system

characterization [2]. A diagram showing the multiple antennas at the transmitter and receiver ends is shown in Figure 4.1.

The use of MIMO antenna systems at the transmitter and the receiver sides requires extra attention compared to their single antenna counterparts. While single antenna performance metrics are still required for MIMO antennas, several extra performance metrics have been devised to assess the performance of the new multi-antenna configuration in MIMO antenna systems. The design of MIMO antenna systems for handheld portable devices is a challenging task, as the antennas are usually placed in close proximity. This will increase the field coupling as well as port coupling. These two issues need to be minimized to achieve low correlation levels that affect the channel capacity and low port isolation that will affect the efficiency of the antenna system. In addition, it is essential to ensure that MIMO antennas share the same ground as they are not isolated, and not sharing the same system ground will cause errors in the levels detected as discussed in [3].

It has been more than a decade now since MIMO antenna systems have entered the real world. They are currently used in a large number of wireless applications covering cell phones, wireless routers and access points, base stations, tablet and laptop PCs among other specialized uses. In this chapter, we will provide an overview of the advancement of such MIMO antenna systems within their various application categories and highlight some of the current and future challenges particularly with the dawn of 5G standards.

4.2 MIMO antenna system performance metrics

In addition to the conventional performance metrics of single antenna systems including resonance, bandwidth, gain and efficiency, MIMO antenna systems have several additional metrics that needs to be checked to be able to obtain the anticipated performance improvements. Although the details of such metrics have appeared in several previous works such as [2,4,5], we will briefly list them here:

(a) Port isolation:
High port isolation is important to maintain in order to have good radiation efficiency. Any energy lost in coupling to adjacent ports will affect the overall radiation efficiency of the antenna system. Port isolation is defined as the $-|S_{ij}|$ between elements ***i*** and ***j*** in a multi-antenna system (or it is the negative of the transmission coefficient). A port isolation value of 10 dB is the least acceptable value for MIMO antenna systems [2], but the higher the better.

In small form factor wireless terminals, antennas might be placed in close proximity and port coupling becomes high (low port isolation). Several techniques for isolation enhancement have been introduced in literature, and can be applied based on the coupling mechanism. Details of such methods can be found in [2].

(b) Total active reflection coefficient:
In a multi-port antenna system, the presence and activity of adjacent ports will affect the port under consideration in terms of its effective bandwidth and efficiency if not properly designed. To assess such interactions, the total active reflection coefficient (TARC) was introduced in [6]. TARC (Γ) is defined as the ratio between the square root of the sum of the reflected powers by its incident power counterpart. For an N-element antenna, TARC is found using

$$\Gamma = \frac{\sqrt{\sum_{i=1}^{N} |\boldsymbol{b}_i|^2}}{\sqrt{\sum_{i=1}^{N} |\boldsymbol{a}_i|^2}} \tag{4.1}$$

where $\boldsymbol{a}_i$ and $\boldsymbol{b}_i$ are the incident and reflected waves, respectively. They are found from the measured S-parameters of the system. TARC values range between 0 and 1, where a 1 means the incident power was totally reflected and not radiated and 0 means the incident power was totally radiated and nothing was reflected. TARC is usually represented in dB.

For the special case of a two port MIMO antenna system TARC can be evaluated by fixing the angle of the incoming wave of one port, and then sweeping the relative angle ($\boldsymbol{\theta}$) with the adjacent port to see the effect of such angle change along with the mutual coupling (S_{ij}) on the effective bandwidth and efficiency of the multi-port antenna system. This can be found using [7]:

$$\Gamma = \frac{\sqrt{\left(\left[\left|S_{11} + S_{12}e^{j\theta}\right|^2\right] + \left[\left|S_{21} + S_{22}e^{j\theta}\right|^2\right]\right)}}{\sqrt{2}} \tag{4.2}$$

(c) Branch power ratio:
The branch power ratio (BPR) is another important metric for MIMO antenna systems that shows the ratio between the lowest and highest power levels within the antenna system. In an ideal multi-antenna system, the BPR should be 1 (0 dB), i.e., all ports are receiving the same power (power divided among all ports equally). The BPR (***K***) can be found using [8]:

$$K = \frac{P_{\min}}{P_{\max}} \tag{4.3}$$

(d) Field correlation:
The most important performance metric for MIMO antenna systems is the correlation coefficient (CC) or the envelope correlation coefficient (ECC). This metric shows the amount of *channel correlation* in a wireless communication link. When the MIMO channels are not correlated, the anticipated increase in the channel capacity can be achieved. Usually we use ECC because we deal with power signals. Ideally, we want an ECC value of "0". While in reality, there is a none zero correlation between the wireless channels due to the multipath environment power distribution as well as the radiation patterns being partially overlapped. The radiation patterns of the MIMO antenna system are the ones used in the evaluation of the CC(ρ) or ECC(ρ_e) since they directly affect the channel between the transmitter and the receiver. In an isotropic[1] channel, they can be evaluated according to [9]:

$$|\rho_{12}|^2 = \rho_e(ECC) = \frac{\left|\iint\limits_{4\pi} \left[E_1(\theta,\phi) * E_2(\theta,\phi)\right] d\Omega\right|^2}{\iint\limits_{4\pi} \left|E_1(\theta,\phi)\right|^2 d\Omega \iint\limits_{4\pi} \left|E_2(\theta,\phi)\right|^2 d\Omega} \tag{4.4}$$

where $E_i(\theta, \varphi)$ is the complex 3D radiated field pattern for antenna ***i***, Ω is the solid angle and the integral is evaluated over the 3D sphere. The expression in (4.4) is based on the assumption of having an isotropic wireless environment and this should always be highlighted. The relation between CC and ECC is not exactly the square function, but they follow approximately a square one in a Rayleigh distributed narrow band environment [10]. The channel power distribution should be taken into account for more accurate ECC evaluation, and a more generalized expression than (4.4) can be found in [2].

Several works in literature advocate the use of the S-parameter equation in [11] for evaluating the ECC/CC due to its simplicity. This equation should not be used anymore due to the following issues. The first is the assumption of using lossless antennas in an isotropic environment. The S-parameters equation for finding the ECC is only valid if the antenna is lossless.

[1]An isotropic or uniform channel is the one that has XPD = 1 and uniform incoming wave distributions, i.e., $P_\phi = P_\theta = 1/4\pi$.

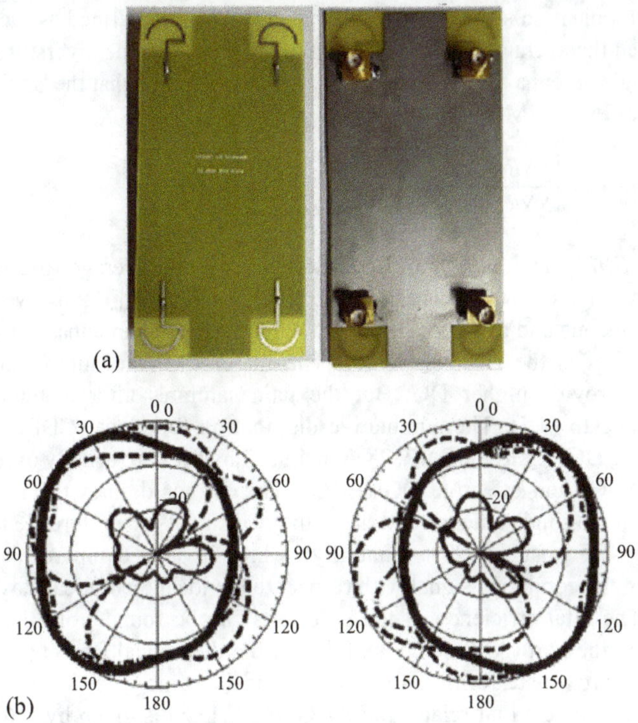

Figure 4.2 GND acting as a reflector for tilting MIMO antenna patterns: (a) 4-element MIMO antenna fabricated prototype and (b) tilted patterns of top antennas. Note the effect of the GND sheet behind them [14] (©2016 Wiley, Reprinted with permission)

Another important issue is the fact that port parameters have nothing to do with channel behavior. This means that the S-parameters equation should not be used to evaluate CC or ECC even for 100% efficient MIMO antenna systems, but rather (4.4) should always be used. Considering similar antennas that are spatially placed in close proximity, a large discrepancy between the S-parameter method and the field method can be identified, which invalidates the S-parameters method. The reader is encouraged to check [3] and [12].

To improve (lower) CC/ECC values in a MIMO antenna system the designer should try to spatially isolate/tilt/separate the radiation patterns. This can be accomplished by using different polarizations, or tilted beams (via beam steering, use of reflectors within the GND, use of parasitic directors, etc.). Some examples can be seen in [13–15]. Figure 4.2 shows one way of achieving tilted patterns for low ECC values.

(e) Diversity gain:
Diversity schemes were devised to combat multipath effects. When the signals are not correlated, their combination at the receiver will provide higher SNR levels, and thus better signal reception. A measure of diversity performance in

a communication system is diversity gain (DG). It is defined as the difference between the averaged SNR of the combined signals in the diversity system and the single antenna system in a specified channel, given that the SNR is above a threshold value. Mathematically it is given by [16]:

$$DG = \left[\frac{SNR^{i}_{div}}{SNR^{m}_{div}} - \frac{SNR^{i}_{\sin}}{SNR^{m}_{\sin}}\right]_{P\left(SNR^{i}_{div} < ref\right)} \tag{4.5}$$

where SNR^{i}_{div} and SNR^{m}_{div} are the instantaneous and average (mean) SNR for the diversity (MIMO) system, respectively, and $SNR^{i}_{\sin}$ and $SNR^{m}_{\sin}$ are the instantaneous and average (mean) SNR for the single antenna system, respectively. Ref is the reference level. The increase in the number of antennas should provide higher DG. But the gain saturates after 8 antennas or so according to (4.5). The cumulative distribution functions (CDF) are used to show the DG versus the SNR. Two main DG metrics are usually used; apparent DG (ADG) and effective DG (EDG). The former defines the improvement between the single antenna and the combined signals at a reference level of 1% on the CDF (marking 99% reliability), while the later compares the improvement of the combined signal with respect to the ideal theoretical Rayleigh CDF [17]. The total efficiency of the two ports can be found from the difference between the port (ANT) CDF and that of the theoretical Rayleigh one.

An approximate formula for antennas of the same type (i.e., similar radiation efficiencies) that relates the ADG to CC/ECC is given by [16]:

$$ADG = 10.5\sqrt{1 - |0.99\rho|^{2}} \tag{4.6}$$

(f) Channel capacity:
The obtained channel capacity ($\boldsymbol{C}$) in bits/sec for a wireless system with $\boldsymbol{M}$ transmitter and $\boldsymbol{N}$ receiver antennas, with no channel state information, Gaussian distributed signals and identity covariance matrix is given by [1]:

$$C = BW\log_2\left(\det\left(\mathbf{I}_N + \frac{P_T}{\sigma^2 M}\mathbf{H}\mathbf{H}^H\right)\right) \tag{4.7}$$

where $\boldsymbol{BW}$ is the channel bandwidth (in Hz), $\boldsymbol{P_T}$ is the equally distributed input power among the elements, σ^2 is the noise power, $\mathbf{I}_N$ is the $N \times N$ identity matrix, and $\mathbf{H}$ is the complex channel matrix. It is clear that in order to obtain higher channel capacity values, we can either increase the $\boldsymbol{BW}$, the $\boldsymbol{SNR}$ or the number of antenna elements.

While increasing the number of antennas at the base station side is easy, having multiple antennas in a small form factor wireless terminal is challenging. Evaluating the channel capacity is an essential step to obtain the increased data rates expected from a MIMO system. The capacity performance is usually done by conducting measurements in a defined environment to come up with its complex channel matrix $\boldsymbol{H}$. Several procedures have appeared in literature such as [817]. Figure 4.3 shows some recent channel capacity curves

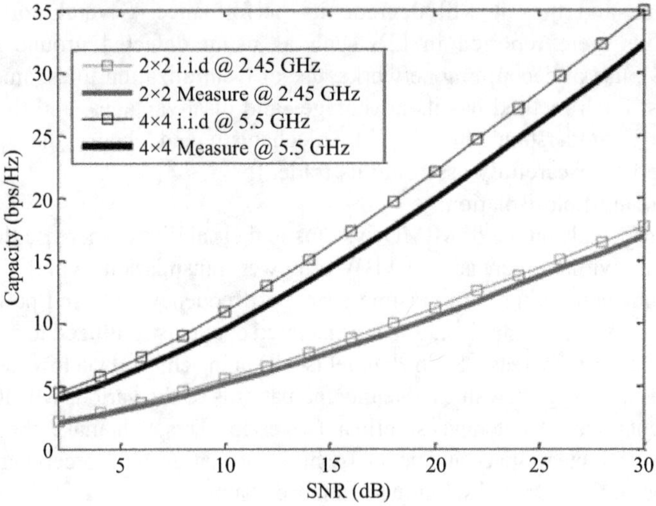

Figure 4.3 Channel capacity curves for the 2×2 and 4×4 MIMO antenna system in [8] (©2016 IEEE, Reprinted with permission)

for a dual-band MIMO antenna system (2.45 and 5.5 GHz) that appeared in [8] and compares the measured actual curves with the theoretical ones based on ideal channel conditions.

4.3 Major MIMO antenna system design challenges

Designing MIMO antenna systems have several challenges, the main ones are:

(a) System integration:
When designing antenna systems for wireless terminals, special care should be given to the system being developed and how the antenna system will be integrated within it. Usually antennas are designed in free space and adjacent components such as the battery, the cover and the LCD screen on mobile terminals are not evaluated. For proper characterization and performance evaluation, the antenna should be placed in its intended package, so that such effects are taken care of. Antenna designers can place the radome (cover), the battery and LCD along with other surrounding parts in a mobile terminal within their model and simulate its behavior before fabrication. Several works have considered this such as [18]. Of course it is a challenge to incorporate all these effects, and close interaction between product design engineers and antenna engineers is required at this stage.

(b) Enhancing port-isolation:
Placing several antennas in close proximity in mobile terminals dictates strong coupling. This coupling as mentioned in Section 4.2(a) degrades the efficiency of the system as part of the power will be wasted in an adjacent

antenna and thus it will degrade its performance. Several port isolation methods were reported in [2], such as using defected ground structures (DGS), use of decoupling networks, use of Neutralization lines, among many others. Each method has its advantages and disadvantages, and the designer needs to understand the coupling mechanism and choose the appropriate method after carefully assessing its tradeoffs.

(c) Enhancing field-isolation:
The major advantage of MIMO systems is their ability to increase the channel capacity without increasing the BW or power transmissions within the system, and thus complying with existing stringent frequency slots and power levels. Thus, only increasing the number of isolated channels is utilized to achieve this large data rate increase. Such channel isolation in achieved via low ECC values. To ensure low ECC values, the antenna patterns of the various MIMO antenna elements should not have significant overlap. This will make the incoming waves separate in space and not cause high correlation at the receiving antennas.

One straightforward solution to enhance channel isolation is to design MIMO antenna systems with tilted beams for their elements with respect to one another. This can be achieved using various techniques. Figure 4.2 shows a 4-element MIMO antenna system where part of the GND plane acts as a reflector to the two top and bottom antenna elements causing a noticeable beam tilting between the two adjacent antennas, thus lowering the ECC values. Pattern-reconfigurable or polarization-reconfigurable antennas can be used as well to ensure low ECC value among adjacent MIMO antenna elements.

(d) Proper characterization:
While the basic antenna performance metrics can be characterized in a laboratory environment using a vector network analyzer and an anechoic chamber, MIMO antenna system metrics such as channel capacity and DG cannot be achieved and measured using such equipment. Special over-the-air procedures [2] should be devised and the channel effects should be accounted for proper evaluation. Reverberation chambers can emulate several wireless environments and can provide accurate measures for the channel capacity and DG [16].

4.4 MIMO antenna system examples

There is a large collection of MIMO antenna systems using various antenna types and for a vast number of applications covering a wide spectrum of frequencies. In this section, we will present few demonstrative examples on the recent advancements in MIMO antenna system development in some recent applications.

4.4.1 Mobile phones and handheld devices

MIMO antenna systems for mobile terminals and cell phones are among the most designed due to the large number of devices required with various dimensions and shapes. Designing MIMO antenna systems for mobile handheld terminals poses stringent requirements for placing several antennas within a small volume. This leads

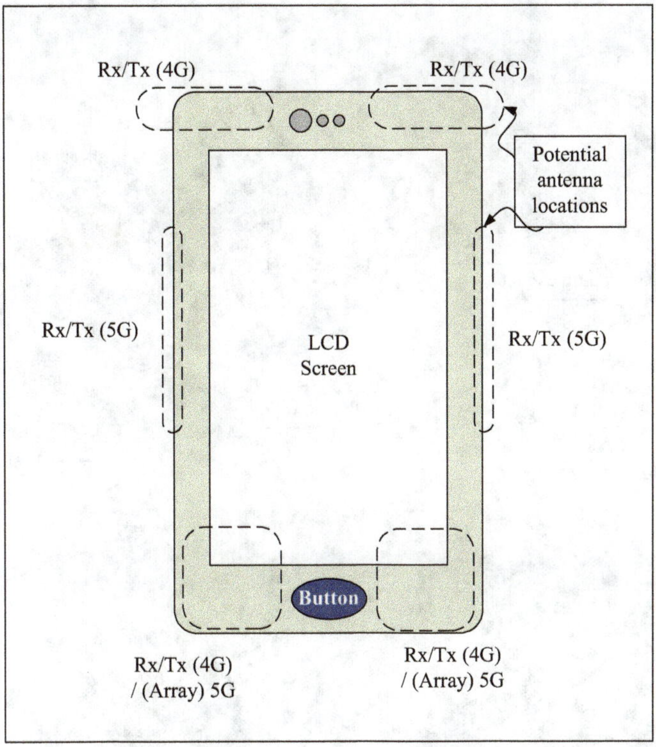

Figure 4.4 Most widely used locations for MIMO antenna systems in current 4G and future 5G bands

to large port and field coupling between adjacent antennas. Figure 4.4 shows the most widely used locations for MIMO antennas in smart phones. The most widely used locations for 4G-based antennas are the top and bottom sides of the cell phone. Most 4G-based-enabled smart phones have 2-MIMO antennas. For the upcoming 5G standard, higher frequency bands are targeted such as the 5.9–8.5 GHz and the mm-wave bands in 27–29.5 and 36– 47 GHz among many others such as 64–71 GHz (several other bands that have not yet been approved are listed in Table 6.1 in [19]). At mm-wave bands, high attenuation will be encountered when signals propagate from the mobile terminal to the base station, and thus antenna arrays are to be used to compensate for such transmission loss. Locations of such antenna arrays are to be placed at the bottom of the phone. For the lower bands targeting 5G systems, the edges of the mobile terminal can be used to stack 8–10 elements for an improved channel capacity.

Several recent works have provided multi-band MIMO antenna system solutions for mobile terminals such as those presented in [20–24]. For covering the low frequency bands of the 4G standard (i.e., 700–900 MHz), the periphery of the mobile phone is usually used to place the antennas such as those in [20]. Two metal

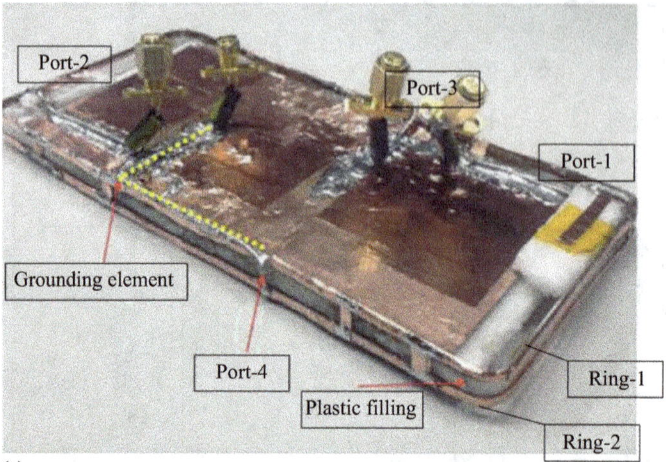

(a)

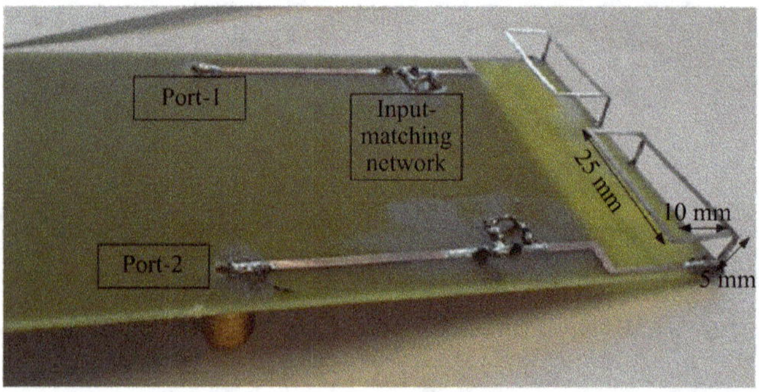

(b)

Figure 4.5 Multi-band MIMO antenna systems: (a) mockup of a 2-element cellular MIMO antenna (ports 1 and 2) and 2-element wireless local area network (WLAN) (ports 3 and 4) one using shorted rings [20] (©2015 IEEE, Reprinted with permission) and (b) 2-element MIMO antenna using 3D coupling elements [21] (©2014 IEEE, Reprinted with permission)

rings separated by 3.5 mm that encircle the GND plane with a total volume of $150 \times 70 \times 7.5$ mm^3 are used. The rings were connected to the GND plane at optimized locations to provide the resonance frequencies of the two element cellular band (820–900 MHz) MIMO antenna and the 2-element WiFi (2,400–2,500 and 5,200–5,800 MHz) MIMO antenna system. The prototype of this design is shown in Figure 4.5(a). Good MIMO performance was measured in three used positions in an anechoic chamber. Another solution for lower bands is the use of 3D folded structures on the top/bottom edges of a mobile phone, with one example shown in [21]. A single antenna had the dimensions of $25 \times 10 \times 5$ mm^3 with raised hollow coupling elements to excite GND currents at the lower 4G bands (750–950 MHz), this

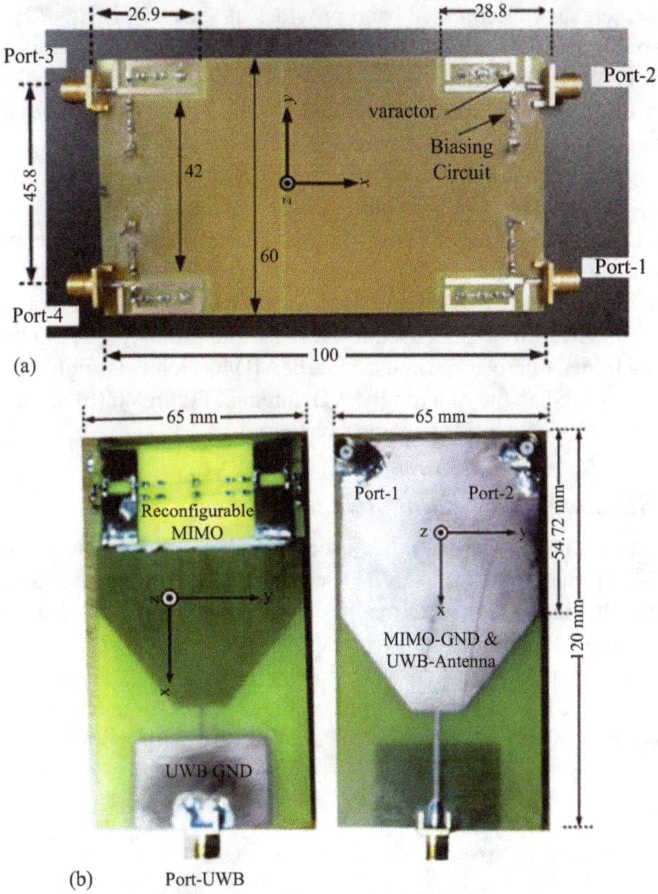

Figure 4.6 Reconfigurable MIMO antennas: (a) 4-element wide BW antenna system using varactor diodes [23] (©2016 Wiley, Reprinted with permission) and (b) integrated MIMO and UWB antenna for CR applications [24]

is shown in Figure 4.5(b). A neutralization line isolation enhancement technique was devised to enhance the port isolation from 6 dB to almost 12 dB in the band of operation. Excited currents generated tilted beam patterns that gave good ECC performance.

4.4.2 Cognitive radio front-ends

Frequency-reconfigurable MIMO antennas and integrated ultra-wide-band (UWB) and MIMO antenna systems for cognitive radio platforms are shown in [23,24]. The four-corners of a standard mobile phone backplane are used in [23] for a 4-element frequency-reconfigurable MIMO antenna system with wide tenability range and wide BW in each step. Each tuned band provides at least 450 MHz of BW, and the reconfigurability covers from 1,610 to 2,710 MHz. Varactor diodes were used for the frequency switching. The patterns were tilted because of the presence of the GND plane (the antennas were surrounded by the GND from three sides). The fabricated prototype is shown in Figure 4.6(a).

An integrated two-element-reconfigurable MIMO antenna covering (755–3,450 MHz) as well as a UWB sensing antenna covering (700–3,300 MHz) was proposed in [24] for CR front-end applications. The UWB sensing antenna provides spectrum sensing, while the MIMO antenna system switches between the available bands via the use of a set of PIN and varactor diodes. A standard mobile terminal backplane of 65 × 120 mm^2 was used to fabricate the prototype. A 3D antenna element type of a folded microstrip was used for the MIMO elements while the hexagonal shape GND was used as the radiating UWB antenna. Figure 4.6(b) shows the fabricated prototype of this integrated antenna system.

4.4.3 USB dongle MIMO implementations

Monopole-based antennas are usually used in UWB MIMO antenna systems as those in [25–27]. Such antennas can be used in mobile terminals supporting UWB operation and standards (some for imaging applications). Smaller-size USB dongle application-based MIMO antenna systems were proposed in [28,29] using slotted patch 3D antennas as well as helix-based ones, respectively. The former targeted WLAN bands in the 2.4 and 5 GHz bands, while the later targeted low 4G bands in the 700 MHz range. Figure 4.7 shows the two designs with proper dimensions.

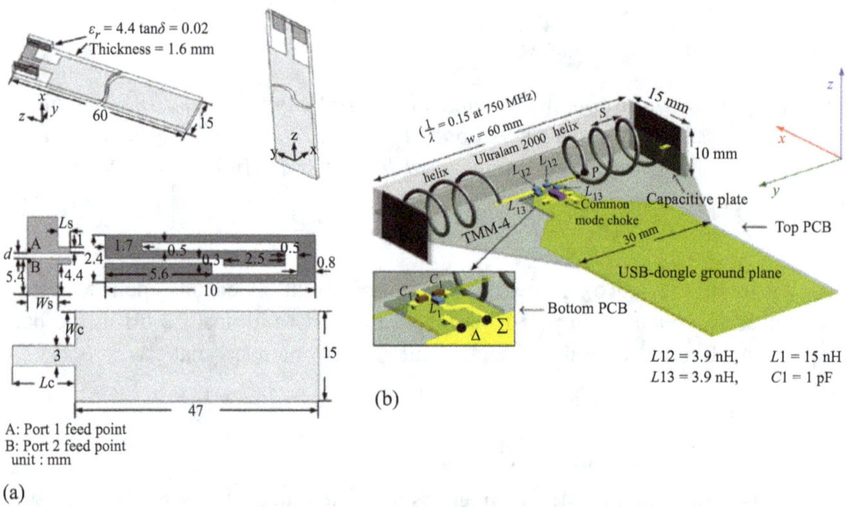

Figure 4.7 USB dongle MIMO antenna system examples: (a) WLAN coverage using slotted patch antennas [28] (©2014 IEEE, Reprinted with permission) and (b) 700 MHz 4G band helix-based antennas [29] (©2013 IEEE, Reprinted with permission)

4.4.4 Wireless access point MIMO implementations

MIMO antenna system implementations for wireless access points targeting WLAN applications have relied on adjacent antennas covering various sectors in 3D space such as those in [30–32]. Two concentric loop antennas were used as the single multi-band antenna element covering 2.4 and 5 GHz WLAN bands in a 3-element MIMO configuration with a circular GND plane of 60 mm radius in [30] as shown in Figure 4.8(a). A 6-port sectoral MIMO antenna system with aperture fed patch elements was proposed in [31] for 5 GHz WLAN applications. Alternate polarizations were used between adjacent antenna elements to lower the ECC (i.e., one vertically polarized and the next is horizontally polarized, and so on). The elements were arranged on a hollow hexagonal pyramid shaped metal with inter-element groves (see Figure 4.8(b)). A 4-beam directional access point design with 4-elements in each of its four sectors was proposed in [32]. Each of the 4-elements

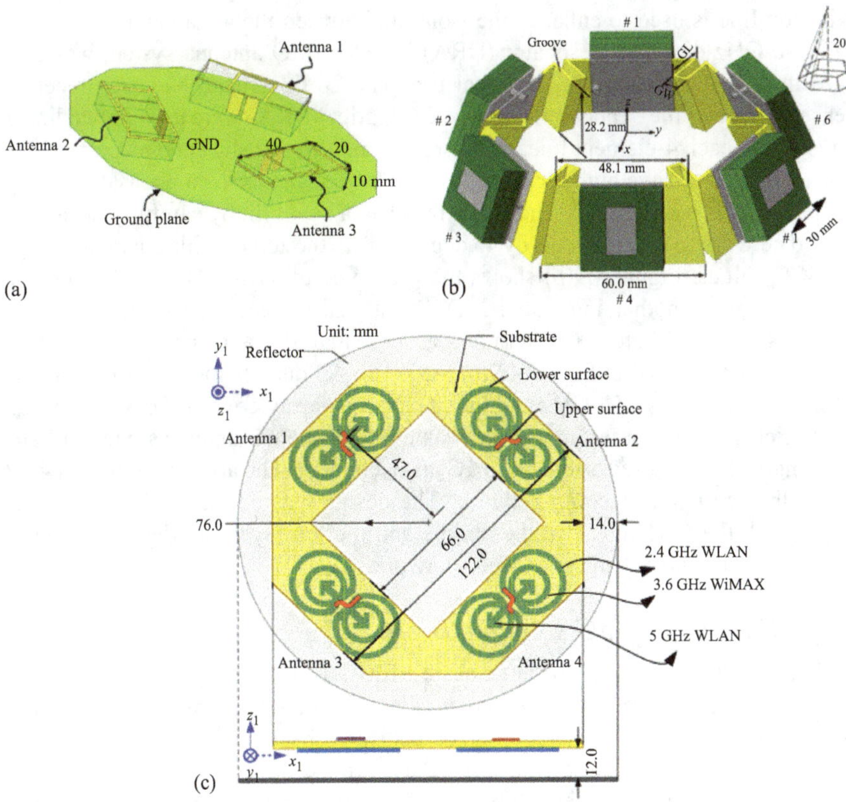

Figure 4.8 *Wireless access point MIMO antenna systems: (a) 3-element loop-based [30] (©2010 IEEE, Reprinted with permission); (b) 6-element aperture fed patch-based [31] (©2014 IEEE, Reprinted with permission); and (c) multi-element multi-band 4-element per one sector (4-sector) design [32] (©2016 IEEE, Reprinted with permission)*

in each of the 4-sectors had two concentric loops for 2.45 GHz WLAN and 3.6 WiMAX bands, and a dipole in its center for 5 GHz WLAN. The MIMO implementation and its single-element geometries are shown in Figure 4.8(c).

4.5 MIMO antenna solutions for 5G-enabled systems

4.5.1 Mobile terminal 5G solutions

Some examples of MIMO antenna systems have recently appeared targeting the 5G potential bands and standard. The first example is a 3.4–3.6 GHz MIMO antenna system with 8-elements that can provide high channel capacities for future 5G terminals targeting low frequency bands. The design integrates the 8-element monopole-based 5G MIMO antenna system on the sides of the mobile terminal main board while its top is used for a 4G 2-element MIMO antenna system. The details of this proposed design are shown in Figure 4.9(a) [33]. Note that a neutralization line is used to enhance the isolation between the 4G antennas.

A 30 GHz dielectric resonator (DRA)-based MIMO antenna system was proposed in [34] for possible 5G terminal use. The 28 GHz band has been recently licensed as one of the 5G potential bands. Cylindrical DRAs were used in two linear arrays where each 4-element linear array represents one MIMO antenna port, where an array is used to have higher gain at such high frequencies to compensate slightly for the path loss encountered. The two arrays had a beam tilt by carefully designing the feed network to tilt the beams by 45 degrees from the zenith. This guarantees very low ECC values. Figure 4.9(b) shows the prototype of such a 5G MIMO antenna system. Finally, another integrated MIMO antenna systems consisting of a two-element 4G MIMO antenna system and a 5G planar slot antenna array operating at 28 GHz is presented in Figure 4.9(c) [35]. The 4G dual-element MIMO antenna system covered 1.87–2.53 GHz and the 5G antenna array covered 26.8–28.4 GHz. A modified two-arm-based monopole was used for the 4G antennas while a half-wavelength slot was used as the basic 5G array element. The array consisted of 4×2 slots with a corporate feeding network. The design was constructed on a 3-layer design, with the GND plane in the middle and the top layer had the 4G elements while the bottom layer had the 5G feed network.

4.5.2 Base station 5G solutions

Massive-MIMO (m-MIMO) or large-scale MIMO is a potential technology to be adopted in 5G standards. This technology relies of hundreds of antennas on the base station side to be able to provide directed dedicated beams for each user, increase the transmitted power via beam steering, and reduce inter-user interference. The idea has been heavily investigated from a signal-processing aspect [36], but so far, very few actual implementations have appeared due to the complexity and high cost of such antenna systems.

One example of a m-MIMO testbed for base-station applications has appeared in [37], where a 100-element patch array arranged in a T-shape pattern was proposed and operated in the 3.7 GHz range with 180 MHz of frequency bandwidth. Figure 4.10 shows a front view picture of the m-MIMO antenna array. The testbed was integrated

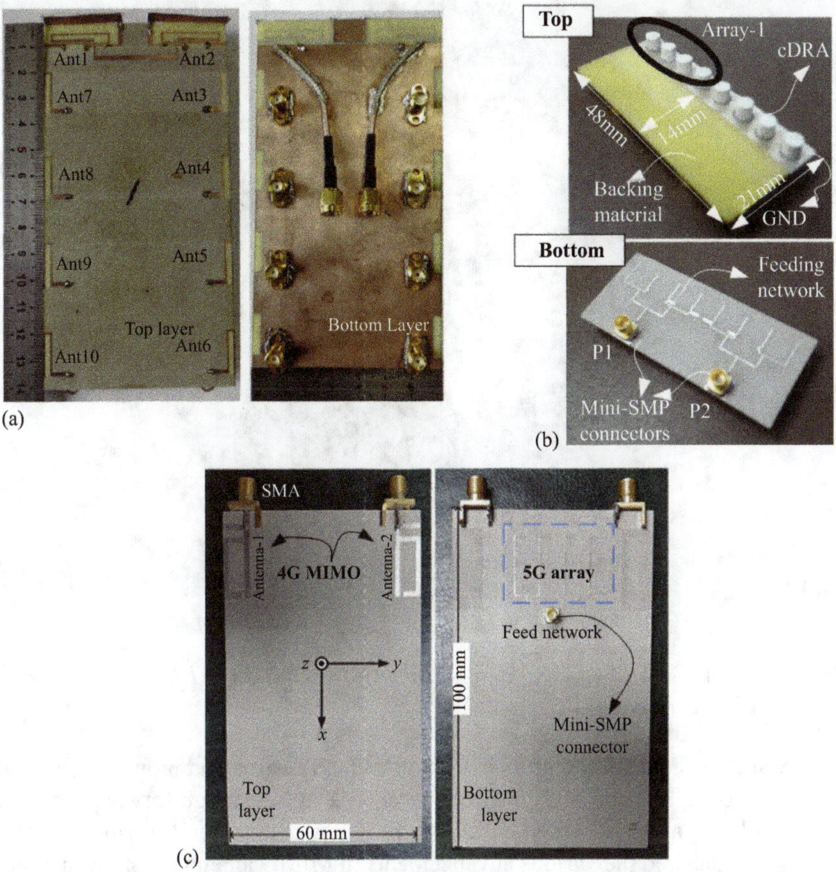

Figure 4.9 5G antenna systems, (a) integrated 4G/5G with 8-element MIMO at 3.6 GHz [33] (©2016 IEEE, Reprinted with permission); (b) 2-element MIMO with each element consisting of a linear array of cDRAs operating at 30 GHz [34]; and (c) integrated 4G/5G with 2-element wideband MIMO and a 5G slot-based array at 28 GHz [35]

with an SDR platform from National Instruments (NI) and was a joint work between NI and Lund University, Sweden.

Another example is found from Xilinx Inc. and BeeCube Inc., where a hexagonal shape base station m-MIMO solution for 4G and 5G bands is proposed [38]. The development platform can support up to 256×256 MIMO configurations. The bands targeted for 5G are the below 6 GHz ones similar to that developed in [37].

4.6 Conclusions

MIMO technology is going to be utilized in future generation wireless terminals due to its ability to provide higher data rates within the limited spectrum and power levels. The design of antennas for such MIMO systems has drawn a lot of attention

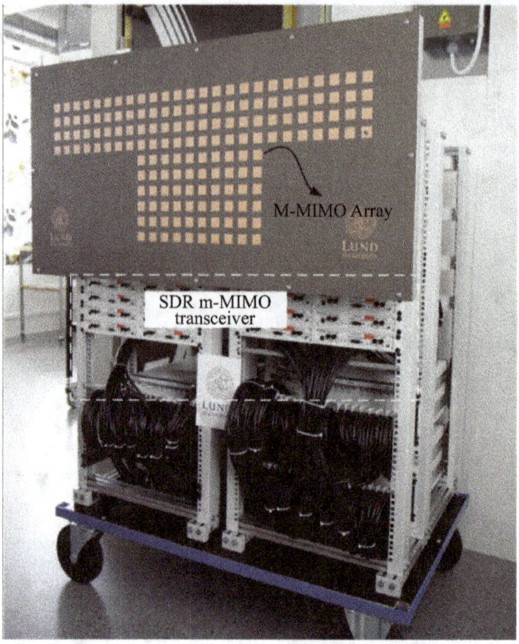

Figure 4.10 m-MIMO antenna array testbed [37] (©2014 IEEE. Reprinted with permission)

in the past decade and will continue to be a driving research and industrial thrust in the future. Designing MIMO antenna systems is not a straightforward task and requires a lot of attention and the evaluation of new antenna performance metrics. This chapter showed the various advancements in MIMO antenna designs that took place within the fourth-generation wireless standard as well as showed several potential solutions for design issues for the upcoming fifth-generation (5G) standard. The use of millimeter-wave is one of the main features for 5G, and integrating MIMO solutions within that band is of great potential. This chapter also gave several recent examples for 5G MIMO antenna solutions.

References

[1] A. Goldsmith, *Wireless Communications*, New York, Cambridge University Press, 2005.
[2] M. S. Sharawi, *Printed MIMO Antenna Engineering*, Norwood, MA, Artech House, 2014.
[3] M. S. Sharawi, "Printed MIMO Antenna Systems: Current Misuses and Future Prospects," *IEEE Antennas and Propagation Magazine*, Vol. 59, No. 2, pp. 1–10, April 2017.
[4] M. S. Sharawi, "Printed MIMO Antenna Systems: Performance Metrics, Implementations and Challenges," *Forum for Electromagnetic Research*

Methods and Application Technologies (FERMAT), Vol. 1, pp. 1–11, January 2014.

[5] M. S. Sharawi, "Printed Multi-Band MIMO Antenna Systems and Their Performance Metrics," *IEEE Antennas and propagation Magazine*, Vol. 55, No. 5, pp. 218–232, October 2013.

[6] M. Manteghi and Y. R. Samii, "Multiport Characteristics of a Wide-Band Cavity Backed Annular Patch Antenna for Multi-polarization Operations," *IEEE Transactions on Antennas and Propagation*, Vol. 53, No. 1, pp. 466–474, January 2005.

[7] S. W. Su, C. T. Lee and F. S. Chang, "Printed MIMO-Antenna System Using Neutralization-Line Technique for Wireless USB-Dongle Applications," *IEEE Transactions on Antennas and Propagation*, Vol. 60, No. 2, pp. 456–463, February 2012.

[8] S. Soltani, P. Lotfi and R. D. Murch, "A Port and Frequency Reconfigurable MIMO Slot Antenna for WLAN Applications," *IEEE Transactions on Antennas and Propagation*, Vol. 64, No. 4, pp. 1209–1217, April 2016.

[9] R. G. Vaughan and J. B Andersen, "Antenna Diversity in Mobile Communications," *IEEE Transactions on Antennas and Propagation*, Vol. VT-36, No. 4, pp. 149–172, November 1987.

[10] J. N. Pierce and S. Stein, "Multiple Diversity with Non-independent Fading," *Proceeding of the IRE*, Vol. 48, No. 1, pp. 89–104, January 1960.

[11] S. Blanch, J. Romeu and I. Corbella, "Exact Representation of Antenna System Diversity Performance from Input Parameter Description," *Electronics Letters*, Vol. 39, No. 9, pp. 705–707, May 2003.

[12] M. S. Sharawi, A. Hassan and M. U. Khan, "Correlation Coefficient Calculations for MIMO Antenna Systems: A Comparative Study," *International Journal of Microwaves and Wireless Technologies*, Cambridge University, Vol. 9, No. 10, pp. 1991–2004, December 2017.

[13] S. Park and C. Jung, "Compact MIMO Antenna with High Isolation Performance," *Electronics Letters*, Vol. 46, No. 6, pp. 390–391, March 2010.

[14] A. T. Hassan and M. S. Sharawi, "Four Element Half Circle Shape Printed MIMO Antenna," *Microwave and Optical Technology Letters*, Vol. 58, No. 12, pp. 2990–2992, December 2016.

[15] J. F. V. Valdes, M. A. G. Fernandez, A. M. M. Gonzalez and D. A. S. Hernandez, "Evaluation of True Polarization Diversity for MIMO Systems," *IEEE Transactions on Antennas and Propagation*, Vol. 57, No. 9, pp. 2746–2755, September 2009.

[16] K. Rosengren and P. S. Kildal, "Radiation Efficiency, Correlation, Diversity Gain and Capacity of a Six-monopole Antenna Array for a MIMO System: Theory, Simulation and Measurement in Reverberation Chamber," *IEEE Proceedings on Microwave and Antenna Propagation*, vol. 152, No. 1, pp. 7–16, February 2005.

[17] P. S. Kildal, *Foundations of Antenna Engineering*, Norwood, MA, Artech House, 2015.

[18] S. Zhang, K. Zhao, Z. Ying and S. He, "Adaptive Quad-Element Multi-Wideband Antenna Array for User-Effective LTE MIMO Mobile Terminal," *IEEE Transactions on Antennas and Propagation*, Vol. 61, No. 8, pp. 4275–4283, August 2013.

[19] International Wireless Industry Consortium, "Evolutionary & Disruptive Visions Towards Ultra High Capacity Networks," *IIWPC Ultra High Capacity Networks, White Paper Version 1.1*, pp. 1–89, April 2014.

[20] K. Zhao, S. Zhang, K. Ishimiya, Z. Ying and S. He, "Body-insensitive Multimode MIMO Terminal Antenna of Double-ring Structure," *IEEE Transactions on Antennas and Propagation*, vol. 63, No. 5, pp. 1925–1936, May 2015.

[21] A. Cihangir, F. Ferrero, G. Jacquemod, P Brachat and C. Luxey, "Neutralized Coupling Elements for MIMO Operation in 4G Mobile Terminals," *IEEE Antennas and Wireless Propagation Letters*, Vol. 13, pp. 141–144, 2014.

[22] S. Shoaib, I. Shoaib, N. Shohaib, X. Chen and C. G. Parini, "Design and Performance Study of a Dual-Element Multiband Printed Monopole Antenna Array for MIMO Terminals," *IEEE Antennas and Wireless Propagation Letters*, Vol. 13, pp. 329–332, 2014.

[23] R. Hussain and M. S. Sharawi, "Wide-Band Frequency Agile MIMO Antenna System with Wide Tunability Range," *Microwave and Optical Technology Letters*, Vol. 58, No. 9, pp. 2276–2280, September 2016.

[24] R. Hussain and M. S. Sharawi, "Integrated Reconfigurable Multiple-Input-Multiple-Output Antenna System with an Ultra-wideband Sensing Antenna for Cognitive Radio Platforms," *IET Microwaves, Antennas and Propagation*, Vol. 9, No. 9, pp. 940–947, September 2015.

[25] J. F. Li, Q. X. Chu, Z. H. Li and X. X. Xia, "Compact Dual Band-Notched UWB MIMO Antenna With High Isolation," *IEEE Transactions on Antennas and Propagation*, Vol. 61, No. 9, pp. 4759–4766, September 2013.

[26] G. Srivastava and A. Mohan, "Compact MIMO Slot Antenna for UWB Applications," *IEEE Antennas and Wireless Propagation Letters*, Vol. 15, pp. 1057–1060, 2016.

[27] S. K. Dhar and M. S. Sharawi, "A UWB Semi-Ring MIMO Antenna With Isolation Enhancement," *Microwave and Optical Technology Letters*, Vol. 57, No. 8, pp. 1941–1946, August 2015.

[28] W. J. Liao, S. H. Chang, J. T. Yeh and B. R. Hsiao, "Compact Dual Band WLAN Diversity Antennas on USB Dongle Platform," *IEEE Transactions on Antennas and Propagation*, Vol. 62, No. 1, pp. 109–118, January 2014.

[29] A. Krewski, W. L. Schroeder and K. Solbach, "2-Port DL-MIMO Antenna Design for LTE-Enabled USB Dongles," *IEEE Antennas and Wireless Propagation Letters*, Vol. 12, pp. 1436–1439, 2013.

[30] S. W. Su, "High-Gain Dual-Loop Antennas for MIMO Access Points in the 2.4/5.2/5.8 GHz Bands," *IEEE Transactions on Antennas and Propagation*, Vol. 58, No. 7, pp. 2412–2419, July 2010.

[31] W. Han, X. Zhou, J. Ouyang, Y. Li, R. Long and F. Yang, "A Six-Port MIMO Antenna System With High Isolation for 5-GHz WLAN Access Points," *IEEE Antennas and Wireless Propagation Letters*, Vol. 13, pp. 880–883, 2014.

[32] Y. Pan, Y. Cui and R. L. Li, "Investigation of a Triple-Band Multibeam MIMO Antenna for Wireless Access Points," *IEEE Transactions on Antennas and Propagation*, Vol. 64, No. 4, pp. 1234–1241, April 2016.

[33] Y. L. Ban, C. Li, C. Y. D. Sim, G. Wu and K. L. Wong, "4G/5G Multiple Antennas for Future Multi-Mode Smartphone Applications," *IEEE Access*, Vol. 4, pp. 2981–2988, June 2016.

[34] M. S. Sharawi, S. K. Podilchak, M. T. Hussain and Y. M. M. Antar, "Dielectric Resonator Based MIMO Antenna System Enabling Millimeter-Wave Mobile Devices," *IET Microwaves, Antennas and Propagation*, Vol. 11, No. 2, pp. 287–293, January 2017.

[35] R. Hussain, A. T. Alreshaid, S. K. Podilchak and M. S. Sharawi, "Compact 4G MIMO Antenna Integrated with a 5G Array for Current and Future Mobile Handsets," *IET Microwaves, Antennas and Propagation*, Vol. 11, No. 2, pp. 271–279, January 2017.

[36] F. Rusek, D. Persson, B. K. Lau, *et al.*, "Scaling Up MIMO," *IEEE Signal Processing Magazine*, Vol. 30, No. 1, pp. 40–60, 2013.

[37] J. Vieira, S. Malkowsky, K. Nieman, *et al.*, "A Flexible 100-Antenna Test-bed for Massive MIMO," *IEEE Globecom Conference Workshop*, Texas, USA, pp. 287–293,

[38] Online, Available from: http://telecom.cioreview.com/news/xilinx-and-megabee-launch-prototyping-platform-for-5g-massive-mimo-antenna-systems-nid-5344-cid-39.html. [Accessed March 2017].

Chapter 5
Reconfigurable leaky-wave antennas

Yingjie Jay Guo[1], Debabrata K. Karmokar[1], and Trevor S. Bird[1,2]

5.1 Introduction

Leaky-wave antennas (LWAs) are travelling-wave antennas that radiate energy as a wave travels along a guided wave structure [1–5]. Owing to the leakage phenomenon of a leaky structure, it has a complex propagation constant consisting of a phase constant denoted by β, and an attenuation constant denoted by α. The value of attenuation or leakage constant depends on the leakage abilities of the structure. If the leakage per unit length is high, the value of α is large. A large value of α means that most of the power is radiated from the portion of the structure close to the feed, thus making a short effective radiating aperture and producing a wide beam. On the other hand, if the value of α is small, the structure leaks energy gradually as it travels away from the source. This results in a long effective aperture and a narrow beam [6]. The direction of the main beam of an LWA is determined by its phase constant β. The phase constant and the free-space wavenumber (k_0) are directly related to the frequency. From early microwave days, leaky wave structures were popular due to their inherent beam-scanning ability [2–5], in which the direction of the maximum beam varies with frequency over a hemisphere, and the beamwidth is determined by the leakage rate of the travelling wave structure.

In order to obtain a desirable beam-scanning range, a wide-frequency sweep is generally required. Scanning an antenna beam via frequency sweeping can be useful for some applications, such as frequency-scanning radars. On the other hand, most communications systems operate in a predefined frequency band determined by the spectrum regulators. In most cases, the band is too narrow to get a desired beam scanning by sweeping frequency. Therefore, scanning the beam at a fixed frequency is highly desirable for most communications systems. Reconfigurable LWAs are a potential candidate for such fixed-frequency beam scanning applications since, by using simple electronic circuits, beam scanning can be realized with LWAs by electronically tuning the circuits typically as part of the LWA

[1]Global Big Data Technology Centre (GBDTC), University of Technology Sydney (UTS), Australia
[2]Antengenuity, Australia

boundaries. In this chapter, we describe the classification of LWAs, their operating principles, recent development of frequency-scanning LWAs, various beam-scanning LWAs at fixed operating frequency, and the scope for future research.

5.2 History of LWAs

LWAs have been known since the 1940s [1]. The first such structures were based on slotted waveguide as illustrated in Figure 5.1. The slot can be tapered, with the waveguide being hollow or dielectric loaded [6,7].

The feasibility of microstrip LWAs was first demonstrated in the 1970s [8,9]. Soon after, better analysis methods became available for calculating leaky mode dispersions. Microstrip LWAs gained strong research interest owing to their planar low-profile configuration. Radiation from higher-order modes of a microstrip was explained in detail [10,11]. Later on, other types of LWAs based on microstrip technology were reported [12,13], and new methods to excite higher-order microstrip modes were investigated. When substrate integrated waveguide (SIW) was invented, it received significant interest since they are widely accepted as a replacement of bulky waveguides. LWA designs based on SIW have been popular in the last decade. Different types of LWAs based on SIW are described [14–16].

In general, it is difficult for a LWA to radiate in the broadside direction because there is an open stopband when the main beam is close to the broadside. The development of composite right/left-handed (CRLH) LWA solved the problem of broadside scanning [4,17]. Various results have been reported on CRLH LWAs based on microstrip and SIW based structures [18,19]. Other LWAs based on different structures and materials have also been developed to overcome the problem of broadside radiation [20–23]. Owing to the frequency band limitation of most communications systems, LWA designs at fixed frequencies have attracted strongest research interest in recent years. Most such structures are based on waveguides, microstrip lines and SIWs [24–29].

5.2.1 Basic operating principle

In the case of a linear surface and its aperture, under the right conditions, the surface can support surface waves when the propagation constant $|\beta| > k_0$ and leaky

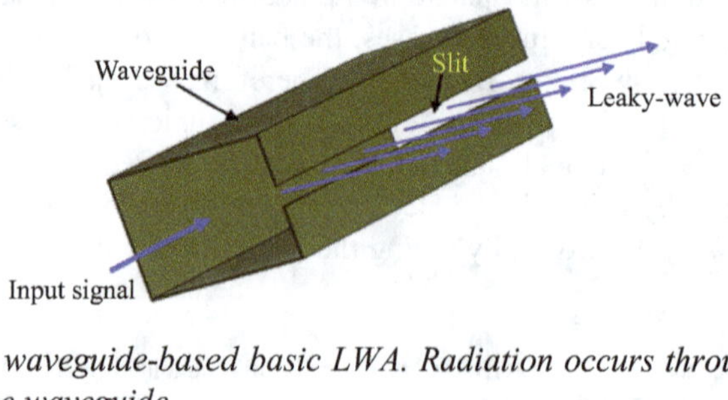

Figure 5.1 A waveguide-based basic LWA. Radiation occurs through the slit of the waveguide

waves when $|\beta| < k_0$, which are fast waves. In this latter case, the leaky mode wavenumber is complex and given by

$$k_z = \beta - j\alpha \tag{5.1}$$

where α is the attenuation constant and β is the phase constant [4,30]. The power which leaks along the structure of a LWA as the wave propagates is accounted for by the attenuation constant α.

The direction of the main beam of an LWA measured from the broadside (x-direction in Figure 5.2(a)) is given by the following approximate expression [7, 31]:

$$\theta(f) = \sin^{-1}\left[\frac{\beta(f)}{k_0(f)}\right] \tag{5.2}$$

A 1-D LWA is usually fed at one end so that the wave propagates along the structure. A basic LWA fed at one end is shown in Figure 5.2(a) [4]. This type of LWA usually produces a single beam. If the travelling wave is reflected from the far-end due to mismatch, a beam is produced in the $-\theta$ direction. This reflected beam will have lower directivity due to the loss of power when travelling forwards. Other beams such as forward bidirectional and broadside beams can be achieved by feeding at the middle of the structure (Figure 5.2(b)). A broadside beam can be achieved from bidirectional LWAs [4].

5.2.2 Classification of LWAs

LWAs can be classified into different categories based on the mode properties, the geometry of the structure and their operating principles. In terms of structural properties they can be divided into uniform and periodic, and one-dimensional (1D) and two-dimensional (2D) LWAs. 1D LWAs can be further classified into different categories, such as uniform and quasi-uniform LWAs [4,30].

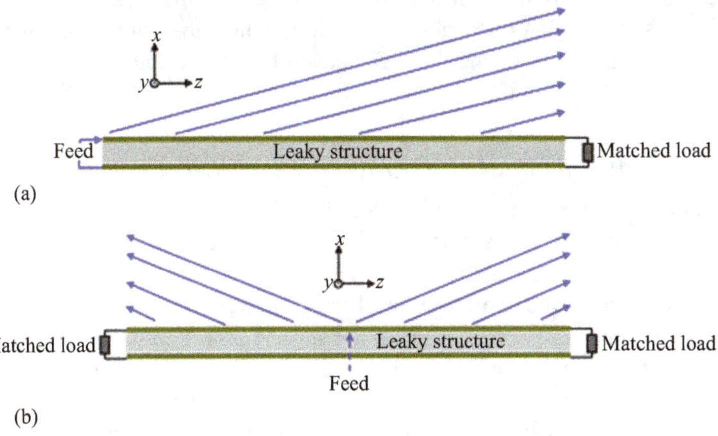

Figure 5.2 Modes of operation for an LWA: (a) unidirectional and (b) bidirectional

5.2.2.1 One-dimensional uniform LWA

The structure of a 1-D uniform LWA is along a uniform line. The wave in a 1-D uniform LWA travels in a particular direction usually away from the feed of the structure. This LWA structure supports a fast wave, i.e., the phase constant (β) is less than the free space wave number (k_0) and lies in the range of $0 < \beta < k_0$.

5.2.2.2 One-dimensional quasi-uniform LWA

The structure of a 1-D quasi-uniform LWA is similar to the 1-D uniform LWA. However, the geometry of the structure is not uniform. Rather, a periodic structure is used in the antenna. The periodicity of the structure is much smaller compared to the wavelength and hence there is negligible effect of the periodic structure on the radiation. Similar to the uniform LWA the structure also supports a fast wave, i.e., radiation occurs from the fundamental mode.

5.2.2.3 Two-dimensional LWA

One-dimensional LWAs usually produce a fan shaped beam such as that from a half-width microstrip LWA. The creation of pencil beams would need a 2D LWA that has a two-dimensional surface. This type of LWAs is usually fed at the centre. The energy of a 2D LWA travels towards the radial direction from the feed point. A pencil beam towards the broadside or a conical beam around the broadside can be easily produced from a simple 2D LWA.

5.3 Passive frequency-scanning LWA structures

Waveguide-based LWA structures have been around since 1940s [1]. With the advancement of microwave technologies came different guided wave structures. LWAs based on new microwave technologies such as microstrip, CRLH materials, metamaterials, and SIW have all been reported. Many of these have been utilized in LWA designs. The LWAs described in the following subsections are basically the main radiating structures of the fixed-frequency beam-steering LWAs presented in this chapter.

5.3.1 One-dimensional (1-D) Fabry–Pérot LWA

A parallel-plate waveguide loaded with a partially reflective surface (PRS) and a grounded high-impedance surface (HIS) can be used for LWA designs. The structure works as a Fabry–Pérot (FP) cavity in which radiation occurs through the top PRS. A 3D view of a short section of the FP LWA is shown in Figure 5.3 with the corresponding transverse equivalent network (TEN). The structure consists of a cavity-backed parallel waveguide which is loaded with two printed-circuit boards (PCBs). Each PCB is designed using periodic metallic dipoles printed on the substrate. The separation between two PCBs is H as shown in Figure 5.3.

Figure 5.4(a) and (b) shows the dispersion diagram and, the corresponding transverse electric field distribution in the LWA, respectively. These show that

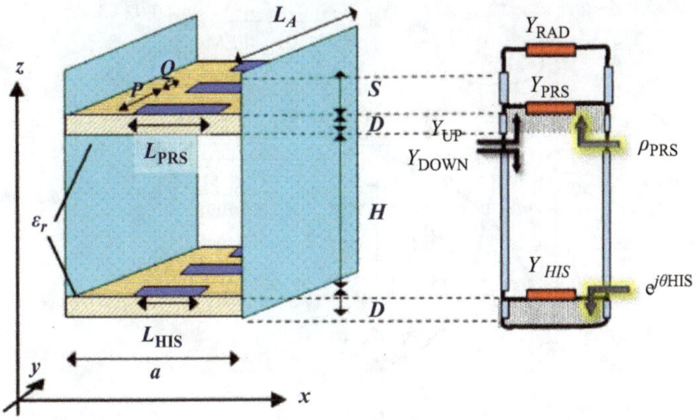

Figure 5.3 Three-dimensional view of a Fabry–Pérot LWA (left) and the corresponding transverse equivalent network (right) [32]

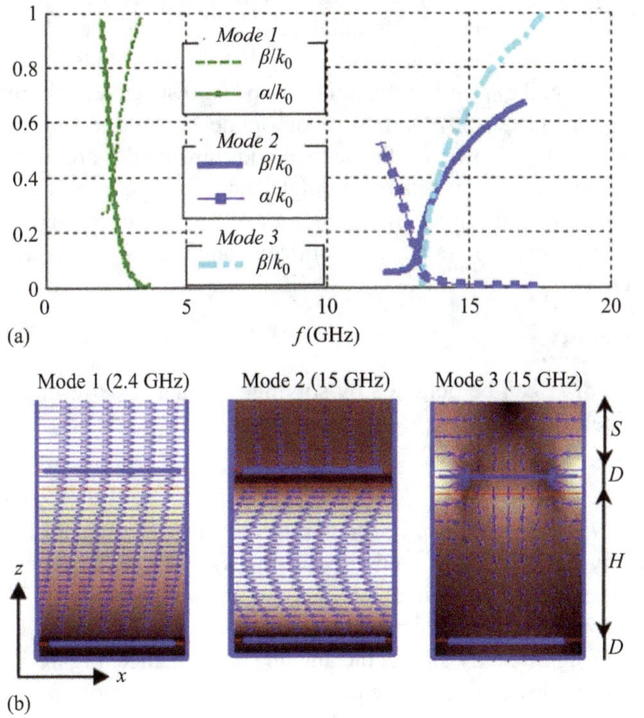

Figure 5.4 (a) Dispersion diagram of three different natural modes in the LWA and (b) electric field (transverse) distribution in the cross-section corresponding to each mode [32]

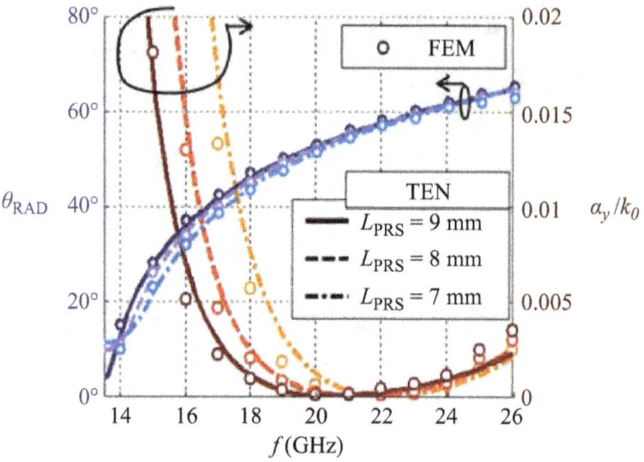

Figure 5.5 Dispersion curve of the LWA in Figure 5.3 with different lengths of PRS dipole when the length of the HIS dipole is set to 9 mm [32]

three different modes are present in the structure, namely, modes 1, 2 and 3. Mode 1 is a horizontally polarized channel-guided mode supported by the host parallel-plate waveguide. The operating frequency of the antenna can be adjusted by changing the height (H) between the PCBs in the design. Modes 2 and 3 are the perturbed horizontal TE_{01} and vertical TE_{10} modes, respectively. In the operating band (15 GHz for this design) the first mode (mode 1) is in the slow-wave region, which is non-radiating. However, the second and third modes are in the fast-wave region. The structure is designed for a single mode of operation by using a horizontally polarized feed. Since the longitudinal direction of the LWA is the y-direction (see Figure 5.3 (left)), the complex longitudinal propagation constant can be written as

$$k_y = \beta_y - j\alpha_y \tag{5.3}$$

The value of k_y can be obtained by solving the transverse resonance equation (TRE) given below [32]

$$Y_{\mathrm{UP}}(f, k_y, L_{PRS}) + Y_{\mathrm{DOWN}}(f, k_y, L_{\mathrm{HIS}}) = 0 \tag{5.4}$$

In the TEN (Figure 5.3 (right)) the two printed circuits, namely HIS and PRS, are modelled by $Y_{\mathrm{HIS}}(f, k_y, L_{\mathrm{HIS}})$ and $Y_{\mathrm{PRS}}(f, k_y, L_{\mathrm{PRS}})$. A detailed discussion on the method to obtain these two equivalent admittances can be found in [33,34].

Both the top PRS and bottom HIS dipoles control the radiation of the LWA, and their length significantly affects the antenna performance. Figure 5.5 shows the leaky-mode (LM) dispersion characteristics of the antenna for fixed HIS dipole length and variable PRS dipole length. It is clear that L_{PRS} has a very small effect on the beam angle (θ_{RAD}). However, it has a great effect on the antenna leakage rate (see the normalized leakage curves (α_y/k_0) in Figure 5.5). When the value of L_{PRS} is approximately half a wavelength, the PRS dipoles resonate and behave like

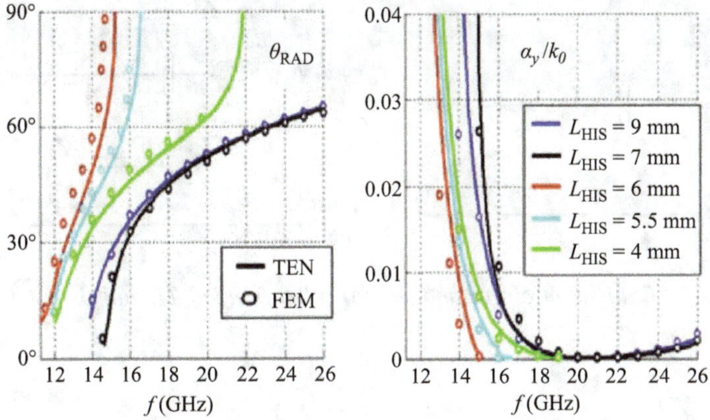

Figure 5.6 Dispersion curve of the LWA in Figure 5.3 with different lengths of HIS dipole when the length of the PRS dipole is set to 9 mm [32]

an effective perfect electric conductor (PEC). In this circumstance, it prevents leakage from the structure though the wave is fast ($\beta_y < k_0$).

The LM dispersion characteristics when L_{PRS} is fixed and variable L_{HIS} are shown in Figure 5.6. As mentioned previously an FSS based on a metal-backed dipole, performing as a HIS, is at the bottom of the LWA structure. The effective height of the resonant cavity can be changed by the HIS, as the length (L_{HIS}) of the dipoles used in the HIS controls the reflection phase of the wave propagating throughout the waveguide. The effect of the LM cut-off frequency on L_{HIS} is shown in Figure 5.6 (left). It can be seen that the cut-off frequency can be varied by varying L_{HIS}. The leakage rates for different values of L_{HIS} are shown in Figure 5.6 (right). It is seen in Figure 5.6 that when the cut-off frequency decreases, i.e., with higher effective aperture, the leakage rate reduces and the pointing angle increases. It is evident from the discussion that by tuning the length of the dipoles used in the HIS, the pointing angle can be changed.

5.3.2 Composite right/left-handed transmission line and LWA

It is well known that the equivalent circuit of a pure lossless conventional right-handed (RH) transmission line (TL) consists of a series inductor and a shunt capacitor. In the case of a pure lossless left-handed (LH) TL, it consists of a series capacitor and a shunt inductor. A lossless CRLH TL has a series impedance (consists of a series connection of an inductor and a capacitor); and the shunt combination of a capacitor and an inductor [35]. The equivalent circuit models of lossless LH and CRLH TL models are shown in Figure 5.7. In a CRLH unit cell, the RH elements are present normally from the portion of a TL and the LH elements are added artificially [4]. If Z is the per-unit-length impedance and Y is the per-unit-length admittance of a CRLH structure the propagation constant of the CRLH TL is given by [35]

$$\gamma = \alpha + j\beta = \sqrt{ZY} \tag{5.5}$$

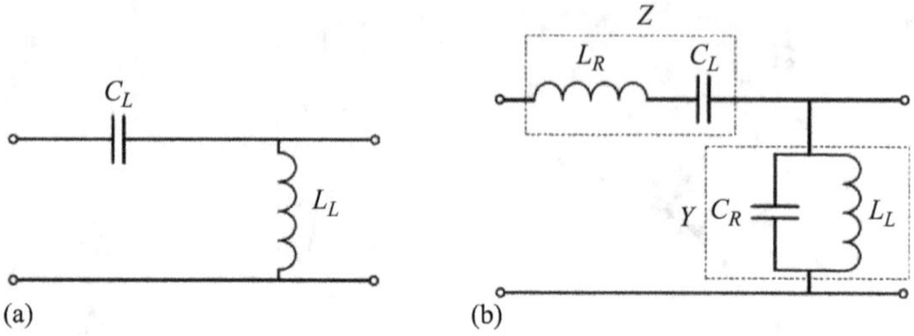

Figure 5.7 Equivalent circuit model: (a) a left-handed TL and (b) a CRLH TL

where

$$Z(\omega) = j\left(\omega L_R - \frac{1}{\omega C_L}\right)$$

and

$$Y(\omega) = j\left(\omega C_R - \frac{1}{\omega L_L}\right) \tag{5.6}$$

As the period of a CRLH structure is small compared to a wavelength this type of structure normally falls into the quasi-uniform category. The radiation occurs from a CRLH structure by the fundamental harmonic ($n = 0$ space harmonic) [4]. The LWAs having these properties are called CRLH LWAs.

The phase constant of a CRLH TL can be calculated using (5.5) and (5.6) and be expressed as [36]:

$$\beta_{\text{CRLH}} = \sqrt{\left(\omega L_R - \frac{1}{\omega C_L}\right)\left(\omega C_R - \frac{1}{\omega L_L}\right)} = \sqrt{\omega^2 L_R C_R + \frac{1}{\omega^2 L_L C_L} - \left(\frac{L_R}{L_L} + \frac{C_R}{C_L}\right)}. \tag{5.7}$$

A typical dispersion diagram of a CRLH unit cell is shown in Figure 5.8 for both balanced and unbalanced conditions. The balanced condition occurs when ($L_R C_L = L_L C_R$), and the structure exhibits the unbalanced condition when $L_R C_L \neq L_L C_R$. The dispersion diagram shows that it has an antiparallel phase (v_p) and group (v_g) velocity in the LH region and they are parallel in the RH region, i.e., $v_p v_g > 0$ in the RH region and $v_p v_g < 0$ in the LH region [35,36]. The CRLH TL supports both slow-wave and fast-wave in its fundamental mode operation, and they can be used as an LWA in their fast-wave region as shown in Figure 5.8 [36]. It noticeable from the dispersion diagram that a CRLH LWA can scan its beam continuously, from the backward to the forward direction, in its balanced condition. As mentioned before, a CRLH LWA can be implemented by adding series capacitances (e.g., by introducing interdigital or gap capacitors in the radiating microstrip) and a shunt inductance (e.g., using metallic vias). A prototype of such an antenna, a CRLH LWA, is shown later in this chapter.

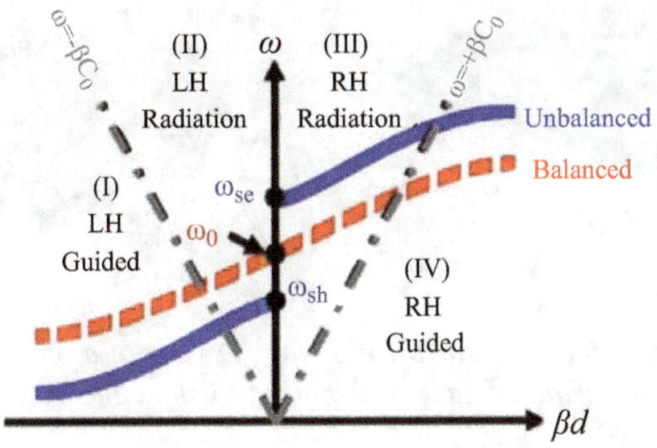

Figure 5.8 Typical dispersion diagram of a CRLH unit cell [36]

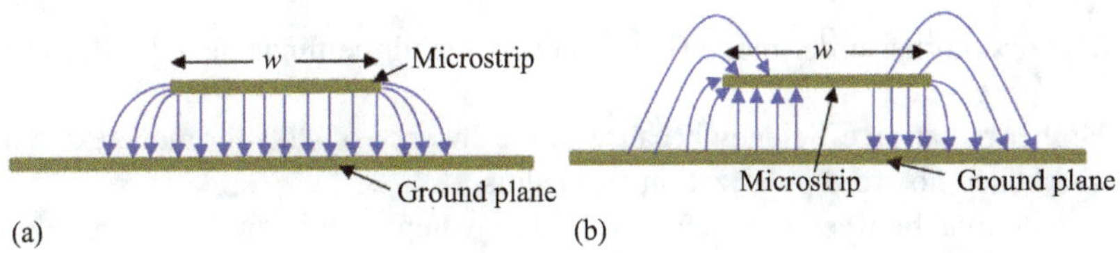

Figure 5.9 Electric-field distributions of two different modes of a microstrip line (substrate omitted): (a) fundamental/dominant mode (EH_0) and (b) first higher-order mode (EH_1)

5.3.3 Half-width microstrip LWA

The field in a fundamental (dominant) mode of microstrip is tightly bound within the substrate between the top line and the ground plane as shown in Figure 5.9(a). In this mode, the microstrip guides an input signal through the structure, i.e., works as a TL. However, higher-order modes are possible some of which radiate leaky waves. Different methods have been developed to excite a microstrip line in its higher-order modes. The first higher-order mode of a microstrip line radiates leaky waves from both of its edges and has a phase reversal and an electric field null at the centre of the microstrip as shown in Figure 5.9(b). Since there is no electric field at the centre of the microstrip in the first higher-order mode, shorting the microstrip to the ground plane will have no effect on the mode.

One easy and effective way to excite a microstrip line in its first higher-order mode is to use a shorting wall along the centre of a microstrip line. The method to excite a microstrip line in its first higher-order mode using this technique was proposed in [12]. Since the microstrip is symmetric with respect to the centre shorting wall, the width of the microstrip can be reduced to half of its original one, thus resulting in a half-width (HW) microstrip LWA (HW-MLWA). A 3-D view of a HW-MLWA is shown in Figure 5.10, where Figure 5.10(a) shows the structure with substrate. In Figure 5.10(b) the substrate is omitted, to show the connection between the microstrip and the ground plane by a conducting wall, called a septum.

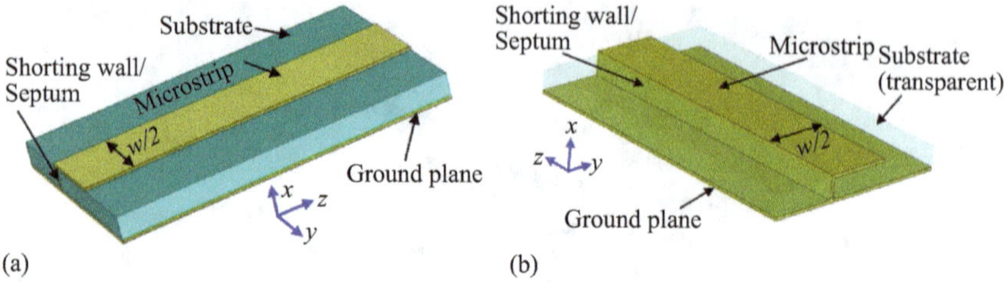

Figure 5.10 Half-width microstrip LWA: (a) 3D view of a basic structure and (b) substrate is made transparent to show the shorting wall/septum

Excitation of a microstrip line in the first higher-order mode using the HW technique has several advantages [12]. These include,

- The first higher-order mode (EH_1) is excited and the fundamental mode (EH_0) is suppressed
- Higher radiation efficiency because of the absence of slots for mode excitation and hence no cross-polarization from slots
- Interactions between the elements are less when used in an array
- Pure guided mode improves radiation efficiency

5.4 Reconfigurable LWAs

As discussed previously, LWAs have inherent frequency based beam-scanning properties. In most cases, a wide range of frequency sweeping is desired for the beam scanning. This frequency-dependent beam scanning may be suitable for some applications such as frequency-scanning radars. When applied to communication systems, however, frequency-dependent beam scanning is very unlikely to be used. The main reason behind this is the predefined frequency band for most wireless communication systems. Although antenna beam scanning can bring significant benefits to a communication system, frequency-dependent beam scanning is of little use.

In the past decades, some research has been conducted and is continuing on fixed-frequency beam-steering antennas. In the following sections, a few beam-steering antennas operating at a fixed frequency are described briefly. Detailed analyses and discussions on each LWA are available in the related reference papers.

5.4.1 1-D FP-reconfigurable LWAs

As discussed in Section 5.3.1, beam scanning is possible from a waveguide-based LWA composed of a PRS and a HIS. A reconfigurable LWA made out of such a 1-D FP cavity consisting of a PRS and a tunable HIS was described in [26]. The antenna is a similar parallel-plate waveguide to that discussed in Section 5.3.1

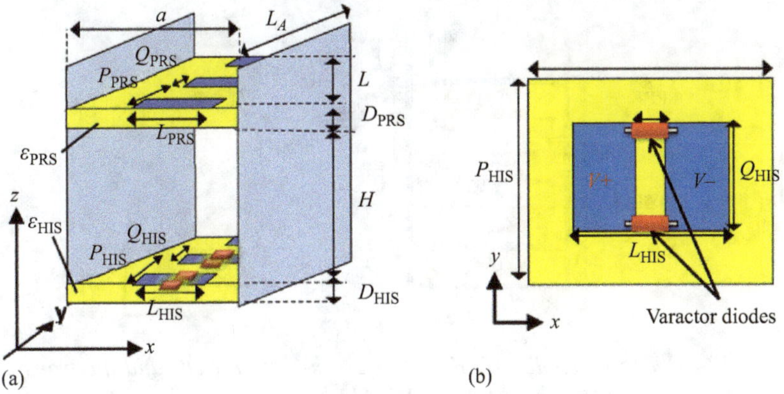

Figure 5.11 1-D reconfigurable FP LWA: (a) 3D view of a short portion the LWA and (b) portion of the HIS in a unit cell. The dimensions of the unit cell and the antenna are available in [26]

where the top plate is a PRS and the bottom plate is a HIS which can be tuned. The HIS is made of electronically tunable patches, where the cavity resonance condition can be changed by tuning the HIS.

Both the PRS and HIS are made of periodic resonant patches as shown in Figure 5.11(a). The HIS and PRS are separated by a distance (H) making a 1-D FP resonant cavity. As shown in Figure 5.11(b), the printed periodic patches in the HIS are loaded with varactor diodes. As discussed in Section 5.3.1 the leakage rate can be determined by adjusting the patches on the top PRS, and the pointing angle can be changed by tuning the resonant length on the HIS patches. These operating principles of a passive 1D FP LWA are used in the reconfigurable FP LWA design. The transverse resonance method (TRM) is also used for the analyses and in designing the reconfigurable LWA. As discussed before, this is a simple yet accurate method to obtain the complex propagation constant of the LWA. Using this method, the LM of the structure can be represented as a function of the main parameters. The dispersion curves obtained using this method not only gives a clear understanding of the operating principle, but also provides a theoretical foundation of a beam-steering LWA.

5.4.1.1 Analysis on the reconfigurable FP LWA

A TEN, which is the key feature of the TRM, is used to analyse the antenna. The TEN needs to be designed carefully so that it models the cross-section of the antenna accurately. The cross-section of the LWA and the TEN for the reconfigurable LWA are shown in Figure 5.12. The most important parts of the TEN are the equivalent admittance of the PRS [Y_{PRS} in Figure 5.12 (right)], and the equivalent admittance of the HIS [Y_{HIS} in Figure 5.12 (right)]. Since the HIS of the antenna design consists of varactor diodes, the admittance of the HIS includes the varactor junction capacitance (C_j) and so the admittance is a function of C_j. Suppose there are N segments in the profile.

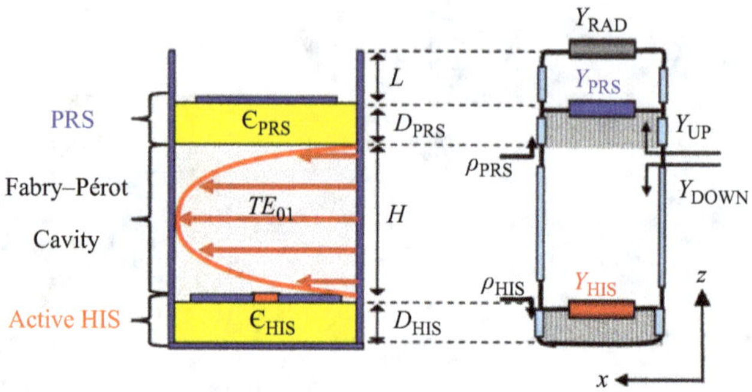

Figure 5.12 Reconfigurable 1-D FP LWA cross-section (left) and the transverse equivalent network (right) [26]

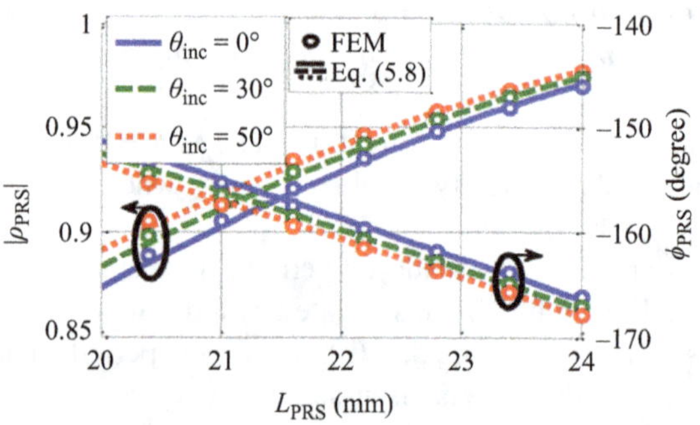

Figure 5.13 Phase and magnitude of the reflection coefficient of an incident plane wave on the PRS with different angles as a function of L_{PRS} [26]

The top PRS is fixed and, using the analytical pole-zero expression given below for the PRS admittance (Y_{PRS}), the length L_{PRS} of the conducting patches can be efficiently determined with:

$$Y_{PRS}(k_y, L_{PRS}) = j\frac{L_{PRS}\prod_{i=1}^{N}\left[L_{PRS} - L_{PRS_{zi}}(k_y)\right]}{\prod_{i=1}^{N}\left[L_{PRS} - L_{PRS_{pi}}(k_y)\right]} \tag{5.8}$$

where the longitudinal wavenumber k_y determines the location of the poles and zeros.

The wavenumber k_y is related to the LM angle (incident or radiating) by

$$\sin\theta_{inc} = \text{real}(k_y)/k_0. \tag{5.9}$$

The phase and magnitude of the reflection coefficient are shown in Figure 5.13 at the design frequency of 5.6 GHz for different incident angles of a plane wave with the variation of L_{PRS}. As discussed before, the value of L_{PRS} needs to be chosen to control the radiation efficiency by tuning the reflectivity of the PRS. In this design, the value of L_{PRS} is selected to be 22 mm to obtain a reflectivity greater than 0.9 for all scanning angles.

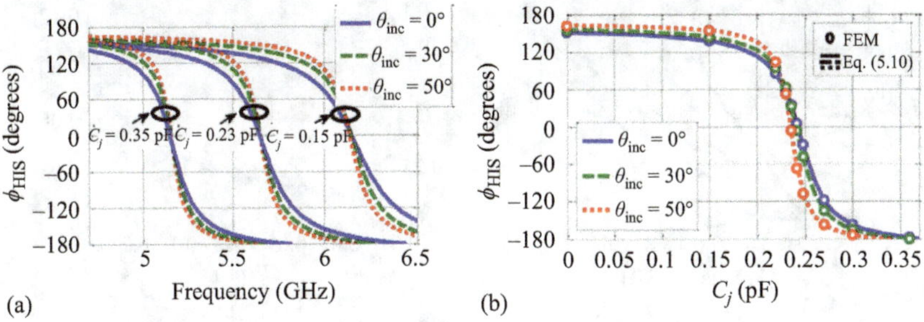

Figure 5.14 Phase of reflection coefficient for a plane wave with different incidence angles on the PRS: (a) variable C_j with frequency and (b) frequency fixed at the design frequency and as a function of C_j [26]

To enable electronic control, the patches on the HIS are divided into two parts along the x-direction as shown in Figure 5.11, leaving a gap (g) of 1 mm between them. Two varactors are placed at the edges of the patches. The pole-zero expression for the electronically tunable HIS is dependent on the junction capacitance (C_j) of the varactor, and can be expressed as

$$Y_{\rm HIS}(k_y, C_j) = j\frac{\prod_{i=1}^{N}\left[C_j - C_{j_{zi}}(k_y)\right]}{\prod_{i=1}^{N}\left[C_j - C_{j_{pi}}(k_y)\right]}. \tag{5.10}$$

As discussed in Section 5.3.1, by varying the physical resonating length of the HIS patches, the boundary condition can be changed. In order to control the TE_{01} LM cut-off frequency, and hence the beam scanning at a fixed frequency, the whole HIS PCB needs to change for each beam position. Unfortunately, this is not feasible for practical applications. The reconfigurable HIS can overcome the problem by varying the effective resonant length electronically as shown in Figure 5.14.

The reflection phase ($\rho_{\rm HIS}$) of the tunable HIS is shown in Figure 5.14(a). The results were obtained using the analytical expression of the pole-zero given in (5.10) for different incidence angles ($\theta_{\rm inc}$) as a function of frequency. It can be seen that the reflection phase provides PMC conditions, i.e., $\varphi_{\rm HIS}$ at 5.15, 5.65, and 6.15 GHz with the junction capacitance C_j set to 0.35, 0.23, and 0.15 pF, respectively. At a fixed frequency, the response of $Y_{\rm HIS}$ (k_y, C_j) can be modelled analytically, using (5.10), as a function of the junction capacitor C_j. The reflection phase as a function of C_j for different $\theta_{\rm inc}$ is shown in Figure 5.14(b). From Figure 5.14(b) it can be observed that the HIS can be changed between a grounded dielectric slab, a PMC sheet and a PEC sheet by controlling the junction capacitance and the incidence angle. For example, when $C_j = 0$ pF and $\varphi_{HIS} = 150°$ the HIS acts as a dielectric slab, with $C_j = 0.23$ pF and $\varphi_{\rm HIS} = 0°$ it works as a PMC sheet, and for $C_j = 0.35$ pF and it behaves as a PEC sheet. By comparing the effect of C_j on the HIS with the analysis given in Section 5.3.1, it is evident that a similar response is obtained with a larger value of C_j and a larger value of $L_{\rm HIS}$. This indicates a direct relation between the effective resonant length of the HIS patches and the junction capacitance C_j.

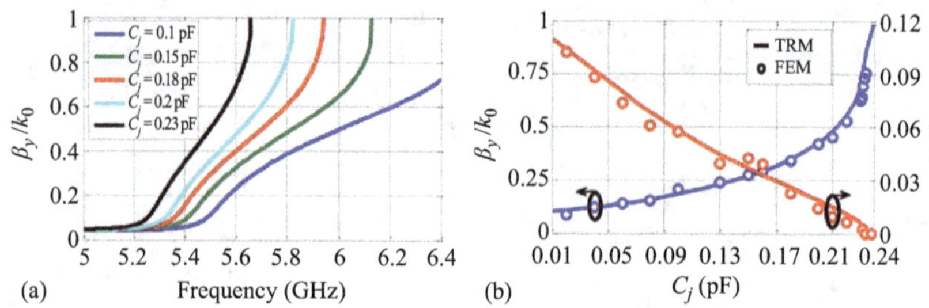

Figure 5.15 Reflection phase for an incident plane wave on the tunable HIS for various incidence angles (θ_{inc}): (a) different value of junction capacitance C_j as a function of frequency and (b) frequency fixed at 5.6 GHz and the reflection phase is plotted as a function of C_j [26]

From this previous discussion, it is clear that the effective resonant length of the HIS can be controlled by controlling C_j, and hence the beam direction of the LWA can be changed at a fixed frequency.

As described in Section 5.3.1, by solving the transverse resonance equation given by (5.4) of the TEN shown in Figure 5.12 (right) the LM complex propagation wavenumber ($k_y = \beta_y - j\alpha_y$) can be obtained. Figure 5.15(a) shows the normalized phase constant (β_y/k_0) as a function of frequency for different values of C_j. It is clear that when C_j increases the cut-off frequency of the TE_{01} mode decreases. It can be noticed that at a particular frequency point, for example, at 5.6 GHz, there is a rise of phase constant for different values of C_j. Figure 5.15(b) represents the normalized phase and attenuation constant (α_y/k_0) as a function of C_j at 5.6 GHz. The value of β_y/k_0 can be varied from near zero (when $C_j = 0.01$ pF) to near one (when $C_j = 0.23$ pF) by controlling the junction capacitance C_j. This indicates that beam scanning at a fixed frequency can be possible from broadside to endfire in the case of a lossless structure, but the analyses do not account for the additional losses from the components in the structure.

5.4.1.2 Antenna prototype and measured results

Figure 5.16 shows the LWA prototype operating at 5.6 GHz, which includes separate images of the parallel plates and the PRS. The length (L_A) of the radiating aperture is $5\lambda_0$. A horizontal coaxial probe is used to excite the TE_{01} LM of the FP cavity. The phase-agile cell used in the HIS and the detailed photograph of the biasing network are shown in Figure 5.17. MGV125-08 varactor diodes are used in the prototype. The diode junction capacitance varies between 0.055 and 0.6 pF with applied DC reverse bias voltage of between 20 and 2 V_{DC} and has a manufacturer tolerance of ± 0.05 pF.

The measured reflection coefficients of the antenna from 5 to 6 GHz for different reverse bias voltages (V_R) are shown in Figure 5.18. The dimensions of the probe were optimized for best input impedance matching for $C_j = 0.15$ pF which is the operating range midpoint. It is seen that as V_R decreases, i.e., C_j increases, the matched band shifts to lower frequencies. Shifting of the matched band to lower

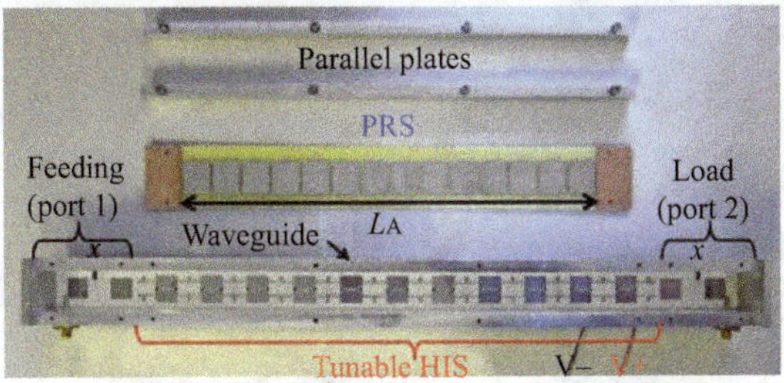

Figure 5.16 Reconfigurable FP LWA prototype [26]

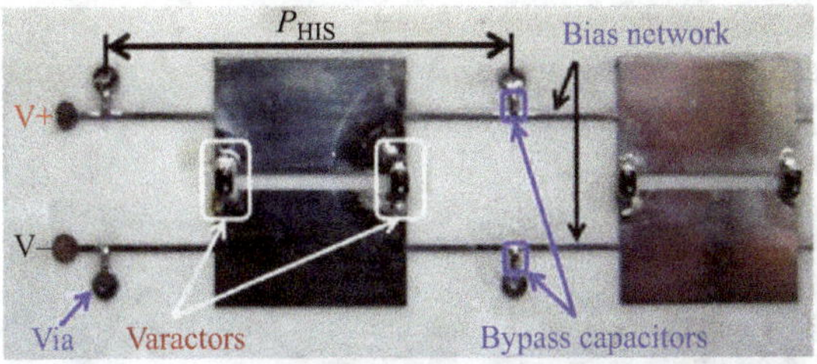

Figure 5.17 Photograph of the tunable HIS cell [26]

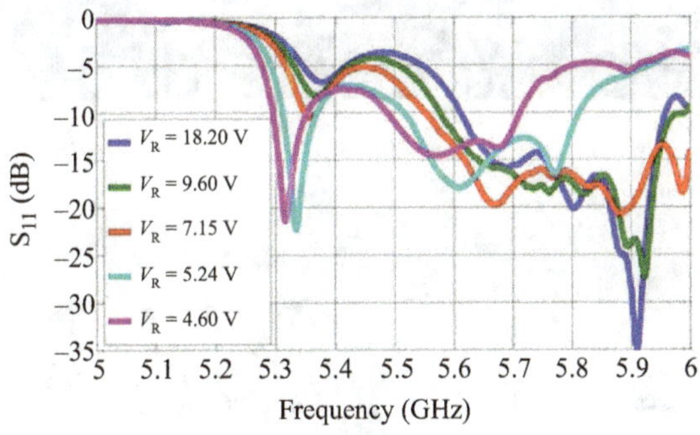

Figure 5.18 Measured reflection coefficient with frequency for different reverse bias voltage of the varactor [26]

frequencies with increase of C_j has a similar effect on the dispersion curves presented in Figure 5.15. The predicted and measured S-parameters of the antenna as a function of V_R at 5.6 GHz are shown in Figure 5.19. Both the reflection ($|S_{11}|$) and transmission ($|S_{21}|$) coefficients vary as V_R changes. Since the probe dimensions were optimized for $V_R = 7.17\ V_{DC}$, which correspond to $C_j = 0.15$ pF, poorer impedance matching can be observed at other operating points.

The experimental set-up for the measurement of the reconfigurable 1-D FP LWA radiation characteristic is shown in Figure 5.20. Figure 5.21 shows the measured normalized radiation patterns of the LWA at 5.6 GHz for different

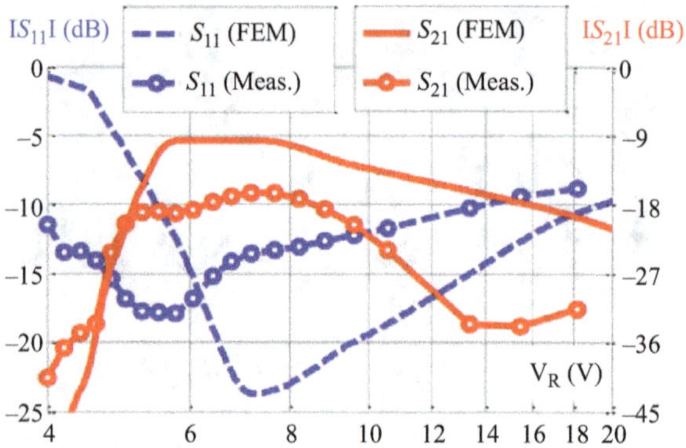

Figure 5.19 Measured S-parameters as a function of reverse bias voltage V_R at a fixed frequency (5.6 GHz) [26]

Figure 5.20 Photograph of the experimental setup of radiation pattern measurement [26]

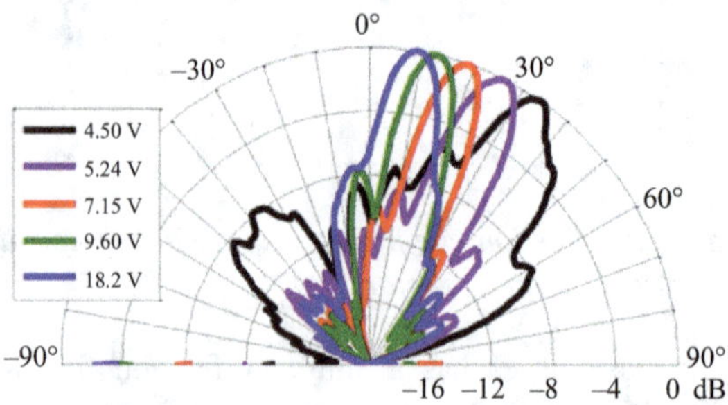

Figure 5.21 Measured normalized radiation patterns at a fixed frequency (5.6 GHz) for different reverse bias voltage of the varactor [26]

reverse bias voltages V_R of the varactor diode. The main beam scans from 9.2° to 34.2° when V_R varies from 18.2 V_{DC} (corresponds to $C_j = 0.06$ pF) to 4.5 V_{DC} (corresponds to $C_j = 0.245$ pF). A comparison between the theoretical and measured beam directions and radiation patterns is given in Figure 5.12(b) and (c),

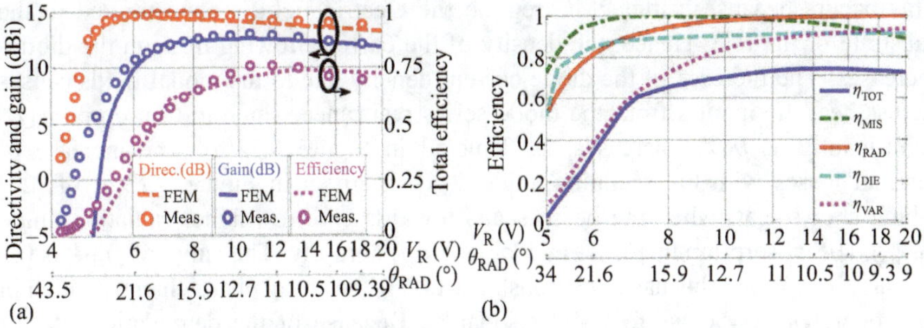

Figure 5.22 (a) Directivity, gain and total efficiency of the reconfigurable 1-D FP LWA at 5.6 GHz and (b) different estimated efficiencies as a function of V_R [26]

respectively, of [26]. The gain, directivity, and total efficiency from experiments are shown in Figure 5.22(a) together with the predicted values. The maximum measured gain is 12.95 dBi when $V_R = 10.6V_{DC}$ which corresponds to $C_j = 0.1$ pF and $\theta_{RAD} = 12°$. A significant drop in the gain curve can be observed when V_R is lower than the optimum point. For example, the measured gain for $V_R = 4.2\ V_{DC}$ is -3.55 dBi ($C_j = 0.26$ pF and $\theta_{RAD} = 35°$). This phenomenon restricts use of large scan angles. The abrupt degradation of gain is due to a higher mismatch loss (Figure 5.19) and a fall of the leakage rate (Figure 5.15(b)) because of the PMC resonance of the HIS with high θ_{RAD}. The gain is stable (greater than 10 dBi) for beam angles below 25° ($V_R > 6V_{DC}$).

The total efficiency ($\eta_{TOT} = G/D$) is greater than 50% and reaches a maximum of 75% within the range when V_R varies between 10 and $14V_{DC}$. For $V_R = 5V_{DC}$, i.e., when $\theta_{RAD} > 30°$, the total efficiency is around 3%. The total efficiency is comprised of the mismatch efficiency (η_{MIS}), LM radiation efficiency (η_{RAD}) and ohmic efficiency (η_Ω) and is expressed as

$$\eta_{TOT} = \eta_{MIS} \cdot \eta_{RAD} \cdot \eta_\Omega \tag{5.11}$$

where η_{MIS} can be computed from the measured reflection coefficient by using $\eta_{MIS} = 1 - |S_{11}|^2$ and η_{RAD} can be estimated using the leakage rate from $\eta_{RAD} = 1 - e^{-2\alpha_y L_A}$. The ohmic efficiency η_Ω is due to dielectric losses (η_{DIE}) and power dissipation in the varactor series resistance (η_{VAR}), so that (5.11) can be expressed as

$$\eta_{TOT} = \eta_{MIS} \cdot \eta_{RAD} \cdot \eta_{DIE} \cdot \eta_{VAR} \tag{5.12}$$

The reason behind the gain drop for large scan angles can be explained by (5.12). The various efficiencies are shown in Figure 5.22(b) as a function of V_R and θ_{RAD}. It can be observed that the mismatch efficiency (η_{MIS}) is greater than 85% throughout the dynamic range of the varactor. However, η_{RAD} is 100% for $V_R = 20V_{DC}$ (radiation angle $\theta_{RAD} = 9°$) and decreases to 52% when the beam points at 30° ($V_R = 5V_{DC}$). As the beam scans from 9° to 30°, η_{DIE} falls from 90% to 50%.

This occurs because in the PMC regime the electric field is concentrated in the substrate of the HIS. Hence, the density of the current flowing through the diodes increases. The increase of the diode current density due to the modal fields causes excess heat dissipation from the diode series resistance which in turn drops η_{RAD} significantly as θ_{RAD} increases, as depicted in Figure 5.22(b). The increase of thermal losses with the scanning angle θ_{RAD} is given in Figure 5.13(c) of [26] which shows that, when the beam scans towards endfire (higher radiation angle θ_{RAD}), HIS absorption increases and $|\rho_{HIS}|$ drops abruptly. The value of $|\rho_{HIS}|$ is 0.3 for $\theta_{RAD} = 34°$ and this has a relationship with HIS's strong phase change, which in turn operates very close to PMC resonance. Because of the degradation of LM radiation efficiency and increase of the varactor resistance losses, the gain drops at large scan angles which restricts beam scanning towards the endfire. However, the 1-D FP LWA can scan the beam continuously within 9° to 30° at 5.6 GHz by varying the varactor bias voltage between 18 and $5V_{DC}$.

5.4.2 *Two-dimensional (2-D) FP-reconfigurable LWA*

As discussed in Section 5.4.1, a 1-D FP LWA can scan its beam electronically along a particular plane, for example, the *x–z*-plane. This section presents a reconfigurable LWA which can scan the beam in both azimuth and elevation [37].

5.4.2.1 Antenna configuration

Figure 5.23 represents the perspective and side views of a reconfigurable FP antenna. Similar to the 1-D FP antenna the structure is also has a resonant cavity consisting of two planar periodic circuit boards at a suitable spacing between them. The antenna shown in the figure is designed for operation at 5.5 GHz, and the area of the aperture is $5.4\lambda_0 \times 5.4\lambda_0$, where λ_0 is calculated at the design frequency. A stacked patch is fed by a probe at the centre and used to excite the antenna. The element is horizontally polarized in the *x*-axis. Each varactor in the HIS is similar to an RLC circuit with a tuneable junction capacitance C_j, which allows tuning the reflection phase of the HIS. The array in the HIS is divided into four independent sectors [sectors S_A, S_B, S_C and S_D as shown in Figure 5.23(a)]. Each sector can be biased by a separate control signal allowing reconfiguration of the HIS varactors in the sector. Simultaneous azimuth and elevation scanning of a pencil beam can be achieved from the sectorized HIS array through electronic control.

5.4.2.2 Dispersion and antenna analysis

In a symmetric centrally fed 2-D FP antenna, the propagation can be an outward cylindrical leaky wave [38]. However, the main objective of this investigation was to design an FP LWA to scan a pencil beam. When this LWA operates at, or around, its resonance it produces a broadside-pointing pencil beam, and when it operates above the resonance, it scans a conical beam. The unit cell of the 2-D FP LWA is shown in Figure 5.24. The dispersion analysis can be easily done for a non-sectorized, infinite symmetrical FP antenna using the unit cell and its TEN. Details of the dispersion analysis are available in references [26,37,39]. Figure 5.25 shows the direction of the beam as a function of junction capacitance. In the

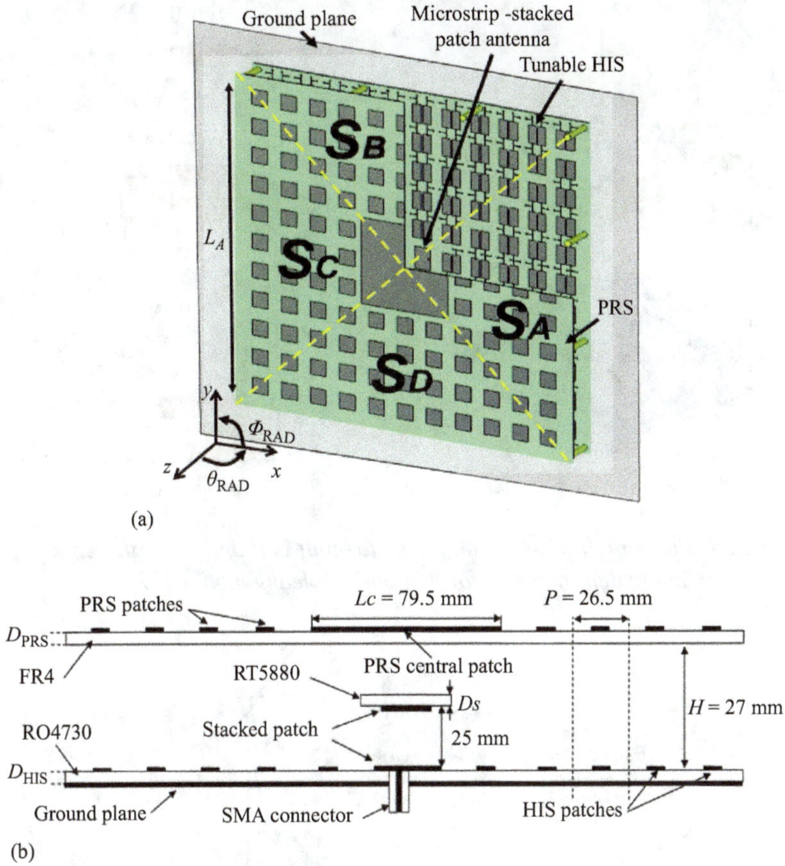

Figure 5.23 Two-dimensional (2-D) reconfigurable FP antenna: (a) 3-D model from CST and antenna side view showing the stacked patch at the centre [37]

scanning region ($C_j = 0.1$ to 0.25 pF) of Figure 5.25 the TM and TE modes propagate and the beam points at an elevation of θ_R. With an increase of the junction capacitance C_j the elevation angle θ_R increases. A cut-off region or electromagnetic bandgap (EBG) appears as C_j becomes greater than 0.25 pF as shown in Figure 5.25. This region prevents LW propagation and hence the structure does not radiate. From the behaviour of the structure, one can summarize that any sector operating in the scanning region (SR) will allow propagation and radiation of leaky waves. The azimuth angle depends on the orientation of the sector, and the elevation angle can be varied by changing the varactor junction capacitance using appropriate reverse bias voltage.

A uniformly polarized 2-D FP LWA produces a cylindrical LW and hence a conical-beam scan in the elevation plane. However, an antenna with four independent sectors can direct the LW to a desired azimuth sector. To do so, the sectors need to tune independently within the SR and EBG regions.

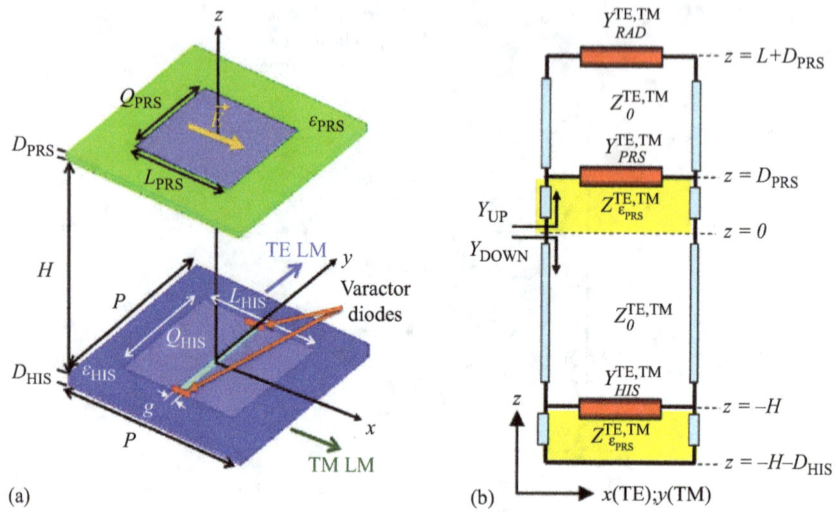

Figure 5.24 Reconfigurable FP antenna: (a) unit cell and (b) transverse equivalent network for TM and TE leaky modes [37]

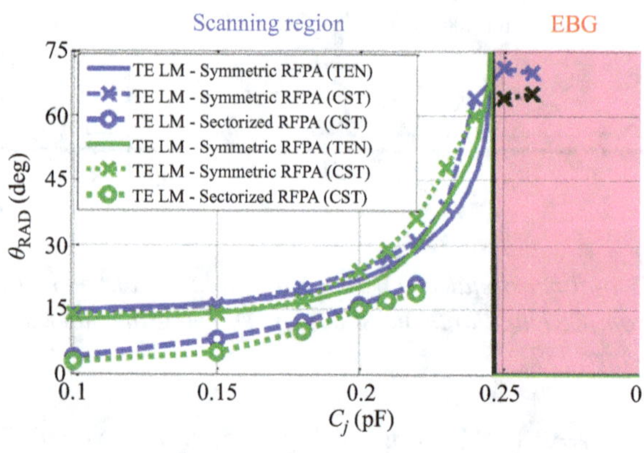

Figure 5.25 Variation of beam direction (θ_{RAD}) with the junction capacitance C_j at 5.5 GHz for TE and TM leaky-modes [37]

The controls of the sectors to achieve the LW in eight different azimuth angles are listed in Table 5.1. The eight azimuth angles φ are 0°, 45°, 90°, 135°, 180°, 225°, 270° and 315°. For example, a pure TM LW in the $+x$-axis will propagate when Section S_A is tuned to SR, while the rest of the sectors are set to EBG. The elevation angle θ_R for the azimuthal direction $\varphi = 0°$ is determined by the varactor junction capacitance C_j for the operating Sector S_A. In a similar fashion, when all the sectors are in the EBG region except S_B a pure TE radiating leaky-wave propagation will be produced along the $+y$-axis. The elevation angle θ_R for the

Table 5.1 Operating regions of the 2-D FP LWA sectors and beam scanning properties

φ	S_A	S_B	S_C	S_D	**Scanning type**
0°	SR	EBG	EBG	EBG	TM scanning
45°	SR	SR	EBG	EBG	Hybrid scanning
90°	EBG	SR	EBG	EBG	TE scanning
135°	EBG	SR	SR	EBG	Hybrid scanning
180°	EBG	EBG	SR	EBG	TM scanning
225°	EBG	EBG	SR	SR	Hybrid scanning
270°	EBG	EBG	EBG	SR	TE scanning
315°	SR	EBG	EBG	SR	Hybrid scanning
Broadside	SC	SC	SC	SC	No scanning

azimuth angle $\varphi = 90°$ is dependent on the operating point of Sector S_B. When two adjacent sectors are in the SR state, and the other two sectors are in the EBG state the azimuthal plane for the beam points between the sectors operating in SR. For instance, if S_C and S_D are in EBG and S_A and S_B are in SR the propagation is around the $\varphi = 45°$ azimuthal direction depending on the SR operating point of each sector.

A cylindrical LW at broadside ($\theta_R = 0°$) radiation can be achieved when all four sectors are set to a splitting condition (SC). By varying the operating point of an active propagating sector a pencil beam can be scanned continuously in the elevation plane for a constant azimuth angle. A scanning range of $\theta_R = [5°, 25°]$ is obtained in the elevation plane when C_j varied between 0.1 and 0.23 pF as shown in Figure 5.25 for the TM ($\varphi = 0°$) and the TE ($\varphi = 90°$) polarization cuts.

5.5 Experimental results

Figure 5.26(a) shows the S-parameter measurement setup of the 2-D FP LWA prototype. The tuneable HIS is shown in Figure 5.26(b), where different parts are marked. Similarly to HIS for the 1-D FP LWA, two varactor diodes (MGV125-20; 0805-2 package) are used in each patch. The junction capacitance C_j of the varactor varies from 0.1 to 1 pF for an applied variable reverse bias voltage from 20 to $2V_{DC}$.

The measured normalized 2-D radiation patterns for the TE, Hybrid and TM operating regimes are shown in Figure 5.27. When Sector B (S_B) is tuned to SR (applied reverse bias voltage of $8.53V_{DC}$ and corresponding $C_j = 0.21$ pF) and all other sectors are in the EBG region (applied reverse bias voltage $2V_{DC}$ and corresponding $C_j = 1$ pF) the TE LW propagates and the corresponding 2-D radiation pattern is shown in Figure 5.27(a). The generated pencil beam points at $\varphi_R = 86°$, which is indeed very close to $\varphi_R = 90°$, and the elevation angle is $\theta_R = 17°$.

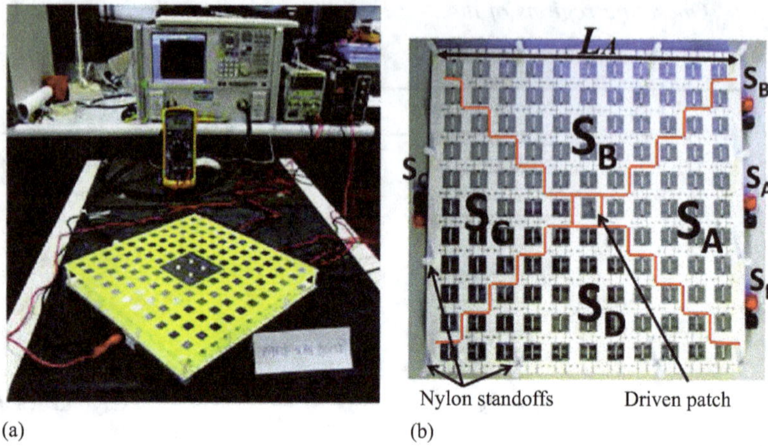

Figure 5.26 Fabricated prototype: (a) S-parameter measurement setup and (b) Tunable HIS with marked sectors [37]

The TM configuration is shown in Figure 5.27(c) when Sector A (S_A) is in the SR region and the rest of the sectors are in the EBG region. As expected, the pencil beam points very close to $\varphi_R = 0°$ (measured direction is 4°) and the elevation angle is $\theta_R = 11°$. In the hybrid configuration shown in Figure 5.27(b) the beam points in an intermediate position between the two sectors. For example, when S_A and S_B are in the scanning region and S_C and S_D are in the EBG region the beam points at $\varphi_R = 30°$ with corresponding $\theta_R = 18°$. Due to the asymmetrical dispersion response between the TM and TE leaky modes (shown in Figure 5.25) a non-identical dispersion response for the same SR operating point is found, for example, for the biasing voltage of $8.53V_{DC}$ $\theta_R = 11°$ and 17° for the TM and TE mode, respectively. To obtain hybrid LM scanning at $\varphi_R = 45°$, the sectors need to be tuned asymmetrically in the SR region to make a symmetric dispersion response, i.e., the propagation constant β should be similar along both axes, in this case, the x- and y-directions.

Measured and predicted normalized radiation patterns for the beam scanning in the elevation plane (different θ_R) are shown in Figure 5.28(a)–(c) for various azimuth direction (φ_R). As it is obvious from the previous analyses, for beam scanning in either the TE or TM regime from a particular sector, the sector needs to operate in the SR region with the rest of the sectors in the EBG region. Once the condition is satisfied, the reverse bias voltage is varied, i.e., thus vary the C_j of the sector operating in the SR region. For hybrid scanning, C_j should be varied simultaneously for both of the sectors operating in SR. Hybrid, TE and TM scanning in the elevation are given in Figure 5.28(a)–(c) with corresponding varactor reverse bias voltages. The broadside beam is shown in Figure 5.28(d) for $\varphi_R = 0°$ (E-plane) and 90° (H-plane) when all four sectors are tuned to SC. Detailed discussions on the gain and efficiency of the 2-D FP LWA are available in [37].

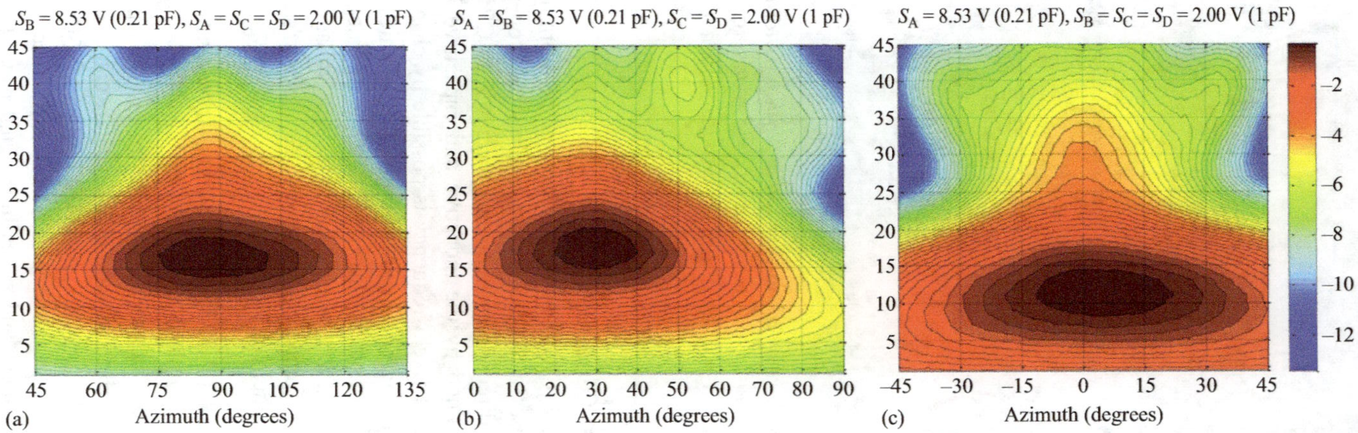

Figure 5.27 Measured normalized 2-D radiation pattern of the FP LWA prototype at the fixed operating frequency (5.5 GHz): (a) TE scanning (beam direction $\varphi_R = 87°$; $\theta_R = 17°$); (b) hybrid scanning (beam direction $\varphi_R = 30°$; $\theta_R = 18°$); and (c) TM scanning (beam direction $\varphi_R = 4°$; $\theta_R = 12°$) [37]

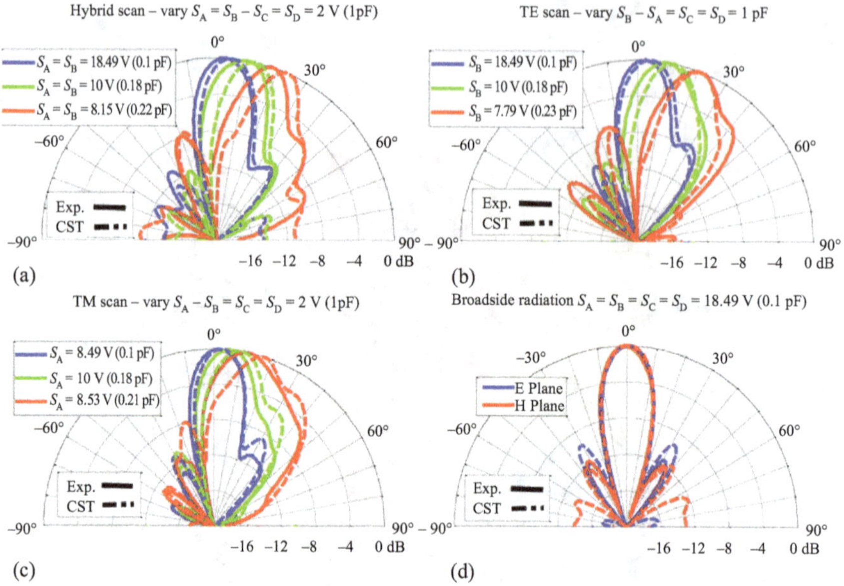

Figure 5.28 Predicted and measured normalized radiation patterns at the fixed operating frequency (5.5 GHz): (a) hybrid scanning; (b) TE scanning; (c) TM scanning; and (d) broadside beam [37]

5.5.1 CRLH-based reconfigurable LWA

As discussed in Section 5.3.2, CRLH-based LWAs scan the beam in the backward direction at low frequencies. Since the structure exhibits LH properties, the beam scans towards the forward direction as frequency increases. In CRLH LWAs, the phase constant β is determined by the inductive and capacitive loading on the structure. Beam scanning at a fixed frequency is possible by changing the inductance or capacitance loading and hence by changing β dynamically in a CRLH structure. The principle of shifting the dispersion curve electronically is shown in Figure 5.29 [24]. It can be seen that by controlling the bias voltage different value of β can be achieved at a fixed frequency. Varactor diodes can be a suitable candidate for the voltage-controlled application.

5.5.1.1 Reconfigurable CRLH LWA for fixed-frequency beam scanning

A varactor-loaded CRLH LWA is proposed in [40] to scan the antenna beam at a fixed frequency, from backward to forward direction. The unit cell equivalent circuit, including the varactor bias network, is shown in Figure 5.30 and the antenna prototype is shown in Figure 5.31. The LH properties are realized in this structure by using shorted-stub inductances and series interdigital capacitances as in [41]. Moreover, the reactance of the left-handed inductance, i.e., the inductance from the

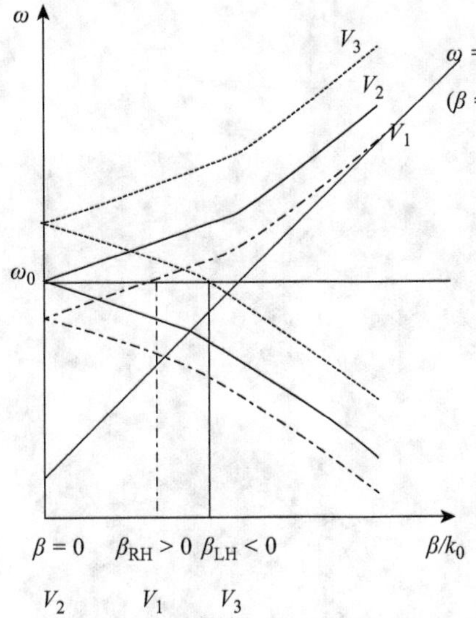

Figure 5.29 Electronic beam scanning principle: vertical shifting of dispersion curve with variable bias voltage of a uniform biased [24]

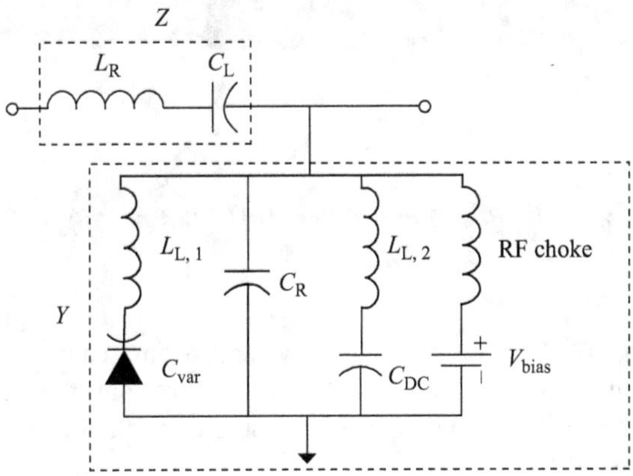

Figure 5.30 Equivalent circuit model of a reconfigurable CRLH LWA unit cell [40]

metallic via, was tuned using a varactor diode. In the unit-cell equivalent circuit the series impedance Z remains similar to the unit cell for the passive CRLH TL unit cell discussed previously and shown in Figure 5.7(b). However, the shunt admittance changes according to the reconfiguration arrangement which includes the varactor diode junction capacitance and the dc bias arrangement. Since the antenna

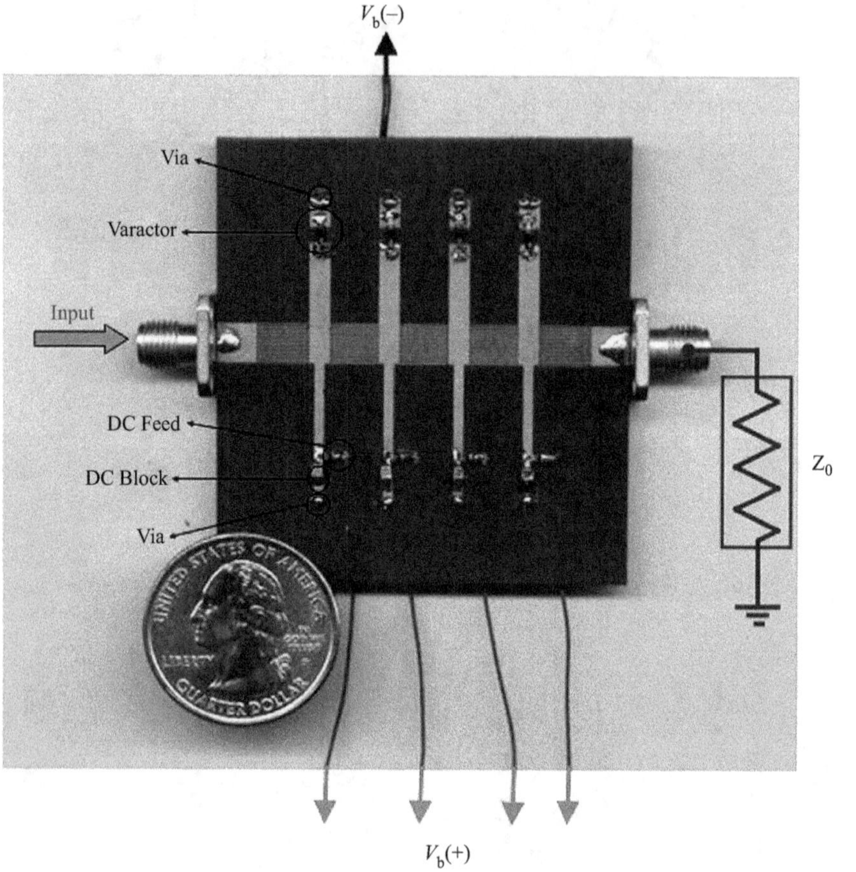

Figure 5.31 Reconfigurable CRLH LWA prototype [40]

is designed for a fixed operating frequency, and beam scanning is achieved by modulating the phase constant dynamically through varying the varactor bias voltage, the direction of the main beam can be expressed as the function of the dc bias voltage as follows [40]

$$\theta(V) = \arcsin\left[\frac{\beta(V)}{k_0}\right] \tag{5.13}$$

where $j\beta = \sqrt{Z'Y'(V)}$ and the values of Z' and Y' can be calculated from the following expressions

$$Z' = \frac{\left(j\omega L_R + \dfrac{1}{j\omega C_L}\right)}{d} \tag{5.14}$$

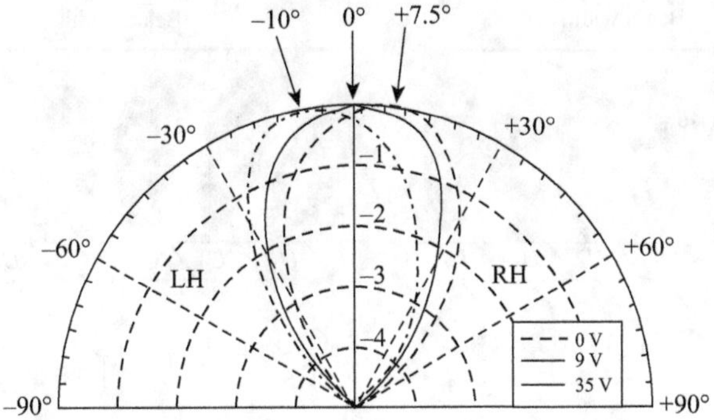

Figure 5.32 Measured radiation patterns of the reconfigurable CRLH LWA prototype [40]

and

$$Y'(V) = \frac{\left(\dfrac{1}{j\omega L_{L,1} + \frac{1}{j\omega C_{\mathrm{var}}(V)}} + \dfrac{1}{j\omega L_{L,2} + \frac{1}{j\omega C_{DC}}} + j\omega C_R\right)}{d} \tag{5.15}$$

where d is the period of the CRLH LWA, C_{var} (V) is the capacitance of the varactor which is tuneable by controlling the bias voltage, and C_{DC} is the capacitance from the DC block.

Metelics MSV34069-E28X varactors were used in the prototype with a variable reverse bias voltage from 0 to 35 V. The antenna was designed for operation at 3.23 GHz. The measured radiation patterns of the reconfigurable CRLH prototype are shown in Figure 5.32. It can be seen from the radiation patterns that the antenna is capable of scanning the beam at a fixed frequency (3.23 GHz) from −10° to 7.5°; when the applied reverse bias voltage across the varactor varies from 35 to 0 V_{DC}.

5.5.1.2 CRLH LWA with tunable beamwidth and radiation angle

The reconfigurable CRLH LWA discussed in Section 5.5.1.1 can scan the beam at a fixed frequency when the bias voltage is varied. Another reconfigurable CRLH LWA is proposed in [24] where both beam direction and beamwidth can be changed electronically. The principle of this beamwidth tuning is shown in Figure 5.33. When the same biasing is applied to all the unit cells in the structure, each unit cell has the same propagation constant, resulting in maximum directivity. On the other hand, for non-uniform biasing each unit cell has different propagation constant depending on the applied biasing voltage across the varactor diode in each unit cell. The radiation patterns of the structure are determined by the response of each unit cell in the structure. It is worth mentioning here that in a passive structure it is almost impossible to change the propagation constant of each unit cell independently once a structure is realized. However, the reconfigurable LWA enables the flexibility to control the propagation constant.

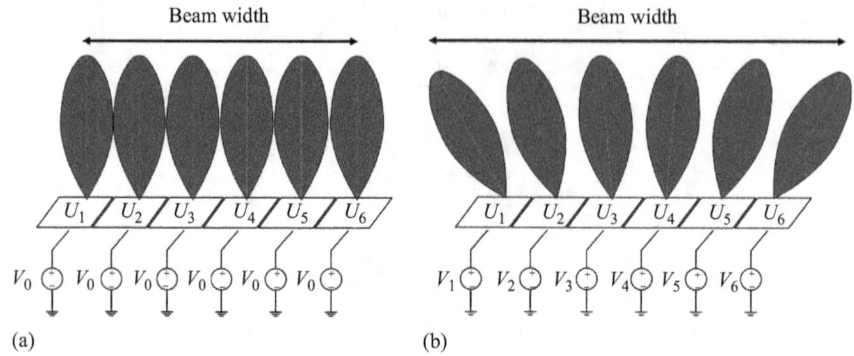

Figure 5.33 Beamwidth tuning principle: (a) when uniform biasing is applied results a narrow beamwidth and (b) when nonuniform biasing is applied results a wider beamwidth [24]

Figure 5.34 shows the reconfigurable CRLH TL unit cell layout and corresponding equivalent circuit model. The reconfigurable CRLH LWA prototype is shown in Figure 5.35. A continuous beam scan from 50° to −49° can be achieved from the prototype. The measured beam-scanning range of the prototype is shown in Figure 5.36 together with the theoretical scanning angle with respect to the varactor reverse bias voltage. Figure 5.37 shows the antenna radiation patterns for both uniform and nonuniform biasing case. In the case of nonuniform biasing the applied voltages on each unit cell is represented on the top of Figure 5.37 and the uniform bias voltages are in the inset. It can be seen that the half-power beamwidth of the nonuniform biasing is wider than with uniform biasing.

5.5.2 Reconfigurable half-width microstrip LWA

As described in Section 5.3.3, one edge of the microstrip line of an HW-MLWA is shorted to the ground plane, and the other edge is free. It is evident from previous investigations that in a fixed-frequency beam steering LWA k_0 is fixed, and in order to achieve beam scanning we need to change the effective phase constant β. From the previous research, it is known that the effective β of a microstrip line can be changed by controlling the reactance profile of an LWA [42]. In an HW-MLWA the open microstrip edge provides great opportunities to reconfigure the edge and hence to control the reactance profile of the microstrip line dynamically. This opportunity is used to design several HW-LWAs to achieve beam scanning at a fixed frequency [25,42,43]. The reconfigurable HW-MLWA proposed in [25] uses 24 reconfigurable unit cells. Each unit cell consists of a gap capacitor, realized by placing a small patch very close to the free edge of the microstrip, and a binary switch to control it. Since there are a large number of unit cells in the antenna design and each unit cell has two different states, to simplify the analysis a multistate macrocell concept is developed for a reconfigurable unit cell. Detailed analysis on the reconfigurable HW-MLWA design is given in the following subsections.

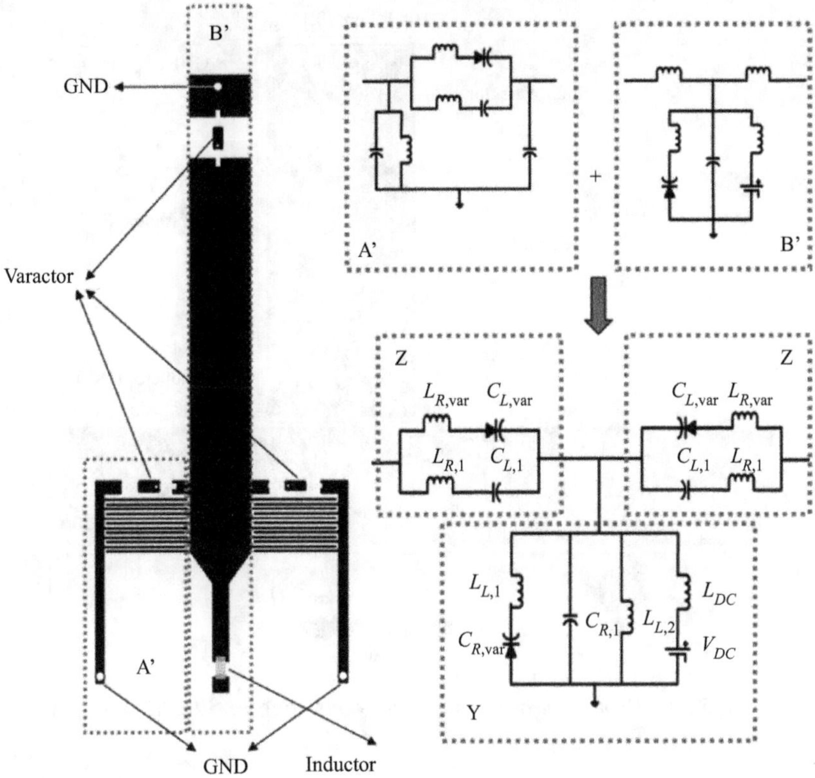

Figure 5.34 CRLH unit cell layout (left) and corresponding equivalent circuit model (right) [24]

5.5.2.1 Analysis of 1-D reconfigurable structure using multistate macrocell

In order to analyse the antenna, a bigger cell is considered, called a 'macrocell'. The concept of 'macrocell' in a binary reconfigurable structure helps to simplify the analysis by reducing the number of configurations. Figure 5.38(a) shows the orientation of 'macrocells' in a generic 1-D reconfigurable periodic structure, and the orientation of unit cells in a 'macrocell' is shown in Figure 5.38(b). It is obvious from the name that each binary reconfigurable unit cell has two states, which can be controlled using a binary switch in the unit cell that has two states either 'OFF' or 'ON' states. For simplicity, the 'OFF' and 'ON' states of a binary switch are denoted here with '0' and '1', respectively. Let's consider a structure with N reconfigurable binary unit cells. For a particular complete binary pattern (CBP) of the whole structure, some of the adjacent patterns can be considered as a group ('macrocell'). Then the CBP can be represented as a repetition of the macrocell in cascade. If M is the number of unit cells in a macrocell, $M << N$. The macrocell presented in Figure 5.38(b) consists of four unit cells, and the states of the binary

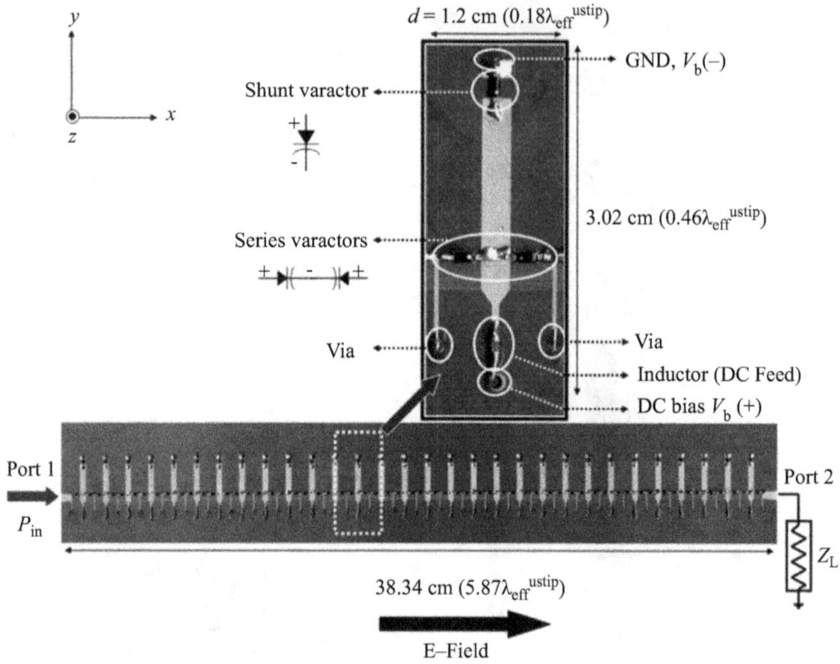

Figure 5.35 Prototype of the structure consisting to 30 unit cells [24]

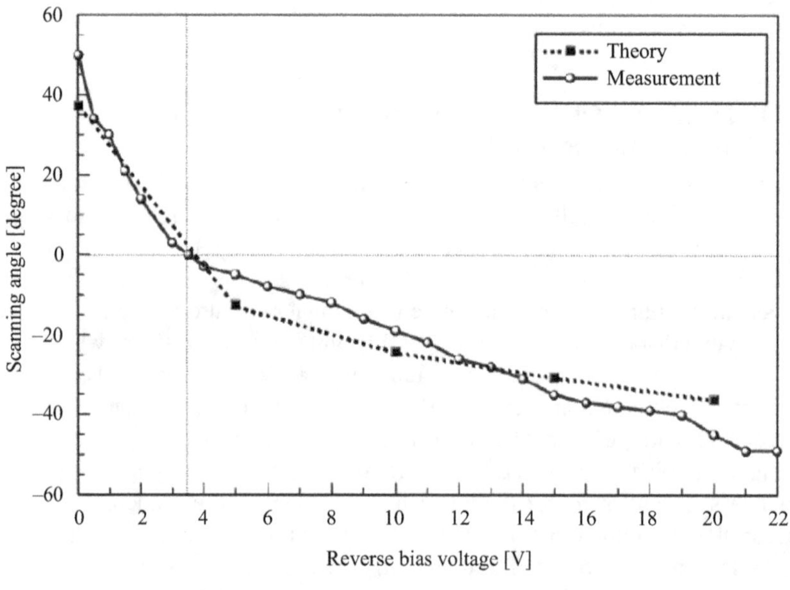

Figure 5.36 Scanning angle (measured and theoretical) with different applied reverse bias voltage [24]

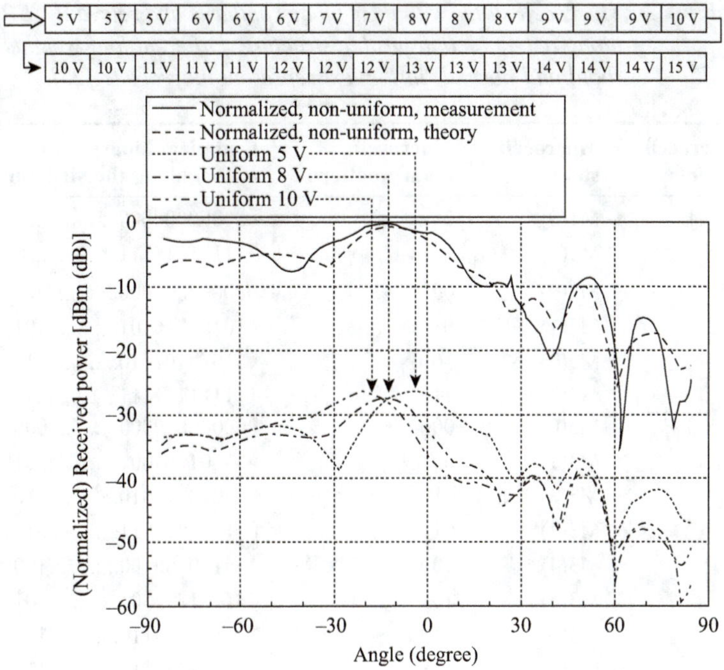

Figure 5.37 Normalized radiation patterns (measured and theoretical) for uniform and nonuniform biasing of unit cells [24]

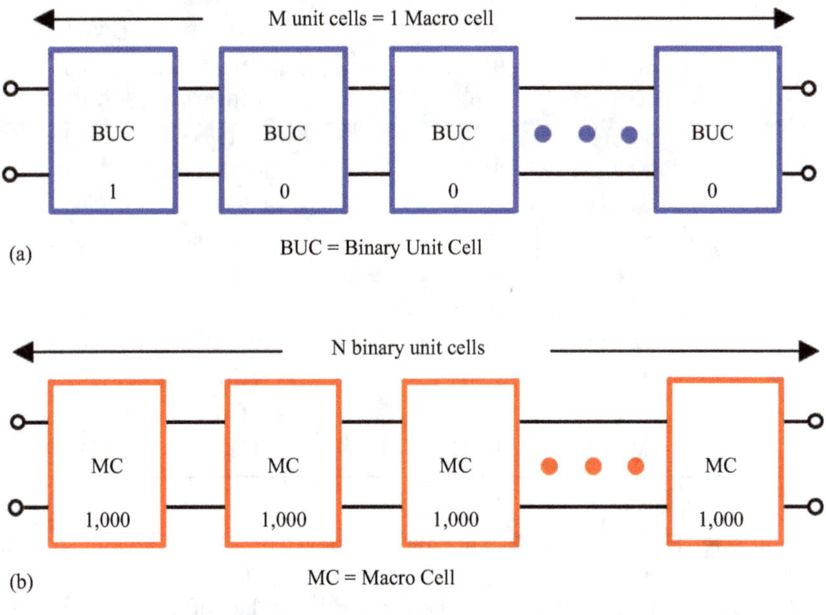

Figure 5.38 Generic 1-D periodic reconfigurable structure: (a) organization of 'macrocells' in a complete structure and (b) organization of unit cells in a macrocell [25]

Table 5.2 Macrocell states with their binary patterns and relevant CBPs along a 1-D reconfigurable structure. Different font weights (bold and normal) in the text represent macrocells [25]

Macrocell order	Macrocell state	Macrocell binary pattern	Complete binary pattern (CBP) along the structure
$M = 1$	M1(0)	0	000000000000..............0
	M1(1)	1	111111111111..............1
$M = 2$	M2(0)	00	000000000000.............00
	M2(1)	01	010101010101.............01
	M2(2)	10	101010101010.............10
	M2(3)	11	111111111111.............11
$M = 3$	M3(0)	000	000000000000...........000
	M3(1)	001	001001001001...........001
	M3(2)	010	010010010010...........010
	M3(3)	011	011011011011...........011
	M3(4)	100	100100100100...........100
	M3(5)	101	101101101101...........101
	M3(6)	110	110110110110...........110
	M3(7)	111	111111111111...........111
$M = 4$	M4(0)	0000	000000000000..........0000
	M4(1)	0001	000100010001..........0001
	M4(2)	0010	001000100010..........0010
	M4(3)	0011	001100110011..........0011
	M4(4)	0100	010001000100..........0100
	M4(5)	0101	010101010101..........0101
	M4(6)	0110	011001100110..........0110
	M4(7)	0111	011101110111..........0111
	M4(8)	1000	100010001000..........1000
	M4(9)	1001	100110011001..........1001
	M4(10)	1010	101010101010..........1010
	M4(11)	1011	101110111011..........1011
	M4(12)	1100	110011001100..........1100
	M4(13)	1101	110111011101..........1101
	M4(14)	1110	111011101110..........1110
	M4(15)	1111	111111111111..........1111

unit cells are 1, 0, 0, and 0. This makes the CBP of the whole structure 10001000……1000.

If M is the order of a macrocell, the macrocell can be defined as a multistate electromagnetic element with a total of 2^M states. In Table 5.2, the CBPs of a reconfigurable periodic structure consisting of 24 binary unit cells are presented for $M = 1, 2, 3$ and 4. The first column of Table 5.2 shows the *i*th order of the macrocell as $M = \mathrm{i}$ and the second column shows the *j*th state of the corresponding order as $\mathrm{M}_{i(j)}$. To represent a complete structure as a periodic macrocell the orders are chosen in such a way that *N/M* is an integer. From all the discussions so far it is obvious that the CBP of a reconfigurable structure consists of periodic macrocell patterns, for example, **1000**1000...**1000** [in Figure 5.38(a)]. However, the binary patterns in a macrocell can be either aperiodic or periodic based on the state of the switch in each unit cell. If we consider the M4(10) and M4(2) macrocell states in Table 5.2, the binary pattern of the M4(10) macrocell state is 1010, which is periodic. On the other hand, the binary pattern of the M4(2) macrocell state is 0010, which is not periodic.

In an LWA design, the complex propagation constant is the key factor. When the macrocell approach is applied in a reconfigurable LWA design using binary unit cells, each macrocell has a corresponding average β and average α, and the direction of the main beam is related to the effective β of the structure as discussed before. With a change of the macrocell state the average reactance profile along the structure changes and hence the phase velocity and the β change. For a fixed frequency, the M1(0) and M1(1) microcell states determine the beam-scanning limits. Other beam positions within the scanning range can be obtained by changing the macrocell states at a fixed frequency, i.e., changing the binary patterns.

The macrocell approach to a 1-D reconfigurable LWA helps greatly to reduce the number of configurations during the analysis because the higher-order macrocells cover some lower-order macrocell with the same electromagnetic (EM) properties. For example, identical EM characteristics can be observed for the M4(10) and the M2(2) states. Moreover, in a structure with a large number of unit cells, the EM properties of the structure do not change significantly when the bits in a macrocell shift their positions. For example, M4(1), M4(2), M4(4) and M4(8) have similar EM properties and hence similar beam directions. From the discussions, it is obvious that a number of states can be omitted and the analysis of a reconfigurable structure consisting of binary unit cells can be simplified using this approach.

5.5.2.2 Antenna design using binary unit cells

The antenna configuration is shown in Figure 5.39. It has a total of 24 reconfigurable binary unit cells. The design and analysis were conducted using CST Microwave Studio. In each unit cell, there is a small patch placed at the open microstrip edge leaving a narrow gap (*g*) of 0.1 mm. The gap *g* between the microstrip and the patch creates a gap capacitor. A binary switch connects the patch to the ground plane by a via in the switch's ON state. The switch can be turned ON or OFF to connect or disconnect the gap capacitor to the ground plane. Since the input impedance of the structure changes with the macrocell states, different feed methods were investigated during the design process to achieve good impedance matching for different macrocell states. It was found that good impedance matching for most of the macrocell states can be obtained with a simple narrow uniform feed

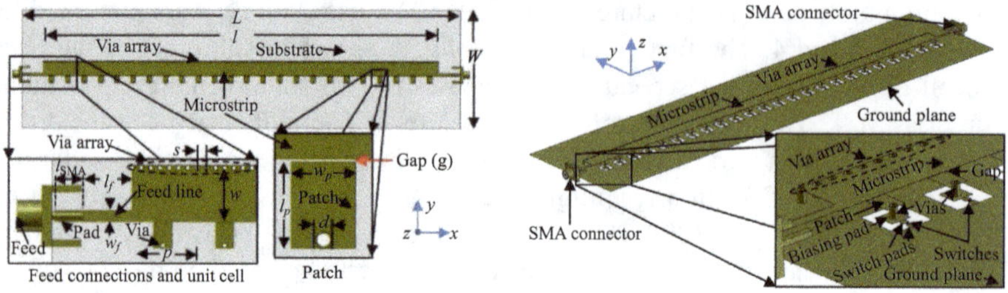

Figure 5.39 HW-MLWA consisting of reconfigurable binary unit cells: (a) top view with detailed feed and unit cell and (b) perspective view omitting the substrate (the structure is not to scale) [25]

line located at the open microstrip edge. The LWA is fed from one side using simple edge feeding technique and the other side of the antenna is terminated in a 50 Ω load to avoid reflected waves.

The antenna was designed on a Rogers RT5880 substrate that has a loss tangent of 0.0009 and dielectric constant of 2.2, and the thickness of the substrate is 1.575 mm. The overall length of the radiating microstrip and the complete structure are $4.8\lambda_0$ and $5.32\lambda_0$, respectively, where the free-space wavelength λ_0 is calculated at the fixed operating frequency of 6 GHz. Detailed geometric parameters of the antenna are given in [25].

5.5.2.3 Analysis of the LWA using macrocell states

From the antenna design we already know that, for the reconfigurable HW-MLWA in Figure 5.39, $N=24$. Different macrocell states were applied to observe the performance of the LWA. From the previous discussion in Section 5.5.2.1 on the selection of the macrocell states of a 1-D reconfigurable periodic structure, we can find that the possible macrocell orders in the current LWA structure are $M=1$, 2, 3, 4, 6, 8 and 12. For instance, $M=3$ indicates that three conjugative unit cells form a macrocell, and hence using 8 such macrocells the whole structure can be represented. In Table 5.2, all the CBPs corresponding to $M=1$ to $M=4$ are already given. It is very easy to notice that all the states generated by $M=1$, 2, and 4 are covered by the macrocell with order $M=8$. Similarly, the states generated by $M=1$, 2, and 3 are covered by $M=6$. A total of $2^8+2^6-2^2-2^1=314$ different states can be obtained from $M=1$, 2, 3, 4, 6 and 8. The states $M=1$ and $M=2$ are removed from the sum since they are duplicates. The antenna characteristics were investigated for the 314 different configurations to validate the multistate microcell concept.

When $M=4$, there are 16 different possible CBPs (as $2^4=16$) as given in Table 5.2. The antenna is not properly matched for all the CBPs applied along the structure. At the operating frequency of 6 GHz, a reflection coefficient

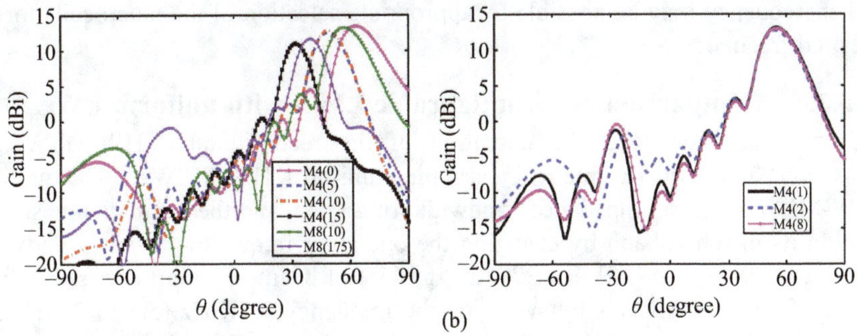

Figure 5.40 Predicted radiation patterns of the HW-MLWA with: (a) six macrocell states and (b) three macrocell states [25]

Table 5.3 Selected five different macrocell states with measured and predicted main beam directions for each state at 6 GHz [25]

Macro-cell state	Complete binary pattern (CBP)	Main beam direction	
		Predicted	**Measured**
M1(1)	111111111111111111111111 (CBP1)	32°	31°
M8(175)	101011111010111110101111 (CBP2)	39°	38°
M2(2)	101010101010101010101010 (CBP3)	49°	47°
M8(10)	000010100000101000001010 (CBP4)	57°	53°
M1(0)	000000000000000000000000 (CBP5)	63°	60°

<−10 dB was observed for 7 different CBPs. Figure 5.40(a) shows the predicted radiation patterns for four selected order $M = 4$ microcell states and two patterns of order $M = 8$. It can be seen that not all the radiation patterns are different. For example, the patterns for M4(10) and M4(5) are very similar because of their similar EM characteristics. Very similar radiation patterns can be achieved by microcells with similar EM properties which can be seen from the radiation patterns shown in Figure 5.40(b) for M4(1), M4(2) and M4(8). The main beam points at the same position for all the three states. The maximum and minimum beam angles are obtained from the M4(0) and M4(15) states. Out of the $M = 1$, 2, 3, 4, 6 and 8 macrocell orders, five different CBPs were selected based on the best input impedance matching and beam position, to observe the beam scanning at the desired frequency. The five selected macrocell states and their corresponding CBPs are listed in Table 5.3, with the beam directions obtained from simulation and measurement. Though the antenna design shown here was for a particular frequency and the macrocell states were selected accordingly. Beam scanning at some

other frequency may be possible by appropriate selection of the macrocells for the desired frequency.

5.5.2.4 Comparison of reconfigurable LWA with uniform LWA

In order to understand the operation of the reconfigurable HW-MLWA, it was compared with a frequency-scanning uniform HW-MLWA. The uniform HW-MLWA has an impedance bandwidth of 2 GHz, and the beam can be steered within its matched band by changing the operating frequency. It was found from the investigations in [25] that 37° (26° to 63°) of beam scanning is possible from the uniform HW-MLWA by sweeping the frequency 1.5 GHz (5.5 to 7 GHz). In uniform LWAs where beam scanning is achieved by sweeping the frequency, the energy radiates at low frequencies from a portion of the leaky structure which is close to the feed point and the main beam points close to the broadside direction. In other words, a short portion of the leaky structure works as an effective radiating aperture. A short effective aperture widens the beam and hence the directivity and gain reduces. With an increase of frequency the energy travels further towards the other end of the structure and the effective radiating aperture become larger. A long effective aperture reduces the beamwidth, and hence the directivity and gain increase. At higher frequencies, the main beam points away from the broadside direction, i.e., close to the endfire. For a better understanding of these characteristics of an LWA, the surface current distributions are shown in Figure 5.41. Figure 5.41(a) and (b) display the surface current distributions of the uniform HW-MLWA at low and high frequencies, when the main beam points at 32° and 63°. It can be seen that the effective radiating aperture is short at the lower frequency and long at the higher frequency. The surface current distributions of the reconfigurable HW-MLWA are shown in Figure 5.41(c) and (d) for the M1(1) (corresponds to CBP1) and M1(0) (corresponds to CBP5) states at 6 GHz, when the main beam points at 32° and 63°, respectively. Comparing the surface current distributions of both the uniform frequency-scanning and the reconfigurable fixed-frequency beam scanning HW-MLWAs it is clear that the effective radiating aperture of a reconfigurable HW-MLWA can be changed by changing the binary pattern applied to the antenna structure.

5.5.2.5 Measured results

The top view of the HW-MLWA prototype is shown in Figure 5.42. In order to prove the concept PEC are used in the simulation to represent the ON state of a binary switch and a small gap is left open to represent the OFF of the switch. For the measurement, a copper tape is used for switches which are ON, and the gap is left open for switches which are OFF. PIN diodes can be used as a switch, and the performance of the antenna was investigated using the actual PIN diode parameters; details can be found in [25]. The predicted reflection coefficients from the full-wave simulation at 6 GHz for five different CBPs are listed in Table 5.4 together with the corresponding measured ones. Predicted and measured gains for these binary patterns are also listed in the same table (Table 5.4). The antenna has a very stable gain within its beam-scanning range. The measured radiation patterns in

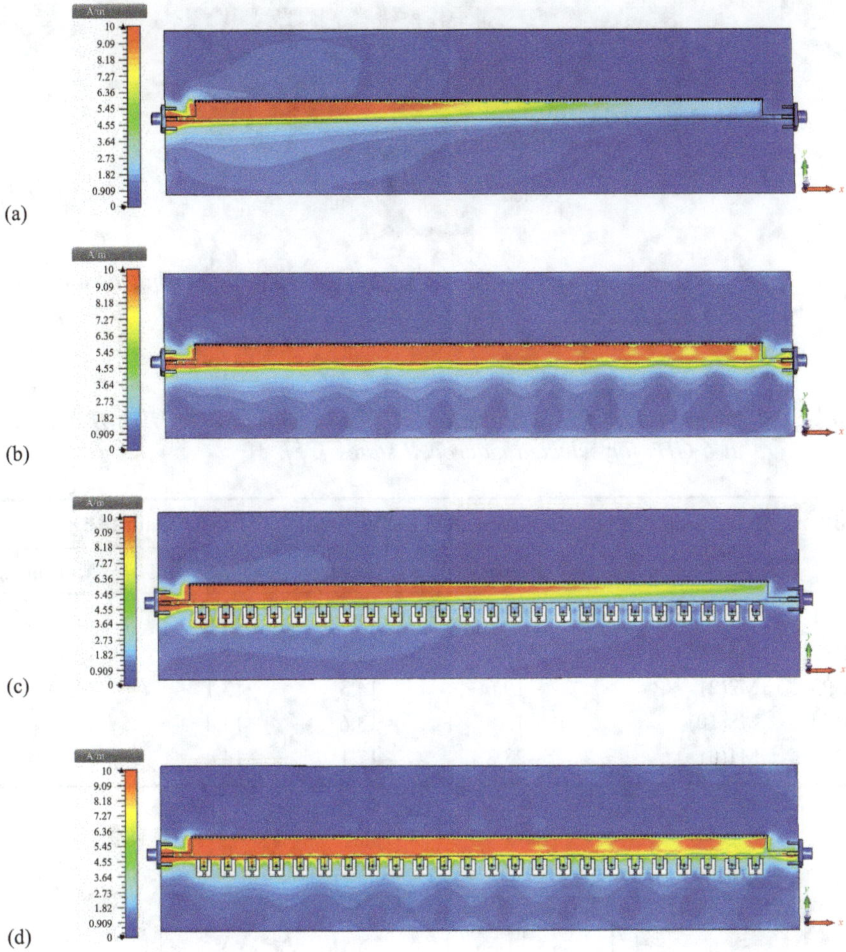

Figure 5.41 Distributions of the surface current (amplitude) of uniform and reconfigurable HW- MLWAs: (a) uniform LWA operating at 5.66 GHz; (b) uniform LWA operating at 7 GHz; (c) reconfigurable LWA operating at 6 GHz (CBP1); and (d) reconfigurable LWA operating at 6 GHz (CBP5) [25]

the H- and E-planes are shown in Figures 5.43 and 5.44, respectively. It can be seen that the antenna produces a beam which is narrow in the H-plane and wider in the E-plane. The antenna has a measured beam-scanning range of 29° at 6 GHz. The measured beam positions for different CBPs are already listed in Table 5.3. The reconfigurable HW-MLWA discussed here represents a very simple antenna configuration to achieve beam scanning at a fixed frequency by using only two values of bias voltage.

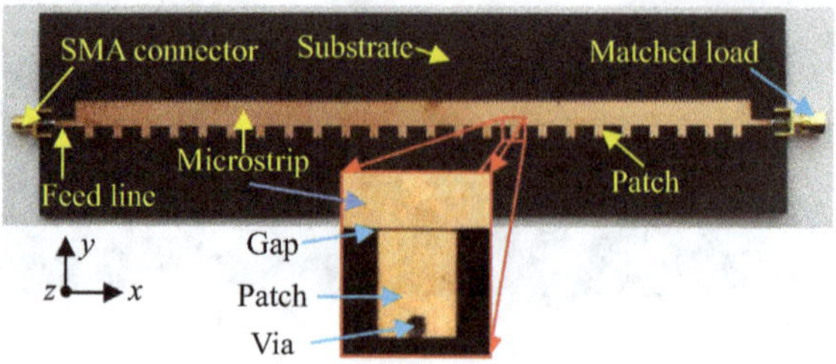

Figure 5.42 Top view of the HW-MLWA prototype consisting of binary unit cells [25]

Table 5.4 Reflection coefficient and gains of the reconfigurable HW-MLWA at 6 GHz for selected macrocell states [25]

CBP	Macro-cell state	$\|S_{11}\|$ (dB)		Gain (dBi)	
		Predicted	**Measured**	**Predicted**	**Measured**
CBP1	M1(1)	−14.2	−11.1	11.8	11.7
CBP2	M8(175)	−14.6	−11.8	11.9	12.1
CBP3	M2(2)	−19.6	−14.3	13.1	12.1
CBP4	M8(10)	−13.8	−13.6	13.4	12.5
CBP5	M1(0)	−23.8	−17.1	14.4	12.9

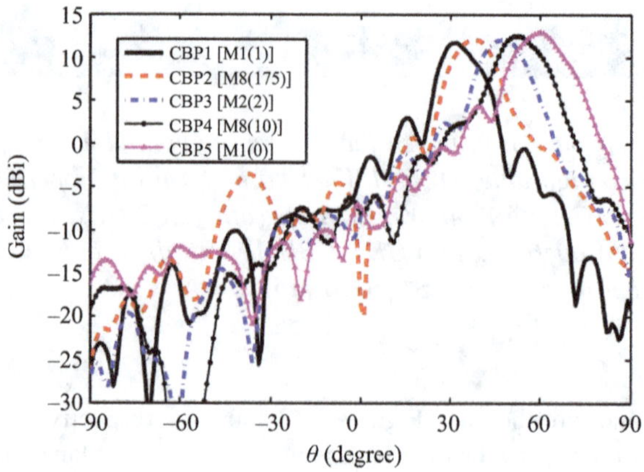

Figure 5.43 Measured H-plane radiation patterns of the reconfigurable HW-MLWA at 6 GHz [25]

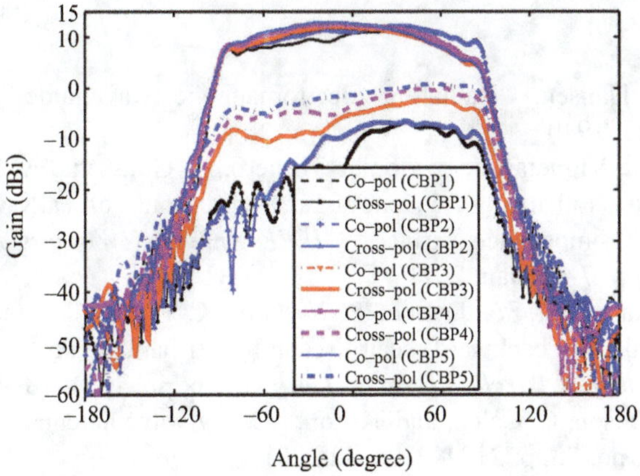

Figure 5.44 Measured E-plane radiation patterns of the reconfigurable HW-MLWA at 6 GHz [25]

5.6 Conclusion

Since the invention of waveguide-based LWAs, this field has been an active research area due to their exceptional inherent beam scanning property. When microstrip-based LWAs were developed, the popularity increased significantly because of the wide-range beam scanning facilities from a simple planar structure. Though much research has been conducted on LWAs, most of it is focused on frequency scanning, i.e., scanning the antenna beam by sweeping frequency. However, the main limitation of LWAs is that a wide range of beam scanning usually requires a broad range of frequency sweep, thus limiting their applications. For narrowband applications, steering the antenna beam at a fixed frequency is highly desirable, especially for wireless communications systems. Some research has been conducted on LWAs for beam scanning at a fixed frequency. One possible approach with reconfigurable LWAs was discussed in this chapter. This includes reference antenna designs as well as their basic working principles. There is significant research scope in the field of reconfigurable LWAs for fixed-frequency operations, especially to improve the beam-scanning range with stable gain. From the discussions in the chapter, it was found that one of the main factors limiting the beam scanning is the losses in the active electronic component used in the design. Components with lower losses in high frequencies may improve the fixed-frequency beam scanning. New methods can be developed to control the phase constant β dynamically for a wider range using less electronic components. Furthermore, the fixed-frequency beam scanning technique can be extended to millimetre-wave or terahertz (THz) on-chip systems with more efficient antenna design for next-generation applications.

References

[1] W. W. Hansen, "Radiating electromagnetic waveguide," U.S. Patent 2.402.622, 1940.

[2] M. Garcia-Vigueras, J. L. Gomez-Tornero, G. Goussetis, A. R. Weily, and Y. J. Guo, "Enhancing frequency-scanning response of leaky-wave antennas using high-impedance surfaces," *IEEE Antennas Wireless Propag. Lett.*, vol. 10, pp. 7–10, Jan. 2011.

[3] D. K. Karmokar, Y. J. Guo, P. Y. Qin, K. P. Esselle, and T. S. Bird, "Forward and backward beam scanning tri-band leaky-wave antenna," *IEEE Antennas Wireless Propag. Lett.*, vo. 16, pp. 1891–1894, Mar. 2017.

[4] D. R. Jackson, C. Caloz, and T. Itoh, "Leaky-wave antennas," *Proc. IEEE*, vol. 100, no. 7, pp. 2194–2206, Jul. 2012.

[5] D. K. Karmokar and K. P. Esselle, "Periodic U-slot-loaded dual-band half-width microstrip leaky-wave antennas for forward and backward beam scanning," *IEEE Trans. Antennas Propag.*, vol. 63, no. 12, pp. 5372–5381, Dec. 2015.

[6] A. A. Oliner, "Leaky-wave antennas," in *Antenna Engineering Handbook*, R. C. Johnson, Ed. New York: McGraw-Hill, 1993, ch. 10.

[7] T. Tamir, "Leaky-wave antennas," in *Antenna Theory - Part 2*, R. E. Collin and F. J. Zucker, Eds. New York: McGraw-Hill, 1969, ch. 20.

[8] W. Menzel, "A new travelling wave antenna in microstrip," in *Proc. 8th Eur. Microw. Conf.*, Sep. 1978, pp. 302–306.

[9] W. Menzel, "A new travelling wave antenna in microstrip," *Arch. Elektron. Uebertrag. Tech.*, vol. 33, no. 4, pp. 137–140, Apr. 1979.

[10] A. A. Oliner and K. S. Lee, "The nature of the leakage from higher modes on microstrip line," in *Proc. IEEE MTT-S Int. Microwave Symp. Digest*, pp. 57–60, Jun. 1986.

[11] A. A. Oliner, "Leakage from higher modes on microstrip line with application to antennas," *Radio Science*, vol. 22, no. 6, pp. 907–912, Nov. 1987.

[12] G. M. Zelinski, G. A. Thiele, M. L. Hastriter, M. J. Havrilla, and A. J. Terzuoli, "Half width leaky wave antennas," *IET Microw., Antennas Propag.*, vol. 1, no. 2, pp. 341–348, Apr. 2007.

[13] D. K. Karmokar, K. P. Esselle, and T. S. Bird, "Wideband microstrip leaky-wave antennas with two symmetrical side beams for simultaneous dual-beam scanning," *IEEE Trans. Antennas Propag.*, vol. 64, no. 4, pp. 1262–1269, Apr. 2016.

[14] J. L. Gomez-Tornero, A. R. Weily, and Y. J. Guo, "Rectilinear leaky-wave antennas with broad beam patterns using hybrid printed-circuit waveguides," *IEEE Trans. Antennas Propag.*, vol. 59, no. 11, pp. 3999–4007, Nov. 2011.

[15] Q. Lai, C. Fumeaux, and W. Hong, "Periodic leaky-wave antennas fed by a modified half-mode substrate integrated waveguide," *IET Microw., Antennas Propag.*, vol. 6, no. 5, pp. 594–601, Apr. 2012.

[16] J. Liu, D. R. Jackson, and Y. Long, "Substrate integrated waveguide (SIW) leaky-wave antenna with transverse slots," *IEEE Trans. Antennas Propag.*, vol. 60, no. 1, pp. 20–29, Jan. 2012.

[17] L. Liu, C. Caloz, and T. Itoh, "Dominant mode leaky-wave antenna with backfire-toendfire scanning capability," *Electron. Lett.*, vol. 38, no. 23, pp. 1414–1416, Nov. 2002.

[18] Nasimuddin, Z. N. Chen, and X. Qing, "Substrate integrated metamaterial-based leaky-wave antenna with improved boresight radiation bandwidth," *IEEE Trans. Antennas Propag.*, vol. 61, no. 7, pp. 3451–3457, Jul. 2013.

[19] C. Jin, A. Alphones, and M. Tsutsumi, "Double periodic composite right/left handed transmission line and its applications to compact leaky-wave antennas," *IEEE Trans. Antennas Propag.*, vol. 59, no. 10, pp. 3679–3686, Oct. 2011.

[20] G. Lovat, P. Burghignoli, and D. R. Jackson, "Fundamental properties and optimization of broadside radiation from uniform leaky-wave antennas," *IEEE Trans. Antennas Propag.*, vol. 54, no. 5, pp. 1442–1452, May 2006.

[21] A. J. Martinez-Ros, J. L. Gomez-Tornero, and G. Goussetis, "Broadside radiation from radial arrays of substrate integrated leaky-wave antennas," in *Proc. 6th Eur. Conf. Antennas Propag.*, pp. 252–254, Mar. 2012.

[22] P. Burghignoli, G. Lovat, and D. R. Jackson, "Analysis and optimization of leaky-wave radiation at broadside from a class of 1-D periodic structures," *IEEE Trans. Antennas Propag.*, vol. 54, no. 9, pp. 2593–2604, Sep. 2006.

[23] D. K. Karmokar, K. P. Esselle, and T. S. Bird, "An array of half-width microstrip leaky-wave antennas radiating on boresight," *IEEE Antennas Wireless Propag. Lett.*, vol. 14, pp. 112–114, 2015.

[24] S. Lim, C. Caloz and T. Itoh, "Metamaterial-based electronically controlled transmission-line structure as a novel leaky-wave antenna with tunable radiation angle and beamwidth," *IEEE Trans. Microwave Theory Tech.*, vol. 52, no. 12, pp. 2678–2690, Dec. 2004.

[25] D. K. Karmokar, K. P. Esselle, and S. G. Hay, "Fixed-frequency beam steering of microstrip leaky-wave antennas using binary switches," *IEEE Trans. Antennas Propag.*, vol. 64, no. 6, pp. 2146–2154, Jun. 2016.

[26] R. Guzman-Quiros, J. L. Gomez-Tornero, A. R. Weily, and Y. J. Guo, "Electronically steerable 1-D Fabry–Pérot leaky-wave antenna employing a tunable high impedance surface," *IEEE Trans. Antennas Propag.*, vol. 60, no. 11, pp. 5046–5055, Nov. 2012.

[27] D. K. Karmokar, D. N. P. Thalakotuna, K. P. Esselle, L. Matekovits, and M. Heimlich, "Reconfigurable half-width microstrip leaky-wave antenna for fixed-frequency beam scanning," in *Proc. 7th Eur. Conf. Antennas Propag.*, pp. 1314–1317, Apr. 2013.

[28] A. Suntives and S. V. Hum, "A fixed-frequency beam-steerable half-mode substrate integrated waveguide leaky-wave antenna," *IEEE Trans. Antennas Propag.*, vol. 60, no. 5, pp. 2540–2544, May 2012.

[29] D. K. Karmokar, D. N. P. Thalakotuna, K. P. Esselle, and M. Heimlich, "Controlling the beam scanning limits of a microstrip leaky-wave antenna," in *Proc. IEEE APSURSI*, pp. 1330–1331, Jul. 2013.

[30] D. R. Jackson and A. A. Oliner, "Leaky-wave antennas," in *Modern Antenna Handbook*, C. A. Balanis, Ed. New York: Wiley, 2008, ch. 7.
[31] A. A. Oliner and D. R. Jackson, "Leaky-wave antennas," in *Antenna Engineering Handbook*, J. L. Volakis, Ed. New York: McGraw-Hill, 2007, ch. 11.
[32] M. Garcia-Vigueras, J. L. Gomez-Tornero, G. Goussetis, A. R. Weily, and Y. J. Guo, "1D-leaky wave antenna employing parallel-plate waveguide loaded with PRS and HIS," *IEEE Trans. Antennas Propag.*, vol. 59, no. 10, pp. 3687–3694, Oct. 2011.
[33] S. Maci, M. Caiazzo, A. Cucini, and M. Casaletti, "A pole-zero matching method for EBG surfaces composed of a dipole FSS printed on a grounded dielectric slab," *IEEE Trans. Antennas and Propag.*, vol. 53, no. 1, pp. 70–81, Jan. 2005.
[34] M. Garcia-Vigueras, J. L. Gomez-Tornero, G. Goussetis, J. S. Gomez-Diaz, and A. Alvarez-Melcon, "A modified pole-zero technique for the synthesis of waveguide leaky-wave antennas loaded with dipole-based FSS," *IEEE Trans. Antennas and Propag.*, vol. 58, no. 6, pp. 1971–1979, Jun. 2010.
[35] A. Lai, T. Itoh, and C. Caloz, "Composite right/left-handed transmission line metamaterials," *IEEE Microwave Mag.*, vol. 5, no. 3, pp. 34–50, Sep. 2004.
[36] M. R. M. Hashemi and T. Itoh, "Evolution of composite right/left-handed leaky-wave antennas," *Proc. IEEE*, vol. 99, no. 10, pp. 1746–1754, Oct. 2011.
[37] R. Guzman-Quiros, A. R. Weily, J. L. Gomez-Tornero, and Y. J. Guo, "A Fabry–Pérot antenna with two-dimensional electronic beam scanning," *IEEE Trans. Antennas Propag.*, vol. 64, no. 4, pp. 1536–1541, Apr. 2016.
[38] A. Ip and D. R. Jackson, "Radiation from cylindrical leaky waves," *IEEE Trans. Antennas Propag.*, vol. 38, no. 4, pp. 482–488, Apr. 1990.
[39] R. Guzman-Quiros, J. L. Gomez-Tornero, A. R. Weily, and Y. J. Guo, "Electronic full-space scanning with 1-D Fabry–Pérot LWA using electromagnetic band-gap," *IEEE Antennas Wireless Propag. Lett.*, vol. 11, pp. 1426–1429, Nov. 2012.
[40] S. Lim, C. Caloz, and T. Itoh, "Electronically scanned composite right/left handed microstrip leaky-wave antenna," *EEE Microw. Wireless Compon. Lett.*, vol. 14, no. 6, pp. 277–279, Jun. 2004.
[41] C. Caloz and T. Itoh, "Transmission line approach of left-handed (LH) materials and microstrip implementation of an artificial LH transmission line," *IEEE Trans. Antennas and Propag.*, vol. 52, no. 5, pp. 1159–1166, May 2004.
[42] M. Archbold, E. J. Rothwell, L. C. Kempel, and S. W. Schneider, "Beam steering of a half-width microstrip leaky-wave antenna using edge loading," *IEEE Antennas Wireless Propag. Lett.*, vol. 9, pp. 203–206, Mar. 2010.
[43] R. Ouedraogo, E. Rothwell, and B. Greetis, "A reconfigurable microstrip leaky-wave antenna with a broadly steerable beam," *IEEE Trans. Antennas Propag.*, vol. 59, no. 8, pp. 3080–3083, Aug. 2011.

Chapter 6

Reconfigurable high-gain antennas for wireless communications

Yingjie Jay Guo[1], Pei-Yuan Qin[1], and Raj Mittra[2]

6.1 Introduction

The last decade has witnessed significant progress in reconfigurable antennas (RAs). The importance of such antennas lies in the agilities of their frequency and radiation characteristics, enabling them to fulfill the multifunctional requirements posed by modern communications and sensing systems. In contrast to traditional antennas whose structures are rigid and designed to fit particular applications, RAs employ flexible physical structures, which can be reconfigured to deliver desired antenna characteristics and optimal system performance.

RAs are poised to play a major role in the upcoming fifth-generation (5G) wireless systems, which relies heavily on cognitive radio technologies to achieve an extremely high data rate by leveraging the combinations of frequency agility, polarization adaptability, pattern diversity, as well as adaptive signal processing. Furthermore, as RAs can be used simply to reduce the number of antennas necessary for wireless systems, they have the great potential to enhance the reliability of wireless platforms by mitigating electromagnetic interference and electromagnetic compatibility problems.

In terms of the parameter that can be reconfigured, RAs can be classified into frequency RAs, polarization RAs, radiation pattern RAs, and compound RAs. Till date, there have been several books and book chapters [1–3] dedicated to it and the number of scientific papers in this research area has been increasing steadily. Most of the published work is concentrated on the reconfiguration of a single radiator due to the ease of validating the design concepts and implementing biasing networks. In this case, it is unavoidable that the gains of those RAs are relatively low. There is no doubt that low-gain RAs can find wide applications, such as multiple-input, multiple-output (MIMO) antenna systems [4]. However, moderate/high-gain RAs are highly desirable for current and future wireless systems to increase the communications range and to achieve high data rates. Meeting the requirement of

[1]Global Big Data Technology Centre (GBDTC), University of Technology Sydney, Australia
[2]EMC Lab, University of Central Florida, USA

reconfiguration of antenna characteristics in an array topology brings new challenges for antenna engineers due to the more complicated biasing networks and feed networks. This chapter overviews the current state-of-the-art reconfigurable high-gain antennas including RA arrays and reconfigurable partially reflective surface (PRS) antennas, and outlines some directions for future research.

6.2 Reconfigurable array antennas

As a type of passive antenna array, microstrip-based Yagi–Uda antennas can be used for the development of RAs. By changing the electrical length of the driven elements using varactor diodes or PIN diodes, frequency RAs can achieve either continuous or discrete frequency tuning [5,6]. In [6], a frequency switchable printed Yagi–Uda dipole sub-array is proposed as an array element for base station antennas. The electrical lengths of the director and driven dipole elements can be changed by controlling the PIN diodes states so that the sub-array can operate in two different frequency bands. The impedance bandwidth ($|S_{11}| \leq -10$ dB) is measured to be 4.9% and 12.9% in the lower (from 2.39 to 2.51 GHz) and higher band (from 3.18 to 3.62 GHz), respectively. The measured gain of the sub-array is 9.2 and 11.9 dBi at 2.44 and 3.4 GHz, respectively. In addition, work has gone into developing beam-steering antennas based on the Yagi–Uda-type array [7–9]. Usually, such antennas have one driven element and several parasitic elements integrated with switches. By controlling the switches, the directive and reflective roles of the parasitic elements can be changed, thereby changing the main beam direction. The driven element can be either a microstrip dipole [7], a microstrip patch [8], or a wire antenna [9]. However, it is found that the RAs aforementioned suffer from low realized gains. In [10], a beam-switching pattern-reconfigurable quasi-Yagi dipole antenna with enhanced realized gains is presented for wireless local area network (WLAN) systems at 5.2 GHz using a novel pattern reconfiguration mechanism. This operating mechanism is well described using a single dipole antenna.

A very thin half-wavelength dipole oriented along the x-axis is shown in Figure 6.1(a). When the currents on the two arms of the dipole are of the same phase, the current distribution is given by $I_{\text{right}} = \vec{x}\, I_0 \cos kx,\ 0 \leq x \leq l$, and $I_{\text{left}} = \vec{x}\, I_0 \cos kx,\ -l \leq x \leq 0$, where $I_0 = $ constant, k is the propagation constant, and l is equal to a quarter of wavelength. Then, the normalized electric-field pattern can be written as

$$F(\varphi) = \frac{\cos\left(\frac{\pi}{2}\cos(\varphi)\right)}{\sin(\varphi)} \tag{6.1}$$

In this case, the maximum beam of the antenna in the E plane (x-y plane) is toward $\varphi = 90°$. If there is a phase difference α between the currents on the two arms:

$$I_{\text{right}} = I_{\text{left}} \cdot e^{j\alpha}, -180° \leq \alpha \leq 180°, \tag{6.2}$$

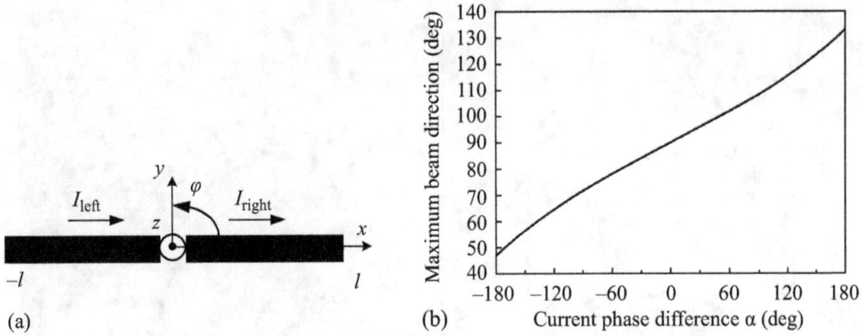

Figure 6.1 Operating mechanism of the beam switching quasi-Yagi antenna [10]: (a) currents on a half-wavelength dipole and (b) E-plane maximum beam direction as a function of the phase difference of the currents on the two arms of the dipole

the normalized electric-field pattern is given by

$$F(\varphi, \alpha) = \frac{\cos\left(\frac{\pi}{2}\cos(\varphi)\right)(1 + e^{j\alpha}) + j\left[\cos(\varphi) - \sin\left(\frac{\pi}{2}\cos(\varphi)\right)\right](1 - e^{j\alpha})}{2\sin(\varphi)} \tag{6.3}$$

When $\alpha = 0$, (6.3) is identical to (6.1). Using (6.3), for each α, the value of φ corresponding to the maximum value of $F(\varphi, \alpha)$ can be calculated, which is plotted in Figure 6.1(b). From Figure 6.1(b), it is seen that the maximum beam direction is almost a linear function of the current phase difference α. For $-180° \leq \alpha \leq 180°$, the maximum beam direction can be changed from $\varphi = 47°$ to $\varphi = 133°$. This beam scanning range becomes to be from $-43°$ to $43°$ with respect to $\varphi = 90°$. For example, when $\alpha = 0$, the maximum beam direction is not tilted, which is at $\varphi = 90°$. When α is equal to 97° and −97°, the maximum beam direction is tilted by 20° and −20°, respectively. Therefore, if the phase of the currents on the dipole can be manipulated, the main beam of the antenna in the E plane can be steered, while the radiation pattern in the H plane (y–z plane, $\varphi = 90°$) does not change with the phase variation.

Based on the above mechanism, a beam switching quasi-Yagi dipole antenna has been developed, as shown in Figure 6.2(a). The antenna consists of a microstrip feed, an impedance transformer, a broad-band microstrip-to-CPS balun, a dipole driver fed by the CPS, and two tilted rows of directive strips. The bottom side is a truncated ground, serving as the reflector. The length of the microstrip-to-CPS balun is controlled by using PIN diodes. The change of the length of the balun allows the currents on the two arms of the dipole to have different phase differences. Different phase differences of the currents lead to different E-plane main beam directions. The antenna operate at three states with the E-plane maximum beam direction toward 20°, −20°, and 0°, respectively. The realized gains of the antenna are between 7.5 and 10 dBi. A photograph of the antenna is given in Figure 6.2(b). The simulated and measured far-field radiation patterns at different

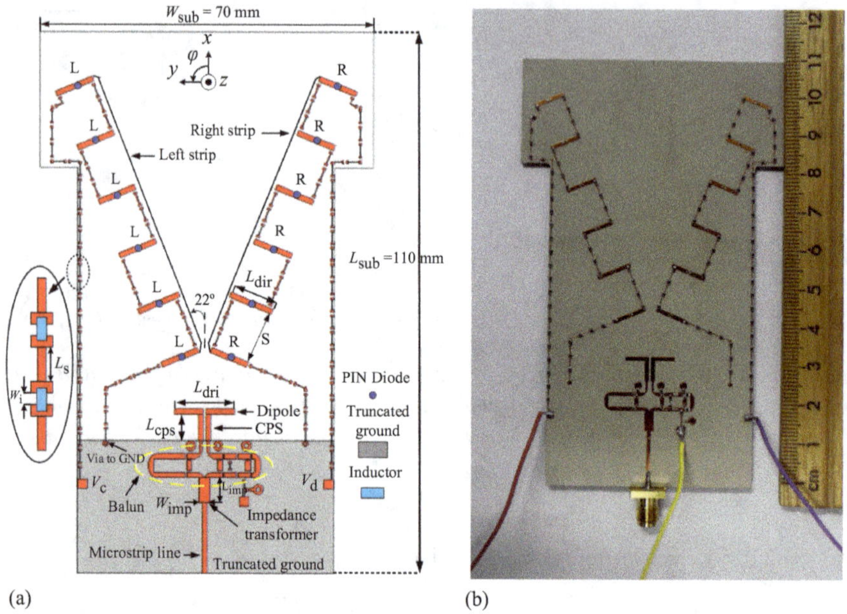

Figure 6.2 (a) Antenna configuration [10] and (b) photograph of the proposed antenna [10]

frequencies are shown in Figure 6.3. The measured main beams for State I are pointed at 0° except for 5.2 GHz where the main beam is toward 2°. For State II, the measured maximum beam radiations are at 20°, 22°, and 19° for 5.0, 5.1, and 5.2 GHz, respectively. For State III, the measured maximum beam radiations are at −17°, −21°, and −20° for 5.0, 5.1, and 5.2 GHz, respectively.

Compared to a dipole antenna, a quasi-Yagi dipole antenna has a reflector and one or several directors which can affect the beam steering range. However, as the dipole driver of the quasi-Yagi antenna has a dominant role in controlling the beam direction, the beam scanning can still be realized by changing the phase difference of the currents on the dipole. But, the actual beam-steering angle is limited by the sidelobe level that increases with the steering angle. The beam steering is realized in a discrete way by using PIN diodes, thus resulting in a switched beam pointing to three directions. It should be noted that the beam could also be steered continuously by using varactor diodes.

As demonstrated in the above two designs, microstrip quasi-Yagi antennas are capable of achieving frequency tuning and beam scanning with enhanced realized gains. However, due to the inherent orientation of the driven dipole, it is very challenging for them to realize polarization reconfigurability.

Phased array antennas can also realize beam scanning with high gains. Usually, phased arrays are not considered to be RAs if the array elements are non-reconfigurable traditional antennas. However, efforts have been made to develop

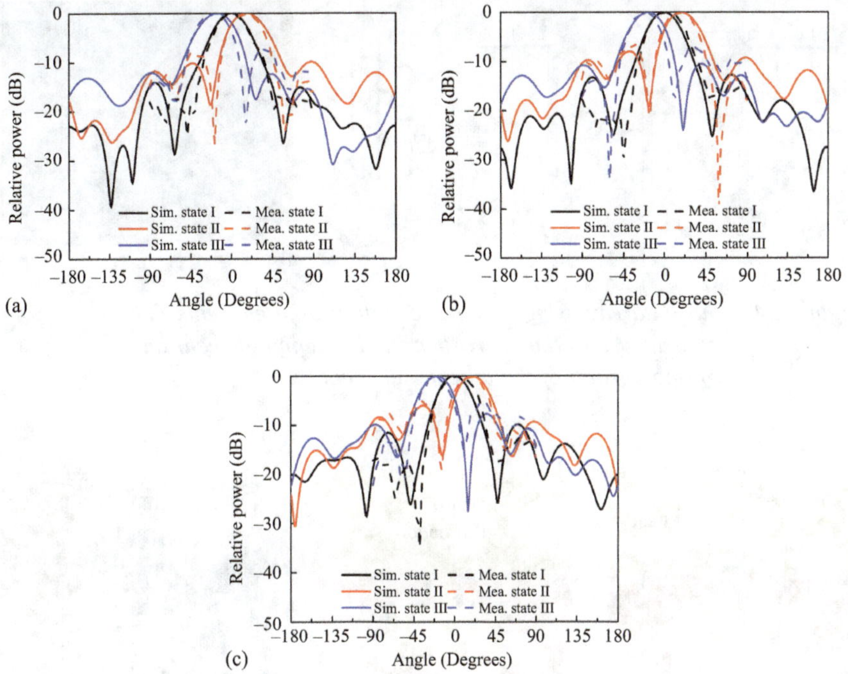

Figure 6.3 Simulated and measured E-plane normalized radiation patterns [10] at (a) 5.0 GHz; (b) 5.1 GHz; and (c) 5.2 GHz

reconfigurable phased arrays using pattern-reconfigurable elements [11,12] or using novel phase shifters based on reconfigurable defected microstrip structure [13]. In [11], a novel phased array is presented to extend array scanning range by using pattern-reconfigurable microstrip Yagi antenna element. The structure of the antenna element is shown in Figure 6.4(a). It consists of driven and parasitic strips, which is able to reconfigure its patterns from broadside to quasi-endfire radiation by controlling the states of the PIN diodes integrated on parasitic strips. More specifically, when all the PIN diodes (K1–K4) are switched on, the main beam in the *H* plane is pointed toward broadside. When K1 and K2 are switched on and the other two are switched off, the main beam is tilted toward positive *y*-axis direction. When K3 and K4 are switched on and the other two are switched off, the main beam is tilted toward negative *y*-axis direction. The photograph of the antenna element is shown in Figure 6.4(b). An array of eight elements has been designed. In order to extend the scanning angle range of the phased array with small gain losses and side lobes, the most appropriate radiation pattern of each array element is chosen for a particular array scanning angle. For example, when all of the elements operate in positive *y*-axis, the array can scan its main beam from 7° to 60° with a gain fluctuation less than 3 dB. The pattern of each active element was measured to synthesize the array pattern. Measured results show that the array can scan its main beam from $\varphi = -60°$ to 60° in *H*-plane with gain fluctuation less than 3 dB.

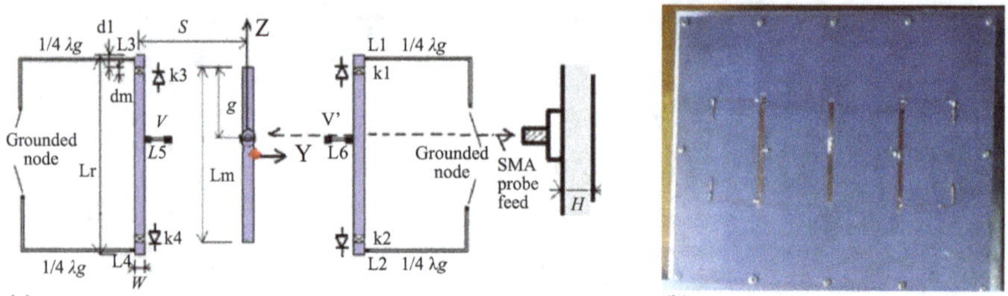

(a) (b)

Figure 6.4 Phased array using pattern-reconfigurable elements [11]: (a) structure of the microstrip Yagi antenna element and (b) photograph of the antenna element

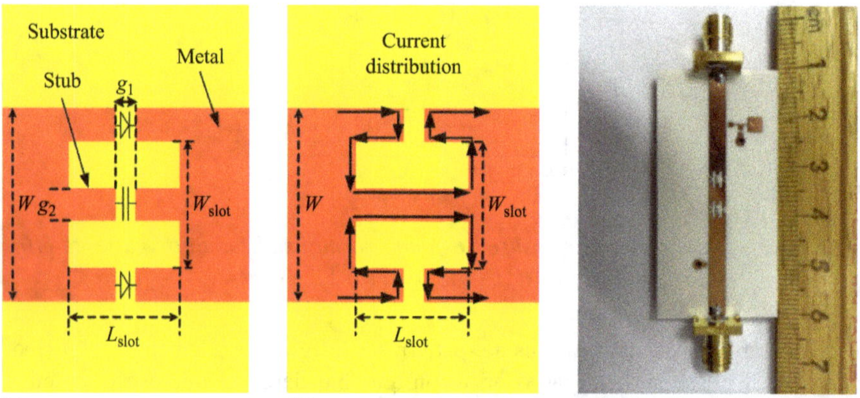

(a) (b) (c)

Figure 6.5 RDMS phase shifter [13]: (a) structure of the RDMS unit; (b) all-off state current distribution; and (c) 2-RDMS phase shifter

The above design does not incorporate a feed network but use active element patterns to synthesis array patterns. In general, a large number of phase shifters are employed in a phased array antenna, and thus their cost, size, and integration method are of great importance [14]. Recently, Can *et al.* proposed compact phase shifters using reconfigurable-defected microstrip structure (RDMS) for phased array antennas [13]. Figure 6.5(a) shows the structure of the RDMS unit based on a 50-Ω microstrip line for the phase shifter design. As depicted in the figure, a rectangular slot with a size of $W_{slot} \times L_{slot}$ is etched on the microstrip line to introduce a defect. PIN diodes are inserted into the edges of the slot area, which enables structural reconfiguration and results in an RDMS unit. Capacitors and metal stubs used for capacitor mounting are placed in the middle of the slot to achieve RF continuity as well as to provide DC isolation. The RDMS unit has two working states, the "All-on" and "All-off" states, when the diodes are turned "on" and "off", respectively.

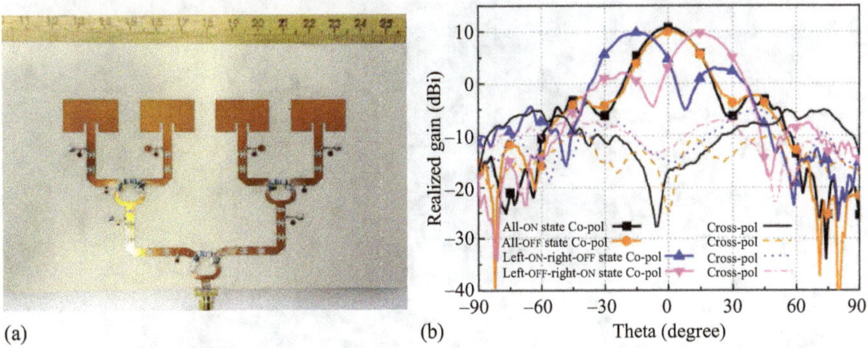

(a) (b)

Figure 6.6 Phased array using RDMS phase shifters [13]: (a) the 4-element phased array prototype and (b) measured far-field pattern of the phased array

In the All-on state, both the diodes and capacitors allow the currents to go through them with minor losses. Numerical simulations show that the RDMS unit behaves like a uniform microstrip line in this state. In the All-off state, the currents across the diodes are blocked, resulting in a different current distribution as shown in Figure 6.5(b). It is noted that in the All-off state, the current path is longer than that in the All-on state, which produces a phase shift. The phase shift value is in proportional to the difference of the electrical length of the current path in the two states, which is affected by W_{slot} and L_{slot}. An RDMS unit with $\{W_{slot}, L_{slot}\} = \{2 \text{ mm}, 3.3 \text{ mm}\}$ is fabricated as an example. It can produce a phase shift of 20° with an insertion loss of 1.1 dB at 5.2 GHz. Larger phase shift values can be achieved by cascading several RDMS units described above. As shown in Figure 6.5(c), a 2-RDMS phase shifter cascading two RDMS units can realize a 40°phase shift.

The above-described phase shifters have been employed in a 4-element phased array feed network as shown in Figure 6.6(a). The array is designed to be able to switch its main radiation beam to 0°, −15°, and 15° in the H plane. In this design, three Wilkinson power dividers are employed to split the power while eight phase shifters are integrated in the feed network to realize 50° progressive phase differences between the array elements at 5.2 GHz. Each phase shifter is composed of two RDMS units with $W_{slot} = 2$ mm and $L_{slot} = 3.6$ mm. For the first-level power divider, there are two phase shifters at each of the 2 branches. For the second level, there is one phase shifter at each of the four branches. Two bias DC voltages labeled by V_1 and V_2 are employed to control the phase shifters. There are 4 different working states of the phased array. When voltages V_1 and V_2 are "+,+", "–,–", "+,–", and "– ,+", the phased array can work in the "All-on", "All-off", "left-on-right-off", and "Left-off-right-on" states. In the All-on and All-off states, the array elements are in phase so that the beam is not tilted. In the left-on-right-off and Left-off-right-on states, phase advances and delays of 50° between the array elements are obtained, respectively. Therefore, beam tilts of −15° and 15° are realized. The measured far-field patterns of the array at 5.2 GHz are given in Figure 6.6(b).

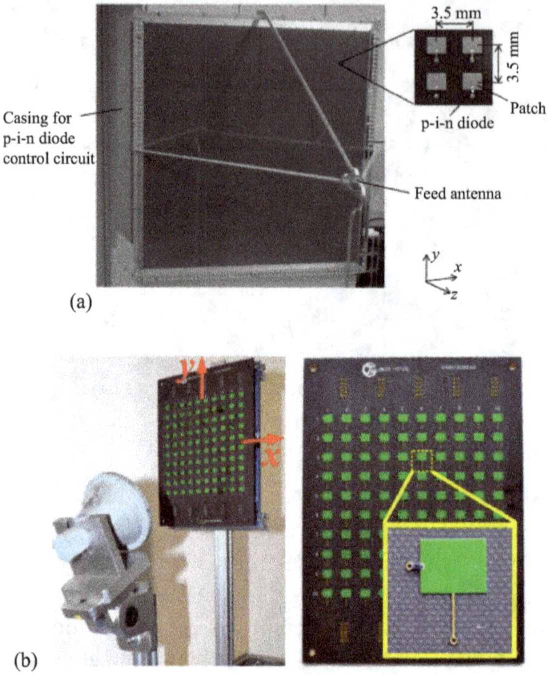

Figure 6.7 (a) 60 GHz-Reconfigurable reflectarray antenna [20] and (b) 1-bit 10 × 10 reconfigurable reflectarray antenna prototype [21]

The disadvantages of phased arrays, however, are their large hardware footprint, high cost and high loss due to the use of transceiver modules and transmission-line feeding networks, especially at millimeter-wave or higher frequencies. Reconfigurable reflectarrays and transmitarrays that employ spatial feeding have been demonstrated as alternative solutions [15,16]. A reflectarray antenna consists of an array of phase shifting elements on a flat surface that provide an appropriate phase response to form a focused beam or a contoured beam when illuminated by a feed [17–19]. Tunable reflectarrays have become promising candidates to achieve both beam steering and high gains due to their attractive qualities, namely, low profile, ease of manufacturing and high efficiency [20–26]. Compared to phased arrays, reconfigurable reflectarrays requires a much lower number of transceivers and do not need transmission line feed networks. For the phase shifting elements of a reconfigurable reflectarray, they usually employ tunable devices to change phase electronically or mechanically. One common way is to us PIN diode to obtain discretized phase shifts. In [20], a 60-GHz-reconfigurable large reflectarray using single-bit phase shifters is presented. The reflectarray consists of 160 × 160 phase shifting elements. The element is a rectangular microstrip patch which is connected to a short circuited stub loaded with a PIN diode, as shown in Figure 6.7(a). By switching ON/OFF of the PIN diodes, the reflection phase changes by 180°. The main beams can scan at ±20° in both the azimuth and elevation planes. Using a similar phase shifting element, a

(a) (b)

Figure 6.8 (a) Dual-frequency reflectarray with the control board [26] and (b) prototype of the reflectarray [29]

Ku-band-reconfigurable reflectarray with 10×10 elements is proposed [21] as shown in Figure 6.7(b). The beams can be scanned within $\pm 50°$ range with a maximum aperture efficiency of 17.9% at 12.5 GHz. In [22], a Ku-band dual-polarized reflectarray based on a new type of polarization turning unit-cell, with single-bit phase resolution is proposed. By using the unit-cell, a linearly polarized incident wave can be transformed into the orthogonal linearly polarized reflected wave in the operating frequencies. Beam-scanning capability is also demonstrated by controlling the PIN diodes. Furthermore, a dual-frequency-reconfigurable reflectarray was proposed to achieve beam scanning at 11.1 and 14.3 GHz [26]. The reconfigurable reflectarray consists of 1600 electronically controllable phase shifting elements. Each one is able to operate at two operating frequencies with 1-bit phase resolution. It successfully addresses the drawback of narrow bandwidth for many conventional reflectarrays. A photograph of the reflectarray is shown in Figure 6.8(a).

By using PIN diodes, the above reflection elements can only provide discrete reflection phase states. A limited number of discretized phase states would introduce phase errors, which can bring quantization losses. To obtain continuous controllable phases, varactor diodes have been employed in elements designs [27–31]. In [27], an electronically tunable reflectarray using varactor diodes is developed. The unit cell can realize over 320° of phase agility at 5.5 GHz. Experimental measurements of the reflectarray operating at 5.8 GHz demonstrate excellent beam-forming characteristics. An aperture-coupled reflectarray element giving a full phase tuning range with a single varactor diode is proposed in [29], as shown in Figure 6.8(b). The element is composed of a microstrip patch fed through an aperture and a microstrip line printed on a different substrate layer loaded with one varactor diode. A 320° phase-tuning range is achieved by changing the biasing voltages. A beam scanning in H plane, a set of multibeam patterns and a cosecant beam shape are accomplished. It is also found that tunable impedance surfaces can be designed as reconfigurable reflectors [31]. As shown in Figure 6.9(a), the impedance surface consists of electrically small square metal patches that are shorted to the ground plane by metal plated vias. All the patches

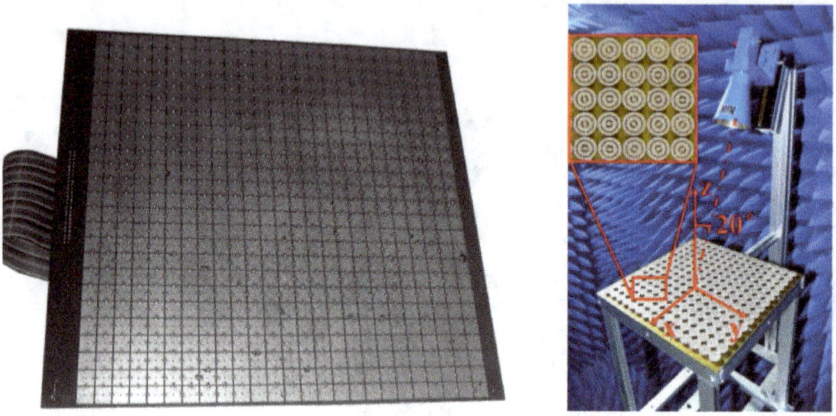

(a) (b)

Figure 6.9 (a) Photograph of the electronically steerable reflector [31] and (b) fabricated RRA prototype with 15 × 15 manually rotated elements [32]

are connected to each other by varactor diodes. By controlling the biasing voltages, a phase variation range of 360° can be achieved, which can electronically steer a reflected beam over ±40° in two dimensions.

One major drawback of using electronic devices (PIN diodes or varactors) to control the phase is the high loss. An alternative approach is to change the phase of the individual reflectarray element mechanically using miniaturized micromotors or actuators [32–34]. In [32], a circularly polarized reconfigurable reflectarray antenna at X band is presented that employs concentric dual split-rings as phase shifting elements as shown in Figure 6.9(b). The elements can be mechanically rotated to obtain continuous 360° phase coverage with negligible element loss. Measured results show that the beam at ($\theta = 20°$, $\varphi = 0°$) direction achieves the maximum gain of 25.6 dB. Well-defined scanned beams within ±60° are obtained with maximum scan loss of 3.7 dB.

Transmitarrays, also known as array lenses, have the capability to achieve high realized gains for beam-forming applications. A transmitarray collimates an incident wave by manipulating the phase value of each phase shifting element. Compared to reflectarrays, where both the feed and the radiation are in the same side, they are at opposite sides for transmitarray. In order to realize beam-scanning, reconfigurable transmitarrays [35–42] utilize electronic phase shift control for each individual element. In [35], a fully reconfigurable 6 × 6 transmitarray based on a balanced guided wave unit cell is presented, as shown in Figure 6.10. A more than 400° phase shift range is realized using bridged-T phase shifters. Well-shaped beams are formed for scanning angles from −50° to 50°. In [36], a reconfigurable millimeter-wave lens-array is proposed employing 2-bit programmable monolithically integrated microelectromechanical systems antenna–filter–antenna unit cells. The array elements have four discrete phase tunings, which make the lens-array

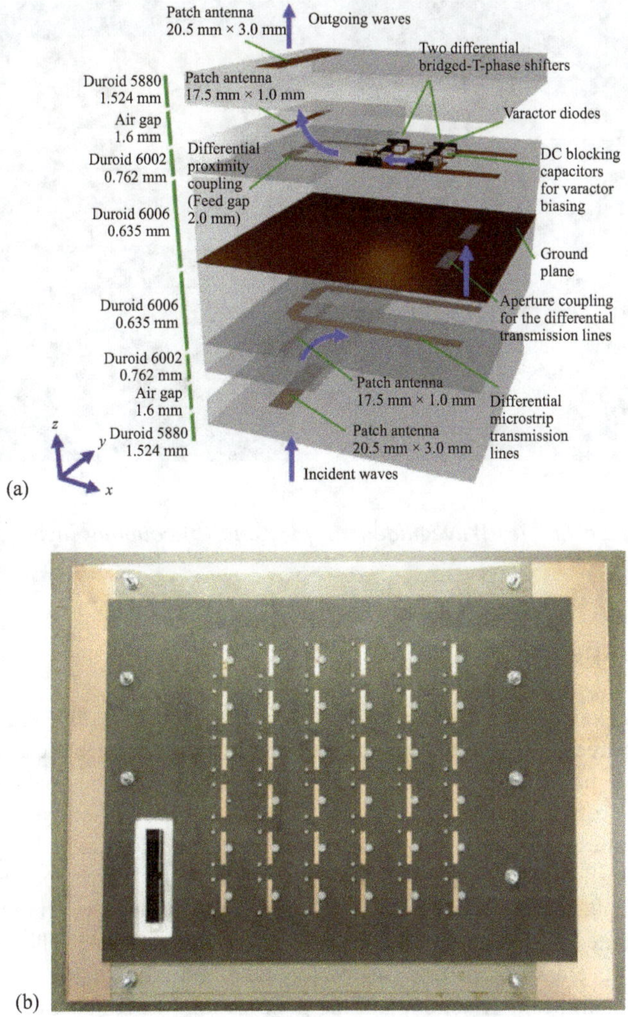

Figure 6.10 (a) Balanced bridged-T guided wave unit cell [35] and (b) assembled 6×6 transmitarray [35]

steer the beam to 40° in either the E or the H plane. A new low-profile, full-space, electronic transmitarray is proposed in [38]. The design employs an integrated leaky-wave waveguide feed to spatially distribute power across a reconfigurable transmitarray aperture, which reduces the profile the antenna architecture significantly. Varactor diodes are used to change the phase of each phase shifting element. The beam steering angle is determined by the superimposition of the phase constants of the leaky wave and the phase shifting elements of the transmitarray. Full-space (both azimuth and elevation) beam steering is achieved at angles up to 45° off broadside. The photograph of the prototype is shown in Figure 6.11.

(a) (b)

Figure 6.11 (a) Waveguide feed [38] and (b) coupling layer [38]

6.3 Reconfigurable PRS antennas

PRS antennas, also known as Fabry-Pérot (FP) antennas, usually employ a low directivity source antenna and a metallic or metallodielectric periodic array, located approximately half a wavelength above the source [43,44]. They have the advantages of low profile, simple construction and high directivity. Compared with array antennas, it does not need a feed network which can become very lossy for large arrays. A significant drawback of the PRS antenna is the narrow operating bandwidth due to the high Q-factor of the FP cavity. Furthermore, traditional PRS antennas radiate fixed radiation patterns with fixed polarizations. This section will present the development of the electronic reconfiguration of the operating frequency, the main beam direction, and the polarization of PRS antennas.

6.3.1 Frequency-reconfigurable PRS antenna

In [45], a frequency-reconfigurable PRS antenna is proposed to address the significant drawback of the narrow operating bandwidth due to the high Q-factor of the FP cavity. The geometry of the frequency-reconfigurable PRS antenna is shown in Figure 6.12, comprising a PRS, a tunable high impedance surface (HIS) and a stacked patch feed antenna. The tunable HIS is placed on the lower surface of the cavity where each cell of the HIS is tuned by changing the voltage applied to the varactor diodes.

Its operating frequency is designed to be tuned from 5.2 GHz to 5.775 GHz to enable use in WLAN applications. Referring to Figure 6.12, the cavity height L_r

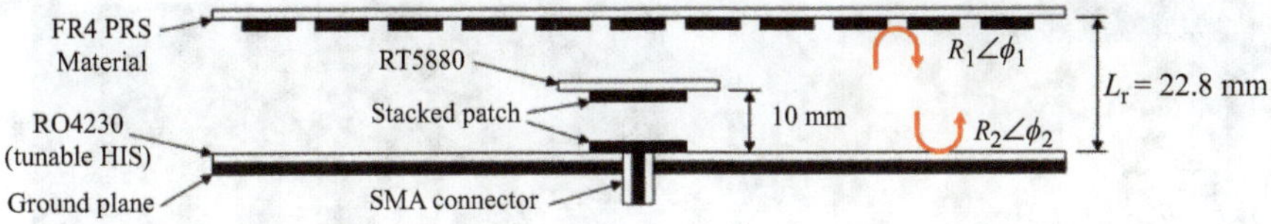

Figure 6.12 Geometry of the frequency-reconfigurable PRS antenna [45]

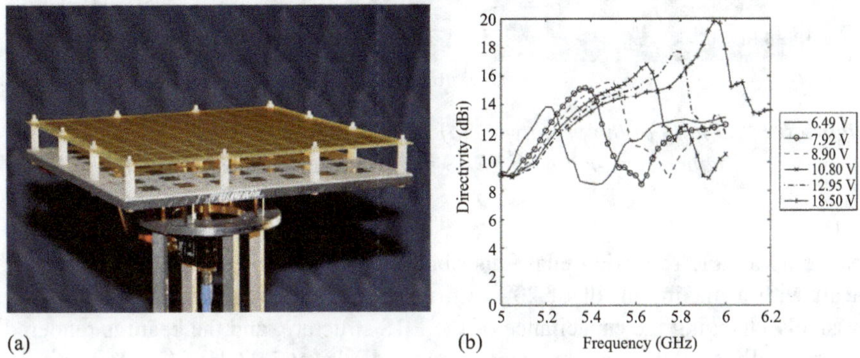

Figure 6.13 (a) Photograph of the frequency-reconfigurable PRS antenna [45] and (b) measured directivity of the frequency-reconfigurable PRS antenna for six different bias voltages [45]

can be expressed in terms of the operating wavelength λ_0, and the reflection phase of the PRS φ_1 and tunable HIS φ_2 as follows:

$$L_r = \left(\frac{\phi_1 + \phi_2}{\pi}\right)\frac{\lambda_0}{4} + \frac{\lambda_0}{2} \tag{6.4}$$

Since φ_2 is a function of the varactor tuning voltage, it is possible to reconfigure the operating frequency of the antenna. Increasing the tuning voltage gives an increase in the antenna resonant frequency, while lowering the voltage reduces the resonant frequency.

A photograph of the fabricated prototype antenna is shown in Figure 6.13(a). The measured directivity versus frequency is plotted in Figure 6.13(b) for six different bias voltages, which correspond to six different junction capacitances of the varactor diodes. It is clear from Figure 6.13 that the measured operating frequency tunes from 5.2 to 5.95 GHz.

6.3.2 Pattern-reconfigurable PRS antenna

In general, a center-fed PRS antenna produces a broadside beam. There are mainly two approaches to tilt the beam at a fixed frequency. The first one is to employ phase-varying metamaterial surfaces as the PRS structure [46–48] in order to

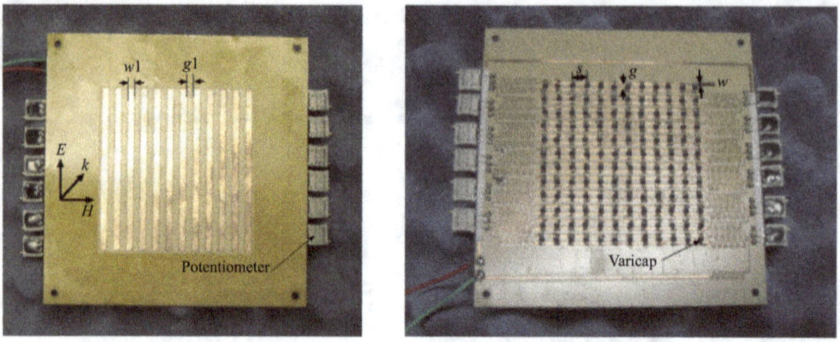

(a) (b)

Figure 6.14 Photo of the prototype: (a) inductive grid of PRS [47] and (b) capacitive grid of the PRS integrated with varactor diodes [47]

realize a variable reflection phase distribution. In [46], a PRS antenna radiating a beam with a maximum tilt of 20° with respect to the broadside is presented by passively changing the capacitance of the PRS structure, and the beam is rendered electronically steerable ±7° by using varactor diodes at 7.9 GHz [47]. As shown in Figure 6.14, the PRS consists of a constant inductive grid and an active capacitive grid integrated with varactor diodes. By changing the biasing voltages, the phase of the input reflection coefficients of the PRS varies, thereby enabling the steering of the radiation beam. Based on a similar structure, a progressive phase variation can also be realized by using tapered inductive grids, leading to a 56° beam tilt with respect to the broadside direction [48]. Another way to accomplish beam steering is to utilize a phased array as the source antenna. In [49], a two-element patch antenna array using a separate Wilkinson power divider and phase shifters is exploited to excite the PRS to steer the beam. It is shown that the beam scanning from −10° to 10° with a realized gain greater than 13 dBi can be achieved at 2 GHz. By using a separate reconfigurable matching network, an overlapped 3% impedance bandwidth is obtained. Additionally, research activities have been conducted to realize beam scanning of a one-dimensional (1D) leaky-wave antenna at a fixed frequency by employing a reconfigurable HIS [50]. The antenna can provide a continuous electronic beam scanning in an angular range from −25° to 25° at 5.5 GHz. This pattern reconfiguration method has been extended to a PRS antenna (2D) enabling beam scanning in both elevation and azimuth angles [51,52]; however, the aperture efficiency of this type of antenna is low as only part of the antenna is used for radiation most of the time. As the reconfigurable leaky-wave antennas mentioned above are described in details in a separate chapter of this book, it is not repeated here.

In [53], a beam-steering pattern-reconfigurable PRS antenna is presented which employs a new non-uniform-reconfigurable PRS structure with a 2-element microstrip patch phased array as the source. The antenna can switch its beam among −15°, 0°, and 15° in an overlapped frequency range from 5.5 GHz to 5.7 GHz (4% impedance bandwidth) with a realized gain over 12 dBi. The reconfigurable PRS structure

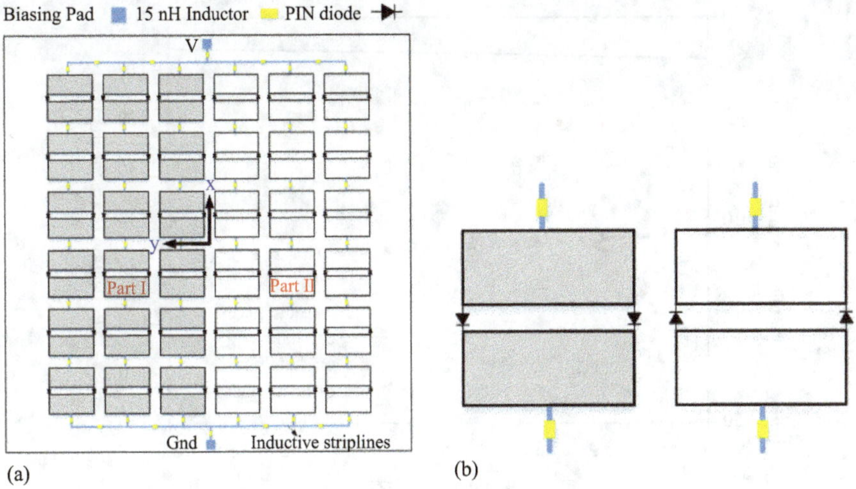

Figure 6.15 Antenna schematics [53]: (a) reconfigurable PRS structure and (b) cell element with PIN diodes and biasing lines

is shown in Figure 6.15(a). It employs 6 × 6 reconfigurable cells and is divided into the left and right parts with respect to the center of the structure. Each reconfigurable cell is comprised of a 20.5 mm × 20.5 mm microstrip patch etched on a 24 mm × 24 mm × 0.8 mm FR4 substrate. A 1 mm slot is inserted in the middle of the patch. PIN diodes are placed at the two sides of the slot. The reflection phase of the cell element can be varied by switching the diodes on and off to produce a reflection phase inconsistency between the two parts of the entire PRS structure, which in turn switches the beam of the PRS antenna. When the PIN diodes are switched off, the two halves of the patch disconnect from each other, which results in a surface with a high reflectivity and a small phase. When the PIN diodes are switched on, a larger phase value is obtained compared to that of the off-state. The simulated reflection coefficients of the cell element for the two PIN diode states are given in Figure 6.16.

As shown in Figure 6.15(b), The PIN diodes in each part are oriented in opposite directions, though the diodes have the same orientation in each individual part. In terms of the DC signal, the PRS cells in each column are connected in series by using very thin (0.15 mm) metallic biasing striplines. In order to mitigate the effects of these biasing lines on the performance of the PRS, the striplines are broken down into small sections and the gaps between them are bridged with surface-mounted RF choke inductors (0402HP, 20 nH). For simplicity, such biasing striplines are named as inductive biasing lines. Subsequently, all the columns are connected in parallel by using the same inductive biasing lines. Due to the reverse orientation of the PIN diodes in the two halves of the PRS, only one DC biasing voltage is utilized to control the diodes. When the voltage V is positive, the diodes in the left part (Part I) are switched on while the diodes in the right part (Part II) are switched off, *and vice versa*. Simulation results indicate that the biasing lines have

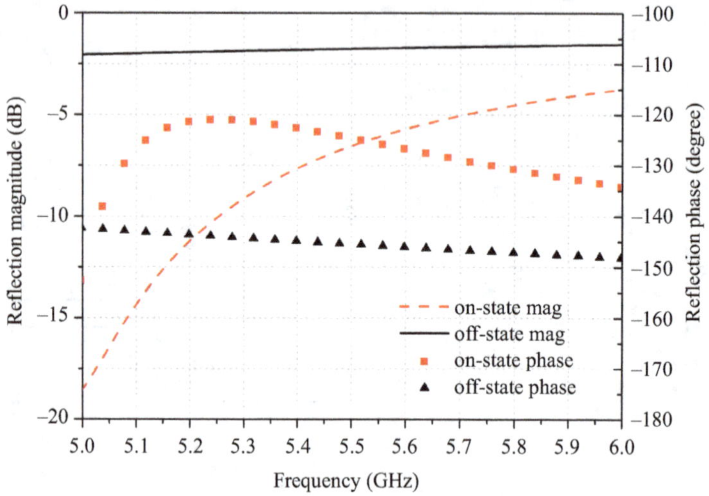

Figure 6.16 Simulated reflection coefficient of the cells [53]

little effect on the performance of the reconfigurable PRS structure. Using this reconfigurable PRS cell element, it is found that an approximate ±5° beam-tilt can be realized for a PRS antenna with a normal patch antenna as the radiating source. Furthermore, the beam-tilt angle is affected by the PRS cell number. It is smaller when the number of cell element is reduced. For a larger number of cell elements, a larger steering angle is achieved but with a lower gain and a higher side lobe level.

In order to increase the tilt range, a phased array antenna is employed as the source (exciter) for the PRS antenna. To obtain phase shifts between the array elements, RDMS-based phase shifters [13] are incorporated in the feed network as shown in Figure 6.17(a). The feed network is comprised of two-phase shifters etched on each branch of a Wilkinson power divider. Each phase shifter consists of three identical RDMS units, as shown in the inset of Figure 6.17(a). Each unit consists of a rectangular slot with a size of $L_{slot} \times W_{slot}$; two gaps etched on the edges for inserting the PIN diodes; and two metallic stubs located in the middle of the slots as mounting pads for the capacitors not only to realize RF continuity but also to provide a DC isolation for the PIN diodes as well.

The phase shift is achieved by controlling the states of the PIN diodes of the RDMS unit to obtain different current paths. The current path for the off-state of the PIN diodes is longer than that of the on-state. For this design, the off-state generates a 30° phase delay with respect to the on-state for a single unit. By cascading three RDMS units, a 90° phase shift can be obtained. If the PIN diodes on both branches of the power divider are "all-on", no phase shift is acquired and this state is referred to "S1". When the diodes on the left branch are switched on and those of the right branch are switched off, the right one can achieve a 90° phase delay with respect to the left one. This state is named "S2" and it results in a tilted beam toward the right side ($-y$ direction). On the other hand, the diodes on the network with a state of left

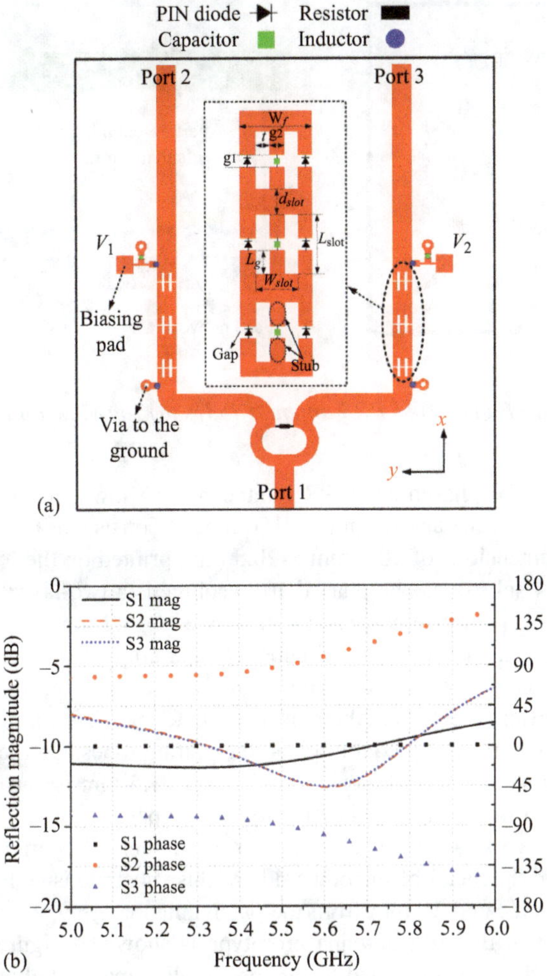

Figure 6.17 (a) Structure of the feed network [53] and (b) simulated reflection coefficient S_{11} of the phase shift feed network [53]

off-state right on-state (named as "S3"), the beam can be titled toward the left side (+y direction). The phased network designed on an 80 mm × 80 mm × 1.524 mm Rogers 4003 substrate was simulated, as shown in Figure 6.17(b). For brevity, only the magnitude of the input reflection coefficient for Port 1 ($|S\_11|$) and the phase difference $\varphi_{\_23}$ between Port 2 and Port 3 are shown. It is observed that $|S\_11|$ is below −10 dB for all the three states across the frequency band ranging from 5.3 to 5.7 GHz. Furthermore, $\varphi_{\_23}$ values for the three states (S1, S2, and S3) remain approximately 0°, 90°, and −90°, respectively, in this frequency band.

Based on the above results, the PRS antenna with a reconfigurable PRS structure and a phased array source has been designed, fabricated, and measured.

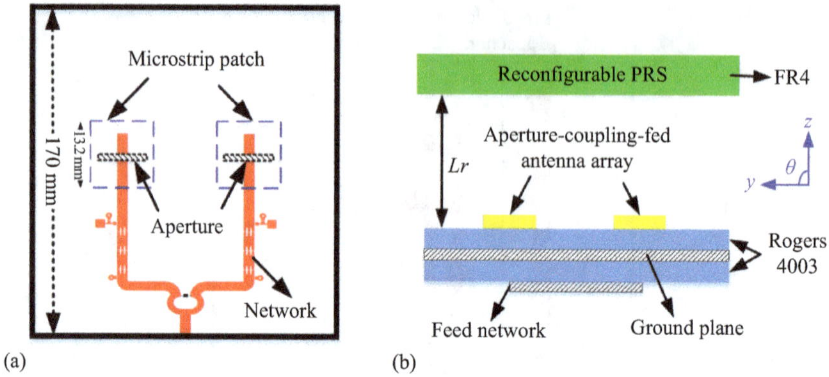

Figure 6.18 Phased array fed PRS antenna [53]: (a) top view and (b) side view

The schematics of the phased array PRS antenna are shown in Figure 6.18. The dimensions of the antenna are 150 mm × 150 mm. It consists of 6 × 6 reconfigurable cells, with dimensions of 20.5 mm × 20.5 mm, printed on the lower side of a 0.8-mm-thick FR4 substrate. The phased array source is a two-layer structure. For the first layer, two square patches, each with a size of 13.2 mm, are placed at one side of a 1.524-mm-thick Rogers 4003 substrate and aligned symmetrically along the *y* direction; hence the beam-scanning would only be realized in the *y–z* plane (H-plane). The spacing between them is chosen to be 12.2 mm, to achieve a compromise between the array coupling and the grating lobes. On the other side of the substrate, two slots with a dimension of 7 mm × 1.5 mm for an aperture coupling are etched on the ground at the position of the patch center. For the second layer, the feed network is printed on one side of another 1.524-mm-thick Rogers 4003 substrate and the metal of the other side of this substrate is etched off. In this design, the PRS is located 30 mm from the patch antenna array.

A photograph of the PRS antenna prototype is shown in Figure 6.19. Nylon spacers and M3 nylon bolts are used to support the PRS material above the ground plane. Visible at the back of the antenna is a low-loss bias tee, used to supply the biasing currents to the reconfigurable PRS unit cells in the measurement.

As discussed previously, the reconfigurable PRS structure has three states corresponding to three radiation patterns. Similarly, the phased array source also has three states; hence, there are 9 states altogether. The simulated beam directions and their corresponding diode states are shown in Table 6.1 for the nine states. As can be seen from the table, the beam can be primarily steered to three directions, namely 5°, 10°, and 15°. The focus of this design is to have a wider steering range of the reconfigurable PRS antenna. Therefore, only the states of 15° beam steering angle are measured. Figure 6.20 plots the simulated and measured input reflection coefficients for the three states. The measured results show that an overlapped impedance bandwidth from 5.5 to 5.74 GHz is achieved and that they agree well with the simulated results. The radiation patterns of the antenna are measured in the H-plane (*y–z* plane) for the three states and are shown in Figure 6.21. It is observed

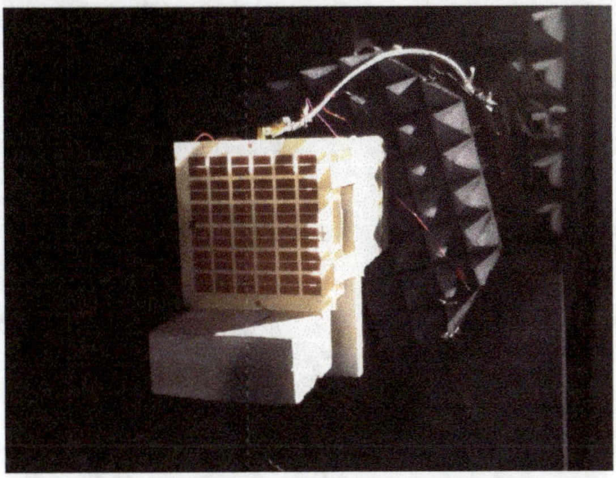

Figure 6.19 Prototype of the proposed antenna [53]

Table 6.1 All states for the proposed antenna and performance at 5.5 GHz

State	PRS state		Array state		Beam direction	Gain (dBi)
	Left-side diodes	Right-side diodes	Diodes on left-arm	Diodes on right-arm		
1	off	off	on	on	0°	16.3
2	on	off	on	off	−15°(−y)	11.7
3	off	on	off	on	15°(+y)	11.7
4	on	off	on	on	−4°	14.2
5	off	on	on	on	4°	14.2
6	off	off	on	off	−9°	14.4
7	off	off	off	on	9°	14.4
8	off	on	on	off	−6°	12.7
9	on	off	off	on	6°	12.7

that a broadside radiation is obtained at State 1. For State 2 and State 3, the beam directions are tilted toward −15° and 15° from the broadside, respectively, in a frequency band that ranges from 5.3 to 5.7 GHz.

6.3.3 Polarization-reconfigurable PRS antenna

Polarization RAs can enhance the performance of radio systems due to their capabilities to mitigate detrimental fading caused by multipath effects, to realize double transmission channels for frequency reuse, and to effectively increase the system capacity and link quality [54,55]. The polarization reconfiguration between linear polarization (LP) and circular polarization (CP) can also enable the antenna to adapt to different electromagnetic environments. For example, the polarization of a reader's antenna in radio frequency identification system needs to be changed between LP (when the orientation of the tag's antenna is known and fixed) and CP

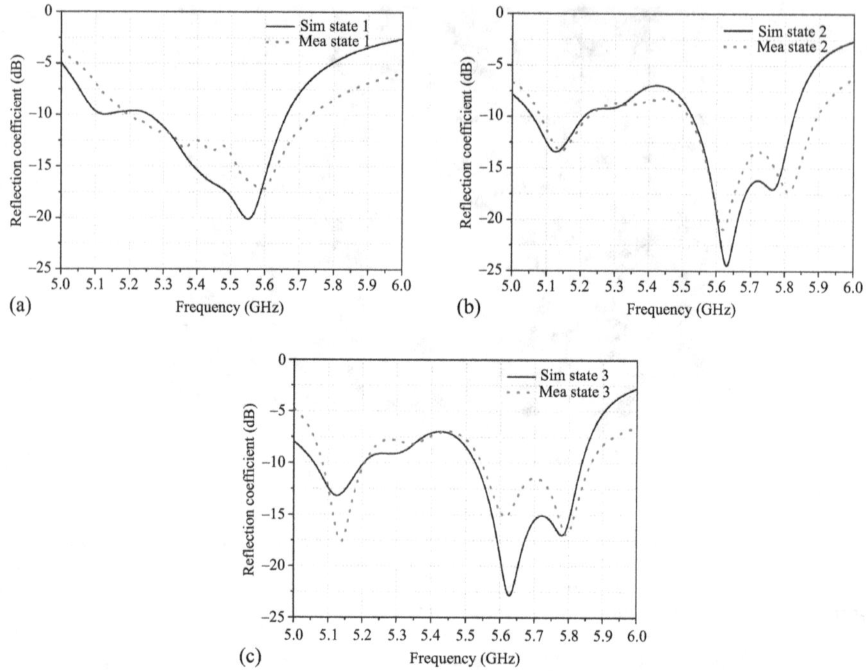

Figure 6.20 Input reflection coefficients of the reconfigurable PRS antenna: (a) State 1; (b) State 2; and (c) State 3 [53]

(when the orientation of the tag's antenna is unknown) [56]. Furthermore, it can be used to receive satellite transmitted TV signals for different countries (linear for Europe and circular for the US/Canada) [57]. Generally, there are mainly two approaches to electronically reconfigure the polarization between LP and CP. One is to introduce reconfigurability on the radiating elements [58–60]. In [58], two PIN diodes are embedded on the two arms of a U-slot patch antenna to switch the polarization between CP and LP. The overlapped impedance bandwidth of is 2.8% with the measured gains varying from 6.3 to 7.5 dB. In [59], a polarization RA employing a perturbed stair-shape slot is designed. The overlapped impedance bandwidth is 37 MHz at the center frequency of 2.46 GHz with a realized gain around 2.9 dB. A coplanar stripline fed dual-loop planar antenna is presented in [60] employing four PIN diodes inserted in the symmetric gaps on the dual-loop. The proposed antenna can electronically switch its polarization between LP and CP with a simulated impedance bandwidth of 5.6%. Another way to accomplish polarization switching is to utilize reconfigurable feed network [61–63]. An RA using dual-feeds incorporating a T-junction power divider and a pair of varactor diodes is presented in [61]. An overlapped impedance bandwidth of 1.4% between LP and two CPs is achieved with a realized gain at maximum 7.2 dB. In [62], an aperture-fed square patch antenna is described using a branch line coupler with four microstrip feeding lines as the feed network. Eight PIN diodes are inserted in the

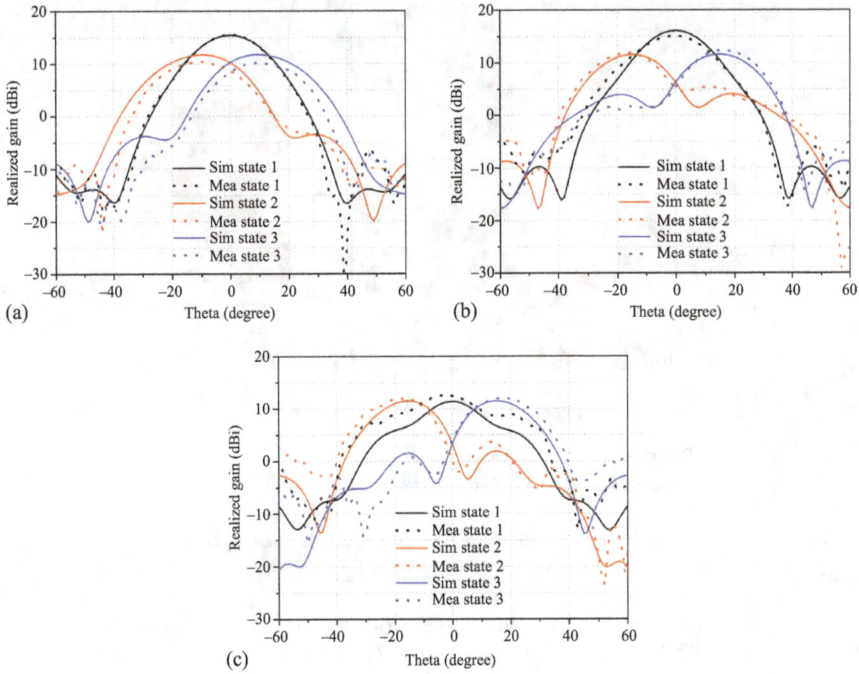

Figure 6.21 Realized gain patterns of the proposed antenna at (a) 5.3 GHz; (b) 5.5 GHz; and (c) 5.7 GHz [53]

feed network to provide quadri-polarization reconfigurability in two orthogonal LPs and two orthogonal CPs. The impedance bandwidth is 3.67% at 2.45 GHz with a 5 dB realized gain. A traditional annular slot antenna can switch its polarizations employing an arrow-shape coupling strip with a PIN diode inserted in each arm. The bandwidth of 4.7% is achieved with measured gains around 4.1 dB [63].

Most of the reported polarization RAs have an overlapped impedance bandwidth of less than 10%. Furthermore, if the bandwidth is considered for the overlapped 10 dB impedance bandwidth and the 3 dB axial ratio (AR) bandwidth, which is more appropriate for practical applications, the reported antennas would have even narrower bandwidth. Therefore, it is noted that to realize a wide bandwidth is quite a challenge for polarization RAs. In [64], a wideband polarization RA is proposed that can switch between LP and CPs. This antenna consists of a patch, a ground plane, and a cross-feeding probe which can be reconfigured by PIN diodes. The overlapped AR bandwidth and the impedance bandwidth is 16.4% with a 7.7 dBic CP gain.

Regarding the realized gain of the polarization RAs, as most of the designs employ a single microstrip patch, the gain is very limited. A polarization RA array is presented in [65]. It can switch between two orthogonal ±45° LPs with a maximum gain of 13.5 dBi. Another method to enhance the antenna gain is to utilize a PRS structure as the superstrate. In [66], a PRS polarization RA is presented using a

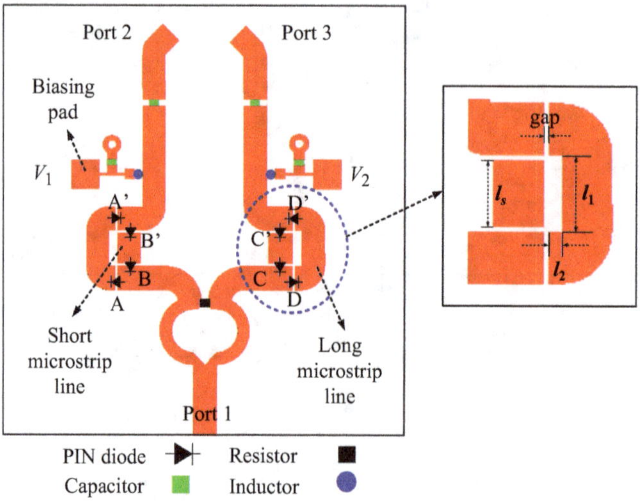

Figure 6.22 Structure of the feed network [67]

microstrip patch antenna with switchable slots as the source. The antenna can radiate either right-hand circular polarization (RHCP) or left-hand circular polarization (LHCP) by controlling the status of the switches. A 1.5% CP bandwidth is achieved with a simulated realized gain of 21 dBi.

Ji *et al.* reported a novel wideband PRS antenna [67], which can electronically alter its polarization between LP, LHCP, and RHCP. Measurement results show that the antenna has an overlapped 10 dB impedance bandwidth and 3 dB axial-ratio bandwidth of 4.7–5.36 GHz (13.1%). The average realized gains are around 9 dBi/dBic, which remain reasonably stable for different polarizations.

To realize polarization reconfigurability between LP and two CP modes, a reconfigurable feed network is designed. The schematic of the feed network is shown in Figure 6.22. It is based on the conventional Wilkinson power divider with a phase shifter inserted on each branch. The details of the phase shifter are shown in the inset of Figure 6.22. The phase shifter is comprised of two separate microstrip lines (a short one and a long one as shown in Figure 6.22) and each line has two gaps etched at the edges for inserting PIN diodes. The long microstrip lines are about $\lambda_g/4$ longer than the short microstrip lines, where λ_g is the guided wavelength at the resonant frequency. By switching the diodes on and off, the length of each branch of the power divider can be varied, which produces the phase difference between the two output ports and in turn realizes polarization reconfigurability. Four groups of PIN diodes are used in this design (A and A′, B and B′, C and C′, D and D′). When diode groups A and D are switched on and the others are off, no phase difference exists between the two branches. This state is referred to as S1. When the diodes A, A′ and C, C′ are switched on, and the others are switched off, the left branch will generate a 90° delay to the right one, which is called S2; When the diodes B, B′ and D, D′ are switched on with the others off, a 90° phase lead can

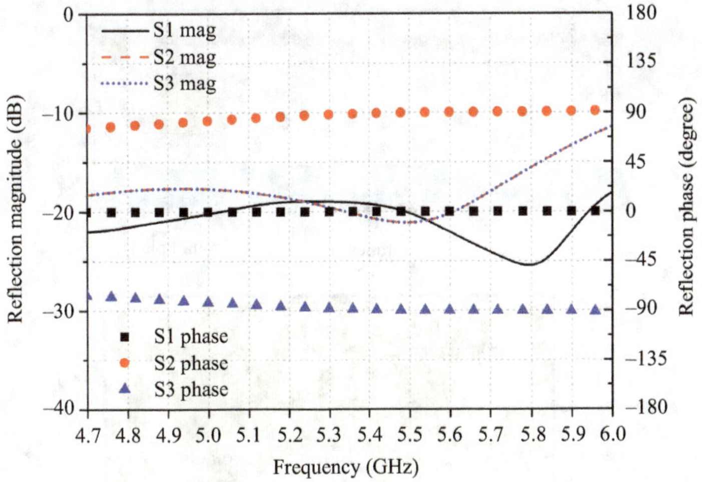

Figure 6.23 Simulated results of the proposed network [67]

be produced at the left branch and this state is referred to as S3. This way, the antenna fed by the proposed network can achieve polarization switching among LP (S1), LHCP (S2), and RHCP (S3).

In order to verify the operation of the feed network, a prototype designed on an 80 mm × 50 mm × 1.524 mm Rogers 4003 substrate was simulated. The biasing of the feed network is also shown in Figure 6.22. The PIN diodes on the two short microstrip lines are oriented in the same directions, and those on the two long microstrip lines are placed in the opposite directions. A capacitor is inserted on the top part of each branch not only to realize RF continuity, but to provide DC isolation for the two biasing voltages (V_1 and V_2) on the two branches. When V_1 and V_2 are connected together and applied to a negative voltage relative to Port 1 that is connected to the ground, the diodes group A and group D on the two long lines are switched on, which results in a 0° phase difference between Port 2 and Port 3 and leads to a LP. A 90° phase difference between Port 2 and Port 3 can be achieved by applying a positive voltage to V_1 and grounding V_2. This way, a RHCP can be obtained, and vice versa.

For brevity, we only report the investigations on the magnitude of the input reflection coefficient for Port 1 ($|S_{11}|$) and the phase difference φ_{23} between Port 2 and Port 3. It is observed from Figure 6.23 that $|S_{11}|$ is below -10 dB across the frequency band ranging from 4.7 GHz to 6.0 GHz for all the three states. Furthermore, the curves of φ_{23} for the three states (S1, S2, and S3) remain reasonably flat with the values approximately 0°, −90°, and 90°, respectively, in this frequency band.

Figure 6.24 shows the configuration of the polarization-reconfigurable PRS antenna. It consists of a PRS structure on the top, a shorted annular patch (SAP) antenna as the source locating above the ground plane, and the proposed feed network underneath the ground plane. The dimensions of the entire antenna are 110 mm × 110 mm ($1.83\lambda \times 1.83\lambda$ at 5 GHz). The PRS material is constructed

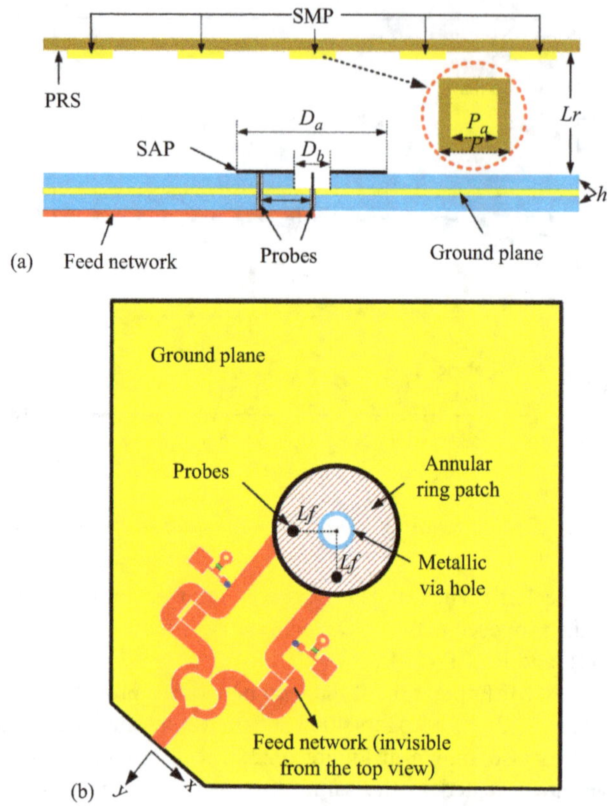

Figure 6.24 Schematics of the proposed polarization-reconfigurable PRS structure [67]: (a) side view and (b) top view of the source

from a 0.8 mm thick FR4 substrate and consists of an array of 5 × 5 square metallic patches. The length of the patch cells (*Pa*) is 18 mm and the periodicity between two adjacent patches (*P*) is 24 mm as shown in the inset of Figure 6.24(a). The cavity height (*Lr*) is set to be 32 mm. An SAP antenna is chosen as the source antenna since it can easily realize a CP when two feed probes with 90° phase difference are employed [68]. In addition, it has more flexibility than a microstrip patch antenna to optimize the impedance matching.

The source is a two-layer structure fed by two probes with a diameter of 1.3 mm, as shown in Figure 6.24(b). For the first layer, an SAP antenna is etched on the upper side of a 1.524-mm-thick Rogers 4003 substrate ($\varepsilon_r = 3.55$). It is designed according to the procedure described in [68]. The diameter of the annular patch (D_a) is 24.6 mm. The patch is shorted to the ground by plating a metallic via hole with a diameter of 11.6 mm (D_b). The metal of the other side of the substrate is reserved as the ground plane. For the second layer, the feed network is printed on one side of another 1.524-mm-thick Rogers 4003 substrate and the metal of the

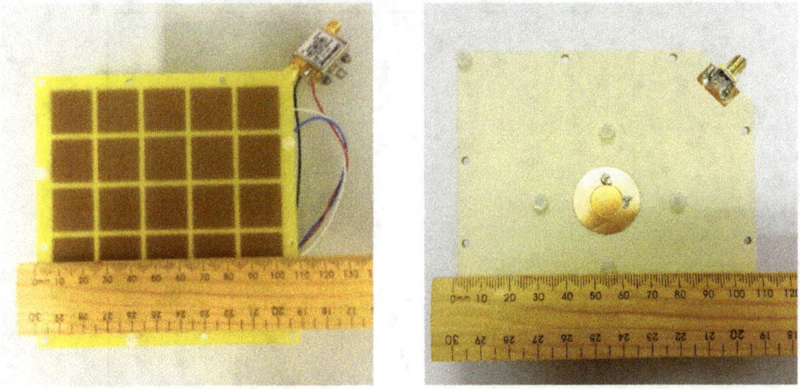

(a) (b)

Figure 6.25 Photographs of the proposed antenna [67]: (a) with the PRS structure and (b) without the PRS structure

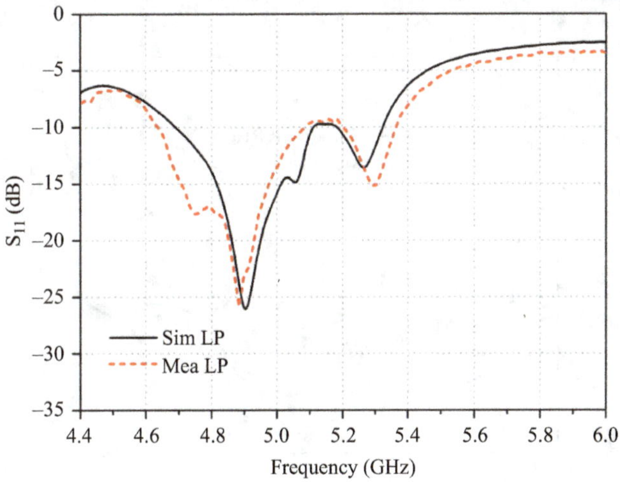

Figure 6.26 Input reflection coefficients of the reconfigurable PRS antenna for LP [67]

other side of this substrate is etched off. The SAP antenna and the feed network are connected by two metal probes with a diameter of 1.3 mm. The probes are located at 7.4 mm (L_f) from the center of the SAP antenna and are 90° sequentially rotated. The probes are isolated from the ground by cutting two circular slots around them with a radius of 1 mm.

A prototype has been fabricated and measured, as shown in Figure 6.25. In Figures 6.26 and 6.27, it is observed that the measured 10 dB impedance

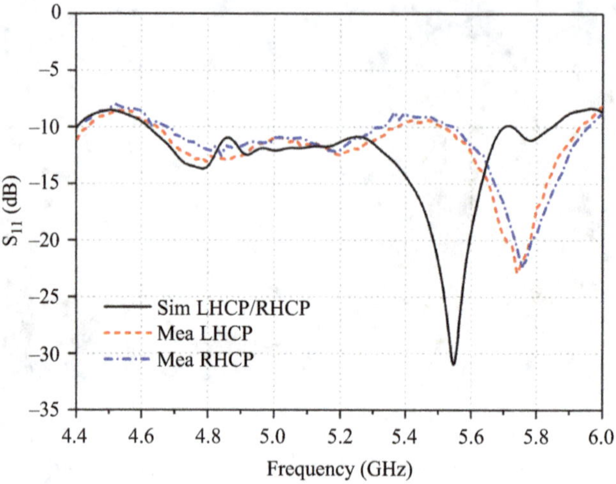

Figure 6.27 Input reflection coefficients of the reconfigurable PRS antenna for CP [67]

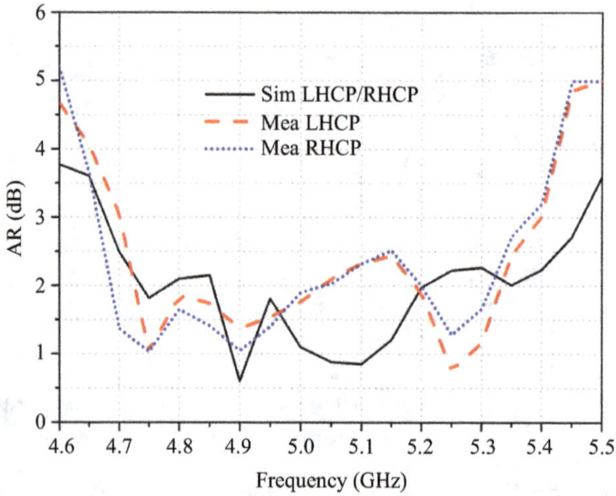

Figure 6.28 Axial ratio of the reconfigurable PRS antenna [67]

bandwidth can cover a frequency band from 4.65 to 5.36 GHz for LP and from 4.63 to 5.95 GHz for LHCP and RHCP. Figure 6.28 gives the simulated and measured ARs for the CP modes. It can be seen that from 4.7 to 5.38 GHz (13.5%) the measured AR is below 3 dB for both LHCP and RHCP. Considering the overlapped 10 dB impedance bandwidth of the three polarizations mentioned above, the overlapped frequency bandwidth ranges from 4.7 to 5.36 GHz (13.1%) for realizing polarization switching between LP and two CP modes.

6.4 Conclusions

In this chapter, an overview of the recent development of reconfigurable high-gain antennas is presented. These antennas range from reconfigurable Yagi antennas, reconfigurable reflectarrays and reconfigurable transmitarrays to reconfigurable PRS antennas. It is demonstrated that PRS antennas are able to reconfigure their frequency, polarization and radiation patterns with high realized gains. Compared with RA arrays, reconfigurable reflectarrays and transmitarrays, it is much easier for reconfigurable PRS antennas to achieve various reconfigurations with a simpler biasing network. In order to fully exploit the diversities of antennas in wireless communication and radar systems, compound reconfigurations of two or three antenna parameters are highly desired. However, there have been no reported products employing such combined RAs. Since PRS antennas have more freedom of parameters (a single feeding source/feeding source arrays and a PRS structure), the above challenge could be addressed by allocating different reconfigurations to different parts. As this work progresses, there is no doubt that fully reconfigurable PRS antennas will deliver significant benefits to practical wireless systems in future.

References

[1] J. T. Bernhard, "Reconfigurable antennas," San Rafael, CA: Morgan & Claypool Publishers, Feb. 2005.

[2] L. Volakis, Antenna engineering handbook, 4th ed., New York: McGraw-Hill, 2007.

[3] Y. J. Guo, and P.-Y. Qin, "Reconfigurable Antennas for Wireless Communications," in "Handbook of Antenna Technologies," edited by Z. N. Chen, Singapore: Springer Science and Business Media, 2016.

[4] C. G. Christodoulou, Y. Tawk, S. A. Lane, and S. R. Erwin, "Reconfigurable antennas for wireless and space applications," *Proc. IEEE*, vol. 100, no. 7, pp. 2250–2261, Jul. 2012.

[5] P.-Y. Qin, A. R. Weily, Y. J. Guo, T. S. Bird, and C. H. Liang, "Frequency reconfigurable quasi-Yagi folded dipole antenna," *IEEE Trans. Antennas Propag.*, vol. 58, no. 8, pp. 2742–2747, Aug. 2010.

[6] Y. Cai, Y. J. Guo, and P. -Y. Qin, "Frequency switchable printed Yagi–Uda dipole sub-array for base station antennas," *IEEE Trans. Antennas Propag.*, vol. 60, no. 3, pp. 1639–1642, Mar. 2012.

[7] S. Zhang, G. H. Huff, J. Feng, and J. T. Bernhard, "A pattern reconfigurable microstrip parasitic array," *IEEE Trans. Antennas Propag.*, vol. 52, no. 10, pp. 2773–2776, Oct. 2004.

[8] M. Donelli, R. Azaro, L. Fimognari, and A. Massa, "A planar electronically reconfigurable Wi-Fi band antenna based on a parasitic microstrip structure," *IEEE Antennas Wireless Propag. Lett.*, vol. 6, pp. 623–626, 2007.

[9] S. Lim and H. Ling, "Design of electrically small pattern reconfigurable Yagi antenna," *Electron. Lett.*, vol. 43, no. 24, pp. 1326–1327, Nov. 2007.

[10] P. Y. Qin, Y. J. Guo, and C. Ding, "A beam switching quasi-Yagi dipole antenna," *IEEE Trans. Antennas Propag.*, vol. 61, no. 10, pp. 4891–4899, Oct. 2013.
[11] Y. Y. Bai, S. Xiao, M. -C. Tang, Z. -F. Ding, and B. -Z. Wang, "Wide-angle scanning phased array with pattern reconfigurable elements," *IEEE Trans. Antennas Propag.*, vol. 59, no. 11, pp. 4071–4076, Nov. 2011.
[12] X. Ding, B.-Z. Wang, and G.-Q. He, "Research on a millimeter-wave phased array with wide-angle scanning performance," *IEEE Trans. Antennas Propag.*, vol. 61, no. 10, pp. 5319–5324, Oct. 2013.
[13] C. Ding, Y. J. Guo, P.-Y. Qin, and Y. Yang, "A compact microstrip phase shifter employing reconfigurable defected microstrip structure (RDMS) for phased array antennas," *IEEE Trans. Antennas Propag.*, vol. 63, no. 5, pp. 1985–1995, May 2015.
[14] R. C. Hansen, Phased array antennas. New York: Wiley, 1998.
[15] S. V. Hum and J. Perruisseau-Carrier, "Reconfigurable reflectarrays and array lens for dynamic antenna beam control: a review," *IEEE Trans. Antennas Propag.*, vol. 62, no. 1, pp. 183–198, Jan. 2014.
[16] P. Nayeri, F. Yang, and A. Z. Elsherbeni, "Beam-scanning reflectarray antennas: A technical review and state of the art," *IEEE Antennas Propag. Mag.*, vol. 57, no. 4, pp. 32–47, Aug. 2015.
[17] G. V. Trentini, "Partially reflecting sheet arrays," *IRE Trans. on Antenna Propag.*, vol. 4, no. 4, pp. 666–671, Oct. 1956.
[18] P. Y. Qin, Y. J. Guo, and A. R. Weily, "Broadband reflectarray antenna using subwavelength elements based on double square meander-line rings," *IEEE Trans. Antennas Propag.*, vol. 64, no. 1, pp. 378–383, Jan. 2016.
[19] Y. J. Guo and S. K. Barton, "Phase correcting zonal reflector incorporating rings", *IEEE Trans. Antennas Propag.*, vol. 43, no. 4, pp. 350–355, Apr. 1995.
[20] H. Kamoda, T. Iwasaki, J. Tsumochi, T. Kuki, and O. Hashimoto, "60-GHz electronically reconfigurable large reflectarray using single-bit phase shifters," *IEEE Trans. Antennas Propag.*, vol. 59, no. 7, pp. 2524–2531, Jul. 2011.
[21] H. Yang, F. Yang, S. Xu, *et al.*, "A 1-bit 10 × 10 reconfigurable reflectarray antenna: design, optimization, and experiment," *IEEE Trans. Antennas Propag.*, vol. 64, no. 6, pp. 2246–2254, Jun. 2016.
[22] M.-T. Zhang, S. Gao, Y. -C. Jiao, *et al.*, "Design of novel reconfigurable reflectarrays with single-bit phase resolution for Ku-band satellite antenna applications," *IEEE Trans. Antennas Propag.*, vol. 64, no. 6, pp. 1634–1641, May 2016.
[23] E. Carrasco, M. Barba, and J. Encinar, "X-band reflectarray antenna with switching-beam using PIN diodes and gathered elements," *IEEE Trans. Antennas Propag.*, vol. 60, no. 12, pp. 5700–5708, Dec. 2012.
[24] R. Pereira, R. Gillard, R. Sauleau, P. Potier, T. Dousset, and X. Delestre, "Dual linearly-polarized unit-cells with nearly 2-bit resolution for reflectarray applications in X-band," *IEEE Trans. Antennas Propag.*, vol. 60, no. 12, pp. 6042–6048, Dec. 2012.
[25] O. Bayraktar, O. A. Civi, and T. Akin, "Beam switching reflectarray monolithically integrated with RF MEMS switches," *IEEE Trans. Antennas Propag.*, vol. 60, no. 2, pp. 854–862, Feb. 2012.

[26] H. Yang, F. Yang, X. Cao, *et al.*, "A 1600-element dual-frequency electronically reconfigurable reflectarray at X/Ku-Band," IEEE Trans. Antennas Propag., vol. 65, no. 6, pp. 3024–3022, Jun. 2017.

[27] S. V. Hum, M. Okoniewski, and R. J. Davies, "Modeling and design of electronically tunable reflectarrays," *IEEE Trans. Antennas Propag.*, vol. 55, no. 8, pp. 2200–2210, Aug. 2007.

[28] S. V. Hum, M. Okoniewski, and R. J. Davies, "Realizing an electronically tunable reflectarray using varactor diode-tuned elements," *IEEE Microw. Wireless Compon. Lett.*, vol. 15, no. 6, pp. 422–424, Jun. 2005.

[29] F. Venneri, S. Costanzo, and G. D. Massa, "Design and validation of a reconfigurable single varactor-tuned reflectarray," *IEEE Trans. Antennas Propag.*, vol. 61, no. 2, pp. 635–645, Feb. 2013.

[30] M. Riel and J. J. Laurin, "Design of an electronically beam scanning reflectarray using aperture-coupled elements," *IEEE Trans. Antennas Propag.*, vol. 55, no. 5, pp. 1260–1266, May 2007.

[31] D. F. Sievenpiper, J. H. Schaffner, H. J. Song, R. Y. Loo, and G. Tangonan, "Two-dimensional beam steering using an electrically tunable impedance surface," *IEEE Trans. Antennas Propag.*, vol. 51, no. 10, pp. 2713–2722, Oct. 2003.

[32] X. Yang, S. Xu, F. Yang, *et al.* "A broadband high-efficiency reconfigurable reflectarray antenna using mechanically rotational elements," *IEEE Trans. Antennas Propag.*, vol. 65, no. 8, pp. 3959–3966, May 2017.

[33] V. F. Fusco, "Mechanical beam scanning reflectarray," *IEEE Trans. Antennas Propag.*, vol. 53, no. 11, pp. 3842–3844, Nov. 2005.

[34] X. Yang, S. Xu, F. Yang, *et al.*, "Design of a circularly polarized reconfigurable reflectarray using micromotors," in *Int. Symp. Antennas Propag. Soc.*, pp. 2155–2156, Jul. 2015.

[35] J. Y. Lau and S. V. Hum, "Reconfigurable transmitarray design approaches for beam forming applications," *IEEE Trans. Antennas Propag.*, vol. 60, no. 12, pp. 5679–5689, Dec. 2012.

[36] C.-C. Cheng, B. Lakshminarayanan, and A. Abbaspour-Tamijani, "A programmable lens-array antenna with monolithically integrated MEMS switches," *IEEE Trans. Microw. Theory Tech.*, vol. 57, no. 8, pp. 1874–1884, Aug. 2009.

[37] J. Y. Lau and S. V. Hum, "A wideband reconfigurable transmitarray element," *IEEE Trans. Antennas Propag.*, vol. 60, no. 3, pp. 1303–1311, Mar. 2012.

[38] J. G. Nicholls and S. V. Hum, "Full-space electronic beam-steering transmitarray with integrated leaky-wave feed," IEEE Trans. Antennas Propag., vol. 64, no. 8, pp. 3410–3422, Aug. 2016.

[39] C. Huang, W. Pan, X. Ma, *et al.*, "Using reconfigurable transmitarray to achieve beam-steering and polarization manipulation applications," *IEEE Trans. Antennas Propag.*, vol. 63, no. 11, pp. 4801–4810, Nov. 2015.

[40] L. D. Palma, A. Clemente, L. Dussopt, R. Sauleau, P. Potier, and P. Pouliguen, "Circularly-polarized reconfigurable transmitarray in Ka-band with beam scanning and polarization switching capabilities," *IEEE Trans. Antennas Propag.*, vol. 65, no. 2, pp. 529–540, Feb. 2017.

[41] A. Clemente, L. Dussopt, R. Sauleau, P. Potier, and P. Pouliguen, "Wideband 400-element electronically reconfigurable transmitarray in X band," *IEEE Trans. Antennas Propag.*, vol. 61, no. 10, pp. 5017–5027, Oct. 2013.

[42] P. Padilla, A. Muñoz-Acevedo, M. Sierra-Castaner, and M. SierraPérez, "Electronically reconfigurable transmitarray at Ku band for microwave applications," *IEEE Trans. Antennas Propag.*, vol. 58, no. 8, pp. 2571–2579, Aug. 2010.

[43] A. P. Feresidis and J. C. Vardaxoglou, "High gain planar antenna using optimised partially reflective surfaces," *IEE Proc. Microw. Antennas Propag.*, vol. 148, no. 6, pp. 345–350, Dec. 2001.

[44] Y. Ge, K. P. Esselle, and T. S. Bird, "The use of simple thin partially reflective surfaces with positive reflection phase gradients to design wideband, low-profile EBG resonator antennas," *IEEE Trans. Antennas Propag.*, vol. 62, no. 2, pp. 743–750, Feb. 2012.

[45] A. R. Weily, T. S. Bird, and Y. J. Guo, "A reconfigurable high-gain partially reflecting surface antenna," *IEEE Trans. Antenna Propag.*, vol. 56, no. 11, pp. 3382–3390, Nov. 2008.

[46] A. Ourir, S. N. Burokur, and A. de Lustrac, "Phase-varying metamaterial for compact steerable directive antennas," *Electron. Lett.*, vol. 43, no. 9, pp. 493–494, Apr. 2007.

[47] A. Ourir, S. N. Burokur, and A. de Lustrac, "Electronic beam steering of an active metamaterial-based directive subwavelength cavity," in *2nd European Conference* on. *Antennas and Propagation*, Edinburgh, Nov. 2007.

[48] A. Ghasemi, S. N. Burokur, A. Dhouibi, and A. de Lustrac, "Inductive-varying grid for highly beamsteerable cavity antennas," *Electron. Lett.*, vol. 49, no. 5, pp. 319–321, Feb. 2013.

[49] T. Debogovic, and J. Perruisseau-Carrier, "Array-fed partially reflective surface antenna with independent scanning and beamwidth dynamic control," *IEEE Trans. Antenna Propag.*, vol. 62, no. 1, pp. 446–449, Jan. 2014.

[50] R. Guzman-Quiros, J. L. Gomez-Tornero, A. R. Weily, and Y. J. Guo, "Electronically steerable 1-D Fabry-Perot leaky-wave antenna employing a tunable high impedance surface," *IEEE Trans. Antenna Propag.*, vol. 60, no. 11, pp. 5046–5055, Nov. 2012.

[51] R. Guzman-Quiros, J. L. Gomez-Tornero, A. R. Weily, and Y. J. Guo, "Electronic full-space scanning with 1-D Fabry–Perot LWA using electromagnetic band gap," *IEEE Antennas Wireless Propag. Lett.*, vol. 11, pp. 1426–1429, 2012.

[52] R. Guzman-Quiros, J. L. Gomez-Tornero, A. R. Weily, and Y. J. Guo, "A Fabry–Pérot antenna with two-dimensional electronic beam scanning", *IEEE Trans. Antenna Propag.*, vol. 64, no. 4, pp. 1536–1541, Apr. 2016.

[53] L. Y. Ji, Y. J. Guo, P. Y. Qin, S. X. Gong, and R. Mittra, "A reconfigurable partially reflective surface (PRS) antenna for beam steering," *IEEE Trans. Antennas Propag.*, vol. 63, no. 6, pp. 2387–2395, Jun. 2015.

[54] S. Gao, A. Sambell, and S. S. Zhong, "Polarization-agile antennas," *IEEE Antennas Propag. Mag.*, vol. 48, no. 3, pp. 28–37, Jun. 2006.

[55] P.-Y. Qin, Y. J. Guo, and C. H. Liang, "Effect of antenna polarization diversity on MIMO system capacity," *IEEE Antennas Wireless Propag. Lett.*, vol. 9, pp. 1092–1095, 2010.

[56] S. Armstrong, Circular polarization vs. linear polarization: Which is the right RFID antenna? [Online]. Available e-mail: http://blog.atlasrfidstore.com/circular-polarization-vs-linear-polarization (Apr. 2013)

[57] A. Kyhle, "Device for switching between linear and circular polarization using a rotatable depolarizer", U. S. Patent, 2684248, Jan. 2014.

[58] P.-Y. Qin, A. R. Weily, Y. J. Guo, and C.-H. Liang, "Polarization reconfigurable U-slot patch antenna," *IEEE Trans. Antennas Propag.*, vol. 58, no. 10, pp. 3383–3388, Oct. 2010.

[59] Z.-X. Yang, H.-C. Yang, J. S. Hong, and Y. Li, "Bandwidth enhancement of a polarization-reconfigurable patch antenna with stair-slots on the ground," *IEEE Antennas Wireless. Propag. Lett.*, vol. 13, pp. 579–582, 2014.

[60] L. Zhang, S. Gao, and Q. Luo, "Polarization reconfigurable loop antenna for satellite communications," in *Antennas and Propagation Conference (LAPC)*, Loughborough, pp. 649–652, Nov. 10–11, 2014.

[61] J.-F. Tsai and J.-S. Row, "Reconfigurable square-ring microstrip antenna," *IEEE Trans. Antennas Propag.*, vol. 61, no. 5, pp. 2857–2860, May 2013.

[62] Y.-F. Wu, C.-H. Wu, D.-Y. Lai, and F.-C. Chen, "A reconfigurable quadri-polarization diversity aperture-coupled patch antenna," *IEEE Trans. Antennas Propag.*, vol. 55, no. 3, pp. 1009–1012, Mar. 2007.

[63] J.-S. Row, W.-L. Liu, and T.-R. Chen, "Circular polarization and polarization reconfigurable designs for annular slot antennas," *IEEE Trans. Antennas Propag.*, vol. 60, no. 12, pp. 5998–6002, Dec. 2012.

[64] K. Wang and H. Wong, "A reconfigurable CP/LP antenna with cross-probe feed," *IEEE Antennas Wireless. Propag. Lett.*, vol. 16, pp. 669–672, 2017.

[65] W. Lin and H. Wong, "Polarization reconfigurable aperture-fed patch antenna and array," *IEEE Access*, vol. 4, pp. 1510–1517, Apr. 2016.

[66] W. Han, J. Ouyang, Z. Guo, Y. Li, and F. Yang, "A single-feed high gain Fabry–Perot antenna with reconfigurable polarization capability," in Proc. Cross Strait Quad-Reg. Radio Sci. Wireless Technol. Conf. (CSQRWC'13), Jul. 21–25, 2013, pp. 279–281.

[67] L. Y. Ji, P. Y. Qin, Y. J. Guo, C. Ding, and S. X. Gong, "A wideband polarization reconfigurable antenna with partially reflective surface," *IEEE Trans. Antennas Propag.*, vol. 64, no. 10, pp. 4534–4538, Oct. 2016.

[68] G. J. K. Moernaut, and G. A. E. Vandenbosch, "Concept study of a shorted annular patch antenna: Design and fabrication on a conducting cylinder," *IEEE Trans. Antennas Propag.*, vol. 59, no. 6, pp. 2097–2102, Jun. 2011.

Chapter 7

Microfluidically reconfigurable antennas

Gokhan Mumcu[1]

7.1 Introduction

Reconfigurable radio frequency (RF) devices are highly attractive for reducing the complexity, cost, and size of RF front-ends that will serve the needs of emerging multifunctional wireless systems, sensor networks, and remote-sensing/imaging applications. A subset of device performance parameters such as size, cost, power-handling capability, frequency-tuning bandwidth, reconfiguration speed, efficiency, frequency-agility (i.e., capability to maintain specified performance across the frequency range), and beam-steering is often considered as the key metrics when evaluating the usefulness of the reconfigurable RF device for a particular given application. Simultaneously excelling in the entirety of these performance parameters has unfortunately not been possible although many technologies have been investigated over the past two decades. These technologies have been primarily based on material loadings (e.g., ferrites [1] and ferroelectrics [2]), semiconductor devices (e.g., varactor [3] and PIN diodes [4]), ferroelectric varactors [5,6], microelectromechanical systems (MEMS) switches, and capacitors [7,8]. Literature review reveals that these technologies offer compact and cost-effective device implementations with high reconfiguration speeds, but suffer in other performance parameters that tend to be more closely related to the RF performance (e.g., power-handling, frequency-tuning bandwidth, efficiency, frequency agility, and beam-steering). For example, material loadings with ferrites and ferroelectrics result in low frequency-tuning range ($<<2:1$) due to the limitations in achievable permeability and permittivity variations. The loss of these materials often hinders the efficiency performance as well. Semiconductor and ferroelectric varactors [9,10] lead to devices with small frequency-tuning ranges and low efficiencies due to the limited capacitance variation and high equivalent series resistances, respectively. Moreover, varactor-loaded devices exhibit a low power-handling capability that is limited by the non-linearities (e.g., third-order intermodulation products) and the thermal conductivity of the materials that form the varactor. The small size of the varactor, although desirable for reducing the RF

[1]Department of Electrical Engineering, University of South Florida, USA

parasitics and increasing the highest frequency of operation, may further exacerbate the power-handling issue by contributing to the generation of thermally hot spots. RF MEMS switches and capacitors have been found very promising for low-loss device implementations, however, they often do not allow for continuous frequency-tuning [11] or high power-handling capabilities [12,13]. Utilization of RF MEMS or semiconductor-based switches and phase shifters for addressing the emerging need of high-gain mm-wave beam-steering antenna arrays seems to be very challenging due to the extreme costs associated with mm-wave components and hardware integration with control and bias circuitries. Consequently, novel alternative techniques and design approaches for reconfigurable RF devices continue to be popular active research areas within the electromagnetics community.

Microfluidics has drawn interest as an alternative technique for reconfiguration of RF devices notably due to its promise for enhancing RF performance in terms of efficiency, frequency-tuning range, frequency-agility, power-handling capability, and beam-steering. The microfluidics-based reconfiguration principles mainly rely on loading a conventionally manufactured RF device (e.g., on printed circuit board (PCB)) with movable fluids in order to change the material properties, achieve capacitive loading, and/or alter the physical length. Although some works utilize plastic/Teflon tubing for housing of the fluids, a more common and desirable approach is to construct *microfluidic channels* by bonding a micromolded polymer (such as polydimethylsiloxane (PDMS) [14]) or micromachined substrate (such as glass or acrylic [15]) that acts as side and top walls of the channel with another planar material layer that acts as the bottom wall of the channel. The planar material layer can potentially be a spin-coated polymer (such as PDMS [16] and benzocyclobutene (BCB) [17]) on a printed RF device to control/minimize the wall thickness in order to achieve significant enhancement in RF interaction between the printed device and movable fluidics inside the channel. *Liquid metals* such as mercury, Galinstan and EGaIn are popular fluid choices due to their conductivity values ranging from 1×10^6 to 3.4×10^6 S/m. Despite being an order of magnitude lower than conductivity of copper, liquid metals have been demonstrated to result in reconfigurable antennas with high radiation efficiencies within the traditional wireless communications spectrum below 6 GHz. Employing liquid metals within structures/channels fully formed from PDMS or other elastomeric polymers have been introduced as a convenient technique for realizing highly efficient flexible and deformable antennas [14,18,19]. Moreover, such antenna configurations allow for frequency reconfiguration through changing the antenna length with substrate stretching [14,20]. This reconfiguration technique is limited by the elasticity of the materials and may be cumbersome due to the number of mechanical parts and very low actuation speeds. Therefore, dynamically changing the shape of the liquid metal volume via pressure or electrically driven actuators such as micropumps or electrowetting has gained more traction as research groups further got interested in the microfluidically reconfigurable RF devices. In these types of antennas, liquid metal volume is often placed in a carrier fluid that has low loss permittivity for micropump-based actuation or in an electrolyte solution with low conductivity for electrowetting-based actuation. The technique has led to the development of compact monopole, dipole, slot, and coplanar waveguide (CPW) antennas and arrays [15,16,21–25] with very wide frequency-tuning ranges (>2:1), high radiation efficiencies (>85%), and polarization reconfiguration capabilities. Movable liquid metals have also been utilized for constructing reflectors and directors of Yagi–Uda type of planar antennas to achieve beam-steering capability [26]. In [27], movable liquid metal volume has acted as a patch antenna that can be repositioned at the focal plane of a microwave lens to achieve low-cost and efficient beam-steering at mm-wave frequencies. Resolving the issues regarding actuation and long-term durability of liquid metals continue to be an active research area along with novel antenna designs that can further take advantage of their reconfiguration capabilities. On the other hand, several research groups have also investigated *liquid-metal free* devices within the concept of microfluidically reconfigurable RF devices. Dielectric liquids such as deionized (DI) water, colloidal dispersions, and acetone have been used for implementing frequency tunable [28,29] and polarization reconfigurable antennas [30] that are potentially promising for accomplishing wide frequency-tuning range, high radiation efficiency, and independent control of multi-band resonances. Metallized plates such as PCBs have been proposed as a long-term replacement of liquid metals within microfluidic channels as well. In this approach, the metallized plate is typically immersed within the microfluidic channel that is completely filled with a very low-loss dielectric solution (e.g., 3M Fluorinert FC-40) and repositioned by fluid flow that can be generated by a micropump unit in a close-loop fluidic cycle. The use of regular high-conductivity metals, selective metallization patters, and repeatable positioning with micropumps have allowed the metalized plate technique to be attractive for a variety of antennas to exhibit high power-handling capability [17] and low-cost efficient beam-steering capability in mm-wave frequencies [31–33].

The goal of this chapter is to present an overview of the microfluidically reconfigurable antennas by discussing antenna examples from the recent literature. The subsequent section briefly summarizes the fabrication and actuation techniques employed by several research groups. As a representative example, fabrication steps followed in for construction of a microfluidics-based mm-wave beam-steering antenna array is detailed [32,33]. The remaining sections group the discussion on the microfluidically reconfigurable antennas under "flexible and stretchable liquid metal antennas", "frequency-reconfigurable liquid metal antennas", "reconfigurable antennas using dielectric liquids", "beam-steerable liquid metal antennas", and "reconfigurable antennas using microfluidically repositionable metallized plates".

7.2 Fabrication and actuation techniques

Plastic and Teflon tubes have found some use in recent past to demonstrate microfluidic reconfiguration concepts when applied to RF devices such as frequency tunable antennas [24], beam-steering antennas [26], and frequency-agile filters [34]. However, the cylindrical cross section of the tubes and their relatively thick dielectric walls were shown to significantly hinder the capacitive loading

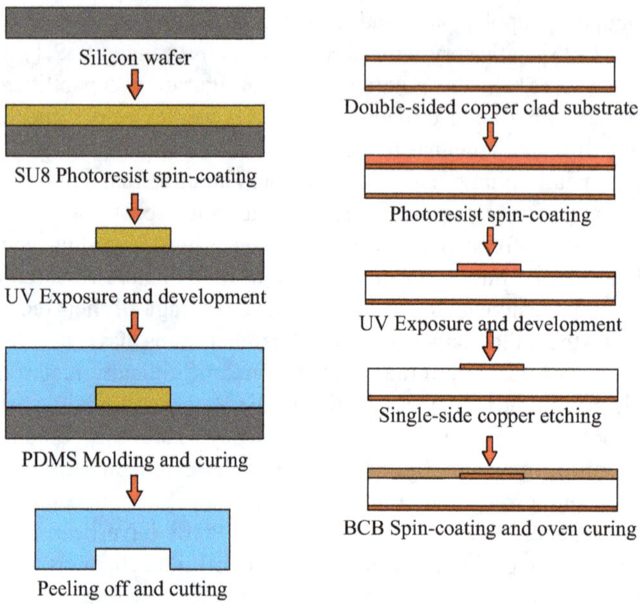

Figure 7.1 Fabrication process for micromolding of PDMS (left) and patterning/ coating of PCB (right). The two structures are bonded via a chemical treatment to form the microfluidically reconfigurable RF device

effect that can be achieved on planar RF devices. In addition, the tubed implementations often resulted in complicated structures. It is therefore not surprising to see that the majority of the recent literature focusing on more advanced fabrication techniques in order to achieve planar and high aspect ratio (width vs. height) microfluidic channels. Laser machining or micromachining of low-loss hard substrates such as acrylic and glass likely stands out as the most promising approach for constructing low loss microfluidic channels with reliable and long service lives. However, the machining costs, time, and experience needed for utilization of hard substrates as microfluidic channels mostly motivated researchers to work with soft elastomeric polymer materials such as PDMS and siloxane in device prototyping. An additional benefit of soft materials has also been their ability to flex and stretch to serve for the needs of conformable and reconfigurable RF devices—assuming the entire device structure is composed of them and RF excitation is carefully designed.

Figure 7.1 presents the basic fabrication steps followed in fabrication of microfluidic channels utilized in the construction of a mm-wave beam-steering antenna array (see Figure 7.2) through micromolding of PDMS. PDMS is known to exhibit a relative dielectric constant of $\varepsilon_r = \sim 2.8$ and loss tangent of $\tan\delta = \sim 0.045$ across the 1–40 GHz band making it suitable for both antenna and RF circuit applications [35]. The mold is obtained by spinning a negative photoresist SU-8

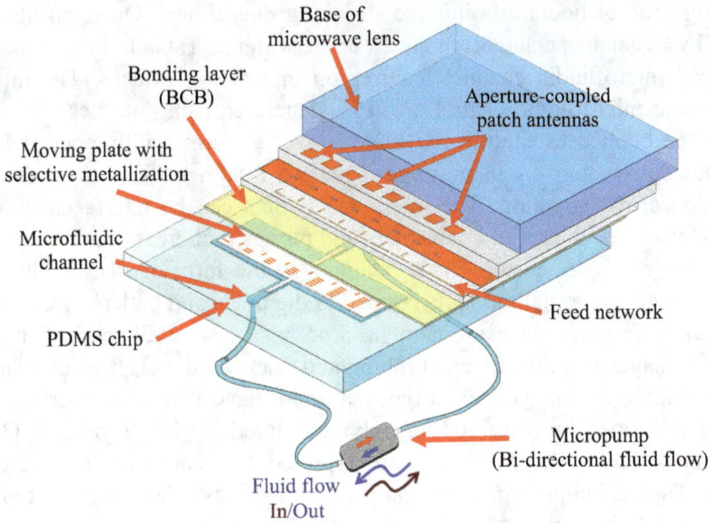

Figure 7.2 Substrate stack-up of a mm-wave 30 GHz focal plane array in which a selectively metallized plate repositionable within a microfluidic channel is utilized to perform switching functionality to activate a desired antenna element (i.e., beam direction) within the array

2075 onto a silicon wafer. SU-8 is suitable for achieving high aspect ratios in microfabrication. After spinning, it is soft-baked in a hot plate at 100 °C temperature for about 15–90 min depending on the desired thickness. The SU-8 layer's thickness defines the height of the microfluidic channel and it is controllable within a certain range (10 μm to several 100 μm) by the spinning speeds provided by the manufacturer. Additional coating can also be pursued over the already soft-baked SU-8 layer to increase or achieve better precision in its final thickness. For the shown array example, the SU-8 thickness is selected as 250 μm to be able to host a 220 μm thick selectively metallized plate within the microfluidic channel. The soft-baked SU-8 is patterned with UV light exposure, subjected to a post-exposure bake at 100°C, developed within a propylene glycol monomethyl ether acetate (PGMEA)-based solution, and hard-baked at 150°C for 20 min to obtain the desired microfluidic channel mold features. The PDMS oligomer and crosslinking prepolymer from the Sylgard 184 kit (Dow Corning) is mixed in a weight ratio of 10:1 and poured onto the SU-8 mold for curing. The quantity of PDMS poured onto the SU-8 mold defines the thickness of PDMS that will be laying over the microfluidic channel. For antenna and RF circuit applications, it is desirable to keep this thickness low to minimize power dissipation in PDMS. However, very thin layers, depending on the width and height of the microfluidic channel, could lead to issues like bulging/bucking as surface tension of liquids tend to dominate in microscale. For the shown array example, the PDMS layer thickness is adjusted to be ~3 mm. After degassing the mixed PDMS in a vacuum chamber, it is cured at room temperature for 24 hours to minimize shrinking due to heat. Once curing is completed, PDMS can be peeled off or cut from the wafer as a stand-alone unit carrying the desired microfluidic channel features on one of its surfaces. The inlets and outlets to the microfluidic channels can be generated using needles. This PDMS layer can be bonded to another flat PDMS layer to form a fully enclosed microfluidic device structure by using oxygen plasma treatment. This also results in a flexible/conformal structure. The inlets and outlets can be interfaced with commercially available microfluidic adapters and pumps to increase durability. For the shown array example, custom-made adapters in the form of PDMS blocks with holes to carry the external tubing is bonded to the inlets and outlets by carrying out an additional oxygen plasma treatment. Epoxy is also used to further prevent potential leakages from the external tubing and microfluidic channel interface. It is also important to note that different types of molds have also been reported. In [20], the mold is formed by laser cutting polymethylmethacrylate layers. In [16], 3D printed microfluidic channel molds are reported for realization of microfluidic channels. This technique can particularly prove itself as a rapid and cost effective fabrication technique as the 3D printing technology continues to advance in terms of process and feature sizes.

In many antenna and device applications, it is preferred to place the microfluidic channel over a planar RF circuit board to achieve enhanced reconfiguration capabilities. Figure 7.1 depicts the fabrication process for the preparation of an example single layer PCB. The patterning of the metallization features such as microstrip lines can be carried out with conventional photolithography by spin-coating photoresist, then proceeding with soft-bake, UV exposure/patterning, developing, metal etching and photoresist removal steps. Subsequently, a thin dielectric layer is spin-coated and cured over the metallization pattern. In several works, the dielectric layer consists of PDMS. However, better efficiency performances and thinner dielectric layers are possible when benzocyclobutene (BCB, $\varepsilon_r = {\sim}2.8$ and loss tangent of $\tan\delta = {\sim}0.002$–0.004 in the 1–40 GHz range) is employed. By being a thermoplastic material, bonding BCB with the PDMS layer (carrying the microfluidic channel) requires an additional processing step. In this step, the BCB-coated board is treated with oxygen plasma and it is placed within 2% v/v APTES (3-Aminopropyltriethoxysilane) solution heated to 70°C for 20 min. The oxygen plasma-treated PDMS layer and the BCB-coated board are later on brought in conformal contact for 15 min in a convection oven at 90°C to complete the bonding process [36]. During this step, a metallized plate can also be placed within the channel. The bonded structure could later be filled with liquid metal and/or dielectric liquids to achieve the intended reconfiguration functionality. Specifically, the shown array example utilizes 6 μm thick BCB layer over a microstrip line-based feed network that consist of 0.8 mm wide gaps. The microfluidic channel carries a selectively metalized plate and all of the remaining volume is filled with low loss 3M Fluorinert FC-40 ($\varepsilon_r = 1.9$, $\tan\delta \sim 0.001$) dielectric liquid. The selectively metalized plate has 1.8 mm long metallization patterns that overlap with these gaps as the plate is repositioned with liquid flow using external actuators such as piezoelectric micropumps. When the overlap happens, the microstrip line appears to be continuous (with insertion loss < 0.1 dB) due to the strong capacitive coupling between the selectively metallized plate and PCB traces thanks to the very thin BCB thickness. The PCB, in general, can be a multilayered structure for advanced functionality. In the shown array example, the board consists of two material layers where the second layer is utilized to realize the 30 GHz aperture-coupled patch antennas feeding the microwave lens. The selective metallization pattern of the microfluidically repositionable plate is adjusted to overlap with the gaps of the microstrip feed line in a way to individually and sequentially excite the antennas to achieve beam-steering functionality.

Actuation of microfluidically reconfigurable RF devices is an ongoing research area with the major goals of compact, efficient and integrated actuation mechanisms. For early concept and prototype demonstrations, manual operation with syringes or automated operation with bulky syringe pumps proves to be useful for certain devices. Since such devices operate with hydraulic pressure, their more compact versions can be readily constructed by utilizing piezoelectric micro-pumps such as the ones utilized in [17] with dimensions of $30 \times 15 \times 3.8$ mm^3, weight of 2 g and less than 200 mW power consumption. Controlling piezoelectric pumps with microprocessors can provide possibility to achieve accurate actuation lengths and actuation speeds. For faster actuation speeds, peristaltic pumps could be utilized, however, achieving positioning accuracy requires a more complicated close loop control approach. Repositioning of liquid metal volumes (possibly embedded within a carrier fluid such as a Teflon solution) has unique challenges associated with their oxidization related issues. Particularly, the nontoxic liquid metals such as Galinstan and EGaIn almost immediately form an oxide skin that drastically changes their surface properties. This causes them to stick on the walls of the microfluidic channels, leave oxidized metal residues, and fail in long-term device operation. Authors in Reference [37] experimentally demonstrated that such residues could greatly be avoided for certain carrier fluid choices and coatings applied over the surface of microfluidic channels. Authors in Reference [38] proposed the idea of placing the liquid metal within an electrolyte carrier fluid to continuously prevent formation of an oxide skin. Moreover, applying voltage potential to the conductive electrolyte fluid allows to electrically displace the liquid metal volume through the principles of electrowetting. Authors in Reference [15] accomplishes actuation of liquid metal through electrical voltage within capillary channels by making use of the large change in surface tension of liquid metals with oxidization. Specifically, in this technique, liquid metal is placed inside a conductive electrolyte and electrical voltage is applied to change its surface properties through removal or formation of the oxide skin. The capillary force of the channel can be used to reposition the liquid metal and formation of oxide skin stabilizes such movement due to the surface tension variation (electrochemically controlled capillary). Although promising for integrated devices, conductive electrolyte is prone to reduce the efficiency of the device particularly for high-frequency approaches. These drawbacks motivate further research for discovering the best actuation techniques for the microfluidically reconfigurable RF devices.

7.3 Flexible and stretchable liquid metal antennas

Liquid metals appear as a desirable choice for implementation of flexible and stretchable antennas due to their high level of conductivities as compared to other cost effective stretchable conductor choices such as conductive rubbers [39,40] and their multilayered variants [41,42]. A dipole antenna that is implemented by filling the microfluidic channels formed in PDMS with a nontoxic liquid metal eutectic gallium – indium (EGaIn) is presented in [14]. EGaIn is an eutectic mixture consisting of 75% gallium and 25% indium. It exhibits a melting point temperature and electrical conductivity of 15.7°C and 3.4×10^6 S/m, respectively. The two rectangular shaped microfluidic channels that define the arms of the dipole antenna are 25 mm in length and have a cross section of 0.5 (width) × 0.15 (height) mm^2. The total antenna length is 54 mm since these two channels are separated by 4 mm to implement the RF excitation. The dipole is excited by directly inserting conductive wires coming from a balanced to unbalanced transition (balun) into the tips of the channels. Due to the elasticity of its materials, the antenna can be reversibly deformed, flexed, and mechanically stretched. Once released, the antenna goes back to its original shape. Stretching the antenna along its length causes a reduction in its resonance frequency as well. At rest, the antenna is characterized to be well impedance matched and operate at 1.85 GHz with a ~11% $|S_{11}| < -10$ dB bandwidth and ~90% radiation efficiency. As the antenna length is increased via stretching, its resonance frequency can be reduced (e.g., 1.6 GHz at 66 mm length). The antenna can remarkably withstand to strains up to 40%. It is also important to note that the immediate oxide formation on the surface of nontoxic EGaIn (or Galinstan) benefits the operation of these kinds of flexible and stretchable antennas. Without the presence of the oxide skin, liquid metals exhibit high surface tensions and tend to withdraw/retract from microfluidic channels exhibiting high width to height aspect ratios. On the other hand, the oxide skin prevents the withdrawal by changing the surface tension and allowing the liquid metal to flow, fill, and remain in a stabilized position within the stretched and/or deformed microfluidic channels.

Figure 7.3 depicts an inverted cone antenna realized by filling microfluidic channels formed within a 1 mm thick PDMS using nontoxic liquid metal Galinstan [18]. Similarly to EGaIn, Galinstan is a eutectic mixture and consists of 68.5% gallium, 21.5% indium, and 10% tin. It exhibits an electrical conductivity of ~3.4×10^6 S/m and a melting point of –19°C. The surface area of the inverted cone antenna is realized from an interconnected grid of 0.1 mm high microfluidic channels to minimize the liquid metal volume and spread it over a large surface area. The antenna operates similar to the well-known volcano and circular disk antennas at lower operation frequencies by acting as a quarter wavelength monopole antenna placed over a large ground plane. On the other hand, the ground plane contributes to radiation at higher frequencies as well. In this design, it is shown that the gap between the cone and the rectangular ground plane determines the upper end of the operation band. The gap is designed as 0.3 mm and the overall antenna footprint is 45.3 × 40 mm^2. The antenna is excited by inserting two pins (soldered on inner and outer conductors of a coaxial cable) into the liquid metal present

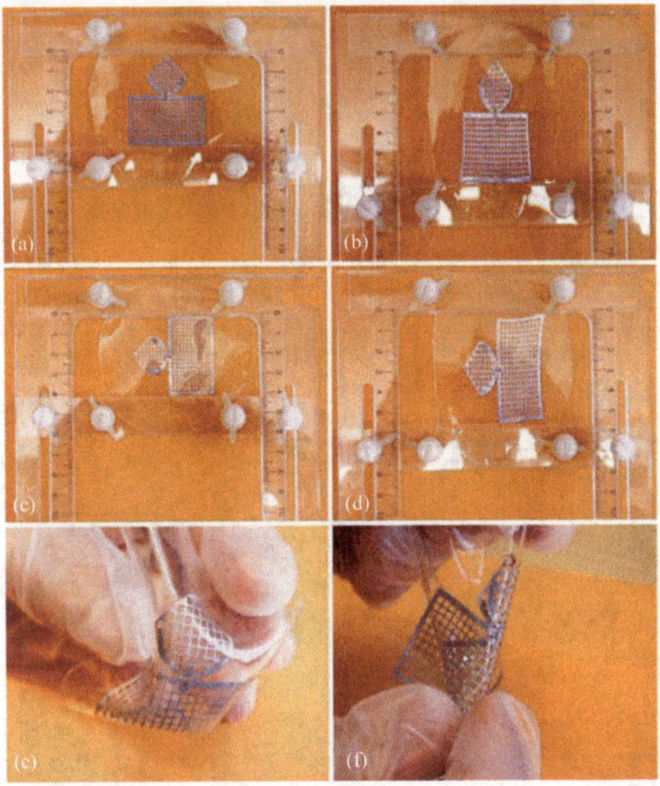

Figure 7.3 *Ultra-wideband inverted cone antenna constructed from liquid metal filled meshed microfluidic channels (after Cheng et al. [18] © IEEE 2009)*

around the gap. This ultra-wideband antenna performs within the 3 GHz – 11 GHz band with an $|S_{11}| < -10$ dB performance. It can be stretched up to 40% and folded. The stretch related elongation in a lateral axis does not degrade the impedance matching performance of the antenna. The antenna exhibits $> 70\%$ radiation efficiency across its bandwidth with near omni-directional radiation patterns.

Patch antenna applications of liquid metals at frequencies below mm-waves are challenging due to their large conductor surface area to thickness ratio. Large aspect ratio microfluidic channel implementations do not only suffer from liquid metals' high surface tension, but also from the low modulus of elasticity exhibited by the PDMS. This may cause the top wall of the channel to collapse and inhibit the flow within the microfluidic channel. Figure 7.4 depicts a flexible patch antenna introduced in [19] and implemented by utilizing serpentine channels separated from each other with walls consisting of a uniformly spaced row of PDMS posts. In rectangular microfluidic channels, minimum pressure to induce liquid flow (i.e., critical pressure) is inversely proportional to the width and height of the channel. Therefore, by keeping the distance among the posts much smaller than the width of the microfluidic

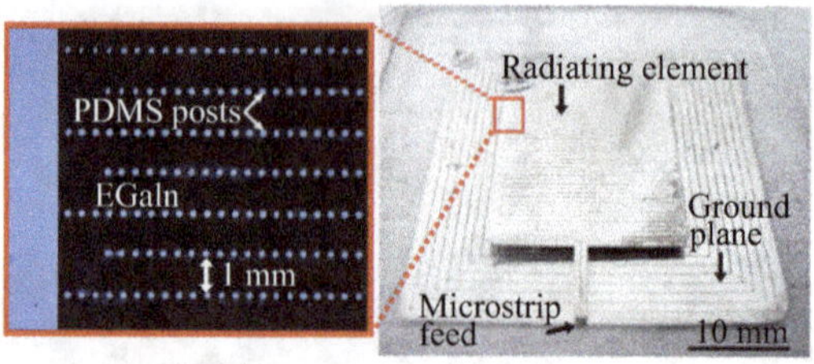

Figure 7.4 Flexible liquid metal patch antenna implemented in PDMS by using serpentine microfluidic channels separated by posts to achieve spreading of liquid metal over a large surface area (after Hayes et al. [19] © IEEE 2012)

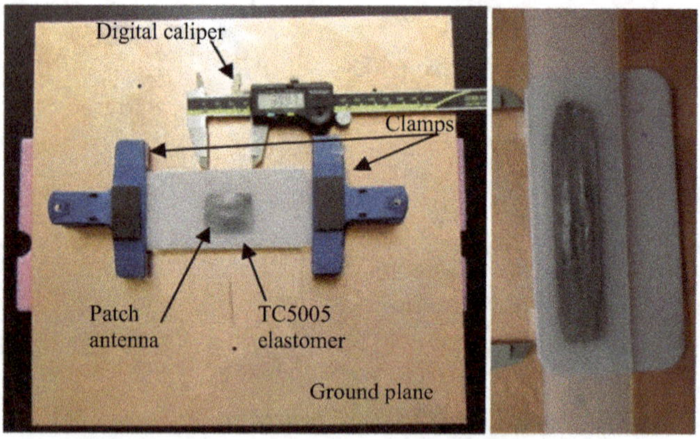

Figure 7.5 Stretchable frequency tunable liquid metal patch antenna implemented in silicone. A plastic isolation layer inserted within the channel helps to spread the liquid metal over a large surface area (after Mazlouman et al. [20] © IEEE 2011)

channel (0.1 mm vs. 1 mm), the flow of liquid metal is favored to follow the path of the serpentine channel during the liquid metal filling/injection stage. Once serpentine channels are filled with EGaIn and the liquid metal inlet is sealed, pressure is applied by pressing on the channels to merge liquids through the posts as well. The same technique is also applied for the realization of the antenna ground plane, resulting in a fully flexible antenna structure. The ground plane is 1 mm below the patch antenna and separated by a PDMS spacer. The microfluidic channels have 0.1 mm height and are located within 4 mm thick PDMS layers. The ground plane and patch antenna footprints are $37 \times 37\ \text{mm}^2$ and $20 \times 24\ \text{mm}^2$, respectively. The antenna is a narrowband design and not well impedance-matched internally. A broadside radiation with 60% radiation efficiency is reported, excluding the impedance mismatch.

Figure 7.5 presents an alternative patch antenna introduced in [20]. To achieve a large frequency-tuning range, reservoir that holds the liquid metal is implemented within a more stretchable Silicone TC5005 elastomer exhibiting a relative dielectric constant around ~2.8 to 3.1. The 0.5 mm thick reservoir is filled by Galinstan, however, a plastic isolation layer with comparable thickness is also inserted within the reservoir to prevent it from collapsing and to reduce its cross section to facilitate spreading of liquid metal over the reservoir's surface area. The ground plane of the antenna is conventionally manufactured and hosts the coupling slot exhibiting a Z-shape for better impedance matching. The silicone elastomer is able to withstand up to 700% stretching, and the antenna is tested up to 300% stretching by utilizing mechanical clamps. The footprint of the free-standing reservoir is 31 × 31 mm^2 and resides within two 40 × 150 mm^2 silicone layers. The ground plane is 5 mm below the patch and 300 × 300 mm^2 in footprint to accommodate a large stretch. The antenna exhibits less than 4% $|S_{11}| < -10$ dB bandwidth that further goes down as antenna length is enlarged. A frequency-tuning range from 1.3 GHz to 3 GHz is experimentally demonstrated. The radiation efficiency of the antenna is 80% and goes down with frequency due to the stretch related thinning of liquid metal and electrical thinning of the ground plane patch separation. More than 65% radiation efficiency is shown to be attainable for 100% or less stretching amounts.

7.4 Frequency-reconfigurable liquid metal antennas

Performance of frequency-reconfigurable antennas often suffer from the limitations of the RF control components (e.g., varactor diodes, PIN switches) employed within their construction. Liquid metals have drawn interest to address such limitations due to their promise in achieving lower loss and large frequency-tuning ratio. Figure 7.6(a) depicts a frequency-reconfigurable dipole antenna implemented by EGaIn filled 0.05 mm high microfluidic channels prepared within PDMS [21]. Each dipole arm consists of two 12.6 mm long and 1 mm wide microfluidic channel segments. These segments are separated from each other two rows of PDMS posts with 0.05 mm diameter and 0.1 mm center to center spacing. The oxide skin of EGaIn stabilizes it to remain outside of the volume between the rows of PDMS posts. In this state (State 1), dipole has its shortest length and operates at its highest frequency (3.6 GHz). It is excited centrally by two electrical leads from a coaxial balun. EGaIn segments in one of the arms can be merged together with pressure to achieve a lower frequency of operation (State 2, 2.6 GHz). Applying pressure can also enlarge the length of the remaining arm to achieve the lowest frequency of operation (State 3, 1.9 GHz). The antenna operates with good impedance matching and bandwidth typical to dipole antennas. A reconfiguration speed on the order of few ms is reported. The reconfiguration mechanism, however, is not reversible and dipole cannot be brought back into the states that operate at higher frequencies.

Figure 7.6(b) presents a slot antenna that can be reversibly reconfigured by pneumatically displacing a small volume of EGaIn over one of it ends [22]. The EGaIn volume is kept inside an air-filled microfluidic channel that is placed within the proximity of the slot antenna. When EGaIn volume is overlapped with the slot, it creates a capacitive loading effect to alter the resonance frequency. The slot

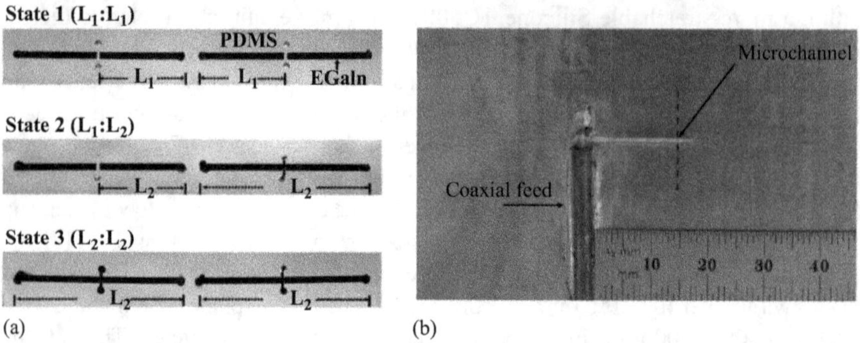

Figure 7.6 Frequency-reconfigurable antennas with discrete tuning states: (a) one-time reconfigurable dipole antenna (after Khan et al. [21] © AIP Publishing LLC 2011) and (b) microfluidically switched slot antenna (after King et al. [22] © IEEE 2013)

dimensions are 21.5×1 mm^2 and a coaxial feed is placed 1.5 mm away from one of the slot ends to achieve the RF excitation. The microfluidic channel is 0.5 mm below the ground plane with 0.5 mm diameter and 17 mm horizontal length. It is fabricated with an alternative technique that relies on weaving sacrificial polymer fibers into a composite textile reinforcement. The waved structure is later infiltrated with epoxy resin and heat treated under vacuum upon solidification of the epoxy to remove the sacrificial materials [43]. The EGaIn volume can be used to switch the operation frequency from 5.34 GHz to 4.6 GHz. Reconfiguration speed is reported as to be below 100 ms. Oxide skin of the EGaIn inhibits repeatability by leaving residues in the channel. However, evacuating the channel using pressurized air and rinsing with ethyl alcohol is reported as a method to clean the oxide skin residues. After this procedure, the antenna can be reconfigured again by inserting the needed EGaIn volume back into the microfluidic channel.

Realizing antenna metallizations from liquid metals is challenging due to the complexity of the feed structure. To alleviate this issue, the frequency-reconfigurable monopole antennas introduced in [23,44] exhibit a coupled feed structure as shown in Figure 7.7(a). Specifically, the microfluidic channel prepared in PDMS is bonded to a 25.4 μm thick liquid crystal polymer (LCP) layer and placed over a PCB that carries the microstrip line feed. The liquid metal within the microfluidic channel is separated from the microstrip line by this thin layer of LCP. However, the thickness is suitable to realize strong capacitive coupling and approximate an RF short condition. The microfluidic channel includes a volume of mercury as the antenna metallization. The remaining volume of the channel is filled by colorless low loss Teflon solution (DuPont AF 2400, 400S2-100). As shown in Figure 7.7(a), retracting the liquid metal volume to reside over the microstrip line prevents that volume from radiating. Consequently, the antenna can be operated over a very broad frequency-tuning range varying from 5.17 to 1.29 GHz. The $|S_{11}| < -10$ dB bandwidth of the antenna varies between 8.3% and 16.9% as it is

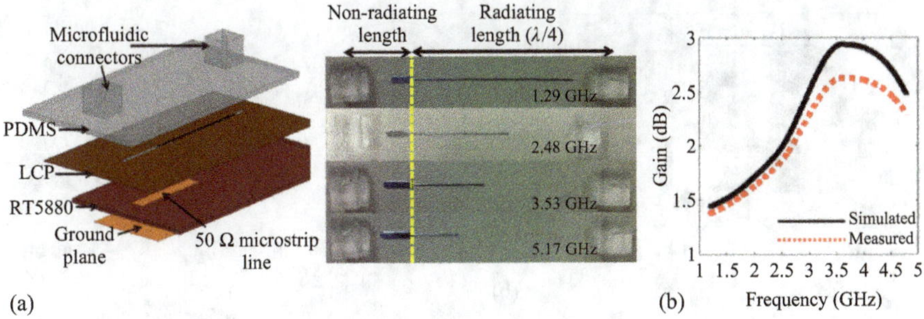

Figure 7.7 Frequency-tunable monopole with a large 4:1 frequency-tuning ratio (a) and its simulated/measured gain performance (b) (after Dey et al. [23] © IEEE 2016)

reconfigured across the frequency-tuning range. The lowest frequency of operation is limited by the maximum continuous length of mercury achievable within a given channel geometry. Through experimental studies performed with different channel geometries, the maximum length achievable is determined to be 50 mm within 0.25-mm high and 0.5-mm wide channels. The highest frequency range is limited by the coupled line resonance of the microstrip line and the retracted liquid metal length. Therefore, a wider microfluidic channel width (i.e., 2 mm) is utilized over the feed line to minimize the non-radiating length of the liquid metal volume. The transition between the two channels exhibiting different widths has a rounded shape to protect the continuity of the liquid metal volume as it is moved from the radiating to nonradiating section or vice versa. Figure 7.7(b) presents the measured and simulated peak gain values of the antenna across its 4:1 frequency-tuning range. The measured gain varies from 1.5 dBi to 2.7 dBi with measured radiation efficiencies > 85%. The ground plane size of the design is adjusted to minimize the pattern variation across the frequency-tuning range. Consequently, the antenna operates with omnidirectional pattern across its entire operation band. The antenna is operated by connecting two piezoelectric micropumps in series to achieve a bidirectional flow (i.e., a bidirectional micropump unit). However, due to the backpressure of the series connected pumps, the reconfiguration time to cover entire frequency-tuning band is measured as 16 s.

The size of external micropump often creates a disadvantage for microfluidically reconfigurable RF devices as the device and pump sizes are typically comparable. As discussed in Section 7.2, integrated actuation techniques being pursued may remove this disadvantage. On the other hand, electrically large reconfigurable RF systems can still benefit from external micropumps if the system size is considerably larger than that of the pump. To demonstrate this possibility, [23] also investigates a 4×1 linear broadside antenna array design by making use of the frequency tunable liquid metal monopole concept. As shown in Figure 7.8(a), by interconnecting the microfluidic channels of individual antennas through tubes or meandered channels, it is possible to actuate the elements of the entire array by

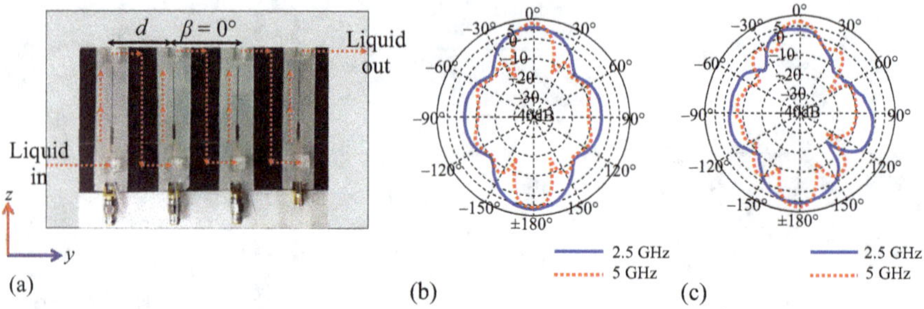

Figure 7.8 Linear array of 2:1 frequency tunable monopoles with interconnected microfluidic channels: (a) prototype; (b) simulated gain patterns; and (c) measured gain patters (after Dey et al. [23] © IEEE 2016)

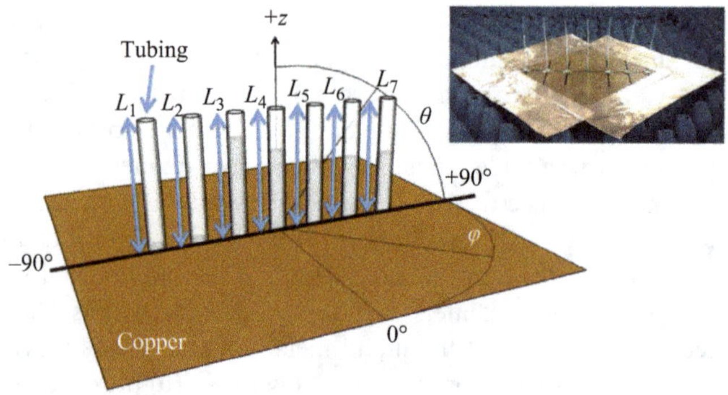

Figure 7.9 Liquid metal monopole array with frequency-tuning, gain adjustment, and beam-steering capabilities (after Morishita et al. [24] © IEEE 2013)

using only a single bidirectional micropump unit. The antenna elements of the array is frequency tunable from 2.5 to 5 GHz. Due to the grating lobe and mutual coupling considerations, the element spacing is selected as 40 mm. Figure 7.8(b) and (c) depicts the simulated and measured gain patterns of the array at the lowest and highest operation frequencies. The measured gains are 6.2 dBi and 8.06 dBi at 2.5 GHz and 5 GHz, respectively. The radiation efficiency of the array is characterized as 65% at 5 GHz and 80% at 2.5 GHz. The reconfiguration time is measured as 20 s.

Figure 7.9 presents another antenna array design pursued with liquid metals [24]. The antenna is constructed from seven 1.02 mm diameter polytetrafluoroethylene (PTFE) tubes vertically erected on a planar ground plane using epoxy. Each tube is connected to a syringe tip from the bottom side of the ground plane. The central tube acts as the driven monopole element whereas the other tubes form the parasitic monopole elements. The configuration is used to construct Yagi–Uda monopole arrays with five elements. The central tube (i.e., driven element) can be filled

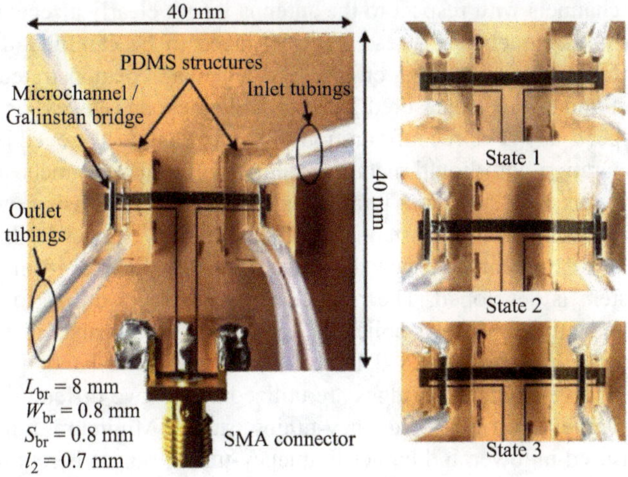

Figure 7.10 Frequency tunable CPW folded slot antenna (after Saghati et al. [16] © IEEE 2015)

with different amounts of EGaIn to achieve frequency-tuning capability. The director and reflector tubes can be placed interchangeably on each side of the driven element to provide beam-steering capability. Since the number of reflectors and directors can also be controlled, it is possible to achieve gain tuning. It is reported that the gain could be adjusted by 1.33 dB when two more additional reflectors are utilized. Ground plane of the antenna is 20×20 cm^2. The inter-element spacing is 2.5 cm, implying 0.2 and 0.32 wavelengths at the operation frequencies of 2.4 GHz and 3.87 GHz, respectively. The measured gain of the antenna is 2.5 to 3 dB lower than the antenna version implemented from copper wires. This is due to the losses associated with PTFE tubing, carrier fluid utilized for EGaIn, syringe tips, and epoxy. The antenna is constructed manually by filling the tubes with syringes and slipping tubing of the center element to the center conductor of SMA connector to achieve the RF excitation.

Authors in Reference [16] applied the capacitive loading of liquid metals for frequency tunability of folded CPW slot antennas. To bring the liquid metal volume close to the antenna surface, a 30 μm thick PDMS layer is spin coated over the metallization plane of the CPW antenna. Two pairs of microfluidic channels prepared in PDMS are subsequently bonded on to the arms of the antenna as seen in Figure 7.10. The microfluidic channel can host a volume of Galinstan that is carried within a low loss Teflon solution. When the Galinstan volume bridges across the antenna slots, a strong capacitive loading effect is created to alter the electrical length of the antenna. The antenna can be operated in three distinct states. In state 1, the microfluidic channels are free of Galinstan and antenna performs at its lowest resonance frequency of 2.4 GHz based on its design geometry. Placing Galinstan on outermost microfluidic channels switches to operation mode to state 2, in which the antenna operates at 3.5 GHz. When the innermost microfluidic channels are filled with Galinstan, antenna operates in state 3 at 5.8 GHz. The locations of the microfluidic channels with respect to the antenna length clearly affects the spacing of the resonance frequencies. These locations need to be carefully designed by considering the capacitive loading effects. The antenna is demonstrated to operate with 3.55 dBi and 2.4 dBi measured gains at its highest and lowest operation frequencies, respectively. The reconfiguration of the antenna is achieved by syringes. Long term operation with liquid metal remaining within the channels suffers from oxidization issues of Galinstan. Evacuating the channels, rinsing with Teflon solution, and refilling with Galinstan is a potential remedy, however, this approach is expected to significantly reduce the reconfiguration time even an automated actuation system is developed. The reconfiguration concept utilized in this CPW antenna has recently also been applied to dual band slot antennas to achieve discrete frequency-tuning within 1.8–3.1 GHz and 3.2–5.4 GHz bands [45]. Specifically, a total of five Galinstan bridges are utilized across two differently sized slot antennas to achieve various frequency-tuning states. Minimum radiation efficiencies measured in lower and higher frequency-tuning ranges is reported as 78% and 82%, respectively. This clearly demonstrates the radiation efficiency advantages offered by the liquid metals.

A reversibly frequency-reconfigurable crossed dipole antenna is introduced in [15] by making use of the electrochemically controlled capillary-based actuation technique. As shown in Figure 7.11, the antenna consists of two pairs of glass capillaries whose ends are supported by acrylic fixture exhibiting six laser-cut reservoirs. The center and outer reservoirs contain the liquid metal EGaIn and electrolyte (1 M NaOH, $\sigma \approx 5$ S/m), respectively. The inner and outer pins of an SMA connector are utilized to center feed the antenna. These pins also contact to a common negative DC terminal. The electrolyte reservoirs are interfaced with positive DC terminals. Applying a positive voltage to the electrolyte removes the oxide skin and causes the EGaIn to withdraw from the capillaries due to the increased surface tension. A negative voltage oxidizes the EGaIn/electrolyte interface and causes the EGaIn to be injected into the capillaries. A small negative voltage stabilizes the position of the liquid metal. Since EGaIn can be independently injected or withdrawn from the capillaries aligned along the x- and y-directions, the antenna offers multiple reconfiguration capabilities. The antenna can generate x-polarized or y-polarized radiation tunable within 0.8 GHz to 3 GHz band. A circular polarization can also be achieved within the 0.8 GHz to 1.6 GHz band by selecting different antenna lengths in the x- and y-axis-oriented capillaries. The antenna size can be changed from 150 to 34 mm. The simulated antenna gains are 0.3 dBi at 1 GHz and 1.7 dBi at 2.5 GHz. The radiation efficiency is impacted with the conductivity of the electrolyte. The reconfiguration speed is similar to a monopole antenna version [46] that is reported to operate with ~3.6 mm/s (withdrawal) and ~0.6 mm/s (injection) reconfiguration speeds for −0.7 V and +.7 V bias voltages.

Figure 7.12 presents a frequency-reconfigurable slot antenna introduced in [25]. The microfluidic channels are constructed from 0.6 mm thick polyimide walls and 0.1 mm thick polyimide floor placed over the antenna plane. The channels are closed from the top by a 1 mm thick polystyrene ceiling. Two microfluidic

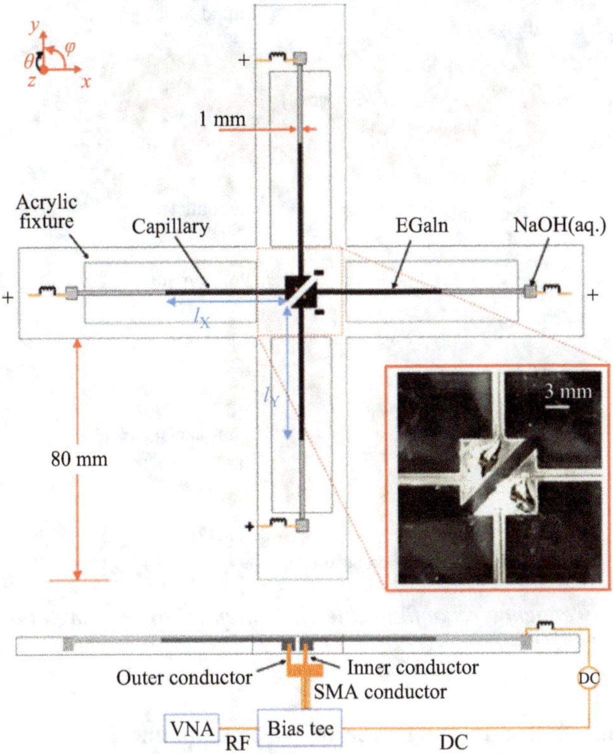

Figure 7.11 Frequency and polarization reconfigurable cross-dipole antenna (after Wang et al. [15] © IEEE 2017)

channels overlap with the 1 mm wide slot antenna arms so that Galinstan within the channel can be used to effectively shorten the antenna length to achieve frequency reconfiguration. To prevent degradation in impedance matching, a third microfluidic channel was placed under the antenna to overlap with the microstrip line feed. Specifically, the Galinstan volume within this channel is used to control the microstrip line stub that crosses over the slot antenna to achieve the impedance matching. The microfluidic channels are coated with PTFE to make them hydrophobic. Galinstan volume is enveloped within a thin layer of NaOH solution to prevent it from oxidization. The remaining volume of the microfluidic channels are free of liquids (i.e., air filled). The microfluidic channels are formed as a cascade of interconnected circular chambers to stabilize the position of Galinstan by making use of the strong surface tension of Galinstan that causes it to minimize its surface volume by withdrawing into the circular chambers. The chambers over the slot antenna are 3.7 mm in diameter and spaced 3 mm apart. On the other hand, the chambers over the microstrip line feed are sized differently by being 2.4 mm in diameter and spaced 2 mm apart. The two Galinstan volumes overlapping with the slot antenna arms are each 27.4 mm long and occupy eight adjacent chambers.

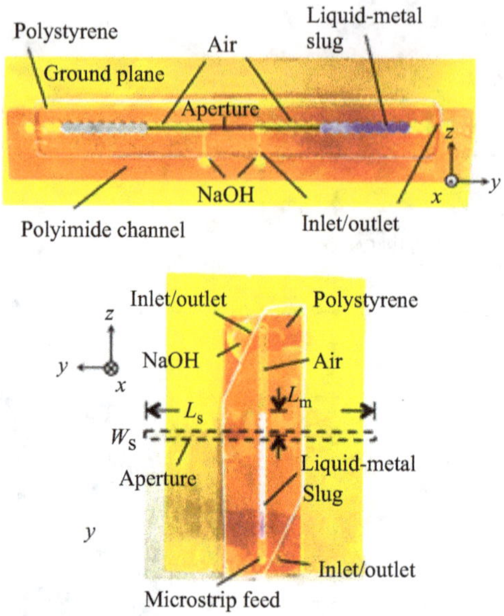

Figure 7.12 Frequency-reconfigurable slot antenna (after Dang et al. [25] © IET 2015)

The Galinstan volume residing over the microstrip line is 32.4 mm long and occupies sixteen adjacent chambers. The antenna length is 60 mm and about half-wavelength at its lowest operation frequency. The Galinstan volumes enveloped with the thin NaOH layers are pneumatically repositioned within the microfluidic channel to achieve the frequency-tuning capability. The antenna achieves frequency-tuning capability from 1.42 GHz to 1.84 GHz as the length is changed from 60 to 30 mm. The $|S_{11}| < -10$ dB bandwidth of the antenna varies from 4 to 5.7%. The gain varies between 4.1 dBi and 4.8 dBi as operation frequency is tuned. The air filled channels are indeed promising for achieving high radiation efficiencies, however, the presented pneumatic actuation is syringe based and manual with reconfiguration speed of 1 s. Implementation of an automated actuation mechanism could significantly benefit the applicability of the presented method.

7.5 Reconfigurable antennas using dielectric liquids

A method to alleviate reliability issues associated with liquid metals is to pursue antenna realizations that solely depend on reconfiguration capabilities achievable by dielectric liquids. Such antenna realizations often switch from one type of material loading (e.g., air or low permittivity liquid) to another type (e.g., high permittivity liquid) to alter the operation of the antenna. The designs, however, potentially suffer from slower reconfiguration speeds due to the need to vary large dielectric liquid volumes and unavailability of readily available low loss high

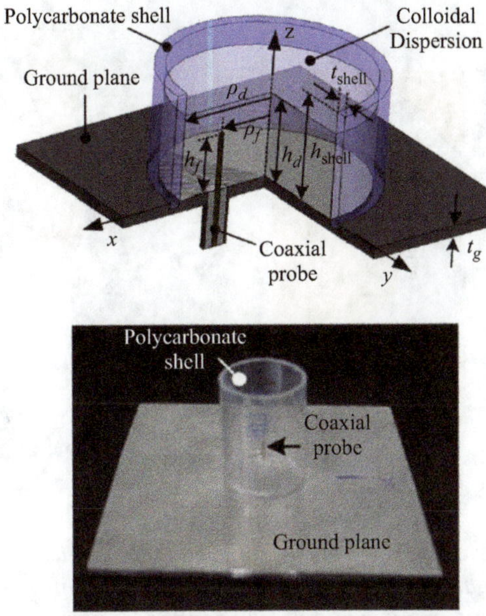

Figure 7.13 Frequency-reconfigurable DRA using colloidal dispersions (after Huff et al. [28] © IEEE 2010)

permittivity dielectric liquids. For example, several designs in literature utilize DI water ($\varepsilon_r = 76.9$, tan $\delta = 0.127$ at 2.4 GHz [30]) as a high permittivity material and report significant radiation efficiency degradation due to its high loss tangent. Therefore, development of low loss high permittivity dielectric liquids is expected to further increase the performance capabilities of dielectric liquid-based microfluidically reconfigurable antennas.

Figure 7.13 presents a frequency-reconfigurable dielectric resonator antenna (DRA) introduced in [28]. To form the antenna structure, a section of polycarbonate tube is mechanically fixed in 1 mm deep annular channel that is centered over a 3.175 mm thick 100×100 mm^2 aluminum ground plane. The tube's radius, height, and wall thickness are 12.7, 30, and 1.6 mm, respectively. The tube is partially filled with a dielectric liquid to operate as a cylindrical DRA in TM_{011} mode. Changing the volume (i.e., height) of the liquid within the tube is utilized as the means of frequency reconfiguration. Since high dielectric constant and low loss are desirable for the construction of the DRA antenna, the dielectric liquid is prepared by dispersing a high permittivity colloidal material barium strontium titanate (BSTO) within nonaqueous hydrotreated naphthenic oil. The BSTO particles are less than 100 nm in diameter and dispersed at a volume fraction of 50%. The mixture is vortexed and sonicated until a homogenized dispersion is achieved. The effective relative dielectric constant of the mixture is estimated to be $\varepsilon_r = 8.64$ based on the Maxwell-Garnett mixing formula. An 8 mm long coaxial probe located 9.5 mm away from the center is designed as the antenna excitation for achieving a

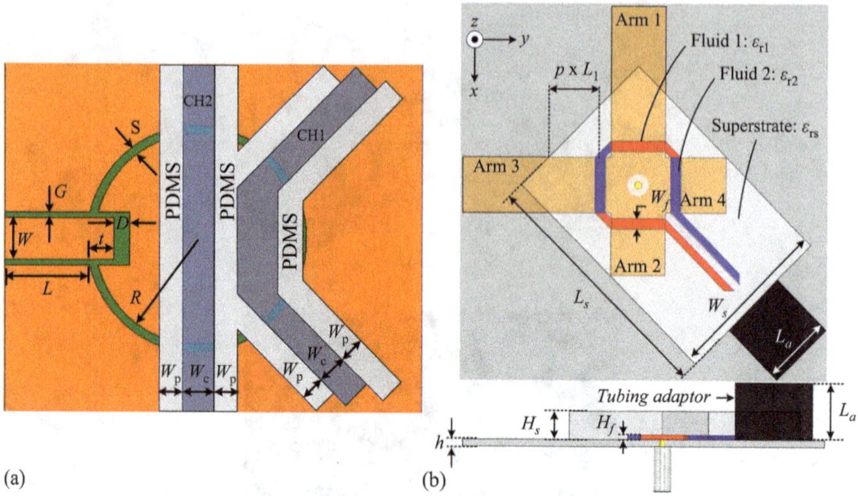

(a) (b)

Figure 7.14 Dielectric liquid-based reconfigurable antennas that utilize DI water as a high permittivity liquid material: (a) frequency-reconfigurable dual band annular slot antenna (after Murray et al. [29] © IEEE 2014) and (b) polarization reconfigurable crossed microstrip patch antenna (after Barrera et al. [30] © IEEE 2014)

resonance frequency of 3 GHz when the height of the colloidal dispersion h_d is 16 mm. The antenna can maintain the impedance matching performance in a single DRA mode for 50% change in h_d. Specifically, varying h_d from 8 to 24 mm tunes the resonance frequency from 4 to 2.5 GHz. The frequency tunability of the design is practically limited by the excitation of higher-order DRA modes and monopole mode of the coaxial probe. Since this antenna prototype is not closed from the top with a lid, its gain measurement is not practical and performed. The antenna is not integrated with an external pumping unit and its frequency-tuning speed is not characterized.

The annular slot antenna introduced in [29] utilizes dielectric liquid loading for independently tuning frequencies of its first and second resonances. As shown in Figure 7.14(a), the antenna is integrated with two microfluidic channels formed by attaching 2.3 mm thick and 2.5 mm high PDMS walls onto the antenna substrate by using PDMS adhesive. A secondary PDMS layer is adhered to the top surface of the walls to complete the fluid/channel housing. The annular slot is fabricated on a 1.524 mm thick FR4 board and fed with a 50 Ω CPW feed. The slot exhibits 7.4 mm radius and 0.6 mm width. The gap between the center conductor of the CPW feed and the center antenna metallization is selected as $D = 1.1$ mm. The center conductor of the CPW also extends into the annular slot footprint by $t = 1.75$ mm to achieve good impedance matching for the first and second resonances observed at 4.2 and 8 GHz, respectively, under the air filled channel conditions. The location of the microfluidic channels are determined by investigating the electric field distributions of the first and second resonances. The straight channel CH2 coincides with the field minimum and maximum of the first and second resonances, respectively. Dielectric liquid loading with CH2 reduces the frequency of the second resonance without impacting the first resonance.

Specifically, when the other channel is filled with air, loading CH2 with acetone ($\varepsilon_r = 23$, $\tan\delta = 0.12$ at 4 GHz) and DI water ($\varepsilon_r = 74$, $\tan\delta = 0.2$ at 4 GHz) reduces the second resonance frequency from 8 GHz to 6.7 and 5.2 GHz, respectively. The frequency-tuning range is 42% and the first resonance frequency is not impacted more than 0.04 GHz (<1%). Due to the field distributions, it is not possible to find a location that coincides with the field minimum and maximums of the second and first resonances, respectively. Therefore, the meandered channel CH1 placed at the field minimum of the second resonance does not directly interact with the field maximum of the first resonance. Consequently, when CH2 is filled with air, loading CH1 with acetone and DI water shifts the first resonance frequency from 4.2 GHz to 3.8 GHz and 3.3 GHz, respectively. The first resonance achieves a 24% tuning range. The measurement demonstrates a 0.5 GHz shift in second resonance frequency. However, this is attributed to slight fabrication error in width of CH1 as simulation studies demonstrate that precise fabrication of CH1 will lead to less than 1% shift in second resonance frequency as well during the reconfiguration of the first resonance frequency. The measured and simulated radiation patterns of the antenna at different operation states agree well with each other. The losses of DI water and acetone significantly degrade the radiation efficiency of the antenna particularly in its second resonance.

Authors in Reference [30] present a polarization reconfigurable crossed microstrip patch antenna that utilizes a dielectric liquid loading mechanism. As shown in Figure 7.14(b), the patch antenna metallization consists of five sections. The center metallization section is excited by the coaxial probe and electrically separated from the other sections (i.e., Arms 1 to 4) with 0.3 mm wide gaps. A 2 mm wide and 1 mm high meandered microfluidic channel implemented in 6.35 mm thick PDMS is placed over the these gaps as depicted in Figure 7.14(b). The antenna footprint is ~ $45 \times 45\ \text{mm}^2$ and realized over a 1.58 mm thick $\varepsilon_r = 2.2$ board. The microfluidic channel carries alternating dielectric liquid volumes with high (DI water) and low (air) relative dielectric constants. Overlapping the high dielectric constant liquid with a gap creates strong capacitive coupling and switches on RF connection with the corresponding arm and center metallization sections. Circulation of the periodic dielectric liquid volume is utilized to activate the patches oriented along the *x*- or *y*-axis to achieve polarization reconfiguration capability. The antenna operates around 2.4 GHz with ~2.4% $|S_{11}| < -10$ dB bandwidth. A similar antenna is also designed and constructed by utilizing PIN diodes as the coupling mechanism. The antenna gains are significantly degraded due to DI water and ohmic losses of the diodes. The radiation efficiencies are 22% versus 10% for the liquid and diode-loaded antenna versions, respectively. Liquid loading is larger in size and exhibits slower actuation speed. The actuation is performed with an external peristatic pump and antenna surface is free of bias lines. On the other hand, the DC bias lines around the PIN diode-loaded antenna causes degradation in antenna performance due to the induced currents. The dielectric fluid-loaded antenna exhibits a better cross-polarization and efficiency level. If the dielectric fluid (i.e., DI water) is lossless, the radiation efficiency is expected to reach up to 70% and 90% for the super substrate choices of PDMS and air, respectively. Hence, the findings of the work clearly demonstrates the need for low loss high permittivity dielectric liquids in order to fully take advantage of the dielectric loading-based microfluidic antenna reconfiguration techniques.

7.6 Beam-steerable liquid metal antennas

Liquid metals have also attracted interest for achieving high-efficiency beam-steering capabilities with low cost and compact assemblies. The liquid metal monopole array [24] discussed in Section 7.4 is an example design that can provide beam-steering capability, however, its beam-steering is limited to two principle directions due to the antenna construction and desire to simultaneously offer other reconfiguration capabilities such as frequency-tuning. An earlier Yagi–Uda antenna design introduced in [26] provides a more advanced beam-steering capability by resorting to a circular geometry. As shown in Figure 7.15, the driven element of the antenna is a circular four section Alford loop that produces an omni-directional radiation pattern in the antenna plane without the presence of the parasitic reflector and director. The driven element is printed on a 0.81 mm thick circuit board exhibiting a relative permittivity of $\varepsilon_r = 3.55$. The antenna is fed by a coaxial probe and the metallization on the back surface of the substrate (shown with dashed lines in Figure 7.15) is patterned to perform a tapered transition from a microstrip line ground plane into the second conductor of a balanced strip line feeding the sections of the Alford loop. The director and reflector of the antenna is

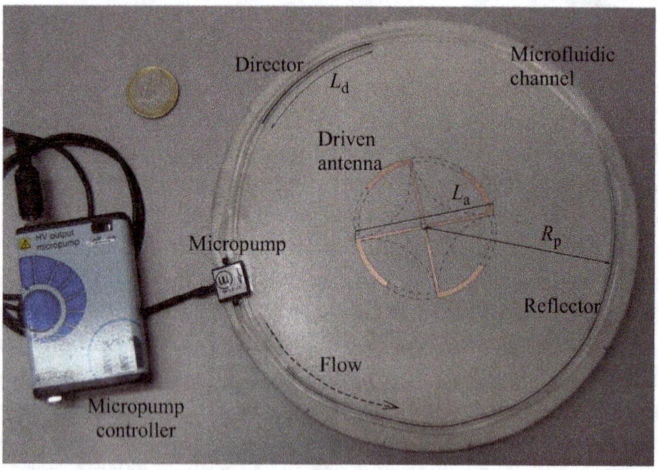

Figure 7.15 Circular beam-steering antenna using liquid metal parasitics (after Rodrigo et al. [26] © IEEE 2012)

implemented from liquid metal mercury volumes within plastic tubing. DI water is used as the carrier liquid and fills up the remaining volume of the microfluidic system. The driven antenna and parasitic element dimensions are related to each other through wavelength considerations. The driven element operates at 1,800 MHz with 4% $|S_{11}| < -10$ dB bandwidth and exhibits a diameter of 58.5 mm. The antenna radius up to the tubing is 80 mm. The tubing is integrated with a micropump and given a double loop shape to avoid the parasitics crossing the pump as beam is steered at least once within the entire antenna plane. The antenna exhibits 90° half-power beam width (HPBW) and 0.55 dB lower gain as compared to the case when the parasitic elements are implemented from copper wires. The gain drop is not only due to the reduction in efficiency associated with the material losses but also due to the directivity decrease associated with plastic tubing-based perturbations. Nevertheless, this performance implies a good overall radiation efficiency especially when the 360° beam-steering capability is considered. The reconfiguration speed is characterized as 2 rad/s.

High-gain beam-steering antenna arrays are challenging to construct in mm-waves due to the size, cost, and performance of RF control devices such as switches and phase shifters. On the other hand, motorized gimbaled antenna systems are characteristically bulky, slow and power hungry. Microfluidic reconfiguration techniques potentially offer an alternative to achieve high-gain beam-steering in mm-waves from low cost and compact assemblies with high radiation efficiencies. To demonstrate this potential, authors in reference [27] introduced a mm-wave beam-steering array concept based on microfluidic reconfiguration of liquid metal volumes. In this concept, the array is located at the focal plane of a microwave lens (i.e., focal plane array, see Figure 7.16(a)). It consists of interconnected microfluidic channels and reservoirs that include a small volume of liquid metal. The remaining volume of the microfluidic channels and reservoirs are filled by a low loss dielectric solution that acts as the carrier fluid of the liquid metal volume. The in-plane dimensions of the liquid metal volume represent a mm-wave patch antenna. The location of this liquid metal patch antenna is varied along the microfluidic channel by utilizing a bidirectional micropump unit as shown in Figure 7.16(b). Since the patch antenna is located at the focal plane of the microwave lens, its repositioning along the microfluidic channel leads to beam-steering functionality. Figure 7.16(c) presents the dimensions of the patch antenna designed to demonstrate the concept at 30 GHz. The microwave lens is an extended hemispherical dielectric lens realized with Rexolite ($\varepsilon_r = 2.56$, tan $\delta = 0.0026$). Since the back surface of an extended hemispherical dielectric lens is also the focal plane, the focal plane array consisting of a passive feed network and microfluidic channel with liquid metal volume is conveniently flush mounted with the lens. Due to the electrically large size of the lens and its ability to minimize reflections at the lens/air boundary; the patch antenna is designed with the assumption that the lens material (i.e., Rexolite) forms a semi-infinite half-space. The substrate stack-up of the antenna consists of 0.25 mm thick microfluidic channel and reservoirs implemented in 2 mm thick PDMS and bonded to 0.05 mm thick LCP (LCP, $\varepsilon_r = 2.9$, tan $\delta = 0.0025$) substrate. This PDMS-LCP assembly is placed over a 0.127 mm

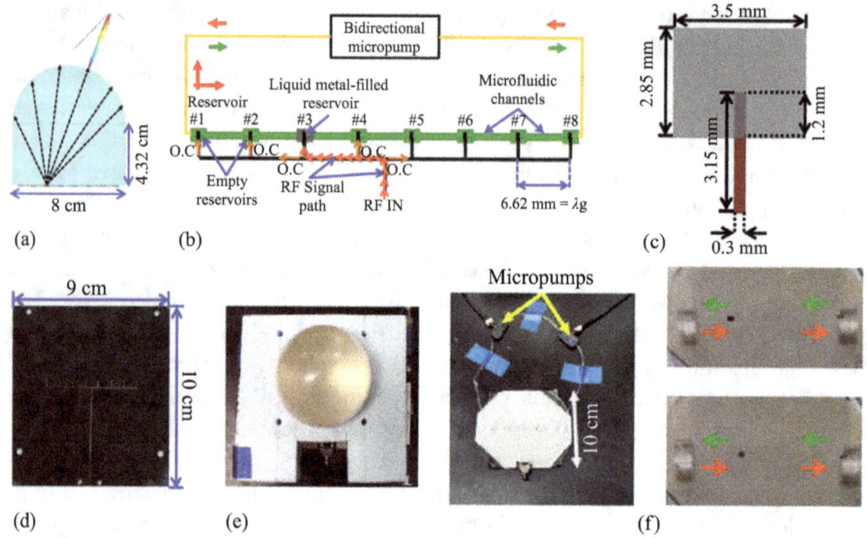

Figure 7.16 30 GHz beam-steering focal plane array using liquid metal volume repositionable within microfluidic channel as a patch antenna feed of extended hemispherical dielectric lens: (a) conceptual illustration of the focal plane array behind the lens; (b) layout view depicting microfluidic channels and reservoirs placed over a resonant corporate microstrip line feed network; (c) layout view of the 30 GHz proximity-fed patch antenna; (d) fabricated resonant corporate feed network (e) array assembly; and (f) microfluidic channel and snapshots of liquid metal patch at different positions (after Gheethan et al. [27] © IEEE 2013)

thick Rogers 5880 PCB ($\varepsilon_r = 2.2$, $\tan\delta = 0.0007$) that supports the passive proximity-coupled microstrip line feed network. Mercury is used as the antenna element and exhibits a volume of 2.5 μl to realize the desired antenna footprint within the given microfluidic channel geometry. The dielectric solution used is Fluorinert FC-77 ($\varepsilon_r = 1.9$, $\tan\delta = 0.0005$). The field of view (FoV) for extended hemispherical lenses are typically $\pm$ 30° due to the lens/air reflection issues for wider scan-angles [47,48]. To construct an 8 element focal plane array with the largest FoV, the HPBW of a stand-alone element should be around $60/8 = 7.5°$.

Consequently, the lens is designed to exhibit 80 mm diameter and 43.2 mm extension length. With these parameters, the stand-alone patch antenna placed at the center of the focal plane provides a pencil beam along the broadside direction with 7° HPBW and 29 dB directivity (from Krauss's approximation). The feed network is all-passive and designed to provide impedance matching conditions as the liquid metal volume is moved among the eight distinct reservoirs. To accomplish this, the lengths of the proximity coupling microstrip lines residing below the reservoirs are taken as half-guided wavelength such that an open-circuit (O.C.)

condition on the feed line is realized within the absence of the liquid metal patch. In addition, these proximity coupling lines are separated from each other by a full-guided wavelength lines to maintain O.C. conditions across the feed network junctions that can potentially direct the RF signal away from the path it needs to take to reach to the liquid metal patch. As seen in Figure 7.16(b), these O.C. conditions allow excitation of the liquid metal patch antenna once it resides in one of the reservoirs. Figure 7.16(d)–(e) present the fabricated feed network and the array assembly, respectively. Figure 7.16(f) depicts the snapshot of the liquid metal patch antenna as it is moved among different reservoirs. The liquid metal patch is well impedance matched in all of the reservoirs with a 3.3% $|S_{11}| < -10$ dB bandwidth. The simulated feed network loss slightly depends on antenna location and its average value is 3.9 dB. Measured gains vary from 24.8 to 21.5 dBi as beam is steered towards different directions. Particularly, the losses related to the lens/air reflection become more dominant as beam is steered further away from the broadside—a typical property of the extended hemispherical dielectric lenses [48]. When compared to the 29 dB directivity expectation, measured gain loss appears as 5.7 dB. Since 1.8 dB of this loss can be attributed to the connector and lens-air reflections; it is assumed that the feed network exhibits a loss close to the expected 3.9 dB value. It is important to note that this loss value is comparable to the performance of a conventional 8 element focal plane array implementation that would utilize a total of 7 single-pole double-throw switches (SP2T). Commercially available Ka-band SP2T switches exhibit ~1 dB insertion loss. Conventional feed network implementation requires the RF signal to pass from three of the SP2T switches before reaching to the excitation antenna. However, the additional interconnects, microstrip line sections, and bends are likely to penalize the insertion loss of the feed network and bring the total insertion loss to 4 dB or more. Therefore, the microfluidic beam-steering focal plane array technique is an efficient technique. Most importantly, the technique is attractive for being low cost as the need for costly SP2T switches and their integration challenges within the feed network is avoided. A disadvantage of the technique is the resonant nature of the feed network since it degrades the bandwidth performance as compared to what could have been achieved by a single stand-alone element. The beam-steering speed can be on the order of few ms depending on the channel geometry and pump choice. The microfluidic channel dimensions and piezoelectric pumps utilized for this particular experimental verification result in 70 ms for moving liquid metal patch from one reservoir to the adjacent one.

7.7 Reconfigurable antennas using microfluidically repositionable metallized plates

Metallized plates repositionable within microfluidic channels through fluid flow are attractive for resolving oxidization and sticking issues associated with liquid metals. The use of regular metals with an order of higher conductivity within the microfluidic channels is also promising for increasing efficiency and operation

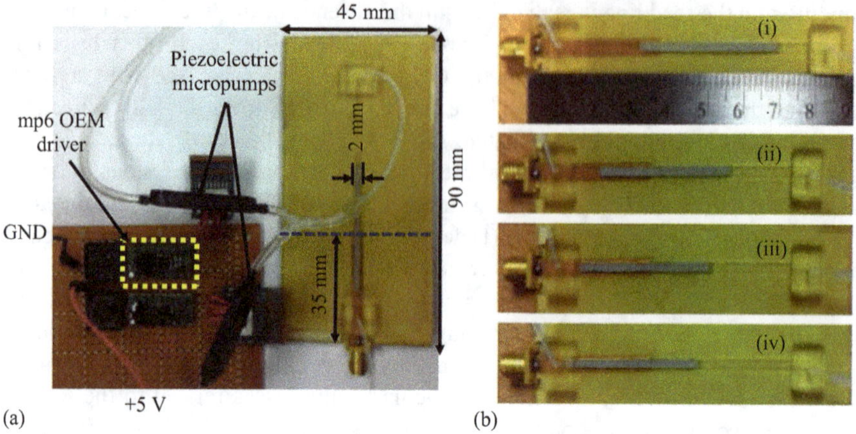

Figure 7.17 Frequency-tunable monopole antenna utilizing a microfluidically repositionable metallized plate within a microfluidic channel as the radiator: (a) antenna assembly (dashed line shows the aligned edges of the ground plane and microstrip feed line) and (b) antenna reconfigured to operate at different frequencies (after Dey et al. [17] © IEEE 2016)

frequency of microfluidically reconfigurable devices. Figure 7.17(a) presents the microfluidically controlled frequency tunable monopole antenna introduced in [17]. The operation principle is similar to the monopole antenna realized from liquid metal [23] and discussed in Section 7.4. The monopole is formed from a metallized plate placed within a microfluidic channel. This plate is realized from a readily available 0.508 mm thick Rogers 4003C PCB ($\varepsilon_r = 3.38$, tan $\delta = 0.0027$) with copper thickness of 17 μm. The microfluidic channel is 0.75 mm thick and fabricated within a 2 mm thick PDMS layer. The antenna is excited through a capacitively coupled 50 Ω microstrip line realized over a 1.524 mm thick Rogers 4003C PCB. The board is spun with a 12 μm thick BCB layer and the microfluidic channel containing the metallized plate is chemically bonded to the BCB surface by using oxygen plasma and APTES treatments (see Section 7.2 for fabrication techniques). The microfluidic channel is filled by low-loss dielectric liquid FC-40 and interfaced with external piezoelectric micropumps using plastic tubing. Two micropumps are connected in parallel to achieve bi-directional flow. The pressure generated by the micropumps is used to push the plate inside the microfluidic channel to change the antenna length, thus tune the operational frequency. Figure 7.17(b) depicts the antenna at four different positions. Switching the piezoelectric micropumps on/off is observed to instantaneously start/stop the plate movement without overshoot. This allows to achieve repeatable reconfiguration based on experimental flow characterization (e.g., 22 mm movement per second).

The antenna design starts by identifying the radiation length (L_{RAD}) required to operate the antenna at the lowest desired frequency. Selecting lowest operation

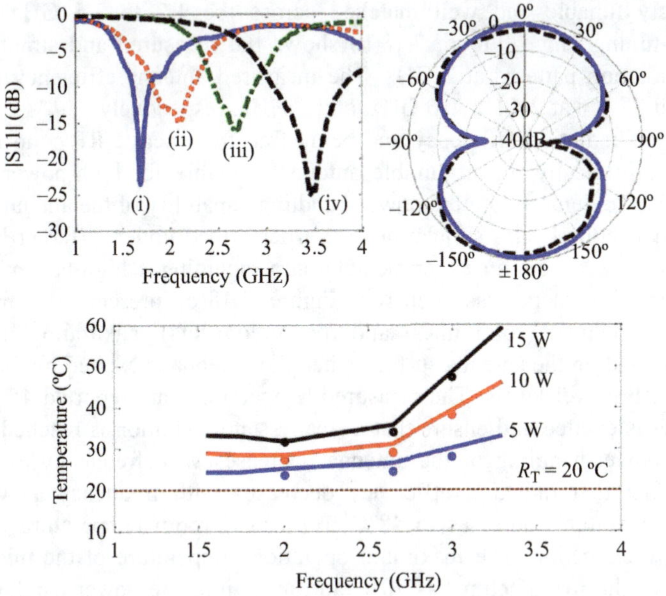

Figure 7.18 Measured performance of the frequency tunable monopole antenna in Figure 7.17: (a) S_{11} at various plate positions; (b) gain pattern at 2 GHz; and (c) maximum temperature observed over the antenna surface under different excitation powers (after Dey et al. [17] © IEEE 2016)

frequency as 1.5 GHz results in $L_{RAD} = 35$ mm within the presence of the antenna substrates. The next design step is to identify the minimum overlap area needed between the feed line and the metallized plate to achieve an RF short. This area is a function of the antenna width (W_{ANT}) and overlap length (L_O) between the feed line and the metallized plate. Designing for the smallest possible L_O is important since retracting the metallized plate over the feed line creates a coupled line resonance that eventually determines the highest operation frequency of the antenna assembly. To keep the fractional bandwidth below 20%, W_{ANT} is selected as 2 mm. Subsequently, by investigating the performance of back-to-back microstrip line to metallized plate transition, $L_O = 5$ mm is selected to keep the insertion loss of this transition below 0.1 dB. Hence, the total length of the metallized plate is finalized at 1.5 GHz as $L_{ANT} = L_{RAD} + L_O = 40$ mm. Ground plane size of the antenna is designed as 45×35 mm^2 to achieve a good impedance match while minimizing the pattern distortions that may occur from operating with an electrically large ground plane at higher frequencies. The highest simulated operational frequency of the antenna is 3.3 GHz before coupled line resonance starts to be clearly visible in the $|S_{11}|$ response. Figure 7.18(a) depicts the measured $|S_{11}|$ performance of the antenna at four distinct metallized plate positions shown in Figure 7.17(b). Specifically, the performance of the antenna prototype is in agreement with the simulation. It is continuously tunable and well matched across the 1.7 to 3.5 GHz with 2:1 frequency-tuning range. Figure 7.18(b) shows the measured and simulated normalized radiation patterns at 2 GHz. The measured antenna efficiency is 92.4%, 88.9%, and 87.4% at 2 GHz, 2.6 GHz, and 3 GHz, respectively.

The high radiation efficiency and being free from active RF control devices make microfluidically reconfigurable antennas suitable for high-power-handling applications. To demonstrate the power-handling capability of the antenna, steady-state thermal multiphysics simulations were performed under different RF power excitations. In addition, an experimental setup including a high-power amplifier and a thermal camera was prepared. Figure 7.18(c) presents the frequency-dependent simulated (solid lines) and measured (dots) maximum temperature values observed on the antenna surface when the antenna is excited by 5 W, 10 W, and 15 W RF power levels. The measured temperatures are reported 45 min after the antenna is excited and ensure that a steady-state condition is reached. Specifically, the power-handling of the antenna decreases with frequency as expected from the fact that radiation efficiency decreases with frequency as well. The antenna temperature increases to 48°C from 20°C room temperature at 3 GHz under 15 W excitation. The maximum operation temperature of the micropumps are given by the manufacturer as 70°C and may limit the power-handling of the antenna to 15 W at the highest operation frequency of 3.5 GHz if no additional heat sinking or cooling is employed. This is much lower than the boiling point of FC-40 (165°C) which is expected to be the first material to get damaged in the antenna construction besides the micropumps. Nevertheless, 15 W also stands as a remarkable power-handling capability considering the limitations associated with conventional solid state or MEMS-based implementations of reconfigurable antennas. It is also important to note that conventional metallization of this microfluidically reconfigurable antenna is a significant contributor to its power-handling capability. For example, replacing the metallized plate with a liquid metal mercury volume decreases the power-handling capability of the antenna by a factor of 2. This is associated with the decrease in radiation efficiency (from 92% to 77% at 2 GHz) since the conductivity of mercury is about 50 times lower than that of copper. Mercury's significantly larger thermal conductivity (compared to the metallized plate's material) is not advantageous in this power-handling scenario, since the antenna metallization (mercury or metallized plate) is embedded within the substrate stack-up and heat extraction is governed by the surrounding materials. Consequently, the metallized plate approach is also very suitable for constructing high power-handling microfluidically reconfigurable RF devices.

Microfluidically repositionable metallized plates have also been utilized in construction of microfluidic beam-steering focal plane arrays. Specifically, the liquid metal volume that acted as a patch antenna feed for the microwave lens (see Section 7.6 and Figure 7.16) has been replaced by a metallized plate repositionable within microfluidic channel with fluid flow [31]. The higher conductivity of metallized plate offers improvement in antenna efficiency as expected. In addition to this improvement, passive feed networks alternative to the corporate resonant feed network mentioned in Section 7.6 have also been investigated to improve the

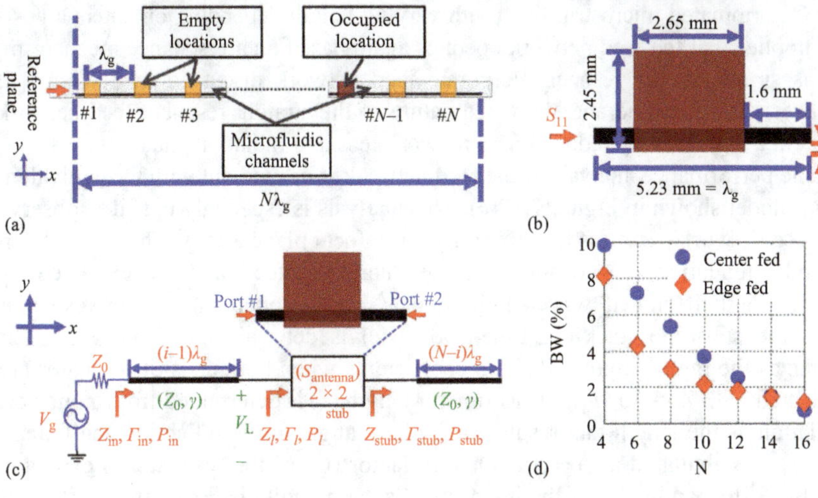

Figure 7.19 Straight resonant feed network of microfluidic beam-steering focal plane array: (a) feed network schematic for edge feeding; (b) schematic of the antenna designed for the 30 GHz feed network; (c) equivalent circuit model of the feed network; and (d) bandwidth performance of the feed network as a function of array size (after Gheethan et al. [31] © IEEE 2016)

feed network performance in terms of insertion loss and undesired radiation leakage. Figure 7.19(a) depicts the straight resonant feed network constructed from cascades of guided wavelength long microstrip lines. The microstrip lines are on a 0.2 mm thick Rogers 4003C substrate and placed under the microfluidic channel that is filled up by a low loss Teflon solution. The microfluidic channel also carries the metallized plate-based patch antenna implemented from a 0.254 mm thick Rogers 5880LZ substrate with 17 μm copper thickness. The channel is 0.4 mm in height and 2.6 mm in width. The metallized plate patch antenna has a footprint of $2.45 \times 2.65\ \text{mm}^2$ as shown in Figure 7.19(b). The 2.6 mm channel width is larger than the corresponding 2.45 mm long dimension of the metallized plate. This is experimentally determined through channel/flow characterization studies to provide a reliable movement using pressure driven fluid flow. The microfluidic channel is implemented within a 2 mm thick PDMS and bonded to 0.1 mm thick LCP before placing on the Rogers 4003C substrate. As seen in Figure 7.19(b), the metallized plate patch is designed to be impedance matched over a guided wavelength long microstrip line by utilizing proximity coupling. An ideal O.C. condition at the right side of the microstrip line gets converted into a perfect 50 Ω matching at the left side due to the presence of the antenna element. Therefore, adding multiples of guided wavelength long microstrip line sections to the ends of this network does not change the impedance matching conditions. Likewise, metallized plate patch can be relocated within the microfluidic channel in single or multiple increments of a full guided wavelength without disturbing the impedance matching conditions. The fact that the antenna is loaded with an O.C. terminated microstrip line (with varying length depending on antenna position) implies that the feed network operates again based on a resonance mechanism. This is similar to the resonant corporate feed network discussed in Section 7.6. However, the simplified and shortened nature of the straight resonant feed network results in a significantly reduced feed network loss and radiation leakage.

The performance analysis of the feed network is carried out with the equivalent circuit model shown in Figure 7.19(c). This analysis is especially useful to observe how feed network performs for arbitrarily large focal plane arrays. The propagation (β) and attenuation (α) constants of the microstrip feed line are extracted by designing a half-guided wavelength long ($\ell = \lambda_g/2$) open-ended line resonator within the substrate stack-up. Line width (w) is determined as the width that minimizes the input/output reflection coefficients when both ends of resonator are terminated with $Z_0 = 50\ \Omega$ ports. Resonator length (ℓ) is determined from a one port simulation as the length that results in 0° phase angle in the reflection coefficient. Hence, β is calculated as $\beta = \pi/\ell$. Quality factor (Q) of the resonator is extracted from the 3 dB bandwidth of the input impedance magnitude. From this, α is evaluated as $\alpha = \beta/2Q$. Scattering parameters of the antenna are extracted from a two-port full-wave simulation. Subsequently, transmission line equations and network analysis are utilized to determine the bandwidth of the system and radiation/line losses for any given number (N) of antenna locations (i.e., N element focal plane array). The analysis predicts that when the metallized plate patch is located at its closest position ($i = 1$) to the excitation (i.e., reference) plane, it exhibits its minimum impedance bandwidth. This is expected since when at this position the open circuited microstrip line length seen by the patch is at its maximum value. The array bandwidth is therefore determined by the bandwidth when the patch is located at position $i = 1$. The analysis also shows that the array bandwidth drops for larger focal plane arrays because longer open-circuited microstrip lines become mandatory to achieve impedance matching. Figure 7.19(d) depicts that the bandwidth can go down from 8% to 1% as N increases from 4 to 16 for the edge-fed straight resonant feed network. Center feeding the straight resonant feed network increases the bandwidth performance since longest open circuited line length is reduced by half. However, the bandwidth and N continues to be inversely proportional to each other. Feed network loss also depends on the position of the metallized plate patch due to the different microstrip line lengths. For an $N = 8$ element array, average loss is 2.83 dB for the edge-fed straight resonant feed network. The feed network loss is almost independent of position in the center-fed straight resonant feed network configuration and evaluated as 1.45 dB. These loss values clearly demonstrate the efficiency advantages of the microfluidic beam-steering focal plane arrays as compared to the conventional SP2T switch-based implementations that are expected to exhibit >3 dB feed network loss.

The performance of the straight resonant edge-fed network is experimentally verified with a 30 GHz array prototype as shown in Figure 7.20(a). Specifically a focal plane array with eight possible antenna locations is designed and fabricated to operate behind an 80 mm diameter extended dielectric hemispherical lens built from Rexolite with 43.2 mm extension length. The array operates with 4% $|S_{11}| < -10$ dB bandwidth.

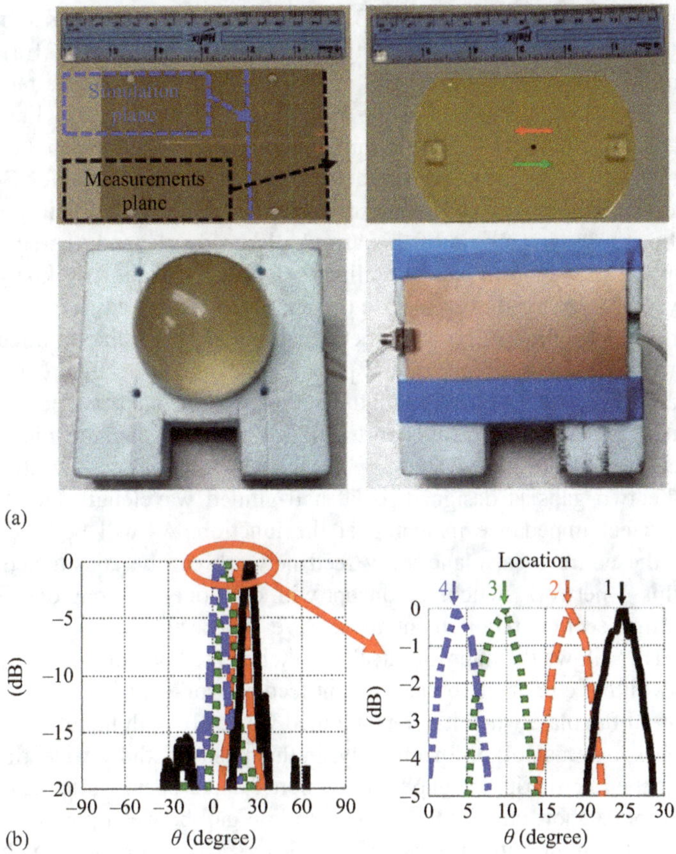

Figure 7.20 Microfluidic beam-steering focal plane array prototype and measured performance with resonant straight feed network: (a) feed network and microfluidic channel with antenna are flushed on back surface of the dielectric lens held by a Styrofoam block and (b) measured radiation patterns as plate is moved among locations 1 to 4 (after Gheethan et al. [31] © IEEE 2016)

This is slightly larger than the 3% bandwidth predicted by the analysis and full wave simulation, however, it is related to the additional length/loss of microstrip line that is added to shift the reference plane to accommodate the edge connector. The radiation patterns measured when metallized plate is moved among positions 1 to 4 agree well with the simulated patterns in terms of HPBWs and beam-steering directions (see Figure 7.20(b)). The measured gains vary from 21.5 dBi to 23.5 dBi as the beam is steered. Accounting for the patch antenna, extension line, connector, and lens losses demonstrates that the feed network loss is less than 3 dB as expected.

More recently, to alleviate the bandwidth issues related to the resonant feed networks, a mm-wave beam-steering focal plane array with microfluidically switched feed network has been introduced [32,33]. A basic T-network that acts as a building block of the focal plane array feed network is shown in Figure 7.21(a) and (b). This technique relies on a "selectively metallized plate" that can be repositioned within the microfluidic channel using fluid flow. The microstrip line exhibits 0.8 mm gaps as open circuits. The 1.8 mm long metallization strips on the plate and the microstrip line are separated from each other through the 6 μm thick BCB layer that acts as the bottom wall of the microfluidic channel (Figure 7.2 in Section 7.2 presents the substrate stack-up of this array). When overlapped, the strong capacitive coupling between the plate metallizations and the microstrip line generates an effective RF short condition to close the open circuits. By adjusting the relative orientations and spacing of the plate metallizations, it is possible to direct the RF power towards different directions and paths. For example, in Figure 7.21(a), the vertical trace of the plate overlaps with the microstrip line gap between port 1 and 2 to direct incident RF signal towards port 2. In this plate position, the microstrip line gap between port 1 and port 3 is an open circuit. Most importantly, the distance between the two gaps is designed to be half-guided wavelength ($\lambda_g/2$) apart to realize a perfect impedance matching at the junction. As will be shown in the following, the entire focal plane array feed network can be constructed by generalizing this T-network switching concept with only one $\lambda_g/2$ long open-circuited microstrip line section requirement to achieve the perfect impedance matching condition. This allows the microfluidically switched feed network to exhibit a very large bandwidth as compared to the resonant feed networks. Figure 7.21(b) presents the position of the plate when it is repositioned vertically with liquid flow to direct the RF power towards port 3. Specifically, in this position, the vertical trace of the plate leaves the microstrip line gap between port 1 and 2 open, while the horizontal trace of the plate short circuits the microstrip line gap between port 1 and port 3. Figure 7.21(c) demonstrates the experimental verification of this switching concept. The microfluidically repositionable selectively metallized plate is located over a microstrip line that exhibits a gap. The switching performance of this assembly is compared to the performance of a continuous microstrip line fabricated under the identical substrate stack-up. Repositioning the selectively metallized plate performs the RF switching functionality as predicted by the concept and numerical simulations. As shown in Figure 7.21(d), within the vicinity of the 30 GHz design frequency, the measured insertion loss of the switched and continuous lines are almost equal to each other implying less than 0.1 dB switch loss. Both lines are also well matched with ~20 dB return loss. This experimental verification also demonstrates the wide bandwidth potential of the designed/fabricated switch. Insertion loss of the switch does not show a significant deviation from that of the continuous line within the 25–40 GHz band.

Figure 7.22(a) presents the layout and key dimensions of the focal plane array prototype that utilizes a microfluidically switched feed network for achieving the beam-steering functionality. The array uses aperture-coupled microstrip patch antennas as the antenna elements. The metallization patterns on the plate are generalized from the aforementioned T-network switching concept. Microfluidically repositioning the plate in increments of 0.6 mm directs the RF power adjacent patch

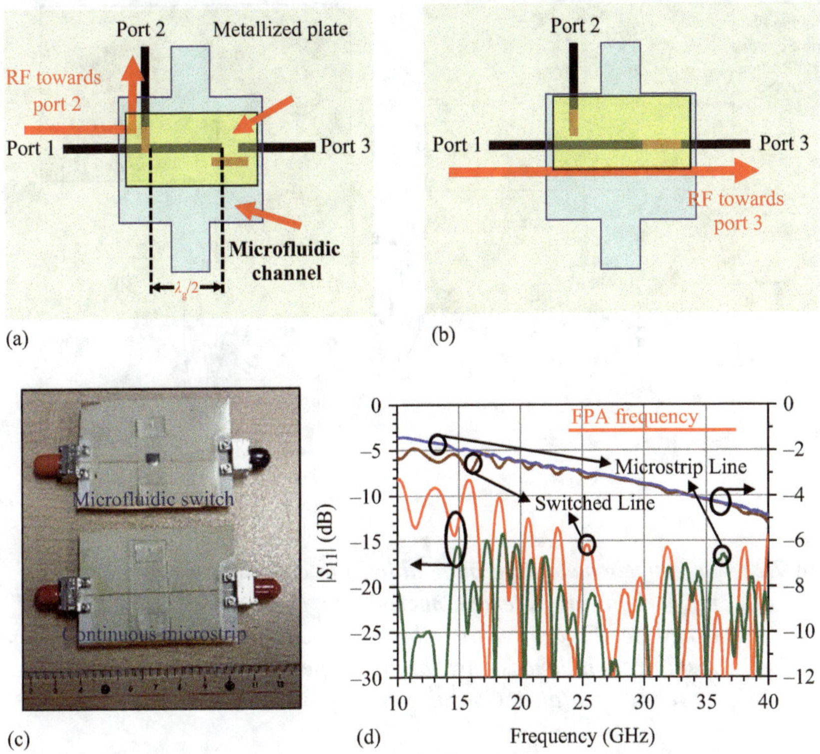

Figure 7.21 *Selectively metallized plate acting as a 30 GHz RF switch in substrate stack-up consisting of 6 μm channel wall (see Figure 7.2): (a) plate is positioned to direct RF power from port 1 to port 2; (b) plate is positioned to direct RF power from port 1 to port 3; (c) experimental verification of the concept in a single throw switch setup; and (d) measured $|S_{11}|$ and $|S_{21}|$ performances depicting switch loss < -0.1 dB at 30 GHz (after Gonzalez et al. [33] © IEEE 2017)*

antenna elements. A 4.2 mm total motion directs the RF power from one end of the array to the other end. This is a significantly reduced motion range as compared to the microfluidic focal plane arrays that use resonant feed networks, as they need to reposition a stand-alone metallized plate-based patch antenna across the focal plane (~40 mm). The prototype shown in Figure 7.22(b) is characterized to operate with 8% $|S_{11}| < -10$ dB bandwidth. As expected, this is identical with the bandwidth of the antenna elements due to the nonresonant nature of the feed network and significantly larger than the 3% bandwidth realized with the resonant feed network-based microfluidic beam-steering arrays. Figure 7.22(c) depicts the measured gain patterns as the beam is steered by repositioning the selectively metallized plate for exciting the antenna elements (the lens is identical to the one utilized in previously

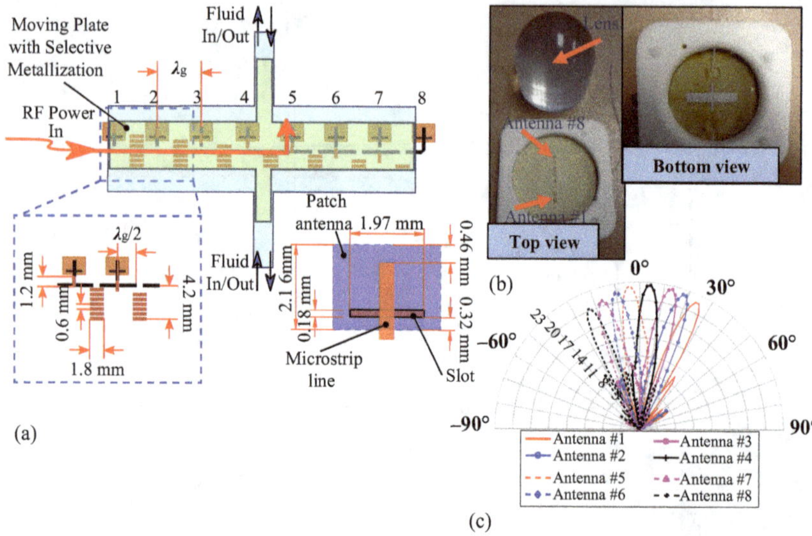

Figure 7.22 Selectively metallized plate utilized to construct a switched feed network for an 8 element microfluidic beam-steering focal plane array (see Figure 7.2 for substrate stack-up): (a) antenna and plate metallization dimensions; (b) prototype; and (c) measured 30 GHz gain patterns (after Gonzalez et al. [33] © IEEE 2017)

discussed microfluidic beam-steering focal plane arrays). The measured patterns exhibit 7° HPBW with >22 dBi gain. Accounting for patch antenna, lens, connector, and extension line losses demonstrate that the feed network loss is less than 3 dB as expected. This performance is again better than what could have been achieved with a conventional implementation that requires 7 mm-wave SP2T switches with bias/control lines. Extending this concept to a 2D beam-steering MFPA is also possible and discussed in [33]. Expected beam-steering speeds are on the order of 20 ms using small piezoelectric micropumps. Microfluidic channel size and actuation characterizations/optimizations are underway to improve the reliability/operation of this system with the small piezoelectric micropumps. In addition, it is important to note that microfluidically repositionable selectively metallized plates have been also recently demonstrated for small top-loaded monopole [49] and frequency-agile RF bandpass filters [50].

7.8 Concluding remarks

The RF performance advantages offered by the microfluidically reconfigurable antennas are likely to attract further interest in the upcoming years. This chapter presented several antenna examples from the recent literature and demonstrated that microfluidic reconfiguration techniques can be applicable in a wide variety of antenna types and achieve different functionalities such as flexibility, frequency-tuning, and beam-steering. The actuation mechanisms and slower reconfiguration speeds appear as the major limiter for the application areas of microfluidically reconfigurable antennas. Therefore, more compact and integrated actuation techniques are highly desired and their development is likely to impact/boost the microfluidically reconfigurable antenna applications. Finally, although this chapter focused on antenna applications, microfluidic reconfiguration techniques have also attracted interest to be utilized in various other RF devices such as frequency-agile filters [34,50–54], frequency selective surfaces [55–57], and imaging systems [58]. This broad scope of interest is therefore expected to keep the research field grow and enable new technologies in future.

References

[1] Keane W.J. 'Narrow-Band Yig Filters Aid Wide-Open Receivers'. *Microwaves*. 1978;17(9):50.

[2] Srinivasan G., Tatarenko A.S., and Bichurin M.I. 'Electrically Tunable Microwave Filters Based on Ferromagnetic Resonance in Ferrite-Ferroelectric Bilayers'. *Electronics Letters*. 2005;41(10):596–8.

[3] Brown A.R., and Rebeiz G.M. 'A Varactor-Tuned RF Filter'. *IEEE Transactions on Microwave Theory and Techniques*. 2000;48(7):1157–60.

[4] Lugo C., Jr., and Papapolymerou J. 'Electronic Switchable Bandpass Filter Using PIN Diodes for Wireless Low Cost System-On-a-Package Applications'. *Microwaves, Antennas and Propagation, IEE Proceedings*. 2004; 151(6):497–502.

[5] Lugo C., Wang G.A., Papapolymerou J., Zhao Z.Y., Wang X.Y., and Hunt A.T. 'Frequency and Bandwidth Agile Millimeter-Wave Filter Using Ferroelectric Capacitors and MEMS Cantilevers'. *IEEE Transactions on Microwave Theory and Techniques*. 2007;55(2):376–82.

[6] Zheng Y., Sazegar M., Maune H., Arshad M., and Jakoby R. 'Compact Tunable Dual-Channel Antenna Based on Ferroelectric Ceramics'. *Electronics Letters*. 2011;47(16):897–8.

[7] Kenle C., Xiaoguang L., Kovacs A., Chappell W.J., and Peroulis D. 'Antibiased Electrostatic RF MEMS Varactors and Tunable Filters'. *IEEE Transactions on Microwave Theory and Techniques*. 2010;58(12):3971–81.

[8] Reines I., Park S.J., and Rebeiz G.M. 'Compact Low-Loss Tunable X-Band Bandstop Filter With Miniature RF-MEMS Switches'. *IEEE Transactions on Microwave Theory and Techniques*. 2010;58(7):1887–95.

[9] Zhang X.Y., Xue Q., Chan C.H., and Hu B.J. 'Low-Loss Frequency-Agile Bandpass Filters With Controllable Bandwidth and Suppressed Second Harmonic'. *IEEE Transactions on Microwave Theory and Techniques*. 2010;58(6):1557–64.

[10] Park S.J., and Rebeiz G.M. 'Low-Loss Two-Pole Tunable Filters With Three Different Predefined Bandwidth Characteristics'. *IEEE Transactions on Microwave Theory and Techniques*. 2008;56(5):1137–48.

[11] Entesari K., and Rebeiz G.M. 'A Differential 4-Bit 6.5–10–GHz RF MEMS Tunable Filter'. *IEEE Transactions on Microwave Theory and Techniques*. 2005;53(3):1103–10.

[12] Shim Y., Wu Z.Z., and Rais-Zadeh M. 'A High-Performance Continuously Tunable MEMS Bandpass Filter at 1 GHz'. *IEEE Transactions on Microwave Theory and Techniques*. 2012;60(8):2439–47.

[13] Blondy P., and Peroulis D. 'Handling RF Power:The Latest Advances in RF-MEMS Tunable Filters'. *IEEE Microwave Magazine*. 2013;14(1):24–38.

[14] So J.H., Thelen J., Qusba A., Hayes G.J., Lazzi G., and Dickey M.D. 'Reversibly Deformable and Mechanically Tunable Fluidic Antennas'. *Advanced Functional Materials*. 2009;19(22):3632–7.

[15] Wang M., Khan M.R., Dickey M.D., and Adams J.J. 'A Compound Frequency and Polarization Reconfigurable Crossed Dipole Using Multi-Directional Spreading of Liquid Metal'. *IEEE Antennas and Wireless Propagation Letters*. 2016;16(99):1–1.

[16] Saghati A.P., Batra J.S., Kameoka J., and Entesari K. 'Miniature and Reconfigurable CPW Folded Slot Antennas Employing Liquid-Metal Capacitive Loading'. *IEEE Transactions on Antennas and Propagation*. 2015;63(9):3798–807.

[17] Dey A., and Mumcu G. 'Microfluidically Controlled Frequency-Tunable Monopole Antenna for High-Power Applications'. *IEEE Antennas and Wireless Propagation Letters*. 2016;15:226–9.

[18] Cheng S., Wu Z.G., Hallbjorner P., Hjort K., and Rydberg A. 'Foldable and Stretchable Liquid Metal Planar Inverted Cone Antenna'. *IEEE Transactions on Antennas and Propagation*. 2009;57(12):3765–71.

[19] Hayes G.J., So J.H., Qusba A., Dickey M.D., and Lazzi G. 'Flexible Liquid Metal Alloy (EGaIn) Microstrip Patch Antenna'. *IEEE Transactions on Antennas and Propagation*. 2012;60(5):2151–6.

[20] Mazlouman S.J., Jiang X.J., Mahanfar A., Menon C., and Vaughan R.G. 'A Reconfigurable Patch Antenna Using Liquid Metal Embedded in a Silicone Substrate'. *IEEE Transactions on Antennas and Propagation*. 2011; 59(12):4406–12.

[21] Khan M.R., Hayes G.J., So J.H., Lazzi G., and Dickey M.D. 'A Frequency Shifting Liquid Metal Antenna with Pressure Responsiveness'. *Applied Physics Letters*. 2011;99(1):013501: 1–3.

[22] King A.J., Patrick J.F., Sottos N.R., White S.R., Huff G.H., and Bernhard J.T. 'Microfluidically Switched Frequency-Reconfigurable Slot Antennas'. *IEEE Antennas and Wireless Propagation Letters*. 2013;12:828–31.

[23] Dey A., Guldiken R., and Mumcu G. 'Microfluidically Reconfigured Wideband Frequency-Tunable Liquid-Metal Monopole Antenna'. *IEEE Transactions on Antennas and Propagation*. 2016;64(6):2572–6.

[24] Morishita A.M., Kitamura C.K.Y., Ohta A.T., and Shiroma W.A. 'A Liquid-Metal Monopole Array With Tunable Frequency, Gain, and Beam Steering'. *IEEE Antennas and Wireless Propagation Letters*. 2013;12: 1388–91.

[25] Dang J.H., Gough R.C., Morishita A.M., Ohta A.T., and Shiroma W.A. 'Liquid-Metal Frequency-Reconfigurable Slot Antenna Using Air-bubble Actuation'. *Electronics Letters*. 2015;51(21):1630–2.

[26] Rodrigo D., Jofre L., and Cetiner B.A. 'Circular Beam-Steering Reconfigurable Antenna With Liquid Metal Parasitics'. *IEEE Transactions on Antennas and Propagation*. 2012;60(4):1796–802.

[27] Gheethan A. A., Jo M.C., Guldiken R., and Mumcu G. 'Microfluidic Based Ka-Band Beam-Scanning Focal Plane Array'. *IEEE Antennas and Wireless Propagation Letters*. 2013;12:1638–41.

[28] Huff G.H., Rolando D.L., Walters P., and McDonald J. 'A Frequency Reconfigurable Dielectric Resonator Antenna Using Colloidal Dispersions'. *IEEE Antennas and Wireless Propagation Letters*. 2010;9:288–90.

[29] Murray C., and Franklin R.R. 'Independently Tunable Annular Slot Antenna Resonant Frequencies Using Fluids'. *IEEE Antennas and Wireless Propagation Letters*. 2014;13:1449–52.

[30] Barrera J.D., and Huff G.H. 'A Fluidic Loading Mechanism in a Polarization Reconfigurable Antenna With a Comparison to Solid State Approaches'. *IEEE Transactions on Antennas and Propagation*. 2014;62(8):4008–14.

[31] Gheethan A.A., Dey A., and Mumcu G. 'Passive Feed Network Designs for Microfluidic Beam-Scanning Focal Plane Arrays and Their Performance Evaluation'. *IEEE Transactions on Antennas and Propagation*. 2015;63(8): 3452–64.

[32] González E., and Mumcu G., editors. 'Low-Loss Wideband Feed Networks for High Gain Microfluidic Beam-scanning Focal Plane Arrays'. *2016 IEEE International Symposium on Antennas and Propagation (APSURSI)*; 2016 June 26 2016–July 1 2016. pp. 645–646.

[33] González E., and Mumcu G., editors. 'Mm-Wave 2D Beam-Steering Focal Plane Array with Microfluidically Switched Feed Network'. *2017 IEEE International Symposium on Antennas and Propagation (APSURSI)*; 2017.

[34] Mumcu G., Dey A., and Palomo T. 'Frequency-Agile Bandpass Filters Using Liquid Metal Tunable Broadside Coupled Split Ring Resonators'. *IEEE Microwave and Wireless Components Letters*. 2013;23(4):187–9.

[35] Cresson P.Y., Orlic Y., Legier J.F., et al. '1 to 220 GHz Complex Permittivity Behavior of Flexible Polydimethylsiloxane Substrate'. *IEEE Microwave and Wireless Components Letters*. 2014;24(4):278–80.

[36] Sunkara V., Park D.-K., Hwang H., Chantiwas R., Soper S.A., and Cho Y.-K. 'Simple Room Temperature Bonding of Thermoplastics and Poly(dimethylsiloxane)'. *Lab on a Chip*. 2011;11(5):962–5.

[37] Koo C., LeBlanc B.E., Kelley M., Fitzgerald H.E., Huff G.H., and Han A. 'Manipulating Liquid Metal Droplets in Microfluidic Channels With Minimized Skin Residues Toward Tunable RF Applications'. *Journal of Microelectromechanical Systems*. 2015;24(4):1069–76.

[38] Gough R.C., Morishita A.M., Dang J.H., Wenqi H., Shiroma W.A., and Ohta A.T. 'Continuous Electrowetting of Non-toxic Liquid Metal for RF Applications'. *Access, IEEE*. 2014;2:874–82.

[39] Agar J., Durden J., Staiculescu D., Zhang R., Gebara E., and Wong C.P., editors. 'Electrically Conductive Silicone Nano-composites for Stretchable RF Devices'. *2011 IEEE MTT-S International Microwave Symposium*; 2011 5–10 June 2011. pp. 1–4.

[40] Mehdipour A., Sebak A.R., Trueman C.W., Rosca I.D., and Hoa S.V., editors. 'Conductive Carbon Fiber Composite Materials for Antenna and Microwave Applications'. *2012 29th National Radio Science Conference (NRSC)*; 2012 10–12 April 2012. pp. 1–8.

[41] Takshi A., and Madden J.D. 'Multilayer Stretchable Conductors with a Large Tensile Strength'. *Journal of Elastomers & Plastics*. 2010;42(4):365–73.

[42] Liyakath R.A., Takshi A., and Mumcu G. 'Multilayer Stretchable Conductors on Polymer Substrates for Conformal and Reconfigurable Antennas'. *Antennas and Wireless Propagation Letters, IEEE*. 2013;12:603–6.

[43] Esser-Kahn A.P., Thakre P.R., Dong H., et al. 'Three-Dimensional Microvascular Fiber-Reinforced Composites'. *Advanced Materials*. 2011;23 (32):3654–8.

[44] Dey A., Guldiken R., and Mumcu G. 'Wideband Frequency Tunable Liquid Metal Monopole Antenna'. *IEEE Antennas and Propoagation Society Symposium*. 2013:1–2.

[45] Saghati A.P., Batra J.S., Kameoka J., and Entesari K. 'A Microfluidically Reconfigurable Dual-Band Slot Antenna With a Frequency Coverage Ratio of 3:1'. *IEEE Antennas and Wireless Propagation Letters*. 2016;15:122–5.

[46] Wang M., Trlica C., Khan M.R., Dickey M.D., and Adams J.J. 'A Reconfigurable Liquid Metal Antenna Driven by Electrochemically Controlled Capillarity'. *Journal of Applied Physics*. 2015;117(19):194901.

[47] Filipovic D.F., Gearhart S.S., and Rebeiz G.M. 'Double-Slot Antennas on Extended Hemispherical and Elliptical Silicon Dielectric Lenses'. *IEEE Transactions on Microwave Theory and Techniques*. 1993;41(10):1738–49.

[48] Filipovic D.F., Gauthier G.P., Raman S., and Rebeiz G.M. 'Off-axis Properties of Silicon and Quartz Dielectric Lens Antennas'. *IEEE Transactions on Antennas and Propagation*. 1997;45(5):760–6.

[49] Dey A., and Mumcu G., editors. 'Small Microfluidically Tunable Top Loaded Monopole'. *2016 International Workshop on Antenna Technology (iWAT)*; 2016 Feb. 29 2016–March 2 2016. pp. 148–149.

[50] Palomo T., and Mumcu G. 'Microfluidically Reconfigurable Metallized Plate Loaded Frequency-Agile RF Bandpass Filters'. *IEEE Transactions on Microwave Theory and Techniques*. 2016;64(1):158–65.

[51] Palomo T., and Mumcu G., editors. 'Highly Reconfigurable Bandpass Filters Using Microfluidically Controlled Metallized Glass Plates'. *2014 IEEE MTT-S International Microwave Symposium (IMS 2014)*; 2014 1–6 June 2014. pp. 1–3.

[52] Saghati A.P., Batra J.S., Kameoka J., and Entesari K. 'A Miniaturized Microfluidically Reconfigurable Coplanar Waveguide Bandpass Filter With Maximum Power Handling of 10 Watts'. *IEEE Transactions on Microwave Theory and Techniques*. 2015;63(8):2515–25.

[53] Shuyan G., Bao Jun L., Wenqi H., Shiroma W.A., and Ohta A.T., editors. 'A Tunable Low-Pass Filter Using a Liquid-Metal Reconfigurable Periodic Defected Ground Structure'. *Microwave Symposium Digest (MTT), 2012 IEEE MTT-S International*; 2012, 17–22 June 2012. pp. 1–3.

[54] Irshad W., and Peroulis D., editors. 'A 12–18 GHz Electrostatically Tunable Liquid Metal RF MEMS Resonator with Quality Factor of 1400–1840'. *Microwave Symposium Digest (MTT), 2011 IEEE MTT-S International*; 2011 5–10 June 2011. pp. 1–4.

[55] Li M., Yu B., and Behdad N. 'Liquid-Tunable Frequency Selective Surfaces'. *IEEE Microwave and Wireless Components Letters*. 2010;20(8):423–5.

[56] Li M., and Behdad N. 'Fluidically Tunable Frequency Selective/Phase Shifting Surfaces for High-Power Microwave Applications'. *IEEE Transactions on Antennas and Propagation*. 2012;60(6):2748–59.

[57] Long S.A., and Huff G.H. 'A Fluidic Loading Mechanism for Phase Reconfigurable Reflectarray Elements'. *IEEE Antennas and Wireless Propagation Letters*. 2011;10:876–9.

[58] Dey A., and Mumcu G., editors. 'High Resolution Surface Imaging Arrays Interrogated with Microfluidically Controlled Metalized Plates'. *2014 IEEE Antennas and Propagation Society International Symposium (APSURSI)*; 2014, 6–11 July 2014. pp. 213–214.

Chapter 8

Flexible and wearable antennas

Muhammad M. Tahseen[1] and Ahmed A. Kishk[1]

8.1 Introduction

Research activities on flexible antennas for the wireless communication have been increased tremendously for their numerous benefits. These flexible antennas are rapidly being utilized in several applications, for example, wearable, defense, biomedical, tracking, and many others. It is expected that these antennas will play a significant role on the internet of things (IoT). The well-known concept of microstrip antenna design is mainly used for wearable antennas, which consists of a radiating patch backed by ground (GND) plane. Usually, to simplify the design, the first trial is based on assuming the metal as a perfect electric conductor (PEC). Although losses of the actual materials will have a significant effect on the antenna gain and it is efficiency, the effect on the resonant frequency may not be that significant. As the antenna is flexible, it is a bendable structure. Depending on the polarization and the bending direction, the resonance frequency will be shifted. There are many different materials and techniques that can be used to design flexible and wearable antennas. Usually, these materials are not of the typical type used by the RF Engineers. Whatever the material used, it is necessary to know the electrical properties of the material at the operating frequency band. In some cases, standard methods of measurements are not applicable. For example, flexible antennas are not easily mounted as an individual element to be measured. Therefore, mounting support similar to the actual working environment is suggested.

In [1], a planar-inverted F-antenna (PIFA) is used for the wearable application, which reduces the effect of radiation on the human body by reducing the radiated power towards the body. As the Global System for Mobile Communication (GSM) transmitted power is reduced, the battery life is increased. This performance is achieved due to matching the feed with the load that reduces the power loss because of reduction in the reflected power. In [2], a FlexPIFA antenna is designed from the flexible dielectric material for the smart clothing application. The transmission range was calculated based on Friis expression. Another PIFA antenna that is integrated with the human body for a mobile phone can be found in [3]. It is

[1]Department of Electrical and Computer Engineering, Concordia University, Canada

found that the best mounting location of the antenna is the shoulder because of the least possible physical blockage of the antenna radiation. Another technique is used to reduce the interaction of the antenna with the host body, which is based on using perfect magnetic conductor (PMC), which does not naturally exist, but we can artificially immolate it. This surface has a high surface impedance, but the conductor surface has very low surface impedance. In order to understand this concept, a normal incident plane wave on a PEC is reflected with a reflection coefficient of -1, which means a phase reversal of 180°. On the other hand, the normal incident plane wave on a PMC is reflected reflection coefficient of $+1$, which means a 0° phase reflection. When the PMC is realized artificially, we refer to it as an artificial magnetic conductor (AMC) surface that must be appropriately designed for a prespecified frequency band. Therefore, AMC surface has a certain bandwidth of operation, unlike the PEC that has frequency independent performance theoretically. There are many advantages of using the AMC in the antenna design. When a PEC is used as a GND plane, the induced surface current on the PEC surface causes undesired radiation, such as high cross polarization, smaller matching bandwidth, and lower gain. On the other hand, due to the high impedance of the AMC surface, surface waves are suppressed. Therefore, the electric antennas mounted on the AMC surface has the wider matching bandwidth, higher gain, and low profile as the electric antenna type can be mounted directly of the AMC. To obtain these benefits in the wearable applications, several AMC-based antennas are designed in the literature.

In [4], a Bluetooth antenna is designed on the low-cost FR4 material for the wearable application. In this design, the AMC used for GND plane. The AMC helped in achieving an adequate matching bandwidth of 4.7%, and a maximum gain of 6.7 dBi. The performance of the antenna near the human body was also studied. A measured 14.2% matching bandwidth has been reported [56], from a PIFA, which was designed on the flexible textile material. In [7], the effect of the human body on the transmission loss has been studied. It was found that the channel loss variation is significant with different body locations and body positions. When the antenna is made of woven conductive fabric and felt materials to be wearable lightweight, a 2 dBi gain loss was observed experimentally between the bended antenna and non-bending case [8]. An array of these antennas is designed on a hat. Similarly, a WLAN band flexible wearable antenna is designed with fleece textile material where the knitted GND plane has been used [9]. In [10], [11,12], wearable antennas are designed on the electromagnetic bandgap (EBG) surface. It has been reported that using the flexible textile-based EBG structure in the wearable antenna design; the antenna size can be reduced more than 30%, while the input matching bandwidth can be increased almost 50%. Furthermore, a 6×6 patch array has also been designed with the textile material using EBG surface where both the radiating patch element as well as the EBG surface are made up of copper tape. The flexible wearable antennas are analyzed near the human body and presented in [13–16]. A flexible dual-band U-shaped slotted patch antenna is designed for wearable application in [17,18], while the antenna is operating in GSM 1900 and WLAN frequency bands. A textile-based flexible wearable antenna is designed by truncating patch corners for circular polarization (CP), presented in [19,20]. The antenna has resulted in a 3-dB axial ratio of 3%. To analyze the effect of the conductive material on the wearable antenna performance, six antennas are designed with different methods, and the presented case study concluded that the choice of the proper material is an important factor in achieving the highest antenna performance [21]. In [22], the antenna performance is compared in terms of before and after using the antenna in the embedded body-worn sensor communication setup. The ultra-wide-band (UWB) monopole and slot antennas are used as an example. The presented results exhibit that a significant degradation in the antenna performance in terms of return loss and radiation pattern has been observed when the antenna is integrated into the body-worn sensor communication system. A novel UWB directional antenna is designed in [23] for the wearable application. A slotted patch is used as the radiating element, which is fed is a y-shaped microstrip fork. A GND plane is used at the back to suppress the backward radiation from the slot and converts the omnidirectional radiation pattern into a unidirectional. A UWB antenna is designed in [24], for bio-medical application specifically analyzing the effect of antenna radiation on the human head and how the performance degrades. The antenna geometry consists of a square patch that is connected to a 50 Ω microstrip transmission line. It has been reported by the presented results that the antenna performance is degraded when used near the region of the human head. To miniaturize the antenna size, a dual-band novel button-sized antenna is designed in [25–27] that has resulted in 5.1% with −10 dB impedance bandwidth in the 2.4 GHz band and 13.5% at 5.2 GHz band. Several printed circuit planar antennas are designed in [28], for wearable and UWB applications. Different WLAN band textile-based antennas are designed for Bluetooth application for linear and CP, in [29]. The antennas are designed with pure fabric materials while even the radiating patches are designed with a conductive flexible fabric material that helps in reducing the degradation in the antenna performance when used in the conformal or bending environment. To overcome the feeding problem in wearable antennas, a WLAN band textile-based aperture-coupled patch antenna is designed in [30] that has shown higher flexibility as well as 63% radiation efficiency when the antenna is embedded in the human body. To achieve polarization diversity in wearable antennas, a dual-polarized antenna is designed with textile material and presented in [31]. The proposed antenna has resulted in almost 7 dBi measured gain as well as a maximum 58% of the radiation efficiency. A wearable antenna designed with flexible protective foam that is suitable for firefighter garments is designed in [32] with 180 MHz measured axial ratio bandwidth. The antenna has resulted in a similar bandwidth even in the bending environment with 62% of radiation efficiency. In [33], three conformal antennas are presented, which are printed on a 0.1 mm thin FR4 material. The three proposed designs are: cross-bowtie dipole, flared dipole, and an asymmetric meandered line dipole. The prominent property of the designs is that all antennas have no GND plane. The meandered line dipole antenna has shown better conformability than the other two antennas when mounted on the human body. The antenna performance on the designed human phantom has also been presented, where the simulation and measured performance has shown good agreement in the UHF band. A WLAN band circular polarized antenna array is designed in [34] with e-textile, for body worn applications. Along these six complementary circular rings, shaped antennas are also designed with e-textile, and their performance is analyzed using them in different positions on the body. In [34], wearable antenna for military application has been presented. In [35], a flexible wearable antenna was fabricated using polydimethylsiloxane material. The complete fabrication process has also been explained, while the simulation and measured results exhibit agreement while the antenna is planar and little degradation in the performance has been observed in the bending environment. A proximity-coupled textile-based wearable antenna is designed to steer the beam at 6 GHz. A U-shaped slot in the patch is used while two switches are connected between the feed line and the patch to steer the beam in three different directions of 0°, 30°, and 331° [36]. An inkjet-printed monopole antenna with AMC surface is designed on a paper substrate for wearable application in [37]. Due to the lightweight, and thin paper thickness, the antenna shows higher flexibility, which is required for wearable application. Moreover, the antenna performance is measured in the presence of human phantom. Wearable antenna design used in a life-saving jacket for sea travellers, has been presented in [38]. A circularly polarized textile-based antenna is presented in [39]. The measured 3-dB axial ration bandwidth of 23% and 10 dB impedance matching bandwidth of 44% was achieved (at 2.45 GHz). The textile antenna was connected to a transponder that was attached to a human body to test the short-range wireless power reception. Different flexible wearable antennas are designed with textile material using conductive thread in [40], in the WLAN band. In [41], four textile-based wearable antennas were designed and measured. The antennas were embroidered with different conductive thread density and techniques. The effect of thread thickness and the gap between meshed intervals, on the antenna resonance frequency and radiation efficiency, has been presented. It has been shown by measuring results that the thread thickness has more effect on the antenna performance than the gap intervals between meshed patch threads. This less dense embroidered meshed patch provides an advantage when the antenna is used in the bending environment in wearable applications, where the drop in the radiation efficiency is not so much compared to the nonbending situation. In [42], dual-band wearable flexible antenna is designed with magneto-electric dipole (MD-ME) concept. The antenna operates at lower WLAN (at 2.45 GHz) and higher WLAN band (at 5.8 GHz), where the radiation efficiency is over 60% in lower, and near 42% in the upper WLAN band. A broadband conformal and the flexible circularly polarized spiral antenna are presented in [43], which provides a 10:1 matching bandwidth with almost stable 6.5 dBi gain, and less than 5 dB axial ratio bandwidth in the whole frequency band. The radiating element was embroidered using a conductive thread on 0.59 mm thick Kevlar fabric, which has almost constant measured dielectric constant of 2.5, and loss tangent of 0.006 in the whole band.

8.2 Wearable antennas for biomedical applications

For continuous monitoring of the patient's health in different ways, temperature, blood flow, heartbeat, pressure, and others, research on the wearable electronic has tremendously increased. Typically, body sensor networks (BSN) are used in these situations. These BSNs are used on, and close to the body to measure the signals and wireless transfer to the external remote wearable communication control system [44]. The conformal wearable antennas are preferred in these situations where the antennas adapt to the surface being mounted on, as well as maintains the reliable wireless signal transfer link. To achieve this objective, wearable antennas are designed with flexible dielectric material, which most of the time is textile materials, flexible polymer thin sheets, and a paper substrate. For designing completely flexible wearable antennas, the radiating geometry should be printed with a flexible conductive material, for example, embroidered with conductive thread on textile materials, and conductive ink or paste.

A compact, low-profile wearable antenna is designed in [45], with EBG structure. The textile denim material is used to print the antenna geometry as well as the elements present in the EBG structure. The EBG structures are widely being used in the antenna applications due to their in-phase reflection coefficients while suppressing the surface waves. As a result, the antenna provides wider matching bandwidth, higher gain, and improves the front-to-back ratio (FBR) in the radiation pattern. The transmission coefficient magnitude and the corresponding phase of the used EBG structure exhibit the reflection phase of 0°, while the wave transmission is less than −10 dB.

From the statistical data presented in [46], a high percentage of women have breast cancer in the United States. However, there is a 99% survival rate if it is early detected while this survival rate drops to 24% if it spreads out and not diagnosed in the early stages. Also, in [46], it is reported that only 61% of breast cancer cases are diagnosed at early stages. To diagnose breast cancer accurately in early stages, a flexible 16-array antenna system is presented in [47], operating in the frequency range of 2 to 4 GHz. In [47], to analyze the complete properties of the tumor, backscattered signals from two polarizations are collected from both the *X*- and *Y*-directions. A spiral single polarized antenna array on a phantom exhibits the dual-polarized spiral antenna array for detecting breast tumor. A model is presented with several homogeneous materials to make the inhomogeneous environment for the numerical solution. Several materials are used in the phantom design, for example, fat, skin, muscles, and glands where the electrical properties of these materials can be found in [48]. It can be seen the flexibility of the antennas, which are highly in demand for wearable applications. The total 16 dual-polarized spiral antennas are fed by SMA connectors. A miniaturized spiral antenna element in the array has a size of 20 mm × 20 mm. The simulated and measured port reflection coefficient magnitude with different antenna positions in the array, for single- and dual-polarized spiral antenna array is given. The results show that the proposed conformal antennas lead to wider measured matching bandwidth from 2 to 4 GHz, making it a suitable candidate for the wearable application.

8.3 AMC-based flexible wearable antennas

Most of the wearable antennas are designed on the microstrip patch principle, which consists of a radiating patch backed by GND plane. The radiating patch, as well as the GND plane, is designed considering that the conductors are a PEC.

As we know that a normal incident plane wave experiences a phase reversal of 180° from PEC, while on the other hand a PMC that is non-real result a 0° phase reflection. The surface that reflects the normal incident plane wave with 0° reflection is also known as AMC surface that shows many advantages when incorporated in the antenna design. The undesired radiations created due to the induced surface current created by the PEC surface cause, higher cross polarization, lower matching bandwidth, and lower gain. On the other hand, due to high impedance AMC surface, surface waves are killed. The antennas designed with AMC surface results, wider impedance matching bandwidth, higher gain, and are low profile. To obtain these benefits in the wearable applications, several AMC-based antennas are designed in the literature. In [49], a printed Yagi antenna is designed for the wearable application, with an AMC surface for end-fire radiations. The antenna aims to operate in the ISM band, while the designed frequency is 2.45 GHz. Two AMC surfaces are used in the design while the radiating printed Yagi antenna is placed above at some distance. The reflection phase of the used AMC surface is provided. The antennas are fabricated on two materials, and their performance is compared in a different case: when the Yagi antenna is backed by PEC, when the antenna is backed by single-layer AMC (S-AMC), when the antenna is backed by double-layer AMC (D-AMC). The antennas and AMC surfaces are designed with FR4, and a flexible material "Latex". The material parameters in terms of dielectric constant and loss tangent of the Latex material are extracted using the T-resonator technique presented in [50] and [51]. The antenna performance in the presence of S-AMC and D-AMC, designed with the flexible Latex material is presented here, while the results with the FR4 material is omitted for brevity, but can be found in [49]. The measurements are provided when the antennas are embedded in the body. The simulated and measured reflection coefficient of the antenna with S-AMC and D-AMC is in good agreement. The radiation patterns are also provided. These results show that the proposed antenna exhibits lower FBR when D-AMC surface is used.

In [52], a dual-band flexible and reconfigurable folded slot antenna is designed for the wearable application. The reconfigurability in the antenna radiation characteristics is achieved with a PIN diode, connected to the stub. The polarization and operating frequency of the slot and stub are different, so a polarization dependent dual-band AMC surface is designed. The antenna is placed on the AMC surface, which acts as a GND plane. The AMC surface improves the antenna radiation performance and reduces the specific absorption rate when the radiating antenna is mounted on the body area network. The antenna operates in the wireless body area network (WBAN) and Worldwide Interoperability for Microwave Access (WiMAX) bands. Detailed information about the antenna can be found in [52].

A compact size broadband wearable antenna backed by AMC is designed in [53]. The antenna has a size of 46×46 mm^2. The AMC surface consists of patch elements. The proposed windmill antenna achieved broadband matching bandwidth. The effect of the different AMC-sized GND plane is also studied. The antenna conformability is considered an essential parameter in wearable applications, so the performance of the antenna is analyzed in the bending environment at 7 and 10 GHz. The proposed antenna provides 63.5% (5.7–11.0 GHz) impedance matching bandwidth, a peak gain of 8 dBi, and FBR more than 15 dB. The proposed broadband AMC-backed windmill-like antenna provides radiation efficiency of 90.5% and 85.5% at 7 GHz and 10 GHz, respectively.

8.4 Inkjet-printed wearable antennas

As mentioned in the previous sections, higher flexibility and conformability in the wearable antennas is desired. Flexible antennas should hold flexible characteristics. For this purpose, flexible substrates and conductive ink play a vital role in fulfilling such requirement. The flexible antennas conform to any hosting surface.

Several flexible antennas exist in the literature, which is designed with flexible dielectric materials: paper, plastic, textile, etc.

In this section, few flexible wearable antennas printed on the paper using inkjet printing technique, are presented. Replacing conventional solid copper with the conductive ink in the antennas provides higher flexibility as well as very good antenna radiation efficiency. There are varieties of conductive inks available which results in a reasonably good conductivity close to the conventional copper, but still not equal. Both copper and conductive ink provides some advantages as well as disadvantages. The antennas printed with conductive ink exhibit higher flexibility and conformability which is helpful in the situations where mounting surface is not planar, while the antenna results in lower radiation efficiency. On the other hand, antennas designed with conventional copper provide higher radiation efficiency, but not suitable in the bending wearable environment.

In [37], an inkjet-printed monopole antenna, placed on an inkjet-printed EBG surface, is presented. The antenna is designed to operate in the ISM band (designed frequency is 2.45 GHz). The antenna and EBG surface is printed on a paper sheet with a thickness of 0.1 mm, dielectric constant of 3.2, and a loss tangent of 0.07. A Dimatix 10 PL cartridge is utilized, while 800 m is kept between paper substrate and printer's conductive ink nozzle. The output print resolution of 1,270 dpi was selected in the printer settings. Cabot conductive ink CCI-300 was used at the temperature of 36 °C, while the paper substrate was maintained at 50°C. The printed shape was centered in a thermal oven for 2 h at 130°C [37]. The simulated and measured reflection coefficient with and without the EBG surface and E- and H-plane radiation patterns exhibit good agreement. Similarly, the simulated and measured gain variation with frequency is also given.

In [59], a patch antenna is designed on the felt material using inkjet printing. The antenna achieved radiation efficiency of 53% at 2.45 GHz. A lightweight, compact, ultra-low profile and UWB flexible wearable slot antenna designs in [54] using inkjet printing. Kapton Polyimide substrate is used, which has a dielectric constant of 3.4. The antenna is fed by co-planar waveguide technique. The simulated and measured reflection coefficients with different bends authenticate that the antenna in the bend environment results has the wider matching bandwidth, which makes the antenna suitable candidate for the wearable application.

Similarly, few other flexible wearable antennas designed with inkjet printing technique are presented in [55–59].

8.5 Textile-based wearable antennas

The extremely flexible and conformal wearable antennas are designed using textile materials. Due to different material compositions, not every textile material is suitable for antenna applications. It is highly recommended to use the textile material which is less elastic, the antenna design as it maintains their effective dielectric constant otherwise the material parameters will be changed when the material is stretched. There is a considerable number of wearable antennas exist in literature, designed with the textile material [9–62], and few of them are highlighted in the following sections.

8.5.1 Single- and multi-layer multi-Bowtie conformal antennas

In this section, single- and multilayer multi-Bowtie textile-based conformal antennas are designed at WLAN band, using conductive thread [60]. The radiating patch is embroidered using silver coated conductive thread. The measured conductivity of the thread is 3.6×10^6 S/m. The dielectric constant of the textile materials is extracted using resonance technique [63]. To achieve flexibility in the design, the solid copper GND plane is replaced with shielded fabric GND [64], as shown in Figure 8.1. The effect of solid GND and the shielded fabric GND plane on

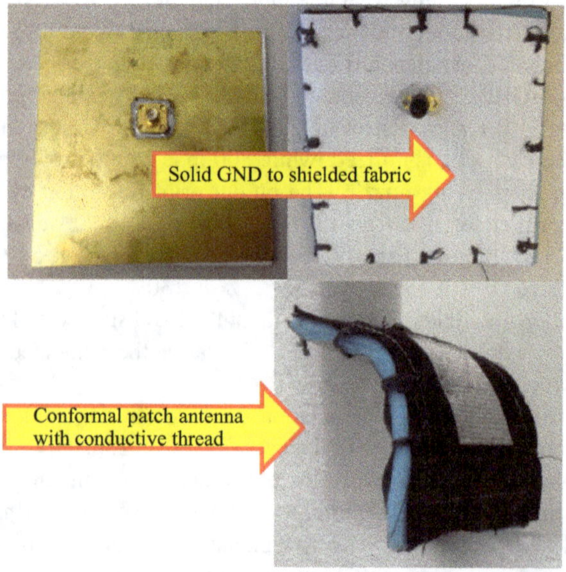

Figure 8.1 Replacing Solid GND plane with shielded fabric GND plane. (Reproduced with permission of IEEE)

the reflection coefficient magnitude is shown in Figure 8.2. Initially, single-layer multi-Bowtie planar antenna is designed at 2.45 GHz and then analyzed its performance in conformal environment. Later, multi-layer stacked Bowtie conformal antenna and single-layer (SL) conformal DRA antenna, has been designed for broadening the antenna bandwidth. The result of the antennas is compared at 2.45 GHz. The Bowtie arrangement to form a single patch makes it more flexible and easily bent without an adverse effect of a possible stretch of the fabric [60]. The antennas are shown in Figure 8.3. The antenna size has similar size of 0.8″ (at 2.45 GHz) with a thickness of 3.6 mm. The performance of the proposed antenna does not vary much within the used 30° surface bending angles. The comparison in terms of voltage standing wave ratio (VSWR), normalized radiation pattern, and the radiation efficiency is shown in Figures 8.4–8.8.

To enhance the operating bandwidth of the conformal antennas, stacked multi-Bowtie and DRA placed on the textile material techniques, are used. The antenna performance is shown in Figures 8.4–8.8. It is evident from these results that the wider matching bandwidth of over 14% is obtained using stacked multi-Bowtie conformal antenna while the designed antenna provides 90% radiation efficiency. To analyze the antenna performance using DRA, a standard RT3010 Rogers material with a dielectric permittivity of 10.2, is used with a thickness of 15 mm for DRA. For prototyping this flexible DRA, it will be stitched on the textile material for supporting at the top layer. The performance comparison of the planar and conformal DRA antenna is shown in Figures 8.4–8.8. The wider impedance bandwidth of over 30% is achieved from the conformal DRA antenna while the

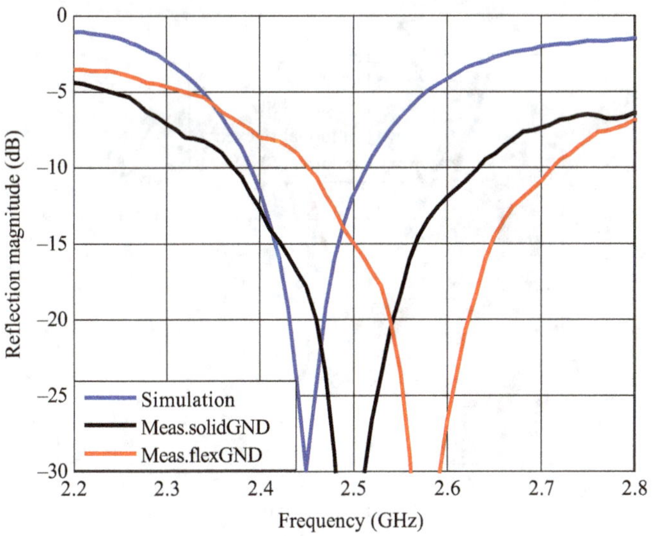

Figure 8.2 Wi-Fi band antenna's simulation and measurement results comparing solid GND and shielded fabric GND plane (Reproduced with permission of IEEE)

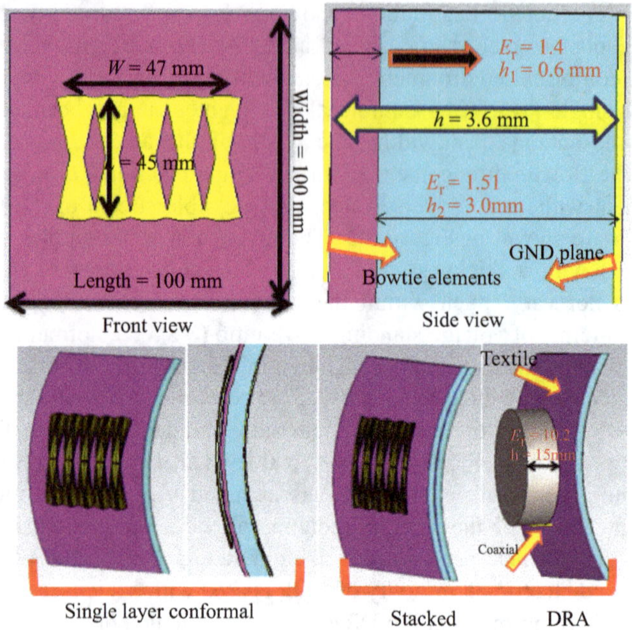

Figure 8.3 Multi-bowtie planar and conformal wide band antennas (Reproduced with permission of IEEE)

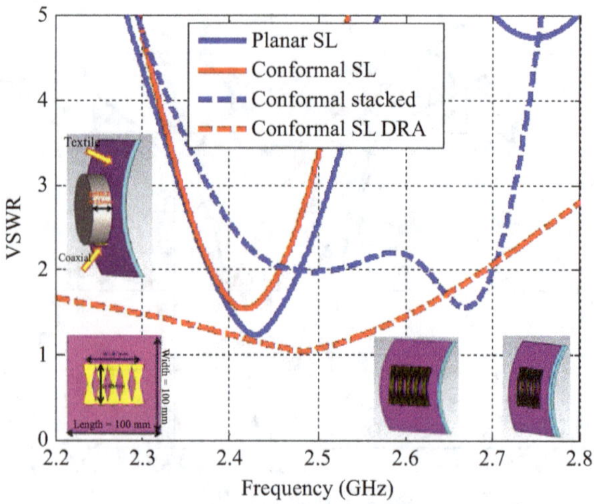

Figure 8.4 VSWR comparison for different proposed conformal antennas (Reproduced with permission of IEEE)

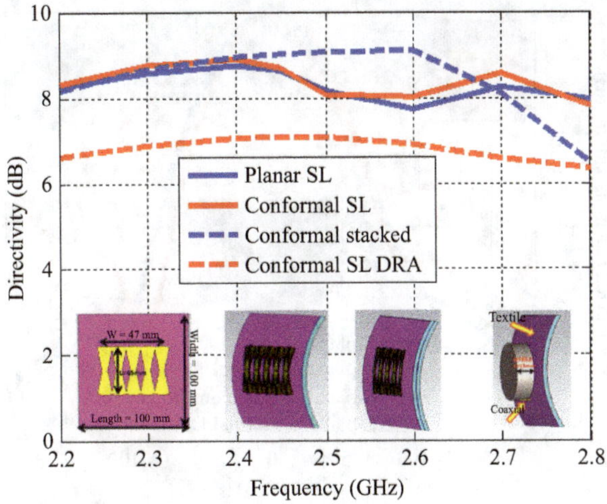

Figure 8.5 Directivity comparison for different proposed conformal antennas (Reproduced with permission of IEEE)

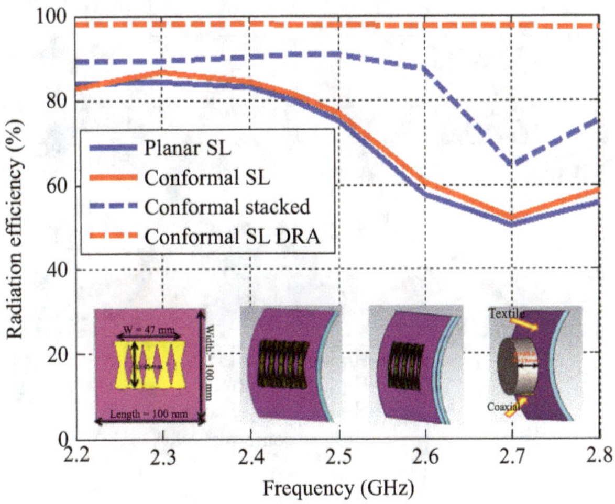

Figure 8.6 Radiation efficiency comparison for different proposed conformal antennas (Reproduced with permission of IEEE)

radiation efficiency of over 95% results in the whole frequency band. The proposed antenna's performance is compared and shown in Table 8.1.

8.5.2 Dielectric resonator antennas for wearable application

Similarly, wearable antennas are designed with DRAs to achieve wider bandwidth and higher radiation efficiency in [61]. In all the antennas, the DRAs are placed on

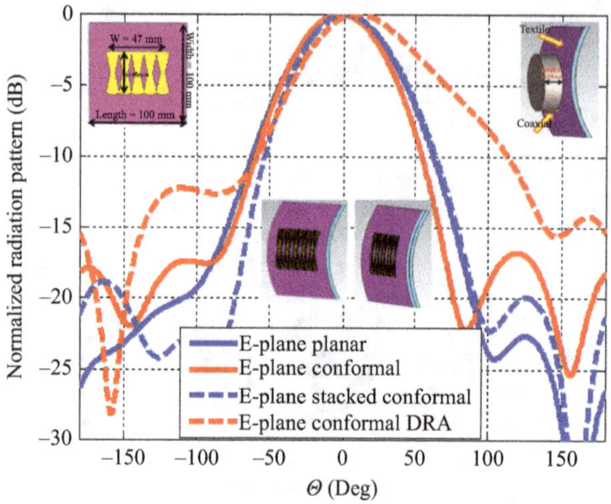

Figure 8.7 E-plane normalized radiation pattern comparison for different proposed conformal antennas (Reproduced with permission of IEEE)

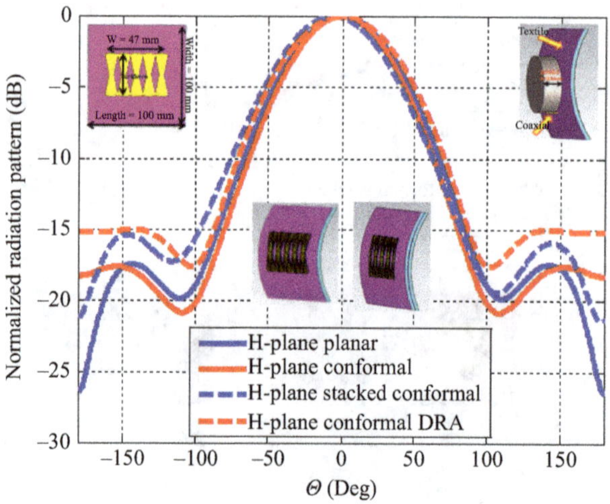

Figure 8.8 H-plane normalized radiation pattern comparison for different proposed conformal antennas (Reproduced with permission of IEEE)

textile material, and excited by coaxial probe and aperture-coupled feed method, as shown in Figure 8.9. The antenna performance in terms of VSWR gain bandwidth, radiation efficiency and the E-H plane normalized radiation patterns are shown in Figures 8.10–8.14. In aperture-coupled method energy is coupled through slot has been used in the GND, as shown in Figure 8.9(b). The central-aperture under the DRA reduces the cross-polarization due to structure symmetry. As the GND plane

Table 8.1 Performance comparison of proposed conformal antennas

	VSWR (3:1%)	Directivity (dB)	η_{rad}(%)	HPBW (Deg)
SL Planar	7	8.62	80	71.3
SL Conformal	6	8.67	81	71.9
Stacked conformal	14.3	9.03	90	66
SL DRA	32.7	7.1	98	76.4

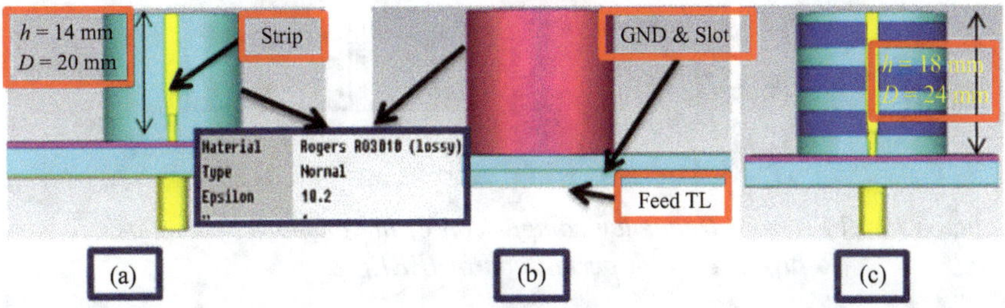

Figure 8.9 Proposed flexible wideband DRA antennas using textile material: (a) Coax-fed DRA; (b) aperture-coupled; and (c) Coax-fed dielectric and textile disk-stacked DRA (Reproduced with permission of IEEE)

is separating the feed and the radiating DRA, the spurious radiations are minimized, which further increase the antenna matching bandwidth. The aperture-coupled antenna provides more than 20% impedance bandwidth, more than 11% 1-dB gain bandwidth and 93% radiation efficiency. The antenna performance is shown in Figures 8.10–8.14. A DRA is designed with the combination of dielectric and felt disks stacking. The diameter and thickness of the DRA are optimized to get wider antenna performance. The antenna model is shown in Figure 8.9(c). The blue layers are the Felt material disks, with an extracted dielectric constant of 1.2 and the loss tangent of 0.02, while the dark green color represents the standard available Rogers material RT 3010, with a dielectric constant of 10.2. The DRA is excited using the tapered transmission line. The antenna performance is shown in Figures 8.10–8.14. The performance comparison of the proposed antennas has also been made and shown in Table 8.2.

8.5.3 Wearable artistic antennas for WLAN-band

A novel concept of transforming complex and artistic geometries into antennas in the WLAN band is presented. Four artistic antennas are designed in the Wi-Fi band and measured. The historical images required to transform into antennas are shown in Figure 8.15, while their corresponding simulated designs are also shown in Figure 8.16. All the antennas are based on the microstrip antenna concept. The antennas are embroidered on the textile material using conductive thread to be

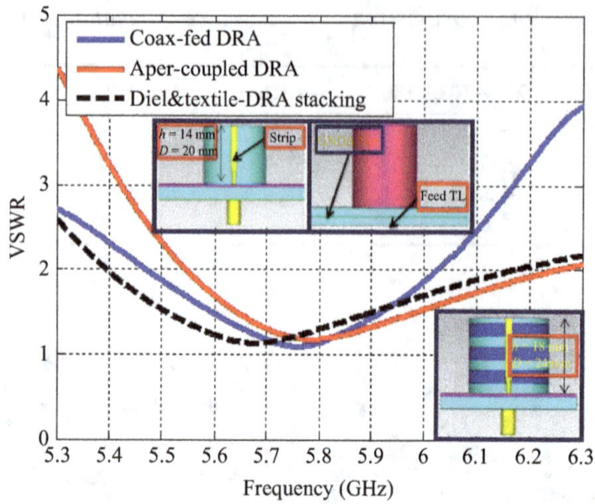

Figure 8.10 VSWR vs. frequency comparison of the proposed antennas (Reproduced with permission of IEEE)

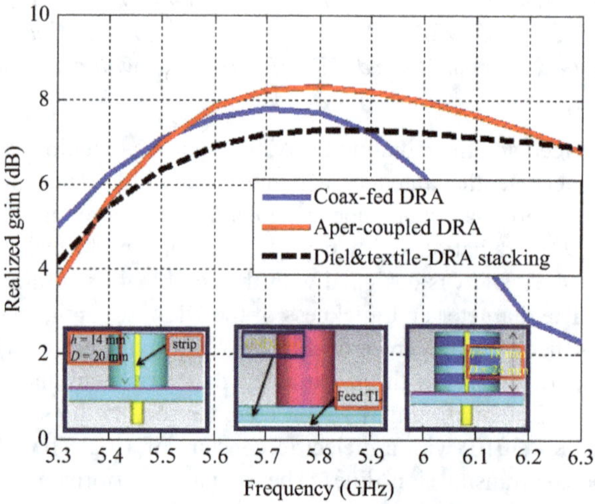

Figure 8.11 Realized gain vs. frequency comparison of the proposed antennas (Reproduced with permission of IEEE)

flexible for wearable applications. The dielectric constant of the sample under test (SUT) is extracted using resonance technique. The solid copper GND plane is made of shielded fabric, which provides the required flexibility. A low-power, short-range Wi-Fi band communication system is designed so that an antenna design has been embedded in each of several dresses for testing the transmission and reception authentication.

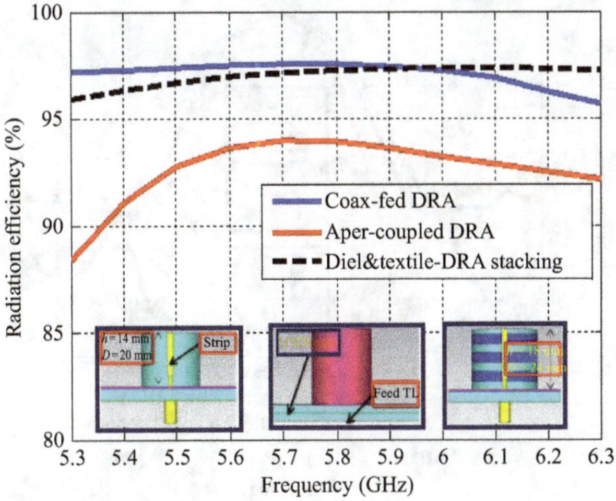

Figure 8.12 Radiation efficiency vs. frequency comparison of the proposed antennas (Reproduced with permission of IEEE)

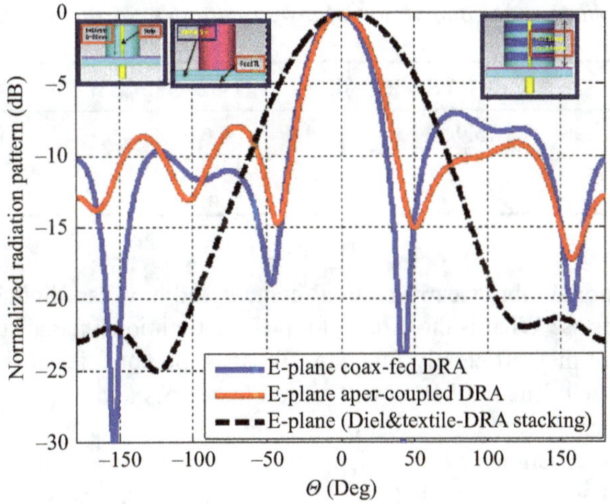

Figure 8.13 E-plane radiation patterns of the antennas at 5.8 GHz (Reproduced with permission of IEEE)

8.5.3.1 Motivation

This work is inspired by well-known techniques of using metal thread handicraft while applying the potential of electrical conductivity to create new digital cloth systems. We want to explore new functionalities of the textile and study wireless transmission for short-range communication. Working at the intersections of arts, culture, science, and technology, the following questions fascinated us: In what

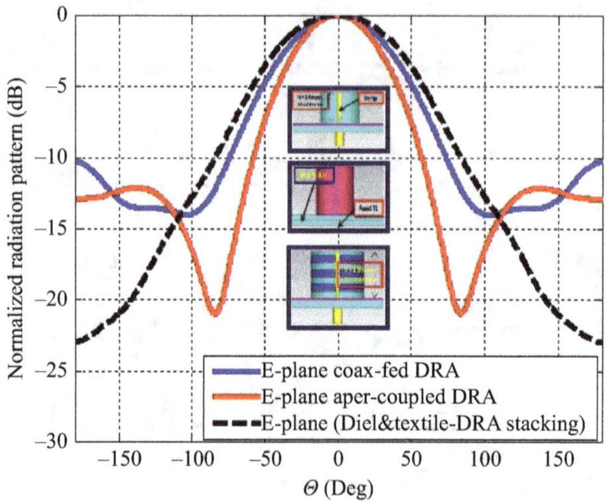

Figure 8.14 H-plane radiation patterns of the antennas at 5.8 GHz (Reproduced with permission of IEEE)

Table 8.2 Performance comparison of proposed antennas at 5.8 GHz

	3:1 VSWR (%)	**1-dB Gain BW (%)**	η_{rad} **(%)**
Coax-Fed	15.9	8.6	97
Aper-Coup	22.4	11.2	93
Diel &Text DRA	27.5	23.3	97

ways do traditional fabrics carry cultural content and how can digital fabrics support new art forms? What is the effect of replacing traditional metal substrates with textile, and will they efficiently function? What is the potential of this concept in today's direction of the IoT? How will wearable devices and the responsive textile environments enhance communication in everyday life and how can this affect the community experience? In addition, can flexible textile antennas be created in the form of imagery and symbols that carry cultural meaning while achieving a high level of functional performance? Our intention is to contribute to knowledge in smart textiles through a variety of ways: in hardware and software design (including the development of textile antennas), in pedagogical and social research on responsive fabrics and through the creative and expressive potential of textiles as a carrier of culture.

Many artistic wearable antennas are designed, but only four of them are presented here. These antennas are designed with the objective of reviving the historic ancient artwork in today's era by making smart textiles to transmit energy wirelessly through antennas. The historic ancient art designs are shown in Figure 8.15 while their corresponding simulated designs are presented in Figure 8.16.

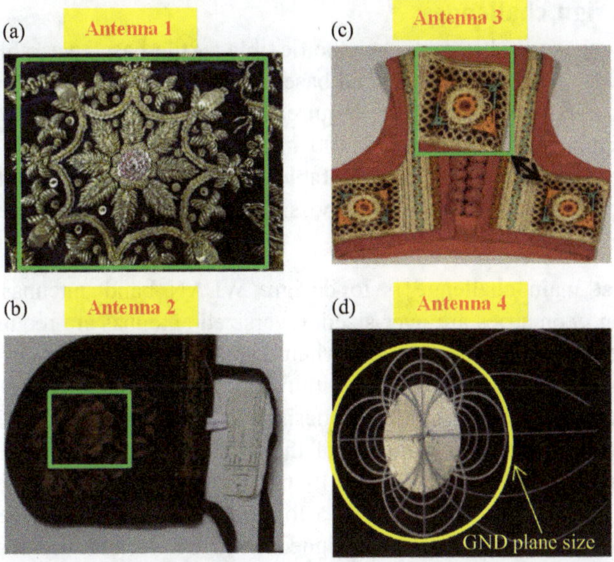

Figure 8.15 Historical images need to transform into functional antennas, where names are given as (a) Antenna 1; (b) Antenna 2; (c) Antenna 3; and (d) Antenna 4, respectively. (Reproduced with permission of IEEE)

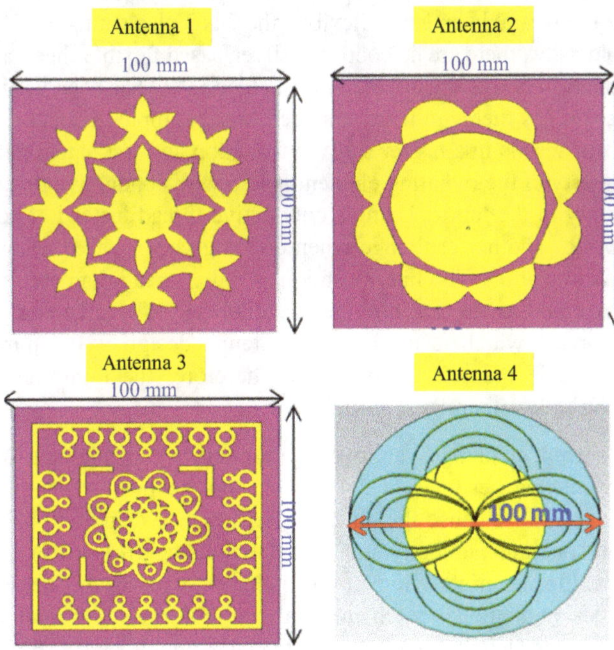

Figure 8.16 The simulated artistic antenna models of the given historical images in Figure 8.15 (Reproduced with permission of IEEE)

8.5.3.2 Design challenges

Certain challenges are observed when the flexible textile-based wearable antenna is designed. These antennas are designed based on the given specification:

"Artistic flexible antennas are required for wearable application, where the required antenna size should be 70–100 mm for better visibility on the dresses (antenna size less than this is not acceptable), while the total antenna thickness of the antenna must remain as minimal as possible (2–3 mm) for higher flexibility and to remain lightweight".

(1) The first main challenge is to design, WLAN band antennas where the radiating geometries are over-sized. Oversized antennas are required to make the antenna images more visible when displayed on the dresses.
(2) The antenna thickness should be within 2 to 3 mm. Minimal antenna thickness results in higher flexibility that is desirable for wearable applications.
(3) The extracted dielectric constant of the SUT remains fixed, after extraction. According to the literature review [9–62], the dielectric permittivity of textile materials are 1.1 to 1.5, which is a low permittivity that is closer to air. As we know, when antennas are designed with low permittivity substrate, they require larger radiating element dimensions to resonate, but as discussed above the antenna size and thickness are already fixed by the design of the artwork. As an antenna designer knows, when most of the parameters are fixed, there is little freedom in designing antennas with highest possible radiation efficiency. As a tradeoff, these artistic antennas demonstrate low radiation efficiency.
(4) To make the wearable antennas flexible, the GND plane at the back should also be flexible. Now, there are no copper or silver-based flexible sheets available for this purpose. To solve this problem, a shielded GND has been used in the proposed design. As there are many shielded fabric samples available, one should be used that results in a higher signal attenuation and less signal penetration.
(5) The geometry of the radiating element is laid with conductive thread resulting in flexibility in the fabric. Using a conductive thread instead of conventional copper brings another challenge when creating vector-based designs for analysis and simulation. The line must equal to the thread diameter, which must also correspond to the machine needle diameter. The diameter of the conductive thread, which is used in the antenna design is 0.3 mm while the machine needle diameter is 0.6 mm. Therefore, the minimum line width should not be less than these dimensions.

With the current challenges, an investigation has been made on a WLAN band circular patch antenna designed at 2.45 GHz, and the results obtained indicates about certain facts those cause the antenna performance degradation. The dimensions of the resonating geometry are calculated using (8.1)–(8.5). Figure 8.17 exhibits that textile materials lying in the dielectric permittivity between 1 to 2, when used to design a circular patch antenna with multiple substrate thickness of 2 to 15 mm, resonates at frequency range between 2 and 3.5 GHz. Because of the fixed extracted dielectric constant of the SUT, a patch size of 45 mm in diameter resonates at the Wi-Fi band when stacked SUT layers make equivalent thickness of

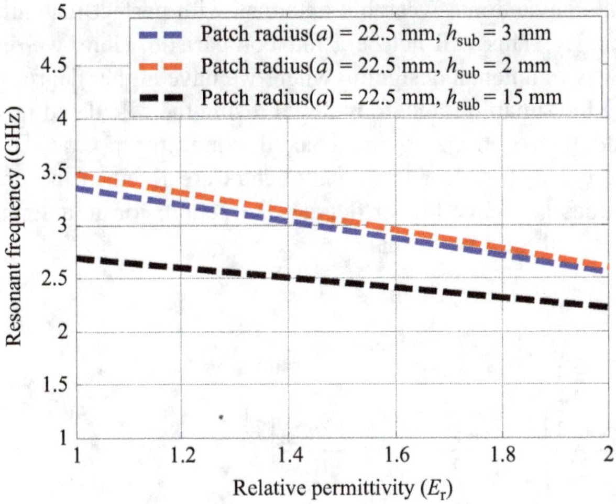

Figure 8.17 The effect of relative permittivity with multiple textile layer thickness of SUT on the resonant frequency of a WLAN band circular patch antenna

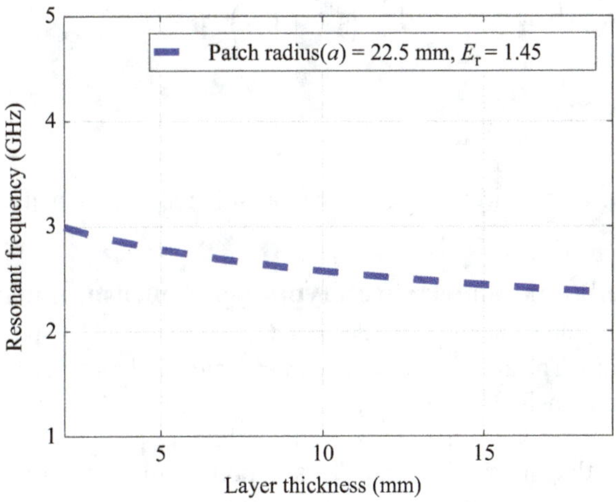

Figure 8.18 The effect of fixed extracted relative permittivity of SUT versus variable textile layer thickness on the resonant frequency of a WLAN band circular patch antenna

15 mm, as shown in Figure 8.18. But we have to stick with the given specifications of the antenna in this particular wearable application where not only the required thickness is 2–3 mm, but the extracted dielectric permittivity of SUT is 1.45, which is an immoveable parameter too. So, due to these obvious performance degrading factors, the performance of the proposed artistic antennas should not be expected to

behave like the conventional wearable antennas with perfect matching and higher radiation efficiency. This will not be a fair comparison. Here, we did not use the conventional way of antenna designing where we have higher degree of freedom in either selecting the suitable standard material, using the calculated resonating patch dimension, and then optimizing the coaxial connector position for impedance matching, whereas majority of these parameters are fixed in the proposed novel antennas that does not leave higher degree of freedom for an antenna designer to design antennas with best performance:

$$W_\mathrm{p} = \frac{c}{(2f_\mathrm{c})}\sqrt{\frac{2}{(\varepsilon_\mathrm{r}+1)}} \tag{8.1}$$

$$\varepsilon_\mathrm{eff} = \left[\frac{\varepsilon_r+1}{2}\right] + \left[\frac{\varepsilon_r-1}{2}\right]\left[1+\frac{12h_\mathrm{sub}}{W_\mathrm{p}}\right]^{-\frac{1}{2}} \tag{8.2}$$

$$L_\mathrm{p} = \left[\frac{c}{(2f_c\sqrt{\varepsilon_\mathrm{eff}})}\right] - 2\Delta L \tag{8.3}$$

$$\frac{\Delta L}{h_\mathrm{sub}} = 0.412\left[\frac{(\varepsilon_\mathrm{eff}+0.3)}{(\varepsilon_\mathrm{eff}-0.258)}\right]\left[\frac{\left(\frac{W_\mathrm{p}}{h_\mathrm{sub}}+0.264\right)}{\left(\frac{W_\mathrm{p}}{h_\mathrm{sub}}+0.8\right)}\right] \tag{8.4}$$

$$L_\mathrm{eff} = L_\mathrm{p} + 2\Delta L \tag{8.5}$$

where W_p is the patch width and L_p is the patch length, calculated at the designed frequency f_c.

8.5.3.3 Simulation models, prototyping, and measurement

Four artistic antennas have been designed in the WLAN band using a CST microwave studio. The antennas have been fabricated and measured. The simulated antennas are shown in Figure 8.16, and referred to as Antenna 1, Antenna 2, Antenna 3, and Antenna 4, respectively. The designed antennas are 100×100 mm^2 in size with the thickness of 2.6 mm. It is important to mention that this antenna size corresponds to the art geometry, which is not related to the RF possibilities.

Due to the complexity of the given geometries, the antenna design has been broken down into individual elements, through a step-by-step (SBS) procedure that helps in understanding which part has more/less influence on the antenna resonance. As an example, Antenna 1 and Antenna 3 are presented with SBS design procedure, as shown in Figures 8.19 and Figure 8.21. The dimensions of the radiating geometry in each step are provided. The reflection coefficient of Antenna 1 from the SBS process is shown in Figure 8.20. It is well evident

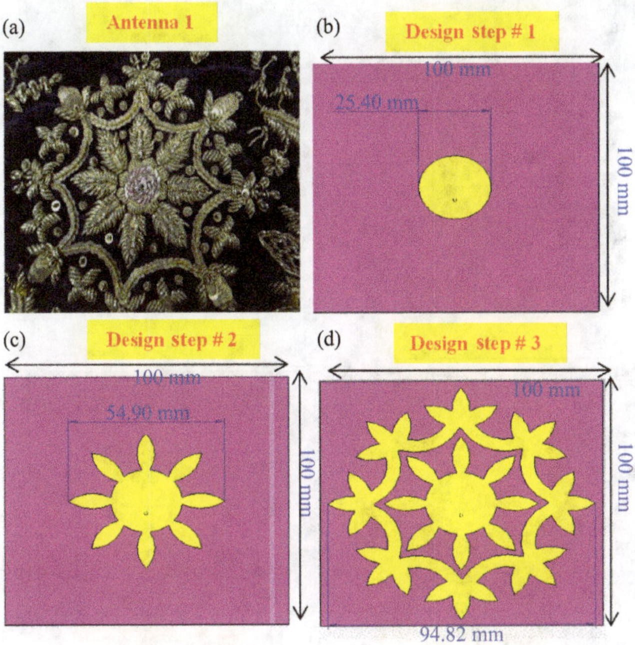

Figure 8.19 A SBS design approach for Antenna 1

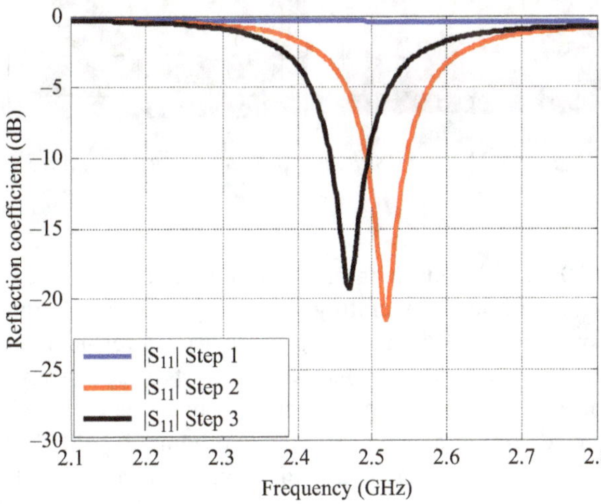

Figure 8.20 Simulated reflection coefficient of Antenna 1 in SBS process shown in Figure 8.19

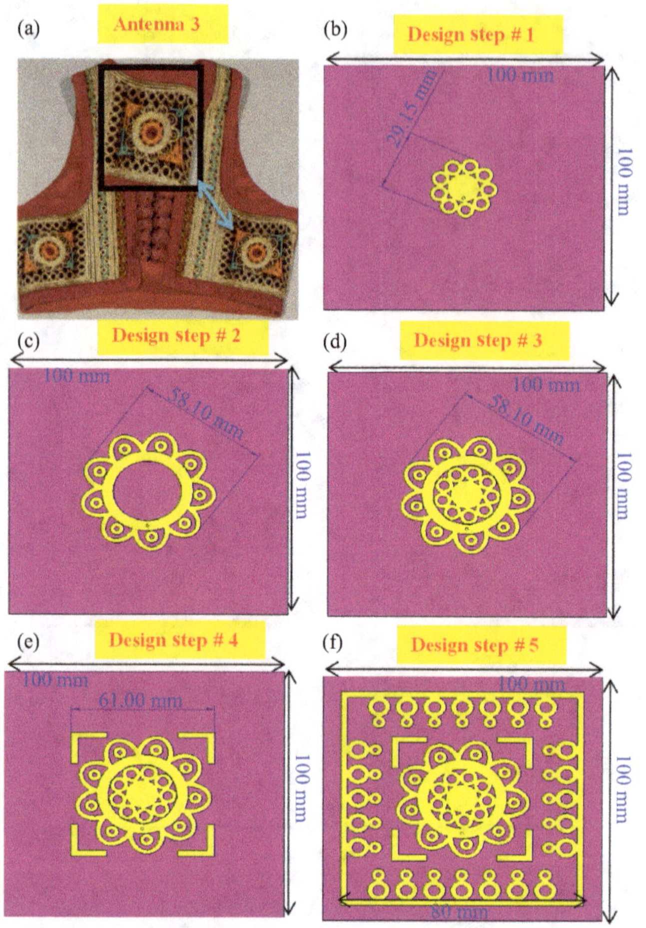

Figure 8.21 A SBS design approach for Antenna 3

from Figure 8.20 that the circular patch designed in Step 1 exhibits a non-resonant behavior within the operating frequency band while due to the coupling with the other parts of the radiating geometries a resonant frequency shift has been observed when design changes from step 2 to step 3. Similarly, for Antenna 3, a five-step design procedure has been used, as shown in Figure 8.21. The reflection coefficient of the Antenna 3 in SBS, is provided in Figure 8.22, which demonstrates that the conventional −10 dB impedance matching level is hard to achieve. Therefore, a −3 dB matching level is considered for this application. From the shown result in Figure 8.22, the smallest size of the radiating antenna in the first two steps is not resonating within the frequency band of interest, while the reflection coefficients in the last three steps are almost similar.

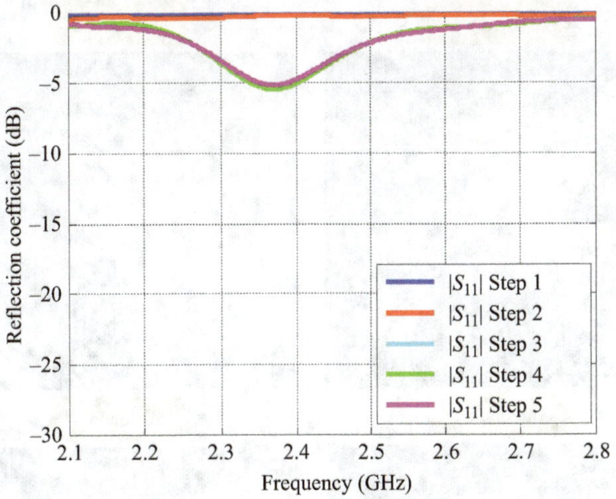

Figure 8.22 Simulated reflection coefficient of Antenna 3 in SBS process shown in Figure 8.21

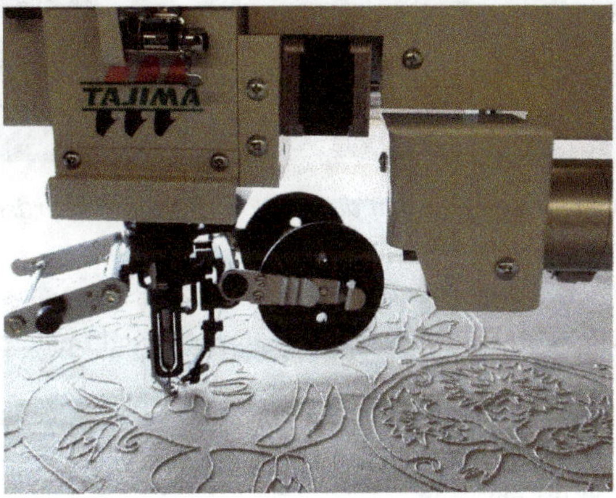

Figure 8.23 A computer-controlled Tajima machine used for laying the proposed artistic antennas

In the upcoming section, the prototyping of these artistic antennas and the comparison between simulated and measured results are presented.

8.5.3.4 Tajima machine used for prototyping

A computer-controlled Tajima machine has been used for laying the artistic antennas. The Tajima laying machine illustrated in Figure 8.23 supports the three-dimensional layering of technical fibers onto the surface of a textile substrate in a seamless way, approximating a flexible circuit board. Unlike commercial

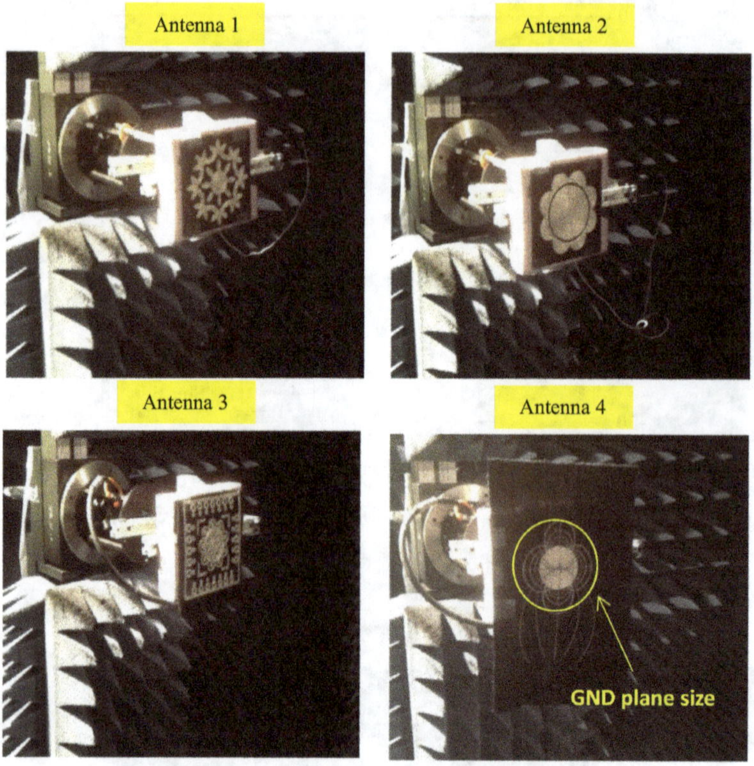

Figure 8.24 Far-field measurement pictures of the proposed artistic textile-based wearable antennas

embroidery machines where the conductive threads would pierce the fabric, the laying machine tethers the conductive thread to only one face of the fabric, isolating the thread from the body and from other areas of conduction inside a garment. The machine is also used for creating flexible circuitry and cabling systems throughout the garment.

8.5.3.5 Simulated and measured results

The antennas are measured at WLAN band. Figure 8.24 exhibits the fabricated artistic antennas in the anechoic chamber for far-field measurements. The artistic antennas are flexible because the geometries are laid on textile material with conductive thread while the conventional solid copper GND is replaced with the shielded fabric GND plane. All the antennas have the same size of 100×100 mm^2. The simulated and measured reflection coefficients are shown in Figure 8.25. Most of the proposed antennas achieved a good -6 dB measured matching bandwidth, while Antenna 3 achieved almost a constant of -3 dB measured impedance

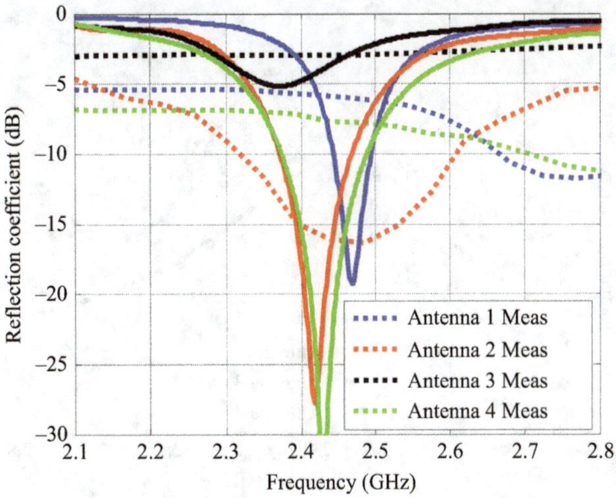

Figure 8.25 Simulated and measured reflection coefficient of the proposed artistic antennas

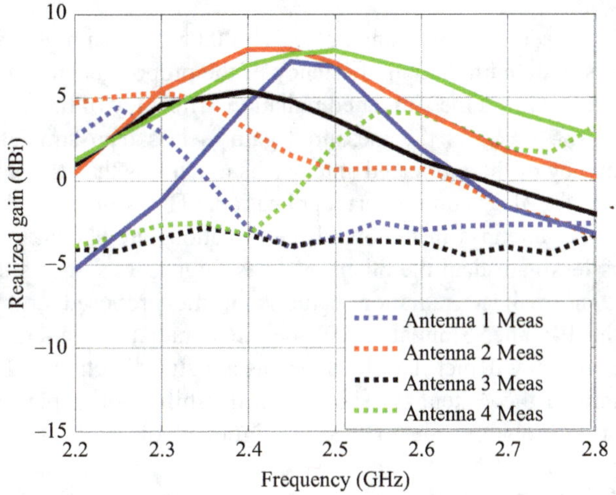

Figure 8.26 Simulated and measured gain of the artistic antennas

matching bandwidth. The simulated and measured antenna gains are shown in Figure 8.26. The disagreement between simulated and measured gain is because of many possible factors: impedance mismatch due to smaller–larger antenna size, possible misalignment between antenna under test (AUT) and probe during the measurements, low conductivity of textile materials, and larger material dissipation

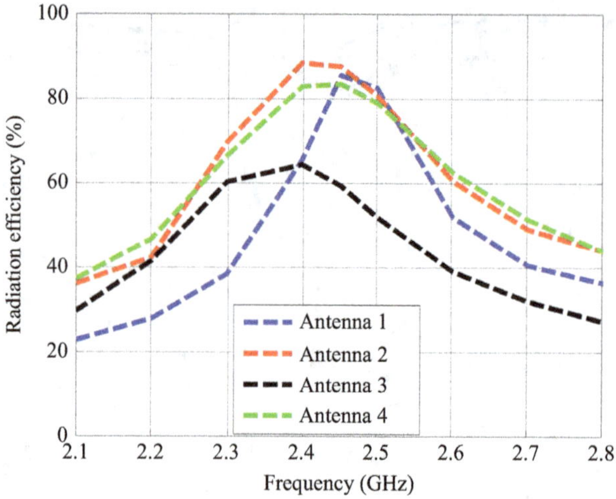

Figure 8.27 Simulated radiation efficiency of the antennas (Reproduced with permission of IEEE)

factor, which is observed in the range of 0.01 to 0.05 or even higher in the textile materials. The simulated radiation efficiency of the proposed antennas is shown in Figure 8.27, which shows the simulated radiation efficiency of the antennas within the range of 60%–82%. As the measured gain is less than simulated gain, the radiation efficiency of the fabricated antennas is much less than the simulation, but still acceptable for the required art application. The simulated 3-D radiation patterns of all the antennas in the broadside directions are shown in Figure 8.28. Antenna 3 has less gain than the other antennas. Figure 8.29 shows the simulated and measured normalized radiation patterns of the proposed artistic wearable antennas in the E-plane. Simulated and measured radiation patterns show some disagreement, but they depict that the antennas are still directional. The measured radiation pattern of the Antenna 3 shows a beam shift at an angle of 20° off the broadside that is probably due to the misalignment of AUT and measurement probe, during the measurements.

The performance degradation is mainly happened due to the challenges described in Section II, along with the possible fabrication errors and measurement errors. These antennas result low radiation efficiency but still enough to make short range communication possible within a room for which these are designed for. The performance of the proposed artistic antennas is compared with previously published in the wearable application, and summarized in Table 8.3. It is important to mention here that the analysis of bending on these antennas, is intentionally not performed because the antennas are embedded on the front side of the dress and remain planar, as shown in Figure 8.30. So, at this location there is no need to perform bending analysis on the proposed antennas.

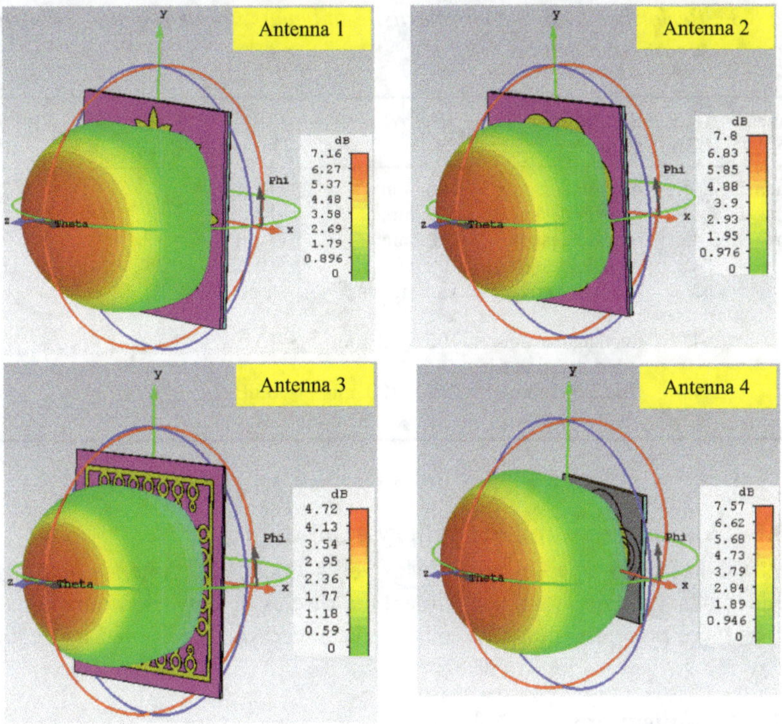

Figure 8.28 Simulated 3-D radiation patterns of the antennas (at 2.45 GHz)

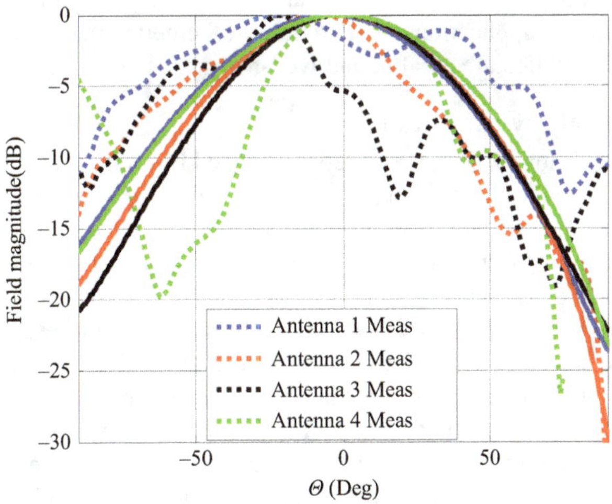

Figure 8.29 Simulated and measured normalized radiation patterns in the E-plane of the proposed artistic antennas (at 2.45 GHz)

Table 8.3 Performance of proposed artistic-antennas versus previous published wearable antennas

Measured	f_r (GHz)	Radiating geometry	Size	$\|S_{11}\|$ Threshold	Gain (dBi)	Sim. XP (dB)	η_{rad} (%)
Antenna 1	2.45	Artistic design 1	100 mm^2	−10 dB	−3	−18	7
Antenna 2	2.45	Artistic design 2	100 mm^2	−10 dB	1.5	−21	18
Antenna 3	2.45	Artistic design 3	100 mm^2	−3 dB	−3	−14	7
Antenna 4	2.45	Artistic design 4	100 mm^2	−10 dB	0.25	−19	13
[65]	2.2	E-Shaped copper patch	18 × 15 cm^2	−6 dB	–	–	low
[66]	2.45	Rectangular embroidered patch	100 × 70 mm^2	−10 dB	2.5	−14	65
[67]	2.45	Gridded embroidered patch	140 × 120 mm^2	−6 dB	0.9	−9	24

8.5.3.6 Applications and art gallery exhibition

Antenna 4 has been given a name "Maxwell's Equation." Antenna 4 has been embedded in several dresses to experiment the wireless energy transformation. The antennas were presented in an art context that successfully achieved the required objectives for which they were designed.

8.5.3.7 Applications

Research into flexible antenna systems makes a significant contribution of scientists working at the forefront of beam-forming antenna systems. Textiles that can effectively communicate with one another can replace traditional hard electronic systems, providing comfort, portability, and new forms of communication. The research has direct applications in the fields of entertainment, fashion, safety, interior design, architecture, health, and well-being.

8.5.3.8 Art gallery exhibition

As a research-creation project, the benefits extended beyond the economic interests of market systems and scientific inquiry. Smart textiles are a dynamic site for social communication just as traditional cloth has served as a system of communication and cultural identity. Activities at galleries, museums and techno-fashion events contribute to the discussions of digital technology and textile arts. Presentations and publications are of interest to students, scholars, and researchers. These antennas were successfully presented in a techno-fashion art gallery "Subtle Technologies Festival, Toronto" on May 12, 2016. During the art gallery event, three models, wore dresses with embedded artistic antennas and LED arrays that display predefined scrolling texts on the LED grid when the antennas are facing each other. Each dress displays different messages depending on the received signal strength from the other two dress antennas as shown in Figure 8.30. When dresses are not in contact with one another (not facing each other) a dotted line scroll across the displays.

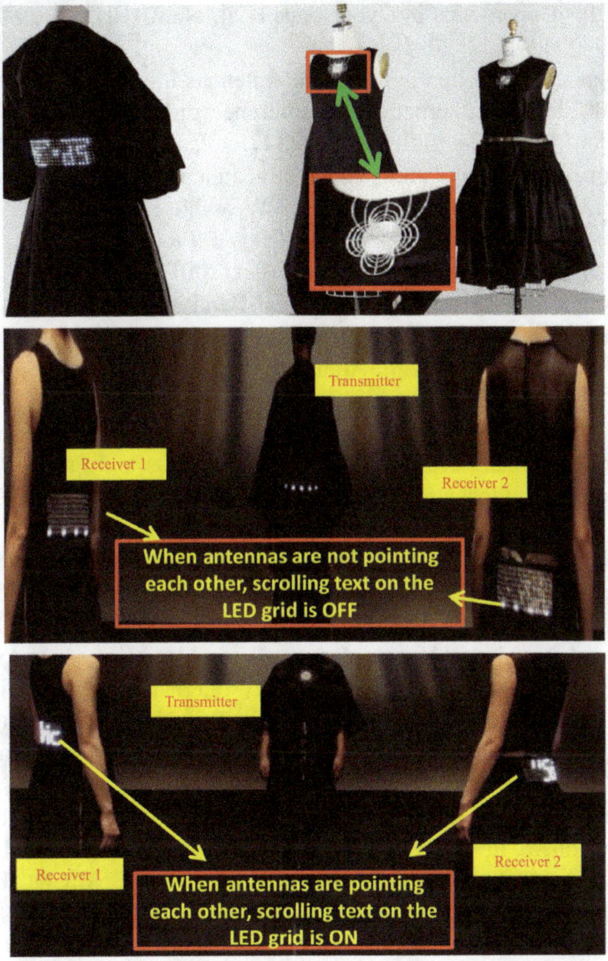

Figure 8.30 Antenna 4 embedded on the dresses of the project "Maxwell's Equation" showing a transmitter communicates with two receivers at a short distance by displaying a scrolling text on the LED grid (Reproduced with permission of IEEE)

References

[1] P. Salonen, L. Sydanheimo, M. Keskilammi and M. Kivikoski, "A Small Planar Inverted-F Antenna for Wearable Applications," in Digest of Papers. Third International Symposium on Wearable Computers, pp. 95–100, San Francisco, CA, USA, 1999.

[2] P. Salonen, M. Keskilammi, J. Rantanen and L. Sydanheimo, "A Novel Bluetooth Antenna on Flexible Substrate for Smart Clothing," in 2001 IEEE International Conference on Systems, Man and Cybernetics. e-Systems and

e-Man for Cybernetics in Cyberspace (Cat. No. 01CH37236), Tucson, AZ, pp. 789–794, vol. 2, 2001.

[3] P. J. Massey, "Mobile Phone Fabric Antennas Integrated Within Clothing," in 2001 Eleventh International Conference on Antennas and Propagation, (IEE Conf. Publ. No. 480), pp. 344–347, vol. 1, Manchester, 2001.

[4] P. Salonen, M. Keskilammi and L. Sydanheimo, "A Low-Cost 2.45 GHz Photonic Band-Gap Patch Antenna for Wearable Systems," in 2001 Eleventh International Conference on Antennas and Propagation (IEE Conf. Publ. No. 480), pp. 719–723, vol. 2, Manchester, 2001.

[5] P. Salonen and J. Rantanen, "A Dual-Band and Wide-Band Antenna on Flexible Substrate for Smart Clothing," in Industrial Electronics Society, 2001. IECON '01. The 27th Annual Conference of the IEEE, pp. 125–130, vol. 1, Denver, CO, USA, 2001.

[6] P. Salonen and L. Sydanheimo, "Development of an S-band Flexible Antenna for Smart Clothing," in IEEE Antennas and Propagation Society International Symposium (IEEE Cat. No. 02CH37313), pp. 6–9, San Antonio, TX, USA, 2002.

[7] P. S. Hall, M. Ricci and T. M. Hee, "Measurements of On-body Propagation Characteristics," in IEEE Antennas and Propagation Society International Symposium (IEEE Cat. No. 02CH37313), pp. 310–313, vol. 2, 2002.

[8] M. Tanaka and Jae-Hyeuk Jang, "Wearable Microstrip Antenna," in IEEE Antennas and Propagation Society International Symposium. Digest. Held in conjunction with: USNC/CNC/URSI North American Radio Sci. Meeting (Cat. No. 03CH37450), pp. 704–707, vol. 2, Columbus, OH, USA, 2003.

[9] P. Salonen and L. Hurme, "A Novel Fabric WLAN Antenna for Wearable Applications," in IEEE Antennas and Propagation Society International Symposium. Digest. Held in conjunction with: USNC/CNC/URSI North American Radio Sci. Meeting (Cat. No. 03CH37450), pp. 700–703, vol. 2, Columbus, OH, USA, 2003.

[10] P. Salonen, Fan Yang, Y. Rahmat-Samii and M. Kivikoski, "WEBGA – Wearable Electromagnetic Band-gap Antenna," in IEEE Antennas and Propagation Society Symposium, pp. 451–454, vol. 1, 2004,.

[11] S. Zhu and R. Langley, "Dual-Band Wearable Antennas over EBG Substrate," in Electronics Letters, vol. 43, no. 3, pp. 141–142, Feb. 2007.

[12] S. Zhu and R. Langley, "Dual-Band Wearable Textile Antenna on an EBG Substrate," in IEEE Transactions on Antennas and Propagation, vol. 57, no. 4, pp. 926–935, Apr. 2009.

[13] P. Salonen and Y. Rahmat-Samii, "Textile Antennas: Effects of Antenna Bending on Input Matching and Impedance Bandwidth," in 2006 First European Conference on Antennas and Propagation, Nice, pp. 1–5, 2006.

[14] P. Salonen, Y. Rahmat-Samii and M. Kivikoski, "Wearable Antennas in the Vicinity of Human Body," in IEEE Antennas and Propagation Society Symposium, pp. 467–470, vol. 1, 2006.

[15] P. Salonen and Y. Rahmat-Samii, “Textile Antennas: Effects of Antenna Bending on Input Matching and Impedance Bandwidth,” in IEEE Aerospace and Electronic Systems Magazine, vol. 22, no. 12, pp. 18–22, Dec. 2007.

[16] C. Cibin, P. Leuchtmann, M. Gimersky, R. Vahldieck and S. Moscibroda, “A Flexible Wearable Antenna,” in IEEE Antennas and Propagation Society Symposium, pp. 3589–3592, vol. 4, 2004.

[17] P. Salonen, Y. Rahmat-Samii, H. Hurme and M. Kivikoski, “Dual-band Wearable Textile Antenna,” in IEEE Antennas and Propagation Society Symposium, pp. 463–466, vol. 1, 2004.

[18] P. Salonen, Jaehoon Kim and Y. Rahmat-Samii, “Dual-Band E-Shaped Patch Wearable Textile Antenna,” in 2005 IEEE Antennas and Propagation Society International Symposium, pp. 466–469, vol. 1A, 2005.

[19] M. Klemm, I. Locher and G. Troster, “A Novel Circularly Polarized Textile Antenna for Wearable Applications,” in 34th European Microwave Conference, pp. 137–140, Amsterdam, The Netherlands, 2004.

[20] M. Klemm, I. Locher and G. Troster, “A Novel Circularly Polarized Textile Antenna for Wearable Applications,” in 7th European Conference on Wireless Technology, pp. 285–288, Amsterdam, The Netherlands, 2004.

[21] P. Salonen, Y. Rahmat-Samii, H. Hurme and M. Kivikoski, “Effect of Conductive Material on Wearable Antenna Performance: A Case Study of WLAN Antennas,” in IEEE Antennas and Propagation Society Symposium, pp. 455–458, vol. 1, 2004.

[22] M. Klemm and G. Troster, “Integration of Electrically Small UWB Antennas for Body-Worn Sensor Applications,” in IEE Wideband and Multi-band Antennas and Arrays 2005 (Ref. No. 2005/11059), pp. 141–145, Birmingham, UK.

[23] M. Klemm, I. Z. Kovcs, G. F. Pedersen and G. Troster, “Novel Small-Size Directional Antenna for UWB WBAN/WPAN Applications,” in IEEE Transactions on Antennas and Propagation, vol. 53, no. 12, pp. 3884–3896, Dec. 2005.

[24] Zhi Ning Chen, Ailian Cai, T. S. P. See and M. Y. W. Chia, “Small Planar UWB Antennas in Proximity of Human Head,” in 2005 IEEE International Conference on Ultra-Wideband, 2005.

[25] B. Sanz-Izquierdo, F. Huang and J. C. Batchelor, “Small Size Wearable Button Antenna,” in 2006 First European Conference on Antennas and Propagation, pp. 1–4, 2006.

[26] B. Sanz-Izquierdo, F. Huang and J. C. Batchelor, “Covert Dual-Band Wearable Button Antenna,” in Electronics Letters, vol. 42, no. 12, pp. 668–670, Jun. 2006.

[27] B. Sanz-Izquierdo, F. Huang and J. C. Batchelor, “Dual Band Button Antennas for Wearable Applications,” in IEEE International Workshop on Antenna Technology Small Antennas and Novel Metamaterials, 2006., pp. 132–135, 2006.

[28] Z. N. Chen, M. J. Ammann, X. Qing, X. H. Wu, T. S. p. See and A. Cai, “Planar Antennas,” in IEEE Microwave Magazine, vol. 7, no. 6, pp. 63–73, Dec. 2006.

[29] I. Locher, M. Klemm, T. Kirstein and G. Trster, "Design and Characterization of Purely Textile Patch Antennas," in IEEE Transactions on Advanced Packaging, vol. 29, no. 4, pp. 777–788, Nov. 2006.
[30] C. Hertleer, A. Tronquo, H. Rogier, L. Vallozzi and L. Van Langenhove, "Aperture-Coupled Patch Antenna for Integration Into Wearable Textile Systems," in IEEE Antennas and Wireless Propagation Letters, vol. 6, pp. 392–395, 2007.
[31] L. Vallozzi, H. Rogier and C. Hertleer, "Dual Polarized Textile Patch Antenna for Integration Into Protective Garments," in IEEE Antennas and Wireless Propagation Letters, vol. 7, pp. 440–443, 2008.
[32] C. Hertleer, H. Rogier, L. Vallozzi and L. Van Langenhove, "A Textile Antenna for Off-Body Communication Integrated Into Protective Clothing for Firefighters," in IEEE Transactions on Antennas and Propagation, vol. 57, no. 4, pp. 919–925, Apr. 2009.
[33] D. Psychoudakis and J. L. Volakis, "Conformal Asymmetric Meandered Flare (AMF) Antenna for Body-Worn Applications," in IEEE Antennas and Wireless Propagation Letters, vol. 8, pp. 931–934, 2009.
[34] T. F. Kennedy, P. W. Fink, A. W. Chu, N. J. Champagne, G. Y. Lin and M. A. Khayat, "Body-Worn E-Textile Antennas: The Good, the Low-Mass, and the Conformal," in IEEE Transactions on Antennas and Propagation, vol. 57, no. 4, pp. 910–918, Apr. 2009.
[35] C. P. Lin, C. H. Chang, Y. T. Cheng and C. F. Jou, "Development of a Flexible SU-8/PDMS-Based Antenna," in IEEE Antennas and Wireless Propagation Letters, vol. 10, pp. 1108–1111, 2011.
[36] S. J. Ha and C. W. Jung, "Reconfigurable Beam Steering Using a Microstrip Patch Antenna With a U-Slot for Wearable Fabric Applications," in IEEE Antennas and Wireless Propagation Letters, vol. 10, pp. 1228–1231, 2011.
[37] S. Kim, Y. J. Ren, H. Lee, A. Rida, S. Nikolaou and M. M. Tentzeris, "Monopole Antenna With Inkjet-Printed EBG Array on Paper Substrate for Wearable Applications," in IEEE Antennas and Wireless Propagation Letters, vol. 11, pp. 663–666, 2012.
[38] A. A. Serra, P. Nepa and G. Manara, "A Wearable Two-Antenna System on a Life Jacket for Cospas-Sarsat Personal Locator Beacons," in IEEE Transactions on Antennas and Propagation, vol. 60, no. 2, pp. 1035–1042, Feb. 2012.
[39] K. W. Lui, O. H. Murphy and C. Toumazou, "A Wearable Wideband Circularly Polarized Textile Antenna for Effective Power Transmission on a Wirelessly-Powered Sensor Platform," in IEEE Transactions on Antennas and Propagation, vol. 61, no. 7, pp. 3873–3876, Jul. 2013.
[40] B. Ivsic, D. Bonefacic and J. Bartolic, "Considerations on Embroidered Textile Antennas for Wearable Applications," in IEEE Antennas and Wireless Propagation Letters, vol. 12, pp. 1708–1711, 2013.

[41] T. M. Nguyen, J. Y. Chung and B. Lee, "Radiation Characteristics of Woven Patch Antennas Composed of Conductive Threads," in IEEE Transactions on Antennas and Propagation, vol. 63, no. 6, pp. 2796–2801, Jun. 2015.

[42] S. Yan, P. J. Soh and G. A. E. Vandenbosch, "Wearable Dual-Band Magneto-Electric Dipole Antenna for WBAN/WLAN Applications," in IEEE Transactions on Antennas and Propagation, vol. 63, no. 9, pp. 4165–4169, Sept. 2015.

[43] J. Zhong, A. Kiourti, T. Sebastian, Y. Bayram and J. L. Volakis, "Conformal Load-Bearing Spiral Antenna on Conductive Textile Threads," in IEEE Antennas and Wireless Propagation Letters, vol. 16, pp. 230–233, 2017.

[44] D. H. Werner and Z. H. Jiang, Editors, "Electromagnetics of Body-Area Networks: Antennas, Propagation, and RF Systems," 1st edition, The Institute of Electrical and Electronics Engineers, Inc. Hoboken, NJ: John Wiley & Sons, Inc., 2016.

[45] Adel Y. I. Ashyap, Z. Z. Abidin, S. H. Dahlan, *et al.*, "Compact and Low-Profile Textile EBG-Based Antenna for Wearable Medical Applications," in IEEE Antennas and Wireless Propagation Letters, vol. 16, pp. 2550–2553, 2017.

[46] American Cancer Society. Facts & figures 2014. [Online]. Atlanta, GA, USA. Available at: http://www.cancer.org.

[47] H. Bahramiabarghouei, E. Porter, A. Santorelli, B. Gosselin, M. Popović and L. A. Rusch, "Flexible 16 Antenna Array for Microwave Breast Cancer Detection," in IEEE Transactions on Biomedical Engineering, vol. 62, no. 10, pp. 2516–2525, Oct. 2015.

[48] M. Lazebnik, D. Popovic, L. McCartney, *et al.*, "A Large-Scale Study of the Ultrawideband Microwave Dielectric Properties of Normal, Benign and Malignant Breast Tissue Obtained from Cancer Surgeries," in Physics in Medicine and Biology, vol. 52, pp. 6093–6115, 2007.

[49] K. Agarwal, Y. X. Guo and B. Salam, "Wearable AMC Backed Near-Endfire Antenna for On-Body Communications on Latex Substrate," in IEEE Transactions on Components, Packaging and Manufacturing Technology, vol. 6, no. 3, pp. 346–358, Mar. 2016.

[50] D. I. Amey and J. P. Curilla, "Microwave Properties of Ceramic Materials," in Proceedings of the 41st Electronic Components and Technology Conference (ECTC), pp. 267–272, 1991.

[51] K.-P. Latti, M. Kettunen, J.-P. Strom, and P. Silventoinen, "A Review of Microstrip T-Resonator Method in Determining the Dielectric Properties of Printed Circuit Board Materials," in IEEE Transactions on Instrumentation and Measurement, vol. 56, no. 5, pp. 1845–1850, Oct. 2007.

[52] S. M. Saeed, C. A. Balanis, C. R. Birtcher, A. C. Durgun and H. N. Shaman, "Wearable Flexible Reconfigurable Antenna Integrated With Artificial Magnetic Conductor," in IEEE Antennas and Wireless Propagation Letters, vol. 16, pp. 2396–2399, 2017.

[53] X. Liu, Y. Di, H. Liu, Z. Wu and M. M. Tentzeris, "A Planar Windmill-Like Broadband Antenna Equipped With Artificial Magnetic Conductor for Off-Body Communications," in IEEE Antennas and Wireless Propagation Letters, vol. 15, pp. 64–67, 2016.

[54] H. R. Khaleel, "Design and Fabrication of Compact Inkjet Printed Antennas for Integration Within Flexible and Wearable Electronics," in IEEE Transactions on Components, Packaging and Manufacturing Technology, vol. 4, no. 10, pp. 1722–1728, Oct. 2014.

[55] H. R. Raad, A. I. Abbosh, H. M. Al-Rizzo and D. G. Rucker, "Flexible and Compact AMC Based Antenna for Telemedicine Applications," in IEEE Transactions on Antennas and Propagation, vol. 61, no. 2, pp. 524–531, Feb. 2013.

[56] S. Genovesi, F. Costa, F. Fanciulli and A. Monorchio, "Wearable Inkjet-Printed Wideband Antenna by Using Miniaturized AMC for Sub-GHz Applications," in IEEE Antennas and Wireless Propagation Letters, vol. 15, pp. 1927–1930, 2016.

[57] B. S. Cook and A. Shamim, "Utilizing Wideband AMC Structures for High-Gain Inkjet-Printed Antennas on Lossy Paper Substrate," in IEEE Antennas and Wireless Propagation Letters, vol. 12, pp. 76–79, 2013.

[58] H. F. Abutarboush, M. F. Farooqui and A. Shamim, "Inkjet-Printed Wideband Antenna on Resin-Coated Paper Substrate for Curved Wireless Devices," in IEEE Antennas and Wireless Propagation Letters, vol. 15, pp. 20–23, 2016.

[59] William G. Whittow, Alford Chauraya, J. C. Vardaxoglou, *et al.*, "Inkjet-Printed Microstrip Patch Antennas Realized on Textile for Wearable Applications," in IEEE Antennas and Wireless Propagation Letters, vol. 13, pp. 71–74, 2014.

[60] Muhammad M. Tahseen, and Ahmed A. Kishk, "Wideband Textile-Based Conformal Antennas for WLAN Band Using Conductive Thread," in EUCAP, Davos, Switzerland, 10–15 April, 2016.

[61] Muhammad M. Tahseen, and Ahmed A. Kishk, "Textile-Based Wideband Flexible Wearable Dielectric Resonator Antennas For WLAN-Band," in ANTEM, Montreal, Canada, 10–13 July, 2016.

[62] Muhammad M. Tahseen, and Ahmed A. Kishk, "Artistic Textile Antennas," in IEEE APS/URSI, San Diego, California, USA, 2017.

[63] S. Sankaralingam, and B. Gupta, "Determination of Dielectric Constant of Fabric Materials and Their Use as Substrates for Design and Development of Antennas for Wearable Applications," in IEEE Transactions on Instrumentation and Measurement, vol. 59, no. 12, pp. 3122–3130, Dec. 2010.

[64] "http://www.lessemf.com/," 776-B Watervliet Shaker Rd, Latham NY 12110, USA.

[65] Y. Rahmat-Samii,"Wearable and Implantable Antennas in Body-Centric Communications," in The Second European Conference on Antennas and Propagation, EuCAP 2007. pp. 1–5, 11–16, Nov. 2007.

[66] Z. Wang, L. Z. Lee, D. Psychoudakis and J. L. Volakis, "Embroidered Multiband Body-Worn Antenna for GSM/PCS/WLAN Communications," in IEEE Transactions on Antennas and Propagation, vol. 62, no. 6, pp. 3321–3329, Jun. 2014.

[67] S. Zhang, W. Whittow, R. Seager, A. Chauraya and J. C. Vardaxoglou, "Non-uniform mesh for embroidered microstrip antennas," in IET Microwaves, Antennas & Propagation, vol. 11, no. 8, pp. 1086–1091, Jun. 2017.

Chapter 9

Wearable technology and mobile platform for wearable antennas for human health monitoring

Vijay K. Varadan[1,2], Pratyush Rai[1], Se Chang Oh[1], Prashanth Shyam Kumar[2], Mouli Ramasamy[2], and Robert E. Harbaugh[3]

Health and long-term care is a growth area for wearable heath monitoring systems. Wearable diagnostic and therapeutic systems can contribute to timely point-of-care (POC) for patients with chronic health conditions, especially chronic neurological disorders, cardiovascular diseases and strokes that are leading causes of mortality worldwide. Diagnostics and therapeutics for patients under timely POC can save thousands of lives. However, lack of access to minimally intrusive monitoring systems makes timely diagnosis difficult and sometimes impossible. Existing ambulatory recording equipment is incapable of performing continuous remote patient monitoring (RPM) because of the inability for conventional silver-silver-chloride-gel-electrodes to perform long-term monitoring, non-reusability, lack of scalable-standardized wireless communication platforms, and user-friendly design. Recent progress in nanotextile biosensors and mobile platforms has resulted in novel wearable health monitoring systems for neurological and cardiovascular disorders. This chapter discusses nanostructured-textile-based dry electrodes that are better suited for long-term measurement of electrocardiography (ECG), electroencephalography (EEG), electrooculography (EOG), electromyography (EMG), and bioimpedance with very low baseline noise, improved sensitivity, and seamless integration into garments of daily use. It discusses bioelectromagnetic principles of origination and propagation of bioelectric signals and nanosensor functioning, which provide a unique perspective on the development of novel wearable systems that harness their potential. Combined with state-of-the-art embedded wireless network devices and printable fractal antenna to communicate with smartphone, laptop, or directly to remote server through mobile

[1]Department of Electrical Engineering, University of Arkansas, USA
[2]Department of Engineering Science and Mechanics, Penn State University, USA
[3]Department of Neurosurgery and Department of Engineering Science and Mechanics, University Park PA, USA

network (GSM, 4G-LTE, GPRS), they can function as wearable wireless health-diagnostic systems that are more intuitive to use.

9.1 Introduction

Public spending on health and long-term care in the Organization for Economic Co-operation and Development (OECD) member countries and BRIICS (Brazil, Russia, India, Indonesia, China, and South Africa) is 6% of the GDP and is projected to increase up to 14%in the next 50 years [1]. Chronic disease diagnosis and treatment are the primary causes for this increase. Patients suffering from chronic diseases need to repeatedly visit the hospital, which can be expensive. As a solution to this, remote POC systems and RPM systems can be used. RPM for POC facilitates the monitoring of a patient's health condition at local or remote places without the need for hospital admission or visit. In case of high-risk patients, it can provide the patient with real-time feedback from a medical center.

Wearable nanosensor systems in the form of smart clothing, equipped with wireless communication technology, provides real-time medical data to health professionals for early diagnosis, planning therapeutic intervention, and following up on the effect of planned therapy. Techniques, such as ECG, EEG, EOG, EMG and electrical impedance tomography, are relevant to POC for cardiovascular disease, neurological disorders, cancer, and stroke. Intelligent wearable sensor systems with simple installation, minimal maintenance, and user involvement can be the best method for ubiquitous health monitoring. They combine high sensitivity of nanosensors with cost effective and lightweight textiles. Long-term, real-time health monitoring is useful in chronic diseases for event detection, onset of critical episodes, and disease management through diagnostics and therapeutics [2]. Unobtrusive wearable health monitoring is found to be effective in prevention and early diagnosis of neurological and cardiovascular disease by non-invasively monitoring a person's vital signs and physiological data [3].

The design of a wearable sensor system is a multidisciplinary effort that involves expertise in various domains including textile engineering, material science, biomedical engineering, wireless communication, cyber-physical systems, and data interpretation algorithms, security and management. Achieving the full potential of wearable sensor systems poses various challenges to the state of the art in all the stated domains.

Smart garments as an embodiment of wearable nanosensors systems can use the textile platform as a substrate to integrate the various components of the system or use the textile platform as an active or passive sensing material. The smallest unit of a textile platform are fibers or filaments the constitutes the fibers. A textile platform is formed by the weaving stitching and fusing of these fibers or filaments. The development of innovative techniques for the integration of fibers to form textiles can result in many textile materials with varying length, cross-section area and shape, and surface roughness. Furthermore, engineering fibers or filaments to introduce intelligent functionality at different levels of fiber, such as material, shape, thin film coating and density, will help incorporate high-density sensor electronics in textile. Thorough knowledge of the physics behind the interaction of textiles and human skin will help design sensors that require constant contact with wearer's skin and without any skin preparation needs.

There are technical challenges in the steps for wireless communication, signal processing and cyber-physical systems. Wearable sensor systems require a system architecture that provides secure, reliable and high fidelity access to patient data for medical professionals. Furthermore, from both the patient and medical professional's perspective, in terms of the adoption of new technologies, system architecture design should carefully consider intuitive user interfaces and simplicity of setup and maintenance of the entire system from the individual components of the wearable to the database solutions for data management. Improvement in microprocessor technology is required to put more signal conditioning and computation on the textile platform. In addition, the wireless communication electronics have to be designed in such a way that they can communicate with other smart wireless devices in the user space such as smartphones and tablet PCs to harness their computation and communication capabilities to eventually upload the data to server, such that the data are accessible to data scientists and medical doctors for interpretation.

From the electronics design perspective, the form factor, battery longevity, and performance in terms of data communication rates and fidelity of data acquired through wearables pose areas of potential improvements. Antennas and wireless communication prove to be important to provide ubiquity to a wearable device in terms of wireless transfer of data, preferably in real-time. Fractal geometries can be used in the antenna design to enable direct printing of the antennas on the garment. Fractal geometries are smooth structures mainly affiliated with Euclidean lines, panes, and spheres. Fractal geometries are naturally found and are then adapted to the engineering design to mimic the structural conformity to achieve the intended functionality. Generally, fractals structures exhibit scale-based symmetry and not translational symmetry and are also its own embodiments. These embodiments are mathematically generated by the iterations of a feedback loop of its nonlinear equations. Fractal geometries can also be tuned to provide better mechanical strength and is paramount to a wearable device design.

The challenges faced in design are unique to the requirements of the intended wearable system. A detailed catalog for the different challenges faced is beyond the scope of this manuscript. However, patterns of design for wearable systems architecture are a burgeoning body of knowledge. The immediate challenges lie in the formulation of consensus standards for data communication, data security and thorough clinical validation of wearable systems in medicine and patient care.

In this chapter, nanotextile-based wireless biosensor systems have been described. Over the past few decades, advancements in pervasive information and communication technologies, coupled with microelectronics and systems development, have provided an opportunity for the integration of electronics with functional textiles. The confluence of these two fields has radically transformed the norms of computing and embedded systems into soft textile interfaces [4].

In Section 9.2, the concept of textile as a health-monitoring platform is explained along with a discussion on newly developed nano and micro scale fiber structures, composite materials and coatings that integrate with textile fabrics to create *smart textiles*. Section 9.3 offers a brief development of the prevalent models used to describe the relationship between the activity of a single neural cell or cardiac myocyte and the characteristic bioelectric signals measured at the level of the skin–electrode interface (EEG, ECG, EOG, and EMG) and a comparison between planar textile electrodes and nanostructured electrodes. In Section 9.5, we describe the bioelectric signals of diagnostic value for both neurological and cardiological disorders.

Section 9.7 discusses about different types of wireless protocols that are used in current state-of-the-art wireless sensor systems. Finally, in Sections 9.8, 9.9, and 9.10, numerous applications have been showcased to demonstrate the cutting edge of textile-based wearable health monitoring technology for neurological and cardiovascular health disorders.

9.2 Smart textile for health monitoring

Current technologies for measuring and recording biopotential signals discussed in Section 9.7 are suitable for bedside monitoring, with the exception of the Holter monitor. POC diagnostics and therapeutics require systems capable of ambulatory and/or remote monitoring. Such systems will allow patients and high-risk individuals to stay in their homes and follow their routine, while continuous monitoring of their neural and/or cardiac functions can be performed remotely. The key to successful adoption of remote health care is invisibility, i.e., sensors that do not interfere with quotidian activities of the individuals and at the same time efficiently monitor parameters critical to neural and cardiac health.

Textile-based sensor systems are flexible sensors that are made of textile or have a suitable texture and flexibility to embed or integrate into textiles of daily use. The resultant functionalized textile is called e-textile or smart textile. They are distinct from wearable computing systems because they emphasize seamless integration of textile with sensors and sensor electronics. Textiles are preferred for the integration of biomedical sensors because they are the most natural materials to use close to the human body. Thus, they facilitate unobtrusive observation, where they simply sense and record physiological signals of the subject without any kind of active interaction with the subject.

Textile-based nano-biosensor systems can be integrated with compact textile integrated wireless electronics, with the help of woven or printed connections, for remote wireless health diagnostics [5]. It eliminates the use of stick-on-glue-based electrodes and can be worn without the help of medical personnel, which is, therefore, a desirable diagnostic system in hospital as well as remote locations.

Smart textile has been an area of focus for space exploration, biomedical, and consumer electronics communities for their potential to significantly augment the body area network (BAN) [6–8], which is also known as internet of things. Xiaoming Tao describes smart textile as a class of smart materials and structures that sense and react to environmental conditions or stimuli. Smart textiles can be divided into three subcategories. Passive smart textile can only sense environmental conditions and stimuli. Active smart textiles can sense and react to environmental conditions and stimuli. Very smart textiles can sense, react, and adapt to environmental conditions and stimuli. In addition to these types, intelligent textiles can cause predictable effects or phenomena by interacting with the environment and the wearer [9–11]. By this definition, nanotextile-based wireless biosensor systems are wearable smart nanosystems.

The smallest units of the textiles are fibers or filaments. Innumerable combinations of these units can result in many textile materials with varying length, cross-sectional area and shape, and surface roughness. The intelligent functionality can be introduced into textiles at different levels. At the fiber level, a coating can be applied or threads can be added to make a composite textile. Fibers of different types can be arranged at random or strictly organized ways in yarns or fabrics to form even 3D structures. These structures can be metalized or functionalized to fabricate a conductive textile electrode and other functional surfaces with micro or nanorod, micro or nanocoil array.

Smart textile (fabric) can be made from materials ranging from traditional cotton, polyester, and nylon to advanced Kevlar with integrated functionalities. However, in the scope of the present chapter, fabrics with electrical conductivity are of interest. There are two kinds of smart textile (fabric) products that have been developed and studied for health monitoring fabric with textile-based sensor electronics [12–16] and fabric that envelopes traditional sensor electronics [17,18]. Pioneering research work, done by Jayaraman *et al.*, showed that weaving can be used to incorporate electrically conductive yarn into the fabric to obtain a textile that can be used as a "Wearable Motherboard." It can connect multiple sensors on the body, such as wet-gel ECG electrodes, to the signal acquisition electronics [12,13]. Later researchers have shown that conductive yarns can be instrumental in fabrication of textile-based sensors made of fabric [14,15] or metallic meshes [16] coated with silver or conductive metal cores woven into the fabric [19].

Naturally occurring fibers have diameters in the order of microns, where the smallest diameter is on silk fiber (10 μm). It is a common conception that textiles made of fibers with diameter in nanometer scale can be deemed as textiles fit for nanosensor applications. These textiles, integrated into the fabric, can serve as different components of smart sensor system.

Based on the degree of integration, the combination of electronics and textiles can be divided into embedded electronics, textronics, and fiberonics. Embedded electronics uses textiles as a platform for building in readily available off-the-shelf electronics (e.g., Phillips illuminative LED shirts and Lifeshirt by VivoMetrics [20]). These can be nanosensor chips made with state-of-the-art nanofabrication techniques. In such smart textiles, the electronics have to be disconnected prior to washing because they cannot endure washing.

Textronics uses electronic components manufactured by using textile materials and textile production techniques. Nanocomposites and nanoparticles can be incorporate in the textile to form sensitive layers and sensor connects. Mazz *et al.*

and Rossi *et al.* have developed polypyrrole on Lycra and carbon-filled-rubber-based printable sensors for the measurement of posture, movement, and respiration. These textronics-based suits have been demonstrated for monitoring rehabilitation, studying ergonomics, virtual reality, and ambulatory monitoring [21,22]. Due to the limitation of materials and fabrication techniques, complex electronic components, such as microprocessors, cannot be fabricated as textronics and should still be embedded in textiles. Research work done by Clemens *et al.* [23] attempts to integrate basic electronic building blocks, such as semiconductor electronics, in yarns for fabrication of transistors on textile. This fiberonics technology can help in full inclusion of microprocessors along with nano-biosensors in future textiles.

Textronics has the potential for creating textiles with new attributes, while keeping them flexible and washable. For textronic technology, knowledge of and access to textile production is necessary. There are many textile production techniques that can be used to build electronic components. A commonly used concept for making textile-based sensors and electronics is weaving or knitting conductive thread into the garment fabric. Jacquard loop weaving can weave conductive yarn into specific patterns for making conductive tracks, contacts, and antennas [24]. Plain and circular knitting, warp knitting, or crocheting can be used for knitting of conductive yarns into making textile electrodes and strain sensors (e.g., GOW trainer [17], Numetrex by Adidas [3,25], Wearable Wellness System from Smartex s.r.l. Pisa Italy [26]). Embroidery of conductive yarns into textiles can be useful in making wearable keyboards [27] and antennas [28].

Smart textile can serve as a platform for electrophysiological sensors that require being in contact with the body. Studies have shown that the textile-based sensor electrodes are as reliable as the conventional silver–silver chloride gel-based electrodes for the detection of ECG signals [29–31].

Textile, as a substrate, can support nanostructures grown on it [32], embedded as composite [33], embedded/mounted as nanomaterials-based devices [34–36], or nanomaterials-based coating and dyes [37,38]. Conductive fabrics can be obtained by weaving conductive yarn into fabric [39], coating conductive layers on fabric surfaces by chemical processes, such as polymerization [40], electroless plating [41,42], and electroplating [43], or physical processes such as vacuum sputter deposition [44,45]. Incorporation of nanofibers in the textile is also possible by drawing out nano-filaments using electrospinning technique [46]. Alternatively, pre-extruded nanofibers can be deposited with the help of electrodeposition. In either case, the nanofibers form a mat or a web that renders the textile substrate as nanotextured. These textile surfaces have a large surface area and surface-to-volume ratio. The large surface area improves the absorption or adsorption property of the textile substrate to make them useful as sensor layer for gas sensors [47], biological sensors [48], chemical sensor [49], biomedical textile [50], water purifier [51], and electrodes for biopotential measurement [52].

Free-standing-aligned nanostructures can be obtained on textile electrode surfaces by using the traditional technique of flocking. It uses electric field or pneumatic force to drive down millions of individual fibers that have a static charge on them. The electric field, in particular, aligns the charge fibers vertically, and static charge ensures that they are apart from each other. The vertically aligned fibers are driven down on to a flexible surface, such as textile or polymer substrate, and pretreated with adhesive for the fibers to get planted [53]. Synthetic long chain polymers, such as polyester, nylon, polyimide, and polyaramid, are melt blown, solution blown, or extruded and spun into fibers on spinneret. The techniques for drawing out the fibers can be modified to obtain fibers with diameter in the order of nanometers (40–2,000 nm). These processes can obtain fibers that are only as wide as the single-layer crystal made of polymer chains [54]. The conventional synthetic polymer fiber-spinning technology has been improved to produce composite fiber. A mixture of two polymers, that are mutually immiscible, can be drawn into fibers by extrusion, such that, one polymer forms long fibers in a matrix of the other. A cross section of such a fiber shows that 60–1,500 islands of one polymer's fibers are distributed in a sea of the other polymer, thus giving the impression of islands in sea [55]. Composite fibers are best suited because they can be flocked as micro-fibers, and then bundled island polymer nanofibers can be released by dissolving the sea polymer (Figure 9.1). This is followed by metallization of the structures with silver by electroless plating method.

The surface of sensor electrodes can have nanoscale and mesoscale free-standing conductive structures. This contributes to increasing the effective surface area of the electrodes, and high aspect ratio nano/mesoscale structures can overcome the obstruction due to rough skin surface and body hair (Figure 9.13). A good skin–electrode interface with these nanostructured sensor electrodes is instrumental

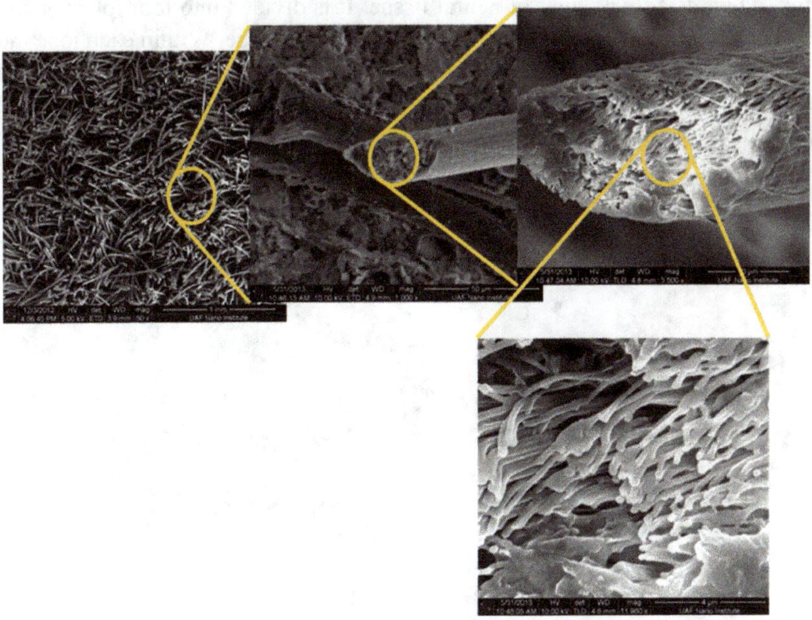

Figure 9.1 Nanostructured electrode surface

in detection of electrophysiological signals emanating from brain and heart to the skin surface. Figure 9.14 shows that the next step would be to understand the principles behind the signals generated by the brain and heart. This understanding should then be extended to the skin–electrode interface to study the effects of nanostructures sensor electrodes on monitoring of signals that are important for diagnosis of neurological and cardiovascular disorders.

9.3 Electrical signals from the brain and heart

The nanotextile-based biosensors for EEG, EOG, EMG, and ECG detect bioelectromagnetic signals generated by brain, muscles, and heart, while being in contact with the skin. In this section, the bioelectromagnetism involved in the origination and propagation of bioelectric signals of interest, i.e., EEG, EOG, EMG, and ECG, have been presented. This will lay the groundwork for electromagnetic theory of skin–electrode interface to explain the superiority of nanostructured electrode over plane and microstructured electrodes. Thus, emphasizing the potential for development of novel systems through a ground-up understanding of the signal sources, i.e., neurons of brain tissue and myocytes of heart tissue, respectively.

The brain is the central organ of the nervous system, which reaches every part of the body. It is responsible for sensory functions such as vision, touch, hearing, taste, and smell. The brain is the center of cognitive functions such as logical thinking, speech, language, and creativity (Figure 9.2).

Most of the volume of brain is made up of cerebrum covered with cerebral cortex, which is a thick layer of neural tissue. It is divided into four lobes, namely frontal lobe, parietal lobe, temporal lobe, and occipital lobe. Within each lobe there are numerous areas, each associated with a particular function. Cerebrum is separated into two hemispheres by a groove called the medial longitudinal fissure. The left and right hemispheres contain almost similar cortical area. However, some areas show lateralization, especially areas related to language are strong on left side and spatiotemporal reasoning are strong on right side [56]. The cerebrum sits

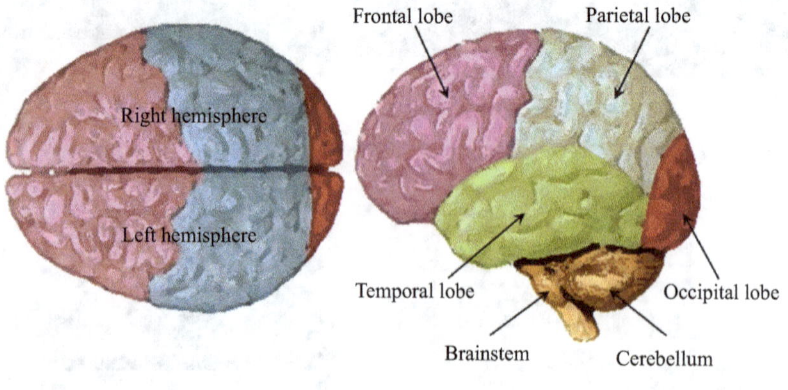

Figure 9.2 Anatomy of brain: cerebral lobes, cerebellum, and brainstem

on top of the brainstem, which is a bundle of cranial nerves that connect the brain to motor and sensory systems in the rest of the body. It also plays an important role in the regulation of cardiac functions, respiratory functions, sleep cycles and facial movements. The cerebellum is situated behind the brainstem and below the cerebrum. It also has a cortical layer with a horizontally furrowed surface called the cerebellar cortex. This part of the brain plays important role in motion control. Though it does not initiate the impulse for motion, it receives and integrates inputs from sensory systems and the spinal cord to fine tune motor activities.

An electroencephalogram (EEG) can be defined as a recorded electric field of the human brain. It can be attributed to the phenomena, which are largely classified into three categories: spontaneous activity, evoked potentials, and bioelectric events produced by single neurons. Spontaneous activity implies all neural activities that occur continuously in a living individual, and it is measured on the scalp or on the brain surface. The respective components are the most prominent features of EEG signals, with an amplitude of about 100 μV on the scalp and 1–2 mV on the brain surface. The signal frequency bandwidth is between 1 and 50 Hz. Evoked potentials arise in response to a stimulus (auditory, visual, electrical, etc.). The relevant EEG signal amplitudes are below the noise threshold. Hence, they are discernible only after averaging the signals in response to a train of stimulus to improve signal to noise ratio. Single-neuron bioelectric events can be recorded using micro/nano electrodes implanted in the brain. It is of particular importance in the monitoring activity of neural clusters to detect asynchronous firing of neurons, which is used as biofeedback by pace-making devices.

A nerve cell has three parts (Figure 9.3): cell body *soma*, numerous short *dendrites,* and single long nerve fiber *axon*. The nerve cell body is similar to that of any other cell with a nucleus, mitochondria, endoplasmic reticulum, and other organelles. The short dendrites receive impulses from one or more neighboring nerve cells and transfer them to the soma.

The effect of these impulses can be excitatory or inhibitory. The axon fiber transfers signal from soma to other nerve cells or muscle cells. The axon communicates with the adjacent nerve cell or muscle cell through synapse. The neural

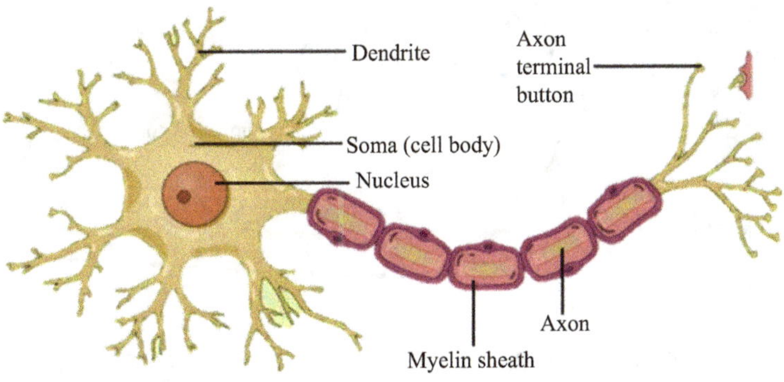

Figure 9.3 Nerve cell structure

impulse passes unidirectional from the axonal presynaptic terminal to the post-synaptic terminals on the cell through chemical neurotransmitters.

The membrane of nerve cells is a bilayer film made of ambiphilic phosphoglycerides. The cell membrane has macromolecular pores, which selectively allow sodium, potassium, and chloride ions to flow through them. The difference between intra cellular ion concentration and extra cellular concentrations results in a resting trans-membrane potential V_{m}, where V_{i} is inner surface potential and V_{o} is outer surface potential of the membrane:

$$V_{\mathrm{m}} = V_{\mathrm{i}} - V_{\mathrm{o}} \tag{9.1}$$

Resting trans-membrane potential is normally negative (−70 mV). This is made possible by ionic concentration gradients of Na^+ and K^+ ions. The extra-cellular concentration of Na^+ is 10 times higher than intracellular concentration, whereas, intracellular concentration of K^+ is 30 times higher than extracellular concentration (Figure 9.4). If a nerve cell is stimulated, V_m can depolarize by change of V_{i} in the positive direction or hyperpolarize by change of V_{i} in the negative direction with respect to V_{o}. Thus, deeming the stimulus as excitatory or inhibitory, respectively [57].

Excitation of a nerve cell is possible only if the stimulation impulse exceeds the threshold potential value of 20 mV, i.e., V_{m} at least changes from −70 to −50 mV. At this point, the ionic permeability of the cell membrane for a sodium ion

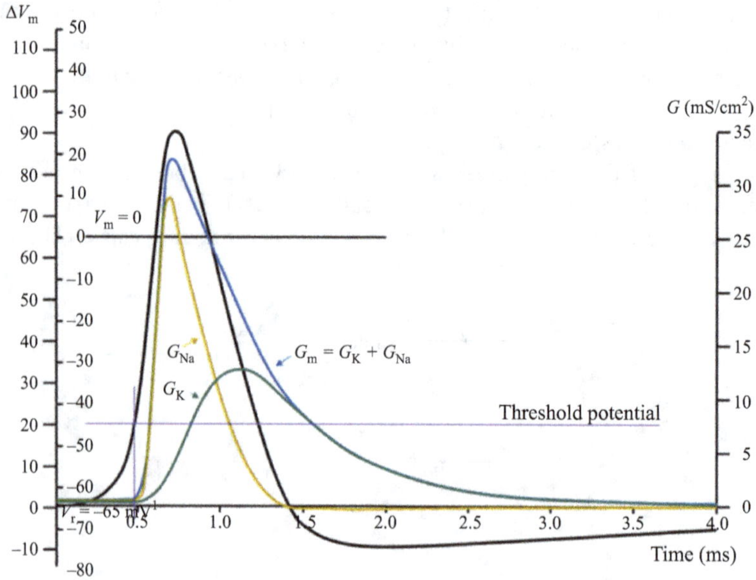

Figure 9.4 Action potential: sodium and potassium conductance (G_{Na} and G_K), their sum (G_m), and membrane voltage (V_m) during a propagating nerve impulse

changes very rapidly to allow flow of sodium ions from outside to inside. This makes V_i more positive to the point where V_m reaches 20 mV, which is followed by a rapid change in the permeability of potassium. This allows potassium ions to flow from inside to outside, thus bringing V_i back to its resting value. The duration of this V_m impulse is ~1 ms. This is followed by restoration of intra- and extra-cellular ionic concentrations by the action of Na–K pump, which is another macromolecular pore in the membrane [57].

The stimulus voltage results in traveling action potential from one nerve to another or to a muscle cell following the path-dendrites-soma-axon-. The action potential is produced by ion transport of Na^+, K^+, and Cl^- through the membrane. It depends mainly on the ratio of ion concentration inside to outside of membrane, voltage across the membrane, and membrane permeability of each ionic species. Under the quasistatic assumption, ion concentration ratio and membrane permeability are represented by Nernst voltage V_n at temperature T (K) in (9.2), where R is the universal gas constant ($R = 8.314$ J/mol·K), $c_{i,n}$ and $c_{o,n}$ are ionic concentrations of the nth species, z_n is moles of electrons transferred during reaction of nth species, and F is Faraday's constant, so the driving force for transportation of nth ionic species is given by $(V_m - V_n)$:

$$V_n = -\frac{RT}{z_n F} \ln \frac{c_{i,n}}{c_{o,n}} \tag{9.2}$$

The EEG signal arises from the field created by localized depolarization, i.e., excitatory postsynaptic potential or localized hyperpolarization, i.e., inhibitory postsynaptic potential. Though the stimulus potential originates at the synaptic terminals and the resultant pulse (current) travels along the neural axon fibers, electrophysiological models consider potential source with a volumetric distribution and a conducting medium that extends continuously in three-dimensional space. They are referred to as *volume source* and *volume conductor*.

Bioelectric activity of nerve cell and muscle cells due to conversion of energy from chemical to electric form gives rise to a non-conservative current. This bioelectric source consists of electric current dipoles formed by charge separation. Hence, the impressed current density $\vec{J}^i(x, y, z, t)$ is similar to volume dipole moment density of the *volume source*, where $\vec{J}^i$ is zero outside of the active cells. An infinite homogenous conductor is a simple approximation of *volume conductor*. Total current density can be given by (9.3). The primary sources $\vec{J}^i$ establish electric field $\vec{E}$ and resultant return current $\sigma\vec{E}$. The return current avoids charge buildup:

$$\vec{J} = \vec{J}^i + \sigma\vec{E} \tag{9.3}$$

Under quasistatic condition, any change in source results in redistribution of charges across membrane. This is expressed mathematically as follows:

$$\vec{\nabla} \cdot \vec{J}^i = \vec{\nabla} \cdot \vec{J} - \sigma\vec{\nabla}^2 V = -\sigma\vec{\nabla}^2 V, \qquad \text{where } \vec{E} = \vec{\nabla} V \tag{9.4}$$

This Poisson's equation in V can be solved to get

$$4\pi\sigma V = -\iiint_{\text{Vol}} \left(\frac{1}{r}\right) \vec{\nabla} \cdot \vec{\mathrm{J}}^i d\, \text{Vol} \tag{9.5}$$

This solution can be extended to inhomogeneous volume conductor, but considering it to be composed of a finite number of homogeneous regions. The head as a volume conductor consists of brain, cerebrospinal fluid, skull, and scalp, not to mention the brain tissue can be divided into gray matter, white matter, and other tissue. Each pth homogeneous region (uniform conductivity σ and unit volume dv_p) has a boundary S_p, which satisfies the conditions of continuity of electric potential V (9.6) and the normal component of current density (9.7). The subscripts 1 and 2 represent either side of boundary S_p:

$$V_1(S_p) = V_2(S_p) \tag{9.6}$$

$$\sigma_1 \vec{\nabla}(V_1(S_p)) \cdot \vec{n}_p = \sigma_2 \vec{\nabla}(V_2(S_p)) \cdot \vec{n}_p \tag{9.7}$$

Following the same steps as shown in (9.3), (9.4), and (9.5) [58], the new expression for V is

$$4\pi\sigma V = -\iiint_{\text{Vol}} \left(\frac{1}{r}\right) \vec{\nabla} \cdot \vec{J}^i \mathrm{d}\, \text{Vol} + \sum_{\mathrm{p}} \iint_{\mathrm{S_p}} (\sigma_{\mathrm{p},2} - \sigma_{\mathrm{p},1}) V \vec{\mathrm{n}}_{\mathrm{p}} \cdot \vec{\nabla}\left(\frac{1}{r}\right) \mathrm{d}\bar{S}_{\mathrm{p}}. \tag{9.8}$$

The first term on the right-hand side is the contribution of the volume source because of the nonelectric energy source $\vec{J}^i$. The second term contribution of a surface source is the summation dipole elements that represent ionic double layers, described by (9.6) and (9.7). Since neural tissue is composed of a very large number of small nerve cells, $\vec{J}^i$ can be summed up as volume dipole moment density function:

$$\vec{J}_i = \frac{\sum_{p=1}^{N} \mathrm{d}\vec{\eta}_{\mathrm{p}}}{\sum_{p=1}^{N} \mathrm{d}v_{\mathrm{p}}}, \quad \text{where } \mathrm{d}\vec{\eta}_{\mathrm{p}} = \int (\sigma_{\mathrm{o}} V_{\mathrm{o}} - \sigma_{\mathrm{i}} V_{\mathrm{i}}) \mathrm{d}\vec{S}_{\mathrm{p}}. \tag{9.9}$$

In an electrophysiological measurement, V can be measured. Thus, (9.8) describes a problem where the field and the volume conductor are known, but the volume source is unknown. Such problems are called inverse problems [59]. This pertains to clinically measured EEG, where the neurologists seek to determine the source of the measured bioelectric signal (EEG). Similar principle applies to other electrophysiological signals such as ECG, EOG, and EMG. Though it is possible to evaluate the source function $\vec{J}^i$ in case of ECG and EMG, it has not been completely feasible in EEG because of the complexity of brain structure and its electrophysiological behavior. Quantitative EEG is largely based on the examination of lead patterns to calculate sensitivity distribution of lead and estimate the statistically most probable source configuration, i.e., neurological conditions. However, clinical

EEG diagnostics is typically based on recognition of typical signal patterns that are known to be associated with neurological conditions [60].

9.4 Cardiovascular anatomy and electrophysiology

The human heart is a muscular organ consisting of four chambers: two upper chambers known as the *atria* and two lower chambers known as the *ventricles*, which are separated by muscular septum into the right and left atria and ventricles, respectively (Figure 9.5(a)). Oxygen-depleted blood from the peripheral organs is returned to the right atrium through the superior and inferior venacava. The contraction of the right atrium then forces blood into the right ventricle through a unidirectional heart valve known as the *tricuspid valve.* The right ventricle then contracts and pumps blood into the pulmonary artery, which takes the blood to the right and left lungs to exchange the carbon dioxide in the blood for fresh oxygen. Meanwhile, freshly oxygenated blood

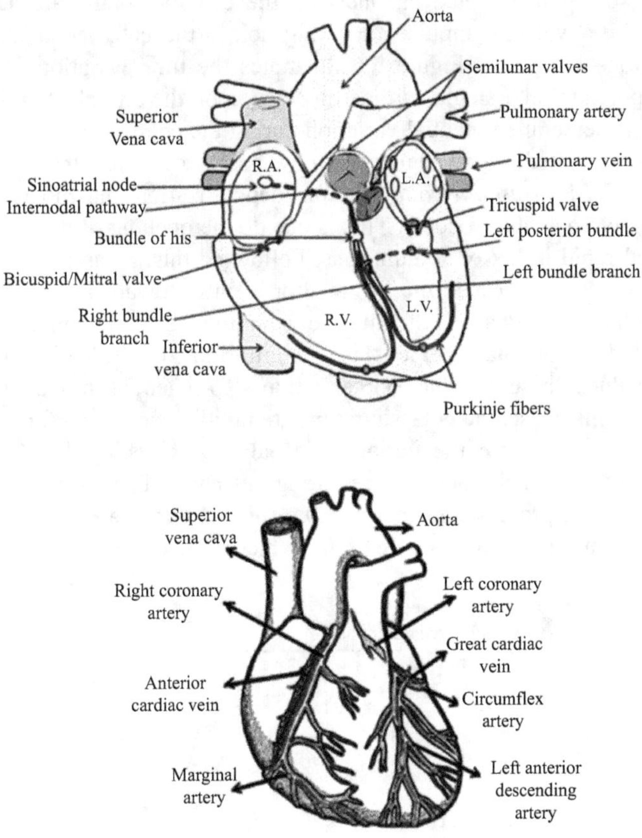

Figure 9.5 (a) A diagram of the internal anatomy of the heart with the impulse conduction pathway and (b) illustration of the blood vessels involved in the circulation of blood to the heart muscles

from the lungs is passed on to the left atrium by the pulmonary vein. The left atrium then contracts and forces the blood into the left ventricle through a unidirectional valve called the *mitral* or *bicuspid valve*. Contraction of the left ventricle forces the blood through the semilunar valve into the aorta, which then branches out to several arteries and supplies fresh blood to all the cells in the body. The major blood vessels involved in the supply of blood to the cardiac tissue are left and right coronary arteries that branch off from the aorta as shown in Figure 9.5(b).

Cardiac muscle cells, or *myocytes*, have a resting potential varying between −80 and −90 mV. The resting (transmembrane) potential (TP) difference is between the intracellular fluid and the extracellular fluid. This potential difference is maintained by the selectively permeable cell membrane that, in turn, maintains the difference in the sodium, potassium, calcium, chloride, and potassium ion concentrations between the two fluids. The TP is regulated predominantly by the sodium and potassium ion concentrations. The cell membrane has voltage-activated channels that transport ions into or out of the cell when triggered by a voltage impulse. As the voltage-gated channels on the cell membrane are activated or deactivated by a voltage impulse travelling across the cell, the transmembrane potential varies with time. Figure 9.6 illustrates the time variation of the transmembrane potential in a single cardiac myocyte upon the arrival of an impulse.

The transmembrane potential variation during the conduction of an impulse is called action potential. Upon the arrival of a positive impulse that shifts the TP above −70 mV, a *threshold voltage*, rapid depolarization and movement of TP towards positive potentials occurs. This is due do the opening of the sodium-gated channels and rapid influx of sodium ions. Following this, potassium and chloride ion channels open and cause the TP to drop a little toward 0 mV. This dip is denoted as Phase 1. In Phase 2, the influx of sodium ions, along with some calcium ions and the efflux of potassium ions, is in equilibrium and the TP is maintained at a constant value. Phase 3 is the repolarization step when TP moves towards the resting TP. In this phase, the potassium ions are rapidly exchanged for sodium ions inside the cells to restore the initial ionic balance at resting TP. Calcium ion channels also reduce their conductance during this phase. Phase 4 is the resting TP condition. The cell processes from Phase 0 result in an increased intracellular calcium ion concentration in the muscle tissue, which initiates the release of energy by

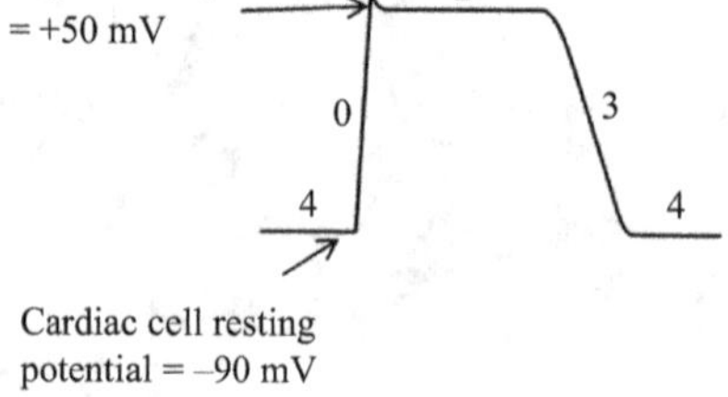

Figure 9.6 Transmembrane potential variations over time for a single cardiac myocyte

breaking down adenosine triphosphate (ATP) molecules and conformational changes in proteins that result in muscle contraction [61].

The heart has a small group of cells, called sinoatrial node, located in the right atrium that are capable of generating impulses periodically. These impulses maintain the contraction rhythm of the heart and their direction of propagation maintains the progression of contraction, i.e., the atria contracts first, then the ventricles. This is often referred to as the natural pacemaker of the heart as it directly regulates the heart rate. Impulses are fed to the sinoatrial node by the *Vagus nerve* and *parasympathetic and sympathetic nervous systems*.

At the beginning of the cardiac cycle, the sinoatrial node generates an impulse which is carried to the *atrioventricular node* or the *Bundle of His* through the internodal pathway, which is made of a fiber of specially modified muscle cells. The impulse is conducted through the internodal pathway when the atria contracts. From the atrio-ventricular node, the conduction pathways split into the left bundle branch and the right bundle branch. The left bundle branch conducts the cardiac impulse along the left ventricle, while the right bundle branch conducts the impulse along the left ventricle. Both bundle branches end in Purkinje fiber cells, which are tree shaped and spread the cardiac impulse along the entire surface of the ventricles. Figure 9.5(a) illustrates the electrical conduction pathway from the sinoatrial node to the Purkinje fibers.

The electrocardiogram (ECG) is a simple non-invasive test to observe the variations in biopotentials originating from the heart, through electrode sensors placed on the surface of the skin. The ECG waveform acquired from a derived Lead II electrode placement system is shown in Figure 9.7, which clearly depicts the

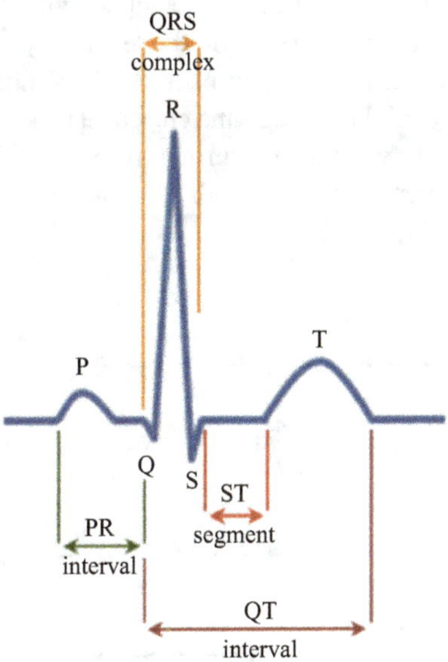

Figure 9.7 ECG waveform with the characteristic P-wave, QRS complex, and T- and U-waves

classical components of the ECG waveform. The waveform characteristics of the ECG include P wave, QRS complex, and T- and U-waves.

The P-wave represents the wave of depolarization that spreads from the sinoatrial node throughout the atria and is usually 0.08 to 0.1 s in duration. QRS complex represents ventricular depolarization. The isoelectric period following the QRS complex is the time at which the entire ventricle is depolarized and roughly corresponds to the plateau phase of the ventricular action potential. The T-wave represents ventricular repolarization and is longer in duration than depolarization. It is followed by a U-wave. Its origin is not well understood. There are three popular hypotheses: (a) late repolarization of Purkinje fibers, (b) late repolarization in the left ventricle, and (c) after-potentials causing variations in normal potentials. In normal subjects, U-waves have the same polarity as T-waves.

9.4.1 The dipole theory for ECG

The ECG is electrical activity observed at the surface of the skin. The most widely followed theory that bridges the origin of cardiac activity and the ECG is the dipole theory [58]. The derivations presented in this chapter are based on the material in [61].

The development of this theory can be divided into three fundamental steps:

1. A model for cell membrane conduction of action potential—**Cable Model.**
2. A representation for the electrical activity propagated from one cell to its neighbor—**Dipole Cardiac Vector.**
3. A model for the transduction of this electrical activity from the heart to the surface of the torso—**Derivation of ECG from Dipole vector.**

The cable model is used to describe the transmembrane potential variations and the currents flowing both inside and outside the myocardial cell. Figure 9.8 illustrates the electrical circuit equivalent of the cell membrane.

V_o and i_o are the extracellular voltage and current at a given instance, respectively. V_i and i_i are the intracellular voltage and current, respectively, at a given instance. M represents the lumped properties over a length Δx. r_o and r_i are the resistances per unit

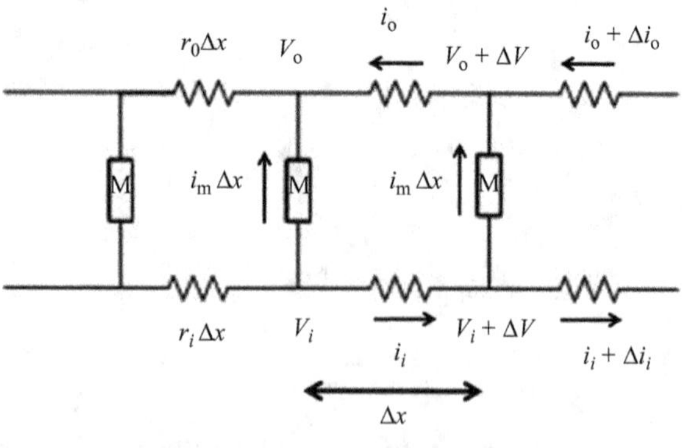

Figure 9.8 The circuit model for currents and voltages at the cell membrane

length of the extracellular and intracellular fluids, respectively. i_m is the current flowing through the membrane per unit length. The TP is given by $V_m = V_i - V_o$ along the membrane. Based on this model, we can arrive at a relationship between the rate of change of TP with respect to position and the ionic current:

$$i_i r_i \Delta x = -\Delta V_i \tag{9.10}$$

$$\frac{\partial V_i}{\partial x} = -i_i r_i \tag{9.11}$$

Similarly,

$$\frac{\partial V_o}{\partial x} = -i_o r_o \tag{9.12}$$

Using Kirchhoff's current law at one of the nodes,

$$\Delta i_i = -i_{\mathrm{m}} \Delta x \tag{9.13}$$

$$\frac{\partial i_i}{\partial x} = -i_{\mathrm{m}} \tag{9.14}$$

$$\frac{\partial V_{\mathrm{m}}}{\partial x} = \frac{\partial (V_i - V_o)}{\partial x} = \frac{\partial V_i}{\partial x} - \frac{\partial V_o}{\partial x} = -i_i(r_i + r_o) \tag{9.15}$$

The TP varies with time as well as spatially.

$$i(x,t) = -\frac{1}{(r_i + r_o)} \frac{\partial V_{\mathrm{m}}(x,t)}{\partial x} \tag{9.16}$$

Substituting for membrane current from (9.14), we attain

$$i_{\mathrm{m}}(x,t) = \frac{1}{(r_i + r_o)} \frac{\partial^2 V_{\mathrm{m}}(x,t)}{\partial x^2} \tag{9.17}$$

At the interface between a depolarized myocardial cell and the neighboring resting cell, there is a transmembrane current that follows to negate the difference in TP between them. This ionic current flow can be depicted as shown in Figure 9.9.

On the assumption that the travelling action potential has a constant velocity (c) and shape within the myocardium, we have

$$\frac{\partial V_{\mathrm{m}}(x,t)}{\partial x} = \frac{\partial V_{\mathrm{m}}(x,t)}{\partial t} \cdot \frac{\partial t}{\partial x} = \frac{1}{c} \cdot \frac{\partial V_{\mathrm{m}}(x,t)}{\partial t} \tag{9.18}$$

$$i = -\frac{1}{(r_o + r_i)} \cdot \frac{1}{c} \cdot \frac{\partial V\mathrm{m}(x,t)}{\partial t} \tag{9.19}$$

This current flowing between a depolarized and a resting myocyte forms a current dipole. The vector associated with this dipole is the dipole moment as shown below in Figure 9.10 by the hypothetical, spherical interface between a depolarized and a resting myocyte.

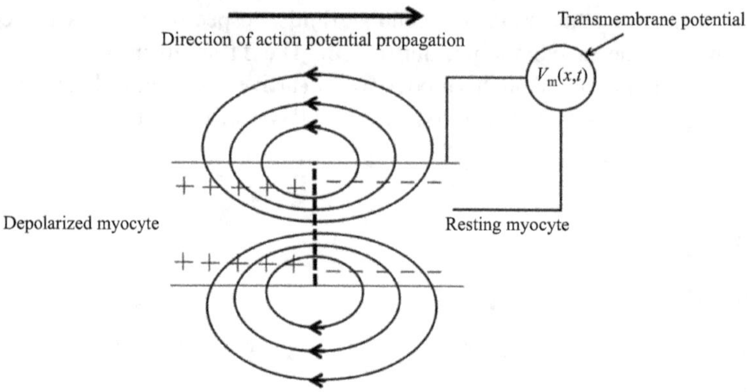

Figure 9.9 Propagation of transmembrane current at the interface between a resting myocyte and a depolarized myocyte

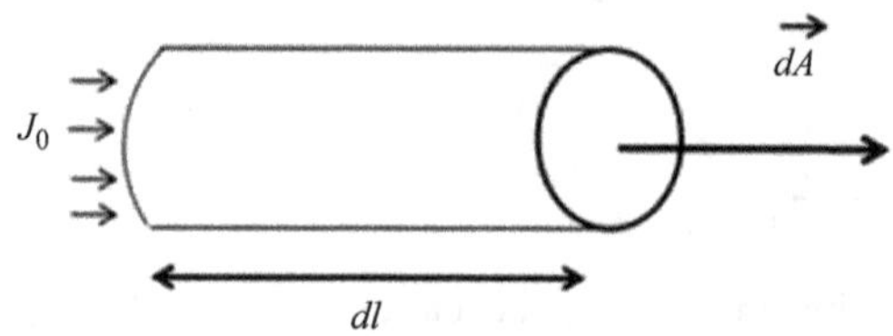

Figure 9.10 Unit dipole produced by the current flowing through a hypothetical spherical interface between a depolarized and a resting myocyte

The magnitude of the dipole moment is given by

$$\vec{m}_o = j_0.\overrightarrow{dA}.dl \tag{9.20}$$

where $\overrightarrow{m_o}$ is the unit dipole moment vector, $\overrightarrow{dA}$ represents the direction of the dipole moment and unit increment in cross-sectional area of cardiac myocyte, and dl is the length of a unit dipole vector. dl and $\overrightarrow{dA}$ are several orders of magnitude smaller that the length spanned by electrodes on the torso; therefore, it is acceptable to represent them as infinitesimal increments.

Values for $\frac{\partial V_m(x,t)}{\partial t}$, r_i, r_o, and dl can be estimated and verified experimentally, yielding a quantitative value for the dipole moment vector.

The heart, as a source of the potential, however, is the resultant summation of all individual dipole moments originating from the cardiac myocyte, given by,

$$M_0 = \iiint J.\overrightarrow{dA}.dl \tag{9.21}$$

9.4.2 Derivation of ECG from dipole vector

According to the dipole theory, the potential difference measured across two points on the torso is the geometric projection of the cardiac vector on the line vector connecting the two points. The aim, therefore, is to derive the relationship between the cardiac dipole vector and the surface potentials at the level of the skin. For the sake of simplicity, the relation between the dipole vector and surface potential at the skin is derived for a single instant of time. By expressing the dipole moment as a function of time, a real-time expression for ECG can be easily derived.

The following assumptions are made in the derivation:

1. The torso is a linear, isotropic, homogeneous, spherical conductor of radius R and conductivity σ.
2. The heart's activity is represented by the time varying Cardiac Dipole vector.

Within the spherical torso, linearity dictates

$$\vec{J} = \sigma\vec{E} = -\sigma\nabla\varphi \tag{9.22}$$

where $\vec{E}$ is the electric field and φ is the electric potential. Since the net charge generation through the cardiac cycle is 0,

$$\nabla\vec{J} = 0 \tag{9.23}$$

From (9.26),

$$\nabla^2\varphi = 0 \tag{9.24}$$

This is a Laplacian equation which is solved in spherical coordinates with the following boundary conditions,

1. No current is allowed to flow out of the body. Therefore, $\frac{\partial\varphi(r,\theta)}{\partial r} = 0$ at $r = R$ (radius of torso)
2. Condition established in the dipole cardiac vector derivation. $M_0 = \iiint J.\overrightarrow{dA}.dl$

The solution that satisfies the boundary conditions is written as a sum of two functions:

$$\varphi(r,\theta) = \varphi_1(r,\theta) + \varphi_2(r,\theta) \tag{9.25}$$

$$\varphi(r,\theta) = \frac{M_0}{4\pi\sigma}\cos\theta\left[\frac{1}{r^2} + \frac{2r}{R^3}\right] \tag{9.26}$$

At the surface of the sphere, i.e., the skin on the torso, $r = R$

$$\varphi(r,\theta) = \frac{3M_0}{4\pi\sigma R^2}\cos\theta \tag{9.27}$$

Figure 9.11(a) and (b) illustrates how the potential at the surface is calculated as a projection of the cardiac dipole vector.

The vector representation of (9.27) is

$$\varphi_A = \vec{M}\cdot\overrightarrow{OA} \tag{9.28}$$

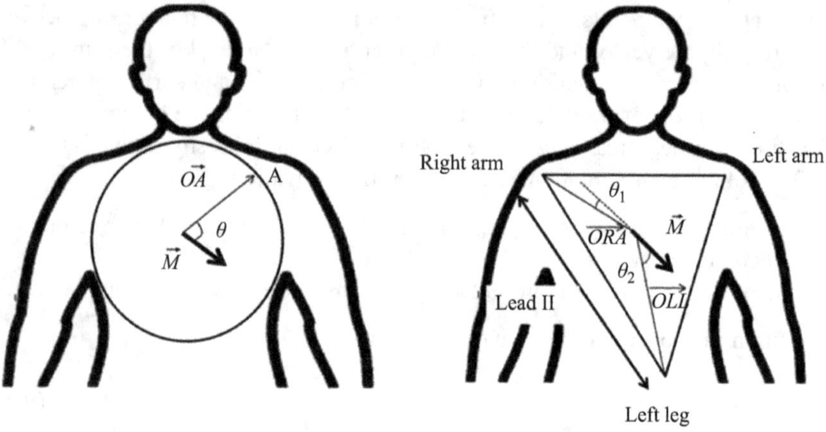

Figure 9.11 (a) and (b) Illustration of how the potential at the surface is calculated as a projection of the cardiac dipole vector

Substituting for $\vec{M}$ from (9.21) and j_0 from (9.17), we get

$$\varphi_A = \iiint \left[\frac{1}{r_o + r_i} \cdot \frac{1}{A_{cr}} \frac{\partial^2 V_{\mathrm{m}}(x,t)}{\partial x^2} . \overrightarrow{dA} . dl \right] \cdot \overrightarrow{OA} \tag{9.29}$$

Lead II refers to the measurement of potential difference between electrodes placed on the right arm and the left leg. Based on this result, the ECG for lead II position is calculated as follows:

$$V_{II} = \varphi_{LL} - \varphi_{RA} \tag{9.30}$$

$$\varphi_{LL} = \vec{M} . \overrightarrow{OLL} \tag{9.31}$$

$$\varphi_{RA} = \vec{M} . \overrightarrow{ORA} \tag{9.32}$$

$$V_{II} = \vec{M} . \left(\overrightarrow{OLL} - \overrightarrow{ORA} \right) \tag{9.33}$$

9.5 Monitoring and diagnosis: neurological signal measurements

EEG is the recording of the electrical activity of the brain along the scalp region by measuring the fluctuation in the voltage induced by the ionic current flows originated from the neurons. EEG is normally measured by placing the electrodes over the skull at defined positions according to the international 10–20 system (Figure 9.12) [62]. The 10 and 20 refer to the distance between adjacent electrodes by dividing the transverse and median planes of the skull perimeters into 10% and 20% intervals. In the international 10–20 system, 19 electrodes are placed over the skull and two electrodes are placed on the ears as reference electrodes (Table 9.1).

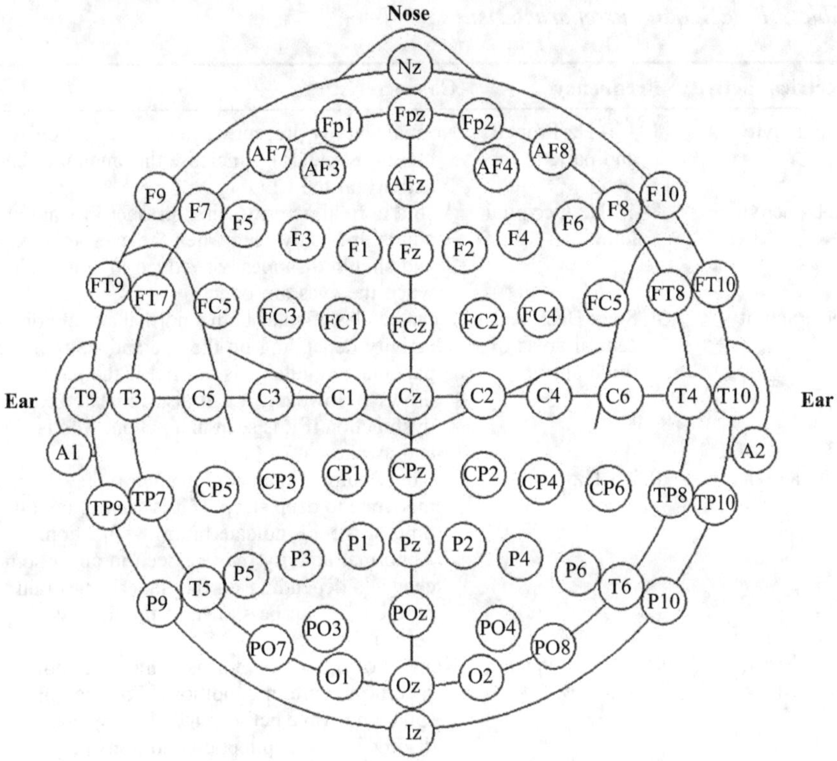

Figure 9.12 International 10–20 placement of electrodes for EEG

The letters F, O, C, P, T in the 10–20 system of placement of electrodes stand for frontal, occipital, central, parietal, and temporal, respectively. The electrodes are placed according to the location of placement and the underlying cerebral cortex. The even numbers represent the right side of the hemisphere and odd numbers represent the left side of the hemisphere. There are two basic methods by which the electrodes are placed: monopolar and bipolar. In monopolar, one side of the amplifier is connected to the reference electrode, and in bipolar, amplifier is connected between the pair of electrodes. The electrical activity is acquired and amplified/filtered to record EEG waveform

Electrooculogram (EOG) is the measurement of the resting potential of the retina. EOG monitors the eye movements by detecting the dipolar current flowing from the cornea to the retina, which also indicates the angular displacement of the eye. Applications of EOG include saccadic movements, smooth pursuit movements, convergence/divergence to record and optokinetic nystagmus. Normally, the electrodes are placed around the eyes with a reference electrode on the forehead. The electrodes are placed on the temple for the lateral movement and the other electrodes are placed vertically, one above and the other below the eye (Figure 9.13) to measure vertical movement of the eyes [62].

Table 9.1 EEG and its characteristics

Electrical activity	Frequency	Characteristics
Beta activity	13–30 Hz Frontal and parietal lobe	Normal activity present when the eyes are open or closed. Some drugs increase the amount of beta activity in the EEG
Alpha activity	8–13 Hz Occipital lobe	Also a normal activity when present in waking adults. It is only seen when the eyes are closed and should disappear or reduce in amplitude when the eyes are open
Theta activity	4–8 Hz Back and central areas of the brain	It can be classified as both a normal and abnormal activity depending on the age and state of the patient. In adults, it is normal if the patient is drowsy. However, it can also indicate brain dysfunction if it is seen in a patient who is alert and awake.
Delta activity	0.5–4 Hz	It is only normal in an adult patient if they are in a moderate to deep sleep. If it is seen at any other time, it would indicate brain dysfunction. Abnormal activity may be seen in all or some channels depending on the underlying brain problem. It can be shown to the depressed person
Spike and wave activity	Random frequency < 60 Hz	Number of other waveforms, which are more specific to certain conditions. For example, spike and wave activity indicates a seizure disorder. Other epileptic conditions may be diagnosed if spikes or sharp waves are seen

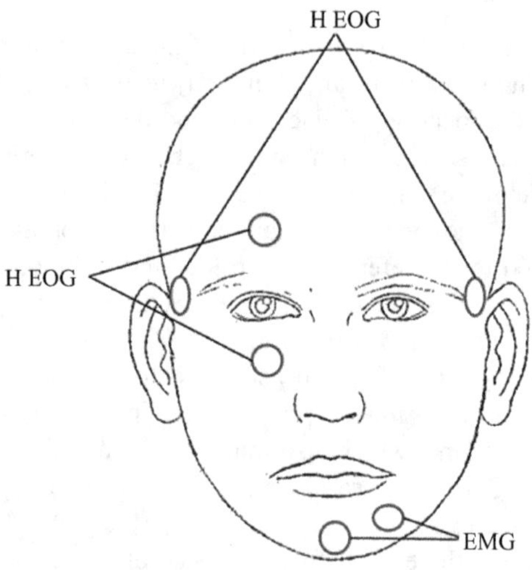

Figure 9.13 Standard electrode placements for EOG for monitoring vertical (V) and horizontal (H) eye movement and EMG for reliable detection of muscle tone REM sleep

EMG depends on the firing action potential of the numerous motors present in the muscles. The electrodes placed on the skin over the muscle detect the electrical activity of the muscles of the underlying tissues (Figure 9.13). It is difficult to correlate the waveform with the specific muscle from which it is generated but this difficulty can be alleviated by the proper placement of electrodes. The muscle fibers that are present near the electrodes will have a greater impact on the waveforms, whereas the muscles at a longer distance will have lesser impact with respect to signal strength. This dependence of the quality of EMG on the distance of muscles and electrodes is mainly caused due to the impedance between the tissues. Therefore, placing the electrodes at a long distance will provide varying and more generalized signals, and placing the electrodes at short intervals gives a signal which is more specific to the muscles over which the electrodes are placed. However, it becomes nearly impossible to identify the specific muscles generating the signal because of the interferences from noise and motion artifacts. The electrodes are generally placed in parallel with the dominant muscles since it minimizes signal cancellation and maximizes the biofeedback sensitivity [63].

Symptoms ranging from structural, biochemical, or electrical abnormalities in the nervous system, especially brain and nerves, which denote abnormality, are termed as a neurological disorder. The two fields of medicine, neurology and neuropsychology, deal with most of the types of neurological disorders. These disorders may vary from loss of concentration to paralysis, where they are also classified into common and rare categories depending on the degree of impact [64]. They are one of the ten leading causes of deaths in the United States. They are followed by infectious diseases such as meningitis and tetanus. Equally prevalent are the degenerative neurological disorders such as Alzheimer's disease and Parkinson's disease [65]. A range of specific disorders can be identified and diagnosed with biopotential signals such as EEG, EOG, and EMG (Table 9.2). Methods to monitor many of these disorders are still in the stage of infancy. For instance, management and/or monitoring of epilepsy involves a complex mechanism because of the diverse nature of the epileptic seizures across the spectrum of the patient population. Treating epileptic seizures has evolved to reach the current state of the art that involves the use of electric and magnetic potential at specific sites in the brain. These deep brain stimulation methods are invasive and are called transcranial direct current stimulation (tDCS) and transcranial magnetic stimulation (TMS). tDCS is one of the modalities of neurostimulation where a low current electric potential is delivered via electrodes to the brain. These electrodes are placed invasively on the affected sites of the brain or over the scalp on the head depending on the requirements identified after medical diagnosis. TMS is another modality of neuromodulation where a change in magnetic field induces the flow the electric potential to the intended site. The aforementioned brain stimulation methods involve two steps: (1) monitoring epileptic seizures through EEG, and (2) using magnetic or electric potential at specific sites to reactivate the neuronal function. These two modalities are used in some of the non-invasive procedures that are simple to administer, however, epileptic seizures are complex in nature, and major improvements in the epileptic management process is required to make it effective.

Table 9.2 Biopotential and neurological disorders

EEG	• Tumors • Stroke • Epileptic seizures • Encephalopathy or delirium • Catatonia • Alzheimer's • Parkinson's
EOG	• Clinical ophthalmology • Parkinson's • Sleep disorders
EMG	• Axillary nerve dysfunction • Centronuclear myopathy • Mononeuritis multiplex • Motor neuron disease • Neuromyotonia • Peripheral neuropathy

The evidence regarding the effectiveness and efficiency of these methods is not well known. There have been animal studies to show the relation between epileptogenesis and excitatory synaptic strength.

9.6 Monitoring and diagnosis: cardiological signal measurements of diagnostic value

The ECG is a fundamental non-invasive method for monitoring the heart's electrical activity by placing electrodes on the skin. ECG provides multiple perspectives of the heart's electrical activity simultaneously [66]. The setup can be a 3-lead system with electrodes placed at corners of the torso section of body (Figure 9.14) as a substitute for the extremities of the limbs (right arm, left arm, and left leg), thus forming an imaginary triangle known as Einthoven's Triangle. They provide a limited view of electrical cardiac activity, but the polarity of these leads is useful for determining the direction of propagation of depolarizing pulse through the cardiac tissue known as the electrical axis. The setup can also be a 12-lead system (Figure 9.14), which uses 10 electrodes, 4 placed at the extremities of the limbs (arms and legs) and 6 placed on the chest. The 6 chest electrodes are called precordial leads that give perspective of electrical cardiac activity in a horizontal plane that is orthogonal to the electrical axis [67]. For chronic disease management, out-of-hospital rehabilitation and diagnostics, such as event detection, require full time ECG recording on ambulatory patients. For the same, the 12-lead ECG setup can be simplified to a 5-electrode system known as the EASI system and simple mathematical transformations exist to

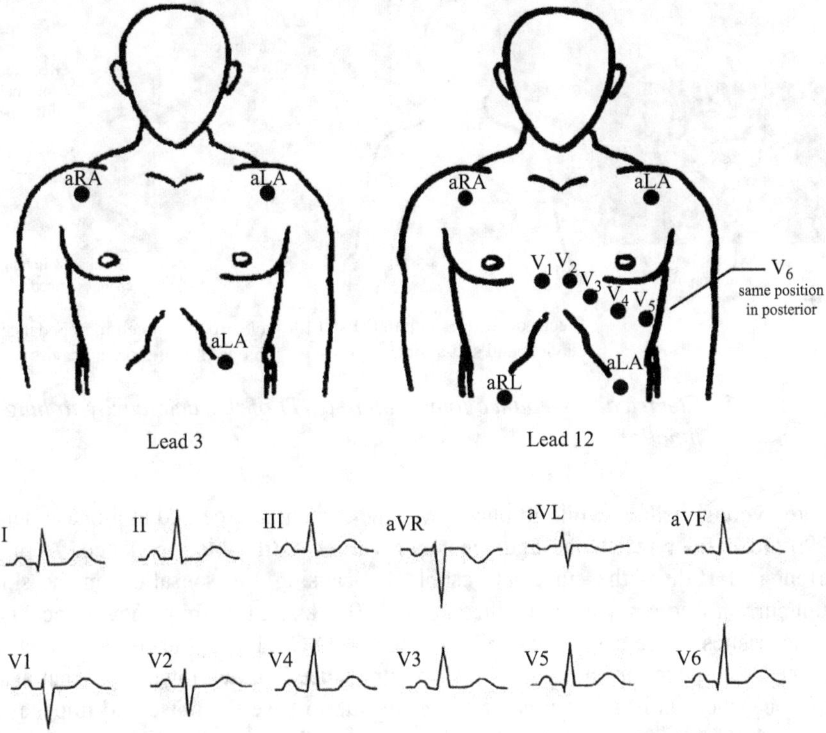

Figure 9.14 Electrode placements for 3-lead and 12-lead ECG, signals from 12 leads [77,79]

derive the 12 lead ECG. These 5 electrodes are on the upper part of body along the sternum and mid-axial region [68,69]. Hence, the ECG panel can be used as an image of the cardiac activity for non-invasive medical diagnosis.

Trans-thoracic impedance is a technique used to measure the change in impedance across the thoracic cavity. It is a type of electrical impedance tomography (EIT) technique. It is important for monitoring pulmonary function, trans-myocardial current, cardiac output, and over all fluid retention of the thoracic cavity. The latter is important in the monitoring of hypertensive patients.

The setup uses 4 electrodes placed in the sub-clavicle, sub-axillar, anterior, or posterior positions (Figure 9.15). A constant current is applied to 2 electrodes, and the resulting voltage is recorded across the other 2 electrodes. The electrode pairs are placed across the thoracic cavity from each other to capture change in conductivity due to ventilation of lungs or cardiac function. This system is also capable of acquiring an impedance image of the thoracic region by using electrode array (16 electrodes or more) placed all around the thoracic cavity. Each set of 4 electrodes acts as a perspective (angle) for scanning the bioimpedance.

The 4-electrode system includes the electrodes, a current driver (source), a voltage-recording unit, and a phase-sensitive demodulator. The phase demodulator

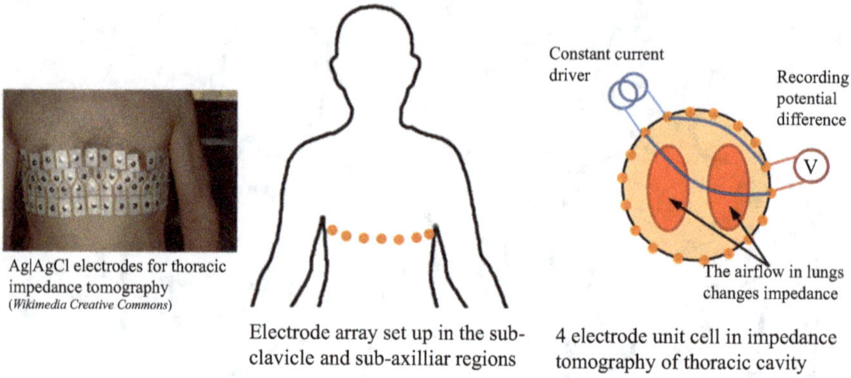

Figure 9.15 Electrical impedance tomography (EIT) of thoracic cavity to lung function

records voltage values while in phase with the current source and at a phase delay of 90° to extract resistance and reactance values of the bioimpedance. Applied current is 1/10th of the current threshold for causing any sensation on the skin. Input current frequency is maintained around 50 kHz. At this frequency, impedance characteristics of the tissue are similar to those at D.C. That is, the current travels in extracellular space but the electrode-skin impedance is much lower than that at D. C. Hence, there is less instrumentation error due to baseline noise and impedance mismatch. However, measurements taken across a frequency spectrum can help in rectifying any phase effects [70].

Early systems, such as the Sheffield Mark 1, used a single impedance measurement circuit and a multiplexer to link with the array of electrodes. More recent systems use devoted circuits for each electrode set. While the former is portable but slow, the latter is fast but bulky. Two types of electrodes are commonly used for this system: silver–silver chloride gel electrodes and the conductive gel filled gold cup electrode.

In theory, the EIT system should be free of impedance at electrode–skin interface. In practice, skin preparation (abrasion) is used for reducing impedance at electrode–skin interface. Still, the system experiences change in impedance at the interface. During thoracic impedance monitoring for longer periods of time, the conductive gel may dry up and increase electrode-skin impedance. Dry textile-based nano-biosensor electrodes have contact impedance less than that of the plain dry electrodes; hence, they can be a good alternative to gel-based electrodes for long-term monitoring.

Cardiovascular disorders can be diagnosed by identifying ECG abnormalities and computing their frequency of occurrence. Apart from the well-defined diagnostic criteria mentioned above, there are a few approaches that are being researched to eventually be included in standard clinical practice. Ventricular arrhythmias are abnormalities in the cardiac conduction physiology or anatomy originating in the ventricles of the heart. From a cardiac patient care perspective, they are the causes of immediate concern because they can lead to fatal outcomes such as sudden cardiac arrest (SCA) leading to sudden cardiac death (SCD) and acute myocardial infarction. The T-wave, QT interval, and the ST segment of the ECG are known to be indicative of the repolarization of the ventricles of the heart during a cardiac cycle. Consequently, among ECG analysis criteria, T-wave alternans (TWA) and T-wave inversion (TWI) have gained significant research interest as means to predict the likelihood of Ventricular Arrhythmias.

T-wave alternans (TWA)—T-wave alternans is the beat-to-beat variation of T-wave morphology and amplitude. Several clinical studies have tried to determine the significance of using TWA analysis to detect abnormalities that may lead to ventricular arrhythmias, as well as establish metrics to perform risk stratification for cardiovascular patients with prior cardiac episodes. The statistical significance of TWA in predicting ventricular arrhythmias has been established in patients across several diagnoses [71]. Studies have also shown the significance of the predictive value of TWA analysis in post myocardial infarction patients [72], risk of SCD [73], congestive heart failure [74], ischemic cardiomyopathy [75], and Chagas disease [76]. Figure 9.16 shows an example of TWA analysis performed in [77].

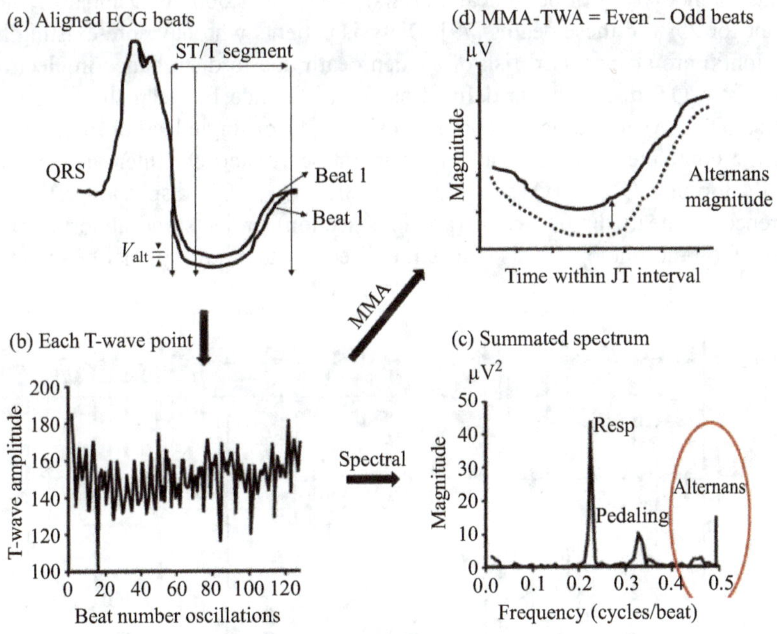

Figure 9.16 TWA Computation: (a) beats are aligned by QRS complexes. At each successive time point in the aligned T-waves (arrows); (b) beat-to-beat oscillations reflect alternans at each time point; (c) spectral analysis applies fast Fourier transformation to yield a power spectrum in which alternans is the peak at the frequency of half the heart rate (0.5 cycles/beat); and (d) MMA analysis uses a nonlinear filter to quantify the maximum difference between the means of "even" versus "odd" beats in an alternating sequence. Figure reproduced with permission, from [77]

T-wave inversion (TWI)–TWI is the reversal in the polarity of the normal T-wave which is upright in most ECG leads. TWI has been associated with cardiovascular as well as cerebrovascular abnormalities [78]. TWI coupled with QT interval changes and dialysate calcium concentrations has also been associated with in increased risk of SCD among patients who have recently undergone hemodialysis [78–81]. Moreover, TWI has gained significant research interest because of its high incidence in young athletes [82] and soldiers [83]. TWI is a known diagnostic criterion for hypertrophic cardiomyopathy (HCM). HCM is known to significantly increase a patient's susceptibility to an SCA leading to SCD when the patient is exposed to exertion through cardiovascular exercises. Consequently, T-wave characteristics are an important criterion used in prescreening for athletes in sports and soldiers in military recruitment. As an example, Figure 9.17 shows the asymptomatic TWI in a Football referee [82].

QT interval dispersion: Sudden cardiac death and fatal arrhythmia have been found as major causes of death among dialysis patients. According to the United States Renal Data System (USRDS) database, the mortality rate among dialysis patients (hemodialysis or peritoneal dialysis) is 230 per 1,000. SCD and arrhythmias account for 25% of these deaths [79]. Dialysis patients who have pre-existing heart condition(s) are at increased risk of sudden death due to disturbance in electrolyte metabolism. QT dispersion is defined as the difference between the longest and shortest QT intervals extracted from ECG signals from single lead or multiple leads up to the complete set of 12 leads. A heart rate-corrected QT interval is given by Bazett's formula $QT_c = (QT_{max} - QT_{min})/\sqrt{RR_{interval}}$. QT dispersion reflects the differences in heart dipole vector (previous section) projects and abnormalities of T-wave loop morphology. This makes it a direct measure of regional heterogeneity

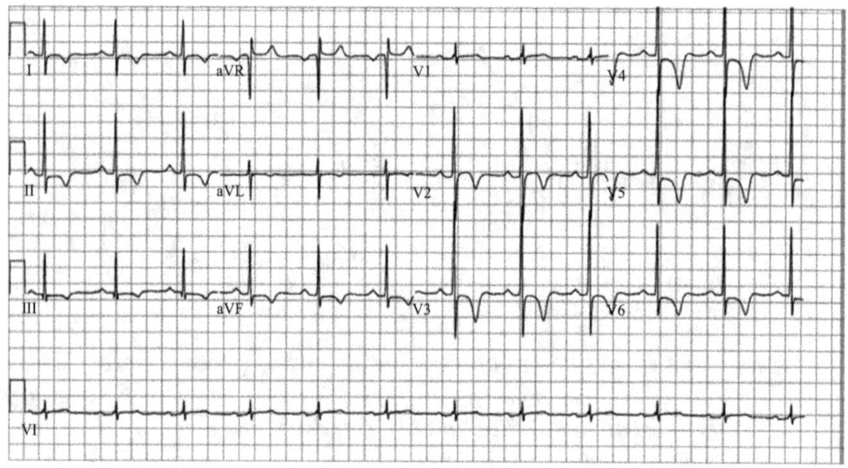

Figure 9.17 T-wave inversion in leads I, II, III, aVF, V2–V6, and ST-segment depression in leads II, aVF, V4–V6 in a 31-year-old asymptomatic professional soccer referee during cardiopulmonary exercise. The subject had no family history of SCD. Reproduced with permission, from [94]

of myocardial repolarization, which has the predisposition to re-entry arrhythmias [81]. It is known that potassium, calcium, magnesium, and metabolic acidosis are important factors for the overall electrical stability of the myocardium to ensure normal cellular excitability, impulse propagation, and regular ventricular recovery. Large amounts of or rapid removal of potassium, low calcium dialysate, intracellular magnesium overload, iron overload, and rapid bicarbonate gain (metabolic acidosis) are the factors that increase QT_c dispersion in dialysis patients [81].

9.7 Monitoring systems

All ECG, EEG, EOG, and EMG monitoring systems in market are mostly defined by the type of signal acquisition and storage system they use. Recording and acquisition of signals is done by a multichannel desktop recording, display, and monitoring system, or its handheld version, or portable data-logging device Holter monitor. A majority of systems can be classified into these two categories. Modern multichannel desktop-recording systems can connect to the physician's office from a remote location with the help of Ethernet connectivity or by wireless network to a nearby workstation for easier workflow. Further advancement has allowed inclusion of automatic triggers that can alert the nursing staff in the hospital but it is still confined to the hospital bed. The Holter monitoring system is the only commercially available multiple lead ambulatory measurement system, and it performs only data logging. A list of noteworthy commercialized ECG recording technologies has been presented in Table 9.3.

In recent years, these monitors have been equipped with event-recording functionality that allows for automatic or manual logging of the time of the event onset, while continuously recording the ECG signals. In addition to this, the Holter can be interfaced with wireless electronics to achieve ambulatory monitoring. This measure has challenges such as shorter battery life and large data volume for transmission.

Table 9.3 Commercial ECG recording platforms

Manufacturer	Product name	No. of channels	Storage	Wireless
Phillips	Page writer TC50 EKG	12 channels of ECG	USB memory stick (up to 16 GB)	No
GE Healthcare	MARS Ambulatory ECG System (SEER 12, SEER light)	3–12 channels of ECG	1 GB internal and optical DVD storage	No
medGadget	TruVue	1 channel ECG, 1Plethysomopragh	N/A	Yes
Imec	Secure Digital Input Output	1 channel ECG	16 GB	Yes
AliveCor®	AliveCor	1 channel ECG	16 GB	Yes
Phillips	EASI (Philips DigiTrak XT)	4 channel ECG	256–512 MB	No

Table 9.4 Commercial EEG, EOG, and EMG recording platforms

Manufacturer	Product name	Number of channels	Storage	Wireless
Philips/Respironics	Alice PDx	21-channel with optional ECG and EEG	1 GB SD card	No
Embla	Embletta X100	12-channel with X100 proxy	128 MB internal memory	No
Compumedics	Somte PSG	16-channel	2 GB Compact Flash	Bluetooth™
Compumedics	Siesta	32-amplified channel	Compact Flash	Siesta's Ethernet radio link
Cleveland Medical	Sleep Scout	9-channel	SD card	2.4–2.484 GHz
ResMed	ApneaLink Plus	4-channel	15 MB internal memory	No
CareFusion	Nox-T3	14-channel	1 GB SD card	No

Most of the EEG recordings for diagnostic purposes are performed in hospitals or other clinical settings. Commercially available ambulatory or out of hospital recording platforms for EEG, EOG, and EMG are primarily intended for sleep studies that need to be performed at home. A brief list is provided in Table 9.4.

From survey of the existing monitoring and diagnostics, it indicates that there are a few systems for ECG, EEG, EOG, and EMG, which facilitate at home and ambulatory monitoring. With the help of post processing of recorded data and/or real-time processing of wirelessly transmitted data at a centralized location, it is possible to perform diagnostics. However, these systems make ambulatory monitoring an added chore rather than a fully automatic process which would be possible with RPM systems. The electrodes need to be replaced after the conductive gel dries up or prolonged exposure to sweat. Applying these electrodes requires the help of a clinical technician and dry electrodes require mechanical appendage such as straps to keep them in place. In addition to this, ECG systems have wire-outs from the electrodes to the recording equipment. Unless tucked and taped as is the case of Holter monitor, the wires limit the movement of patient.

The systems listed in this section do not use or reluctantly use wireless technology. This means that the data needs to be brought to the hospital for post processing and administration of therapy. These intermittent treatments are brief and expensive supervised episodes [84,85]. Therefore, well-defined wireless communication is required in wearable monitoring systems to provide continuous and remote healthcare in an affordable way.

The textile-based nano-biosensors can be combined with flexible wireless sensors to be incorporated in the garment of daily use, e.g., a vest, brassiere, headband, or skull-cap. The sensors can combine with a wireless health diagnostic system made of embedded wireless networking devices, which can communicate with a smartphone. This enables the connection of nanosensors to cloud computing, via smartphone, to fundamentally advance remote cyber-enabled health care.

Wireless devices follow standard communication protocols to interface and connect to each other. The established communication standards are IEEE 802.11 Wireless Local Area Network (WLAN), IEEE 802.15.4 from which ZigBee™ is derived, and IEEE 802.15.6 which is Bluetooth™ and custom radio frequency (RF) transceivers. The choice between wireless protocols for development of wearable wireless health monitoring platform largely deals with complexity of connection, power consumption, and scalability of the application from single to a multiple user system.

WLAN, Wi-Fi-based systems, and Wi-Fi-Bluetooth™ integrated systems have been demonstrated [86], but Wi-Fi consumes as much as four times the power consumed by Bluetooth™ and is hence not an energy efficient solution. ZigBee™ is an attractive protocol as far as connection stability, link layer retransmit in case of data loss, and large range of network configurations are concerned, but the data rate is much lower than Bluetooth™. Moreover, ZigBee™ is not available on standard portable devices. Custom RF transceivers do not have limitations on data rate and consume less power that Bluetooth™ but additional hardware is necessary to connect to standard devices. Thus, Bluetooth™ is ideally suited for pervasive wireless healthcare devices because of the new ultra-low power connection profiles, high data rates of up to ~3 megabits per second (Mbps) and standard availability on all portable electronic devices.

The ultimate goal for all communication architectures is to provide a means of storing the large quantity of real-time patient healthcare data in a remote server and provide emergency warning mechanisms wherein abnormal data is automatically identified and a warning is sent to hospitals, physicians, and the patient as well. Based on these requirements, the solutions proposed thus far have been:

1. Custom RF transceiver sends data to a PC with a plug in RF transceiver, and the PC processes and uploads the data [87].
2. ZigBee™-based approach which is similar to custom RF in that connectivity is to a PC or a customized receiver module that has a Wi-Fi-ZigBee™ combination chip that directly uploads to a server [35]. This has the potential to scale up in setting such as home and local area network (LAN) hotspots.
3. Bluetooth™-based smartphone relay with local signal processing and data relay through Global System for Mobile Communication (GSM)/General Packet Radio Service (GPRS) [88]. This solution truly facilitates wearable health monitoring systems application.

The integration of sensors and the signal conditioning and wireless modules in wearable platforms can be achieved through three strategies.

1. Connecting the sensors through wires to a small module that incorporates signal conditioning, amplification circuits, as well as a wireless transceiver.
2. Connecting each sensor to a small wireless module that acts as an independent node –BAN
3. Integration of various sensors on a garment and using conductive threads and yarn to carry power and signals from the sensors to a small signal conditioning and wireless transceiver module.

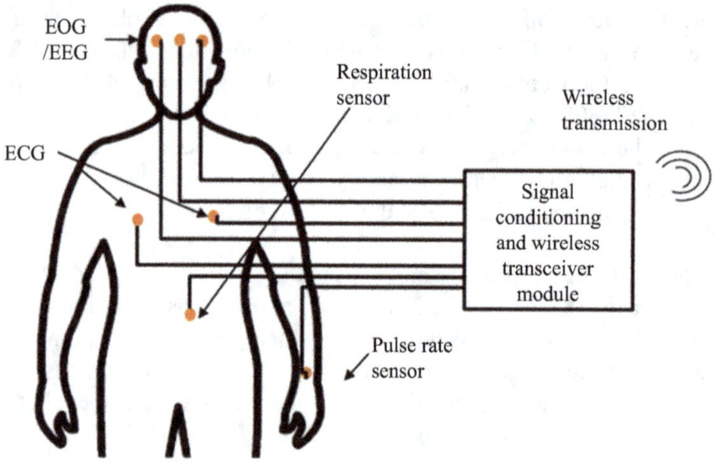

Figure 9.18 Schematic of sensors using wires to connect to a single amplifier, signal conditioning and radio frequency transceiver module

This is the first generation of portable healthcare devices in which systems for POC monitoring consisted of wired sensors that the patient needed to mount manually (Figure 9.18).

Complete health monitoring needs multiple sensors for bioelectric signals like ECG, EMG, EEG, EOG, light absorption-based sensors for pulse oximetry, pressure flow sensors for air flow measurement, and strain sensors for respiration effort measurement. Each sensor has to be wired through regular insulated wire or shielded wires in case of biopotential signals. The wireless communication module uses custom data transmission hardware and software that did not easily integrate into a patient's quotidian life–an individual had to carry a special device for data collection. This approach is not user-friendly. Although a significant improvement in terms of portability is achieved, the sensors used in a majority of these systems are not wearable. Figure 9.18 shows a schematic of these wired sensors.

BAN is a network of independent wireless nodes that span the personal space of a user (Figure 9.19). Various wireless protocols have been used to implement BAN. The state-of-the-art standards in BAN in wearable sensor systems are discussed in detail in [89]. The ultimate objective is to have independent sensors for physiological signals like ECG, EEG, EOG, EMG, pulse rate, blood oxygen saturation (pulse oximetry and photoplethysmography), temperature, and respiration that have wireless transmitters sending data to a single receiving station which may be a personal digital assistant, a tablet personal computer (PC), a PC, a smartphone, or a custom receiver unit. The various wireless communication protocols used are Bluetooth^{TM}, ZigBee^{TM}, and custom communication protocols for low-power transmission of data. The truest embodiment of this concept will have all body worn sensors (slaves) communicating with a single receiver unit (the master) and transmitting time synchronized data in real time. Figure 9.20 presents a schematic of the concept.

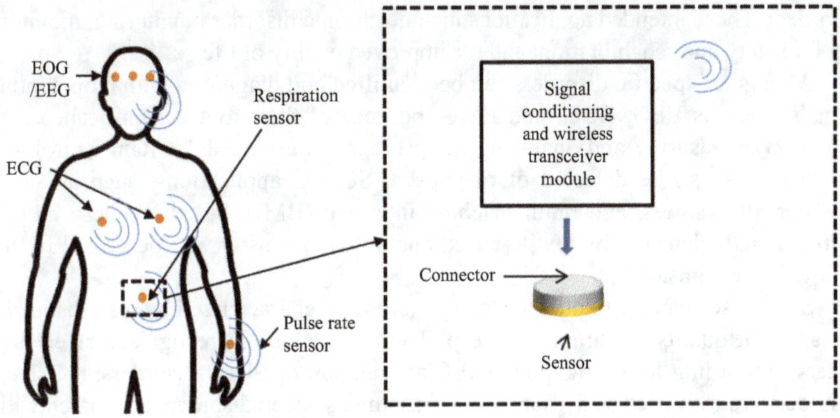

Figure 9.19 Illustration of the concept of body area network of wearable sensors incorporated in band-aid

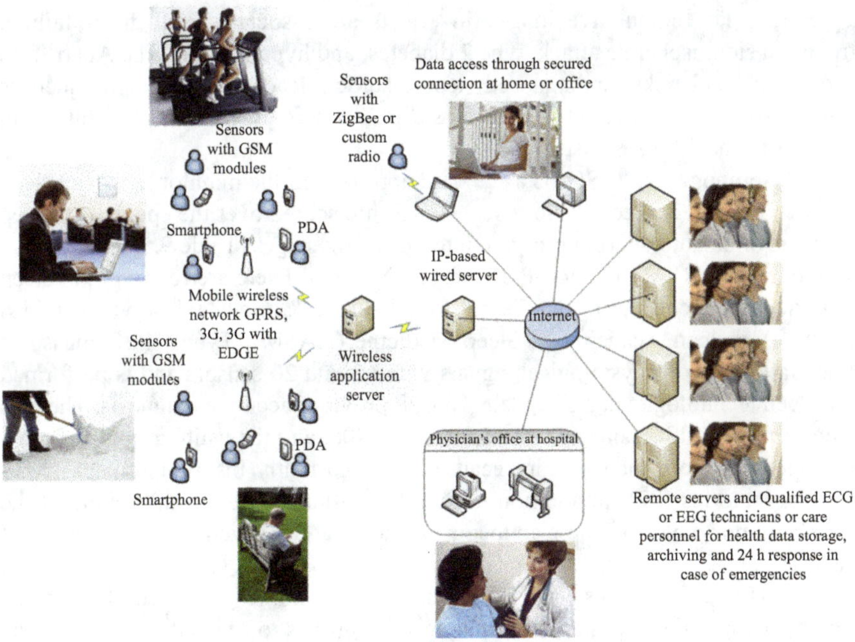

Figure 9.20 Textile-integrated sensor for relaying sensor data over the Internet to a remote server for diagnostics by a doctor

9.8 Neurological disorder monitoring by wearable wireless nano-bio-textile sensors

This section presents textile-based wearable nano-biosensor systems that can measure neurological signals and identify anomalies for diagnosis of targeted neurological disorders. These intended applications include chronic disorder monitoring, monitoring for safety and rehabilitation, and for improved quality of life.

A range of specific disorders can be identified and diagnosed with biopotential signals, such as EEG, EOG, and EMG, because of their rooted significance and origin. By measuring and analyzing these biopotential signals, certain neurological disorders can be detected or diagnosed. Several applications, such as sleep disorder, drowsiness, and brain machine interface (BMI), based on these biopotential signals detected by textile-based analog nano sensors, are described in the following sections.

Sleep disorders are related to sleep patterns and characterized by disturbance in the amount, quality, or timing of sleep. There are about 88 recognized sleep disorders. According to the International Classification of Sleep Disorders, the sleep disorders are classified as dyssomnias, parasomnias, sleep disorders associated with other disorders, and proposed sleep disorders [64]. According to the National Institutes of Health (NIH), 50 to 70 million Americans suffer from sleep disorders and sleep deprivation [90]. The short-term effects of sleep disorder include morning headaches, excessive daytime sleepiness, short-term memory loss, and depression. The cumulative long-term effects are associated with heart failure, stroke/transient ischemic attack, type 2 diabetes, and hypertension [91]. Apart from chronic physical risks, undiagnosed sleep disorders lead to serious consequences from a social perspective. Undiagnosed sleep disorders impose over $22 billion in unnecessary health care costs.

Polysomnography (PSG) is a standard approach to the monitoring of the sleep patterns. These recorded physiological signals are scored over the epoch by a sleep specialist or an auto-scoring program into one of six stages (Table 9.5)–wake, rapid eye movement (REM) and non-REM stages (NREM). There are four types of sleep study devices according to the Center for Medicare & Medicaid Services (CMS) and the American Academy of Sleep Medicine (AASM). Type I PSG measures almost all kinds of physiological signals with around 20 sensors and is performed by a sleep technologist in a sleep lab. Thus, it provides accurate diagnosis. The cost of performing a PSG ranges from $1,000 to $5,000, and the waiting time is from a few weeks to more than a year because of the currently insufficient capacity of sleep laboratories [92]. In addition to that, the patient has to spend a night at the sleep laboratory, which can be an inconvenience that may even affect the results of the test.

The Home Sleep Test (HST) is performed at home in the comfort of the patient's own home and there are no long waiting lists to schedule the exam. The HST records and saves the sleep data to the internal or external memory. Despite its convenience and cost-effectiveness, it has lack of real-time monitoring and limited amount of physiological information. Recently introduced wireless HST devices save the sleep data to a local server wirelessly. Most of them adopt Bluetooth™, ZigBee™, or their own protocols as a wireless communication method over industrial, scientific and medical (ISM) band. The wireless communication is used to build wireless personnel area network (WPAN) or wireless body area network (WBAN) to save or monitor the sleep data. However, the wireless network area is

Table 9.5 Biopotentials in sleep disorders according to the Rechtschaffen and Kales standard

Physiological characteristic	Wake stage	Stage 1 NREM	Stage 2 NREM	Stage 3 NREM	Stage 4 NREM	Stage REM
EEG	Parieto–occipital alpha waves (8–13 Hz) more than 50% of the mixed with fronto-central beta rhythms (>13 Hz)	Alpha waves decrease to less than 50% of the epoch: theta (4–8 Hz) and beta rhythms occur, may have vertex waves	Sleep spindles (approximately 12–14 Hz) and K-complexes lasting at least 0.5 s; delta waves of 2 Hz or less measuring 75 μV or more occupying <20% of the epoch	Delta waves of 2 Hz or less measuring 75 μV or more occupying 20%–50% of the epoch	Delta waves of 2 Hz or less measuring 75 μV or more occupying >50% of the epoch	Theta waves; saw-tooth waves; beta rhythms
EOG	Waking eye movement	Slow eye movement	Silence	Silence	Silence	Rapid eye movement
EMG	Elevated	Elevated but less than in awake stage	Mildly decreased	Mildly decreased	Mildly to moderately decreased	Markedly decreased to absent

limited to the personal area or body area. The real-time monitoring of the physiological signals at remote locations for sleep disorders can be achieved by combining two wireless communication standards or using the mobile communication network. The concept of combining two wireless communication standards is extending WPAN/WBAN to WAN. Therefore, one wireless communication standard is for building WPAN/WAN and the other wireless standards are for WAN.

The sections hereafter describe the wireless HST devices which support the real-time monitoring of sleep data at remote places by using a combination of the wireless communication standards.

The system mainly consists of sensors, a wireless sensor electronic unit, a wireless receiver unit, and a monitoring unit. The system measures 5 biopotential signals to evaluate the sleep stages: 2× EOG, 1× EMG, and 2× EEG. The receiver unit adopts two wireless communication standards—Wi-Fi and ZigBee™. The combination of two wireless standards in the receiver unit allows us to build WPAN and extend the WPAN to WLAN/WAN. The sensor electronic unit sends the sensed 5-channel signals 2× EOG, 1× EMG, and 2× EEG to the receiver unit through a ZigBee™ network. The receiver unit then retransmits to the monitoring unit or remote server through the Wi-Fi network. Figure 9.21 shows the data flow of the system [35].

Dry electrodes, nanostructured textile-based electrodes, or gold nanowire electrodes can be used instead of conductive gel to overcome the drawbacks of wet silver–silver chloride electrodes as mentioned previously, i.e., Figure 9.22 shows both the dry electrodes. The main functions of the wireless sensor electronic unit are data acquisition, data processing, data flow management, and data transmission. The wireless sensor electronic unit consists of amplifier module, data acquisition/process module, and ZigBee™ wireless module.

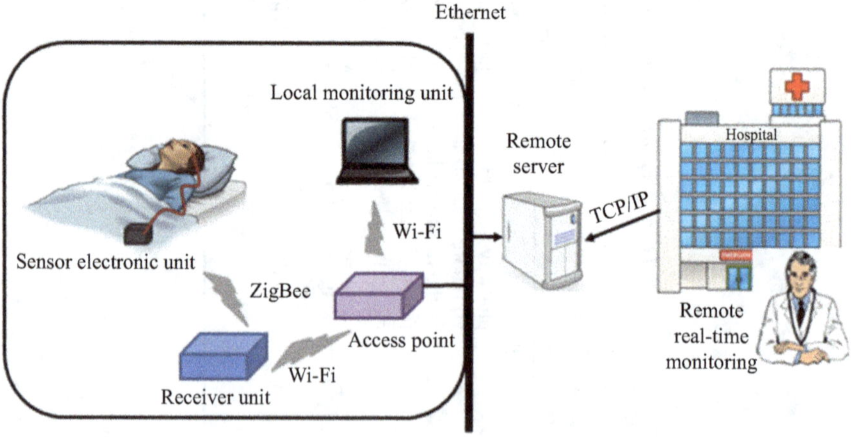

Figure 9.21 Data flow of the wireless telemedicine system for diagnosing sleep disorders

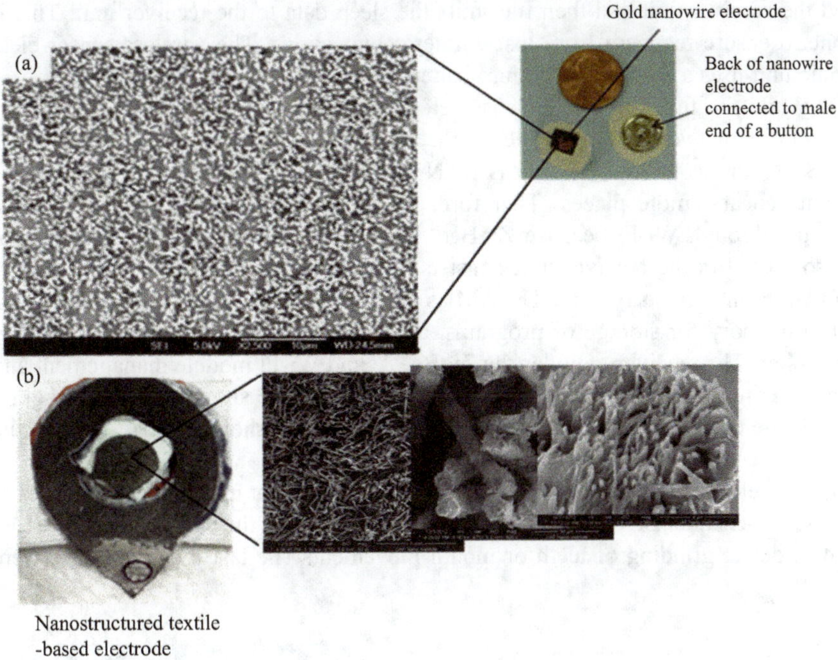

Figure 9.22 (a) Vertically aligned gold nanowire electrodes and (b) nanostructured textile nano-biosensor

The raw physiological signals from the sensors are very weak, ranging from micro-volts to millivolts, and contaminated by noises, especially 50/60 Hz power line interference. Therefore, the signals need to be amplified and filtered to improve the signal to noise ratio. The biopotential signals for sleep studies are differential signals. The body impedance and contact impedance between skin and the electrodes might vary under different skin conditions leading to impedance mismatches. Therefore, an amplifier should have high common mode rejection ratio (CMRR) and high input impedance to maintain signal integrity. The amplifier module has 5-channel for 2× EOG, 1× EMG, and 2× EEG. The gain and bandwidth of each channel was tuned according to the AASM manual [93]. The amplified signals are digitized using an analog to digital converter (ADC) on the microprocessor.

Finally, the signals are transmitted to the receiver unit by ZigBee™ module. The output of the ADC is transferred to the ZigBee™ module through universal asynchronous receive/transmit (UART) interface. The data from the microprocessor are transmitted to the wireless receiver unit through ZigBee™. The data received by the receiver unit are reconstituted as packets with sequential hex code including node number, sampling rate, the number of channel, and the data according to the defined communication protocol. These packets are saved in the data buffer temporarily and transmitted when a data request is received from a PC or a remote server. ZigBee™ interface routinely checks the wireless connection between the sensor electronic unit and the receiver unit and then transmits the sleep data to the receiver unit. This is done to ensure recovery from lost wireless connections. The wireless sensor electronic unit uses a 3.7 V poly lithium battery as a power supply (Figure 9.23).

While the function of the sensor electronic unit is to build the wireless personnel area network with ZigBee™, the role of the receiver unit is to extend the ZigBee™ network to the wireless LAN with Wi-Fi. The sleep data needs to be monitored at remote places. Therefore, the network should be extended to use Internet through Wi-Fi because ZigBee™ is not as prevalent as Wi-Fi. The processor board in the receiver unit consists of two wireless modules, the ZigBee™ and Wi-Fi module, a 16 MB SD RAM as a buffer for sensor data, a 64 MB NAND Flash memory for storage of programs, a 32-bit microcontroller, and an Ethernet controller. The receiver includes the ZigBee™ and Wi-Fi module management and the server management based on TCP/IP [35]. Figure 9.23 shows the block diagram and image of the wireless receiver unit, and Figure 9.24 shows the data flow in the wireless receiver.

Two electrodes were placed on the left lower and right upper side of the eyes to measure 2-channel EOG. One of electrodes was placed on the chin to measure EMG and to detect grinding of teeth or mouth movement. The last two electrodes were

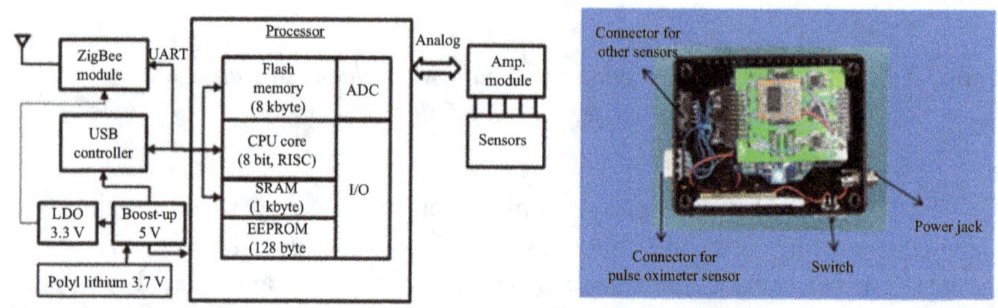

(a) (b)

Figure 9.23 (a) Block diagram and (b) image of the wireless sensor electronic unit

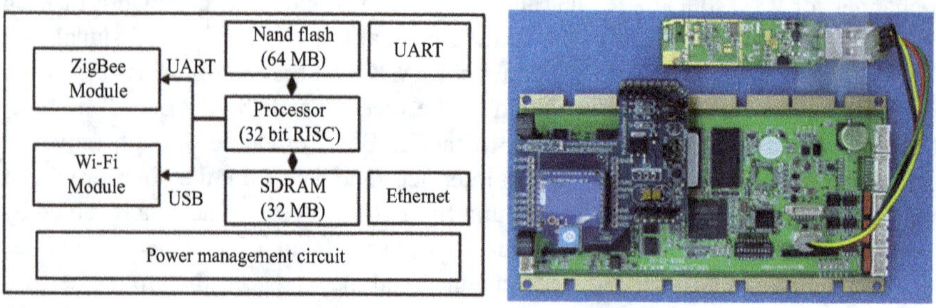

(a) (b)

Figure 9.24 Data flow in the wireless receiver unit

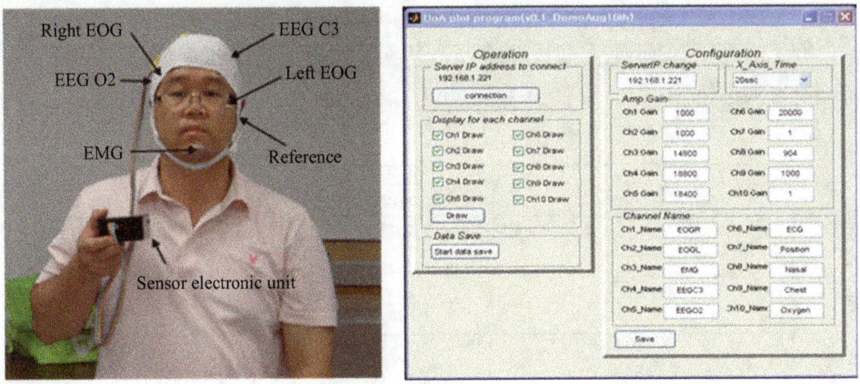

(a) (b)

Figure 9.25 Experimental test setup for the system with ZigBee and Wi-Fi: (a) image of a subject under experiment and (b) GUI of the monitoring utility program

placed on C3 and O2 position on the scalp according to the international 10–20 system. The reference electrode to measure the differential biopotential signals was placed on the left ear lobe. Figure 9.25 shows an image of a subject under sleep study with the proposed system and the GUI of the monitoring utility program developed using MATLAB® (Mathworks, Inc., Natick, MA, USA).

Figure 9.26 shows the measured and recorded biopotential signals using this system. In case of EOG, five different types of eye motions (blinking, left, right, up, and down) movement of eyeballs can be detected to emulate REM and wakeful sleep stage. The left and right EOGs show opposite polarity of slope due to the placement of the electrodes over the reference electrode. EMG reflects movement of the chin. The EEG acquisition performance can be verified by discerning between beta waves and alpha waves. Beta waves are associated with normal waking consciousness and alpha waves with wakeful relaxation with eyes closed. By opening and closing eyes alternately, the changes of brain waves can be observed. EEGs from C3 and O2 positions show that alpha waves are dominant when eyes are closed. EEG from position O2 shows stronger alpha waves than from position C3 even though both the channels of the amplifier are set to the same gain. It verifies that the alpha waves predominantly originate from the occipital lobe during wakeful relaxation with closed eyes.

In a case study on a human subject from 10 p.m. to 4 a.m., 5-channel signals were monitored and recorded in real time [35]. The EEG signals are plotted in Figures 9.27–9.29. Transition between sleep stages can be seen in the EEG data from C3 and O2. This shows that a wireless telemedicine system for diagnosing sleep disorders can address the need for the efficient extension of network to provide remote monitoring in real time. The dry electrodes and a scalable WPAN–WLAN wireless technology will foster more convenient and accurate diagnosis of sleep disorders.

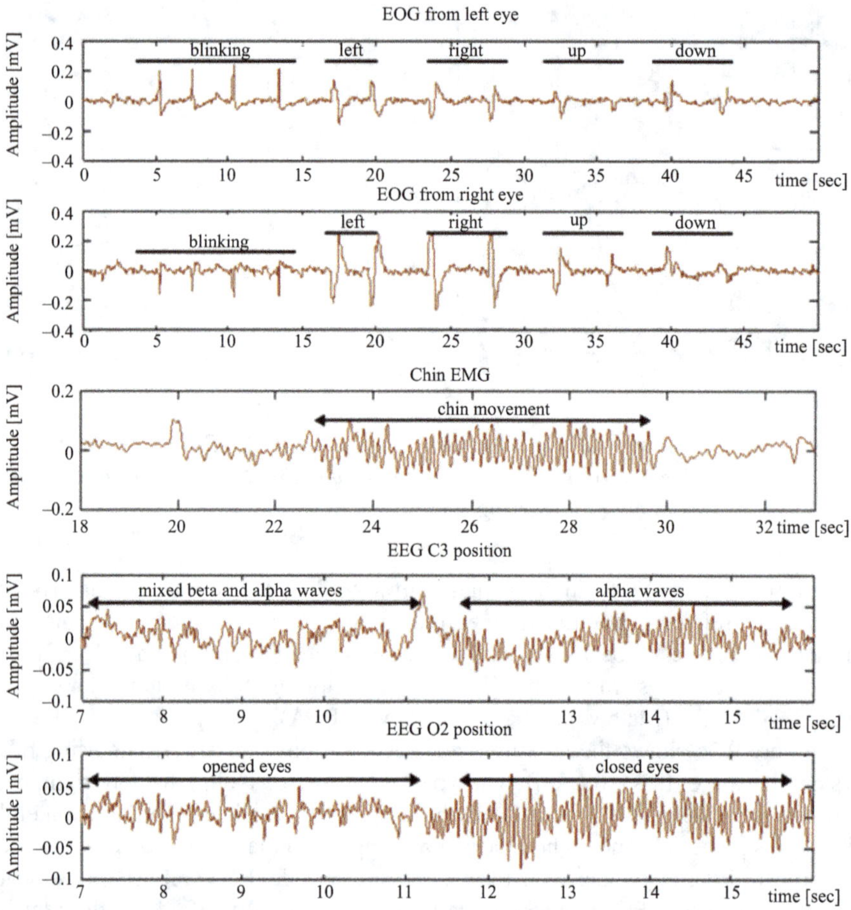

Figure 9.26 Monitored EOG, EMG, and EEG from the wireless telemedicine system for sleep study

Wireless wearable textile-based nano-biosensor systems can also be useful in improving safety. This application addresses a condition that interferes with human performance and can become an occupational hazard. Drowsiness is defined as the onset of sleep or the precursor for sleep. For instance, the transition state from wakefulness to sleep under undesired circumstances (in most cases) is defined as drowsiness. It can also be referred to as an intervening process in concentration and attention. Additionally, fatigue or drowsiness is also referred to as the psychological disability which results from various sources and origins [94]. Drowsy driving is one of the third most common reasons for road accidents. The National Highway Traffic Safety Administration (NHTSA) estimates that approximately 25% of police-reported crashes involve some form of driving inattention—the driver is

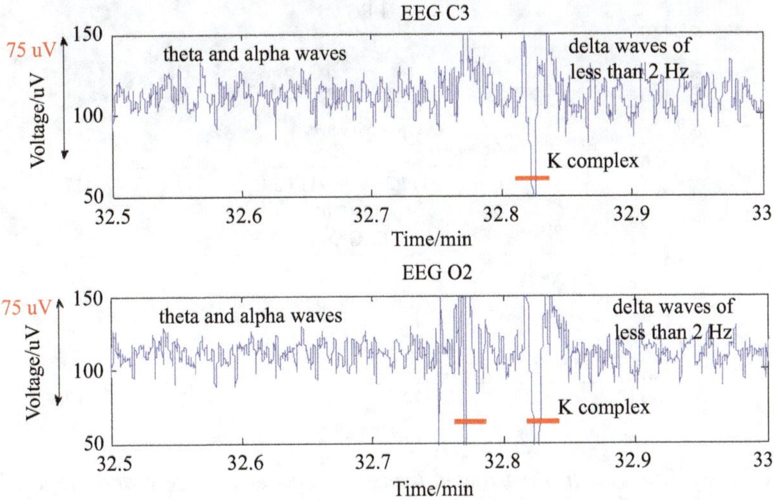

Figure 9.27 *This epoch illustrates the beginning of transition from Stage 1 to Stage 2. Vertex sharp waves are very prominent. In the beginning of the epoch, theta and alpha waves are dominant. In the middle of the epoch, there is a K-complex. At the end of the epoch, delta waves less than 2 Hz are shown*

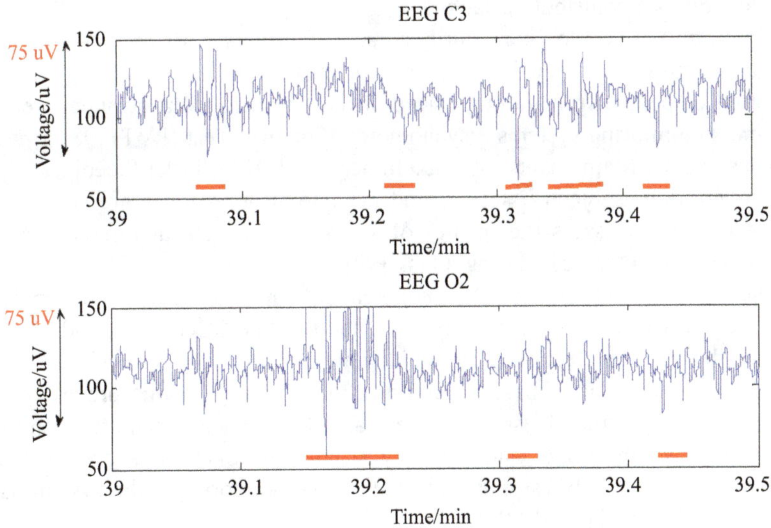

Figure 9.28 *This epoch illustrates Stage 3. Acceptable high amplitude, slow-wave activity occupies between 20% and 50%*

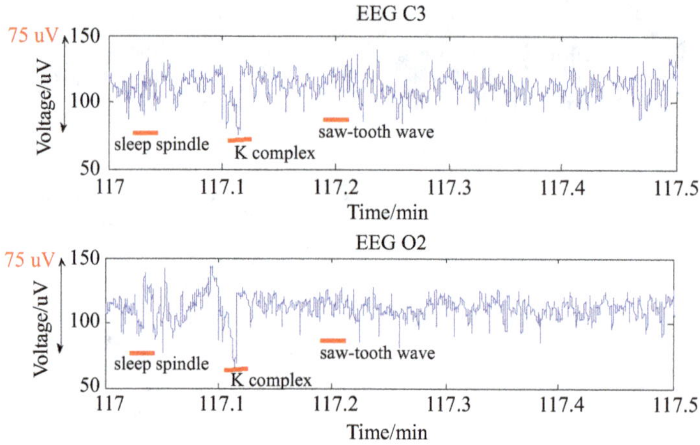

Figure 9.29 This epoch illustrates a transition between Stage 2 and Stage REM. In the beginning of the epoch, there is sleep spindle followed by a K-complex. Following the K-complex are saw-tooth waves which herald the appearance of REM. The last half of epoch shows relatively low voltage, mixed frequency

distracted, asleep or fatigued, or otherwise "lost in thought" [95,96]. The 100-Car Naturalistic Study [97] recorded the activities of 241 drivers over the course of 12–13 months and found that 78% of the crashes and 65% of near crashes had one form of inattention as a contributing factor.

Monitoring of people, like security personnel, drivers, shift-based workers, and custodial workers, who need to stay awake during the normal sleep cycle and prevention of people from falling asleep at the wheel to avert accidents requires drowsiness monitoring systems. Psychomotor Vigilance Test (PVT) [98], Epworth Sleepiness Scale, Maintenance of Wakefulness Test, Karolinska Sleepiness Scale, video monitoring, subjective report, and behavioral approaches are some of the tests performed to assess the quality of sleep from which an inference can be derived about attentiveness during a task [99].

These systems rely more on the biopotential signal approaches to discriminate sleepiness and wakefulness. Primarily, biopotentials like ECG, EOG, and EMG are collected through the electrodes by establishing a skin-electrode contact. Methods like sample entropy and phase synchronization are used in this process of discriminating sleep/wakefulness. However, a real-time and accurate monitoring system to detect drowsiness and classify sleepiness/wakefulness or drowsiness/alertness becomes highly essential for continuous monitoring if the system has to avert road collisions [100–102].

Considering the preceding parameters as the factors of a reliable and a real-time monitoring unit, a flexible headband equipped with sensors and GSM wireless transmitter that can communicate with a receiver at a remote location was built (Figure 9.30). The textile-based nano-biosensor is placed in such a way that it is in

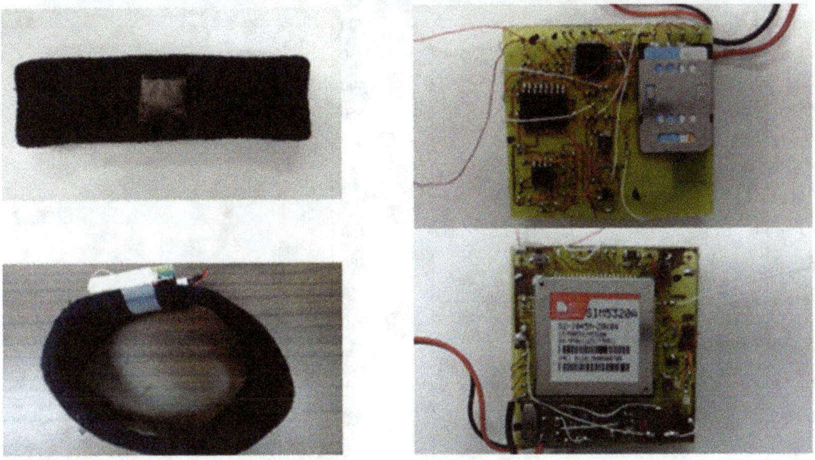

(a) (b)

Figure 9.30 (a) Front and top view of the headband and (b) front and backside of the processing module

contact with the forehead for monitoring frontal cortex of the brain. The sensor was designed to detect raw biopotential signal, which is a combination of EEG and EOG [100]. The amplifier and A/D converter circuit feeds the digitized signal to a GSM module, which can send data to a remote server for real-time monitoring and detection of a person's level of drowsiness. The usage of the GSM module provides an advantage over the communication distance as it covers longer distances as compared to Bluetooth™ and ZigBee™. This attribute is important in application for personnel with high risk or field jobs.

This system correlates the EEG and EOG signals to facilitate interlinking of eye blinks and the discrimination of alpha and beta waves. In the EEG signal, alpha wave dominance is present when a person is relaxed and beta waves are dominant when the person is attentive. Similarly, an assumption that a person falling asleep will have a slower eye blink rate than an attentive or a normal person is also made. Subsequently, the pace at which the eyes blink are measured, where the difference in pace and waveform during slow and normal blinks are analyzed. Correlation of all these features extracted from EEG and EOG is finally used to estimate the condition of the person [100].

The back end of the module consists of software at the receiver end which is programmed to extract the features from the EEG and EOG signals. Specific features from which sleep can be detected are extracted in the software and it pops out a warning signal when the driver falls asleep. The rate of eye blinks from the EOG signal and the peak values from autoregressive power spectral density (AR-PSD) of the brain signal in the alpha frequency band are correlated to estimate the condition of the person. The receiver side software is programmed to perform feature extraction, AR-PSD of the signal, and improve the signal quality to adapt the processing strategy and parametric analysis. These separate results are then correlated in the

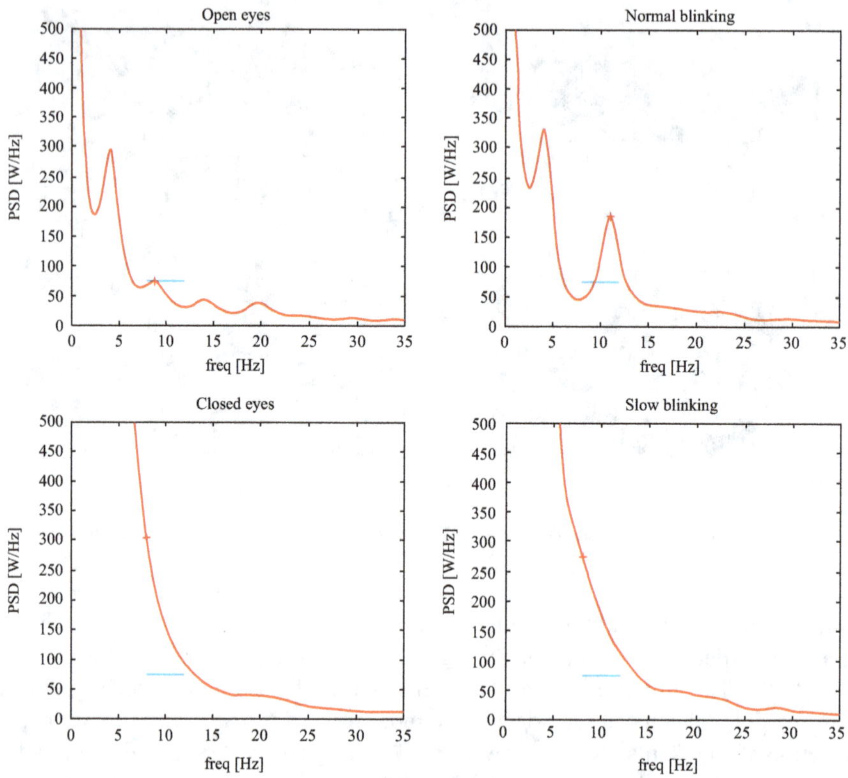

Figure 9.31 AR-PSD values depicting open eyes, normal blinking, closed eyes, and slow blinking [112]

main server program of the software according to the decision criteria, which is based on the initial set of experiments [100].

Initial experiments, as shown in Figure 9.31, are conducted to get ARPSD values fixing the decision criteria. It shows that the measured values clearly indicate that closed eyes have the highest value and open eyes has the lowest value. Similarly, Figure 9.32 shows the average values obtained over a series of test cycles.

Upon comparison of test results of the flexible headband system on a real-time driving simulator and PVT, it was observed that an increase in the number of hours of sleep drastically worsened the reaction time of each subject in any given cycle of PVT test. Similarly, the number of times drowsiness was detected in the period of the PVT test proves that the flexible headband system detects more number of occurrences of drowsiness as the number of hours of sleep deprivation increased. Therefore, a correlation between the PVT and flexible headband system was derived. Hence, it was evident that the flexible headband system equipped with textile-based nanosensor provided effective drowsiness and alertness discrimination. It is also evident from the analysis that the number of drowsiness occurrences increased when the reaction time of the subject increased [110].

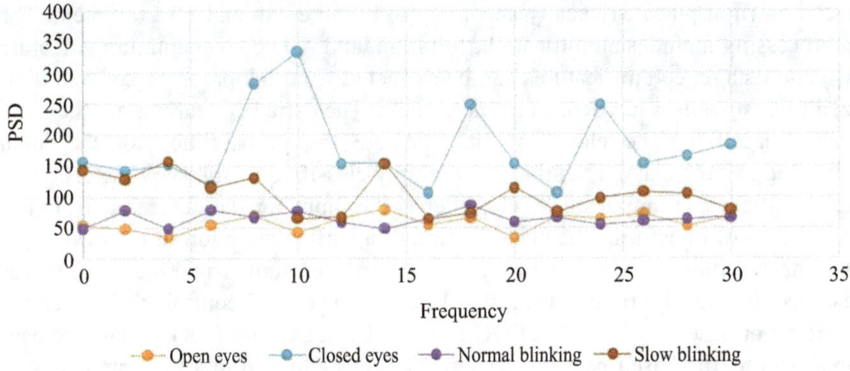

Figure 9.32 Cumulative results of (a) peaks observed during normal blinks, closed eye, and slow blinks from EOG signal and (b) AR-PSD values during open eye, closed eye, normal blinks, and slow blinks from the EEG signal [110]

Wearable nano-biosensor systems can be used in potential applications for interfacing human mind with devices to operate them with thought and other neurological signals as cues. A BMI is a communication system that translates human's thought into signals to control devices such as a computer application or a neuroprosthesis [103]. A BMI enables the brain to communicate with the external world by deciphering the brain's activity. Hence, the assistive devices or systems using a BMI improve the quality of life in disabled people. In addition, a BMI has been proposed to replace humans with robots in the performance of dangerous tasks like explosives handling/diffusing, hazardous materials handling, firefighting, etc.

Previous research projects have demonstrated the feasibility of BMI with the invasive method by implanting the intracranial electrodes in the motor cortex of monkeys [104–108]. Though an invasive BMI can use good quality of brain signals, it is expensive and the implanting surgery may lead to undesirable side effects. A noninvasive BMI using EEG signals are preferable for humans. EEG signals represent the electrical activity of millions of neurons in the brain. EEG has various properties and it can be used as a basis for a BMI: rhythmic brain activity, event-related potentials, event-related desynchronization, and event-related synchronization [109]. Different rhythmic brain activities are shown depending on the level of consciousness. The brain waves are classified according to the frequency band: delta (0.5–4 Hz), theta (4–8 Hz), alpha (8–13 Hz), beta (13–30 Hz), and gamma (30–100 Hz). These rhythms are affected by different actions and thoughts, for example, the thinking of movement attenuates or changes a typical brain rhythm. The fact that the thoughts affect the brain rhythms connote the rhythmic brain activities that can be used for the BMI.

In general, implementation of a BMI based on EEG signals requires measurement of EEG, preprocessing, feature extraction, classification, and device control. To measure EEG signals, textile-based nano-biosensor electrodes are placed on right places, typically according to the international 10–20 system. The preprocessing includes amplification, filtration, and A/D conversion. In the feature extraction stage, certain features can be extracted from the preprocessed and digitized EEG signals in frequency or time domain. The extracted features are the input of the classifier. The classifier can calculate the probabilities for the input belonging to each class. The signals are simply classified by threshold detection by the classifier. The output of the classifier is the input for the device control. The device control transforms the classification to a particular action of the device.

The wearable wireless brain-machine interface, shown in Figure 9.33, was designed to control a robot wirelessly [110,111]. The BMI controls the movement of the robot based on EEG and EOG signals. The EEG and EOG signals are captured with textile-based nano-biosensor electrodes attached to a headband at positions across the forehead. This electrode placement is not a conventional placement meant for EOG measurements. It was chosen because the signal response to the left and right movement of the eyes was strong enough to use as a command signal. The classified rhythmic brain waves are used to control the acceleration and deceleration or stopping of the robot by setting the threshold. The classified EOG signals from the left and right movements of eye balls control the left and right direction changes of the robot.

The wireless transmitter module consists of a high gain biopotential amplifier to enhance the weak microvolt scale biopotential signals, a microcontroller that performs the analog to digital conversion, and a Bluetooth™ module (Figure 9.34).

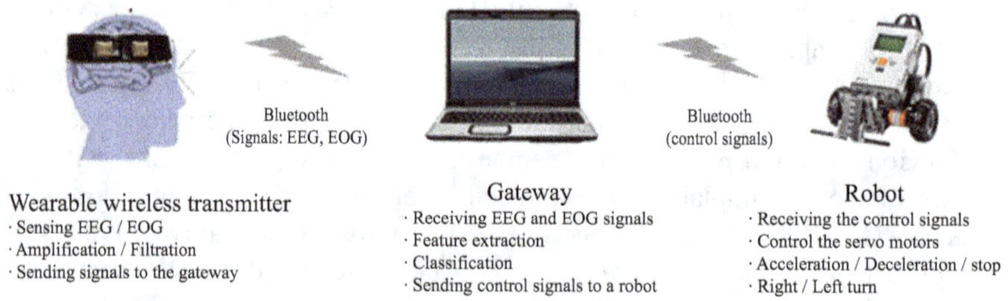

Figure 9.33 Data flow of the wearable brain machine interface

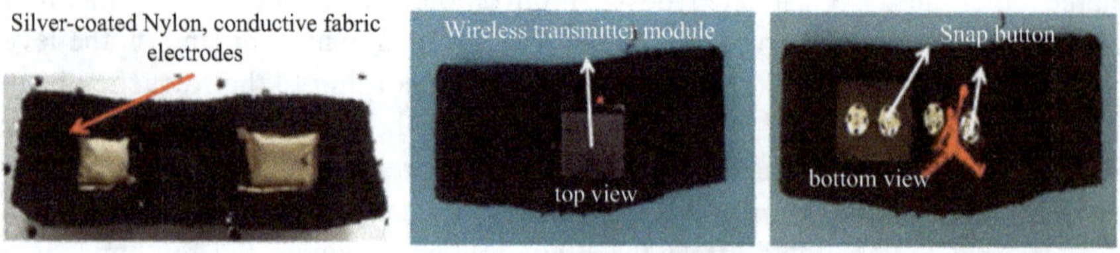

Figure 9.34 Images of the wearable sensing transmitter: (a) conductive textile electrode sensors on the headband; (b) top view of the wearable sensing transmitter snapped on the headband; and (c) snap button male and female between module and the headband

The power source is a 250 mAh poly-lithium battery. The biopotential amplifier is a three-stage amplifier with an overall gain of 65.44 dB and a 3 dB bandwidth of 1.45–40 Hz set using active RC and Butterworth filters [110,111].

The nano-biosensor electrode system measures the differential resting potential of the retina according to the eye movement. If the eye is moved from the center position toward the electrode, this electrode sees the positive side of the retina and the opposite electrode sees the negative side of the retina. With this measurement method, all kinds of the eye movements are measured–left/right, up/down, and blinking movement. In this BMI system, left/right eye motion is used to take decision of a robot turning left or right, hence the other up/down and blinking motions are considered as noise. Thresholds of amplitude and time duration are used to remove the blinking and up/down motion. After removing the undesired eye motion signals, only the left and right waveforms remain. The two waveforms should be classified as left and right motion to send turning commands to the robot. The classification between two eye motions is done by investigating the polarity of waveforms. Left motion shows positive polarity and right motion shows negative polarity. Figure 9.35 shows the classification processes of left/right eye motion. The left and right motion signals are filtered with 5~15 Hz bandwidth to achieve sharp waveforms to process further. After filtering, integration is performed to make the signals smooth. The floor noises are removed by setting the adaptive amplitude threshold. The amplitude threshold is defined by 75% of peak values of waveform. The cleaned waveforms by removing noise floor are used to decide the command signals. As shown in Figure 9.35(e), finally, left eye motion coded to the pulse amplitude 1 and followed by amplitude 5 pulse. Right eye motion is coded to the amplitude 5 pulse and followed by amplitude 1 pulse.

Acceleration and deceleration of the robot are controlled by EEG signals. Among several brain waves, the alpha and beta waves are focused because they represent the mental states of attention and relaxation, respectively. The received EEG signals are transformed into frequency domain through Fast Fourier Transform to investigate the brain waves. Each rhythmic brain wave is defined by the frequency band, and the mental states can be deduced by considering the amplitude change of the frequency band. To define the attention and relaxation level, the threshold are set based on the ratio between the sum of AR power spectral density (PSD) of alpha and that of beta waves. Instead of taking the sum of AR-PSD value, taking the peak AR-PSD value over the frequency band was found to be more robust in noisy conditions. In an experiment to validate the EEG feature extraction algorithm, a subject was asked to be in attention or relaxation mode by an experimenter and the calculated attention level was compared for accuracy with the actual mental states of a subject. The subject was asked to solve a series of quantitative problems to achieve the attentive state. Figure 9.36 shows the experiment result. As shown in the result, when the ratio of peak PSD value of low beta and alpha was used, the accuracy rate was about 95%.

Based on the EOG and EEG feature extraction and classification algorithms, a whole system was implemented to control the robot. The speed of a robot was controlled by attention and relaxation level extracted from EEG signals in the

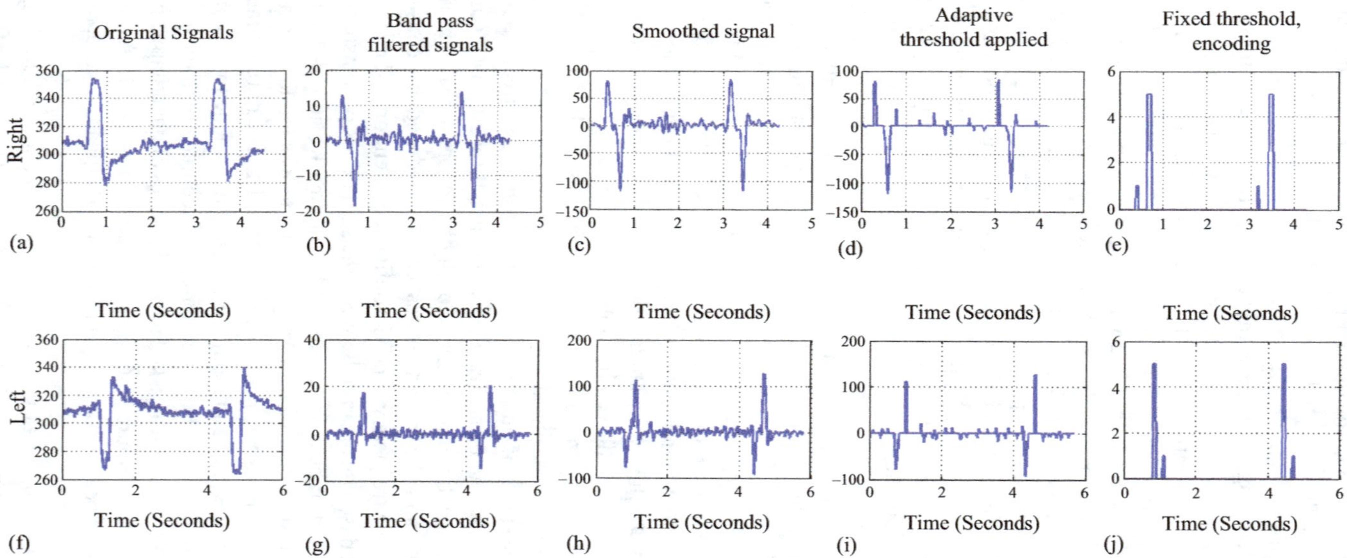

Figure 9.35 Classification process of EOG signals: (a) raw left and right EOG signals; (b) integrated signals; (c) filtered signals; (d) noise-removed signals; and (e) classified signals

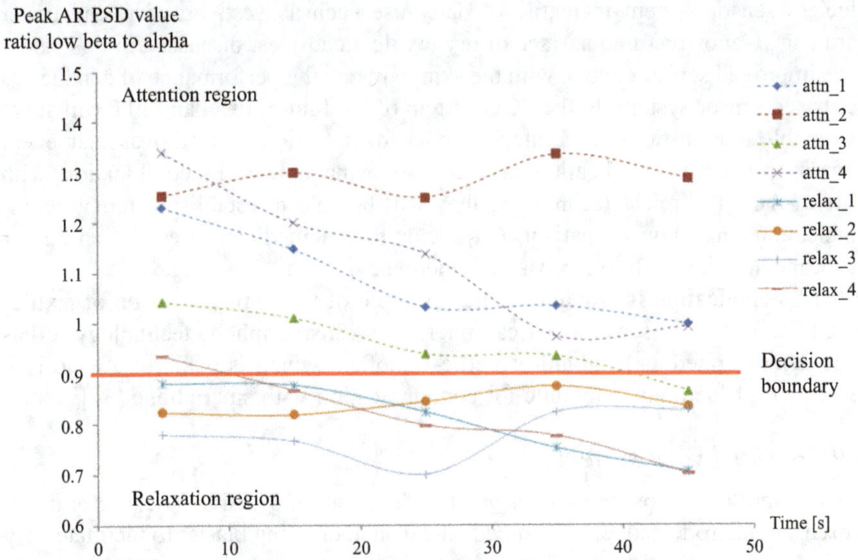

Figure 9.36 Experiment results of the ratio of peak AR-PSD value of low beta and alpha waves

straight section, and the changing direction of a robot was controlled by the left and right of eye motions in the curved section.

9.9 Cardiovascular health monitoring

Coronary heart disease, cardiovascular diseases, and strokes are the leading causes of mortality in the United States as well as around the world. According to statistics by the Centers of Disease Control and Prevention (CDC), cardiovascular or cerebrovascular disease was the most prominent cause of death in the United States. It was the cause of death in more cases than cancer, accidental injury, diabetes, and infectious diseases [84]. Hence, an average American is more likely to die of heart attack or stroke than cancer, accidental injury, diabetes, or deadly infections, such as HIV.

The POC techniques suitable for ambulatory or out of hospital monitoring are non-invasive electrophysiological techniques: ECG and EIT. Both the techniques measure electrical activity and properties of cardiovascular system for long periods of time to detect abnormality. In current medical practice, they are considered preliminary testing techniques. Any abnormalities detected in these tests are confirmed by in-hospital techniques: echocardiogram, cardiac catheterization, computed tomography (CT) scan, magnetic resonance imaging (MRI).

In the present form, ECG and EIT techniques use bulky instrumentation and constrictive conductive gel electrode-based sensor systems that are not suitable for very long monitoring. Smart textile technology incorporates textile-based dry and glueless sensor system in textile of daily use such as vest, brassier, and under garments. Nanostructured surface of dry textile electrodes, discussed in this study, have improved sensor contact with the skin to match the performance of conductive gel-based sensor system. In the POC system of the future, ECG and EIT will serve as reliable diagnostic and monitoring tools for detection of cardiovascular event towards preventive medication and disease management. In combination with state-of-the-art wireless technology, they will be able to establish a remote communication link between patient and doctor for telemedicine prescription and/or quick medical intervention in case of emergency.

This application is a quintessential example of the implementation of textile-based ECG nano-biosensors that can interface with smartphone technology. It has been demonstrated as band-aid named e-Nanoflex, which is made of a nano-biosensor with Bluetooth™ module for communication with smartphone [36].

9.9.1 Hardware system

The e-Nanoflex prototype was designed for acquisition of ECG. The pair of nano-biosensor electrode pads was mounted on a wound dressing plaster to facilitate easy testing. The application was demonstrated with an electrode pair made of gold nanowire electrode on flexible titanium foil as reported in [112] and textile-based nano-biosensors described in Section 9.2 (Figure 9.37).

The ECG is a differential signal, i.e., the signal is perceived as the difference in potential between two points on the skin wherein one of the electrodes acts as the reference for the other. An amplifier was used to enhance the signal strength because the signal strength of ECG, at the level of the skin, is of the order of hundreds of microvolts. Key attributes of amplifier design were large CMRR, small input offset voltage, and low input power consumption. The amplifier is designed in multiple stages. The instrumentation amplifier provides the first high impedance stage with a gain of 10 to avoid saturation of the saturation due impedance mismatch. Analog

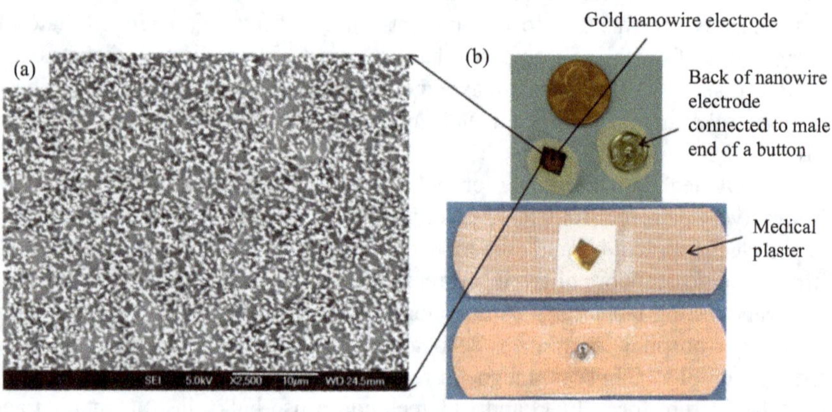

Figure 9.37 (a) Scanning electron microscope image of vertical gold nanowires and (b) the gold nanowire electrode packaged in a medical plaster

filters with a pass band between 0.2 and 70 Hz make sure the mismatch is not seen by the later high gain stages. Two more stages of non-inverting amplifiers are used to further improve the signal quality. The amplified analog signal is then digitized by the ADC on the microprocessor at a fixed sampling rate of 200 Hz. The microprocessor then communicates with the Bluetooth™ module using the UART interface. The Bluetooth™ module continuously sends the data from the microprocessor to the data logging unit and the received data is then stored or processed by the host microprocessor of the smartphone. This application can be extended to a personal computer with a Bluetooth™ receiver [113].

The type of connection between Bluetooth™ devices, also known as profile, is chosen based on the kind of data that needs to be sent from one device to another. For this application, the data is a series of digits which are the output from the microprocessor. The serial port profile (SPP) was chosen because it supports the continuous transmission of data. The module basically broadcasts the SPP service, and any Bluetooth™ device, like a phone or a Bluetooth™-enabled PC, that comes within range can securely connect to this device on that profile and start receiving the data if the correct passkey is entered.

One of the drawbacks of the SPP is that, in case the data transmitted is not received by the data logging device, the data is not retransmitted automatically. This problem has been considered in detail in [114]. The additional retransmit feature has to be implemented in software to improve the reliability of an SPP connection. Alternatively, the introduction of a special profile for health devices, called the health devices profile, by the Bluetooth™ Special Interest Group is expected to be more robust and suitable for this application than SPP.

9.9.2 ECG signal acquisition

This band-aid-type e-Nanoflex can be used for measurement of single ECG leads. Lead I ECG signal obtained from the sensor was plotted on a smartphone in real time as shown in Figure 9.38. The frequency content of ECG is between 0.2 and 70 Hz. The sampling rate was fixed at 200 Hz to satisfy the Nyquist–Shannon sampling theorem. A digital band pass filter with a bandwidth of 0.2 to 70 Hz was implemented on the data logging device to improve the signal to noise ratio. The three-limb lead signals were obtained from commercial wet Ag/AgCl electrodes, *Moore Medical LLC.*, and from gold nanowire electrode as shown in Figure 9.39. The signals from the two electrodes are observably similar.

Band-aid-type e-Nanoflex can be placed at different positions relative to the heart to get different perspectives of the heart (Figure 9.40). Different positions enhance individual components of the ECG waveform, thus making each position relevant to the diagnosis of ECG abnormalities mention in Section 9.6. The band-aid sensor system can potentially be used for monitoring QT interval dispersion, which is prevalent in dialysis patients at risk of SCD. The system is simple to apply and does not interfere with the dialysis setup. Its adaptability to Bluetooth™ for smartphone or PC makes it applicable to in hospital dialysis as well as POC for ambulatory dialysis.

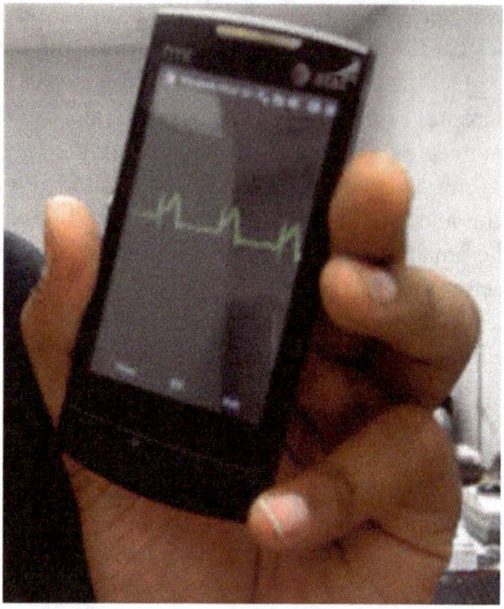

Figure 9.38 Lead I ECG data displayed in real time on a smartphone

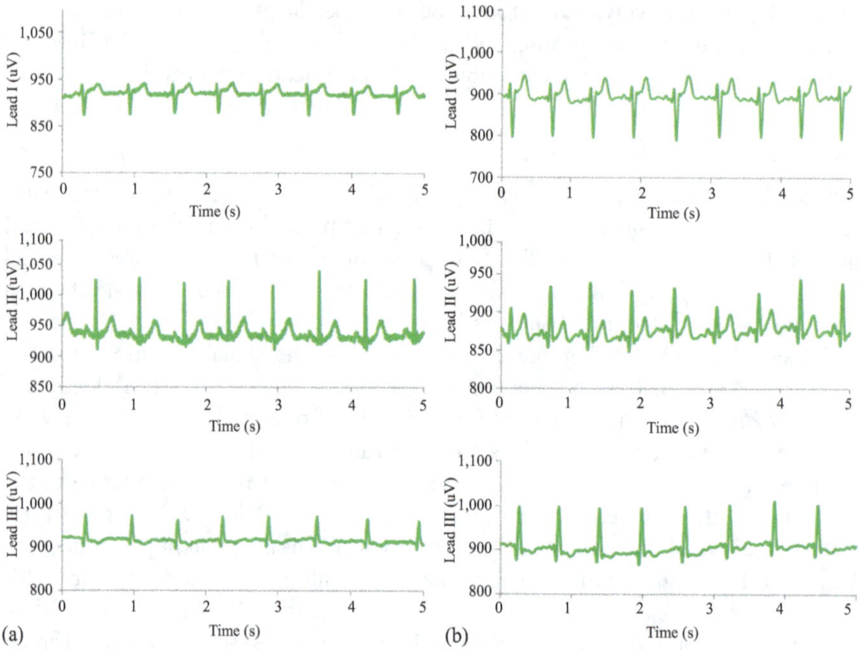

Figure 9.39 Three-limb lead ECG acquired from e-Nanoflex sensor using (a) commercial Ag/AgCl electrode and (b) gold nanowire electrode

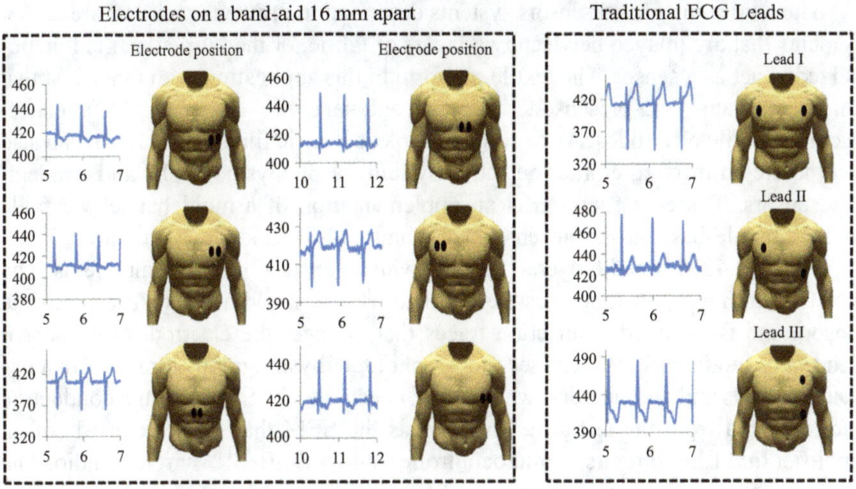

Figure 9.40 Different positions of band-aid-type e-Nanoflex and the ECG signal associated with it

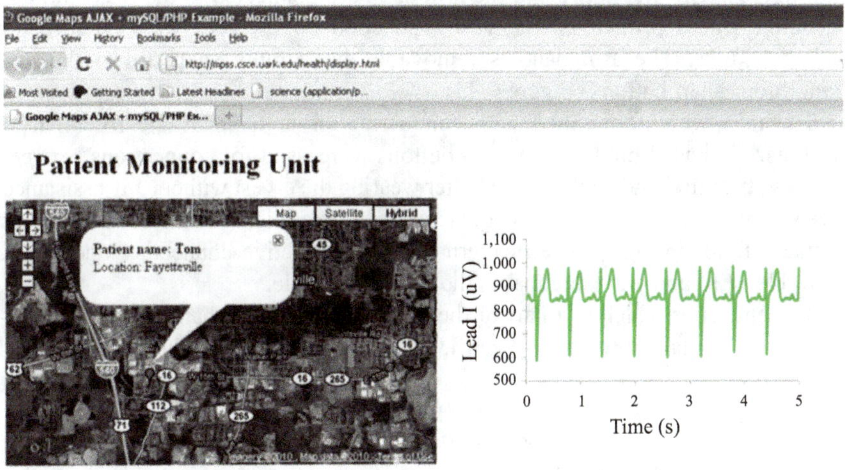

Figure 9.41 Snapshot of ECG data stored from the user and the corresponding location from where the data was sent

The data from the e-Nanoflex sensors can also be sent to a remote server through the 3G network. The data was tagged with the current GPS location and the time. In case of an emergency, a message with recorded data, current location and time can be sent to the emergency response team. The snapshot in Figure 9.41 shows the location of the user and also the data received from the sensor as recovered from the server.

The textile-integrated sensors systems described in previous sections, are sensor elements that are inlayed between two layers of fabric, or the fabric itself is functionalized to act as a sensor. The textile platform in this application is an inner vest that can incorporate nano-biosensors such as gold nanowire electrode [112], nanostructured textile electrode [54], or composite piezoelectric films [115,116]. It can also incorporate an infrared emitter-detection system for plethysmography and temperature sensors. The e-vest system is an implementation of a multichannel wearable wireless textile-based nano-biosensor that monitors ECG and blood pressure.

Figure 9.42 shows the overall system, which consists of four components. The first is a compression inner vest referred to as e-vest, with the textile electrode sensors and the printed connection traces that connect the electrodes to a sensor electronics module (SEM). The second is a photoplethysmography arm band that has Near IR LEDs and photodiodes, which are connected to the SEM through conductive traces printed from the left arm. The third is the SEM that further consists of an amplifier and filter circuits, a microcontroller, and a ZigBee^{TM} wireless radio. The last is a software program running on a PC, which receives and plots the incoming data from the person wearing it.

The photoplethysmography arm band consists of two arrays of Near IR Photodiodes and a central array of three photodiodes. This assembly was described in detail in [29]. The electrodes and the arm band are connected to the SEM through conductive traces printed on the vest. The conductive traces were made using conductive inks. The arm band is removable and simply snaps on to the vest through four snap buttons (Figure 9.43).

The inks were formulated with conductive silver nanoparticle fillers in an elastic acrylic-based binder. The snap buttons were used for connections between the e-vest, the arm band, and the SEM after wearing the e-vest without any assistance. Screen printing, which is textile manufacturing compatible, was used to transfer the ink traces following the desired pattern for the conductive connecting traces. The removable SEM and arm band makes the e-vest washable.

The amplifier which is a part of the SEM consists of 4 channels. Three channels for the bipolar limb leads—Lead I, Lead II, and Lead III. The fourth channel

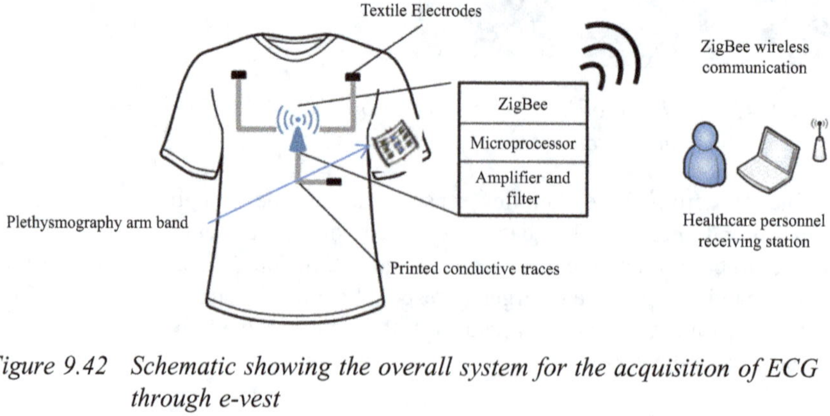

Figure 9.42 Schematic showing the overall system for the acquisition of ECG through e-vest

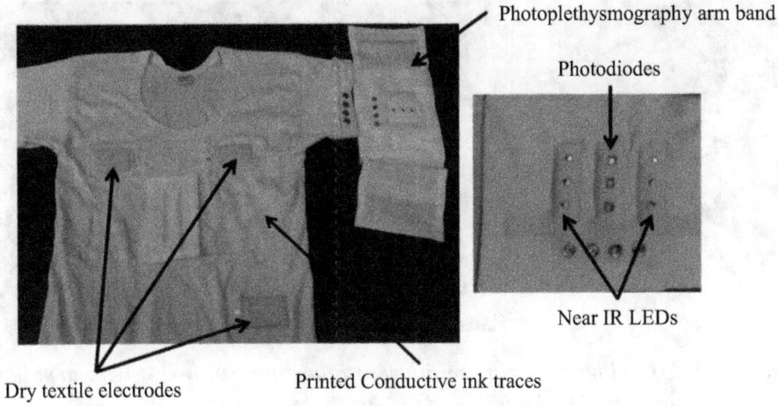

Figure 9.43 e-vest and the photoplethysmography arm band

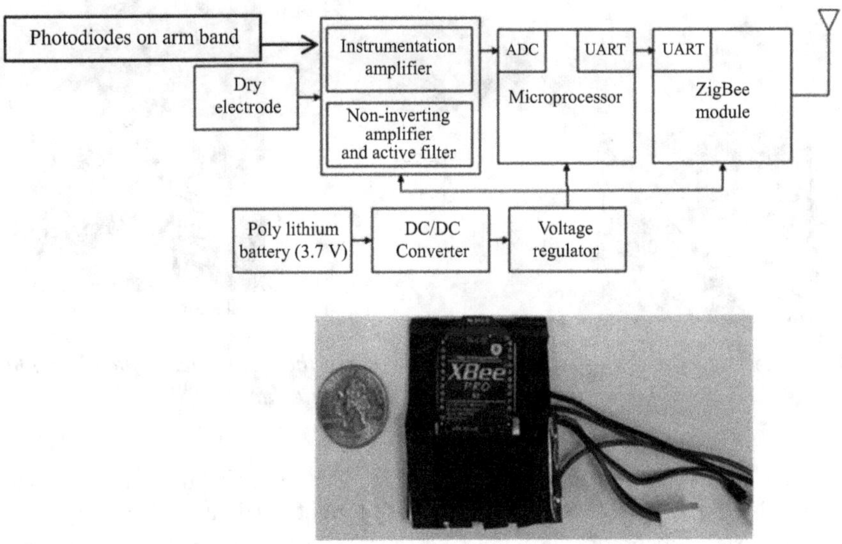

Figure 9.44 (a) Schematic of the SEM and (b) actual SEM used in e-vest system

amplified the potential difference across the photodiode, which detects the reflected IR waves from the brachial artery. The amplifiers used in the SEM had a pass band of 0.2 to 70 Hz and a mid-band gain of 50 dB for the three ECG channels. The gain was increased to 55 dB for the photoplethysmography sensors for a band of 0.2–15 Hz. The amplified signals from the amplifier are digitized using the onboard micro-controller for transmission (Figure 9.44).

The choice of ZigBee^{TM} radio module was motivated because of two desired functions. Firstly, it has to support data rates higher than 9,600 bps, because 4 channels of digitized ECG and BP signal has to be transmitted in real time. Secondly, it provides

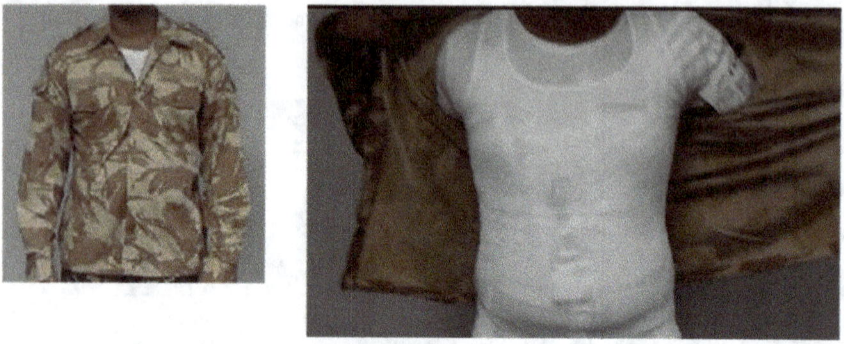

Figure 9.45 *(a) Subject wearing military fatigues and (b) e-vest with arm band worn under the military fatigues*

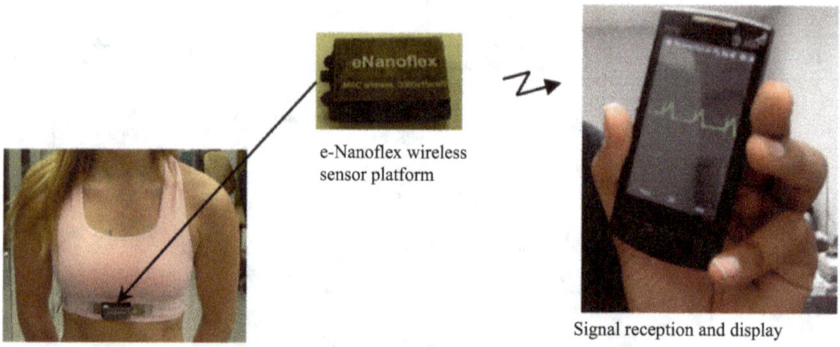

Figure 9.46 *Depiction of the e-bra worn by one of the test subjects, the eNanoflex module and the smartphone display interface*

communication ranges as high as possible for applications in sports, military expedition, and high-risk work environments such as firefighters (Figure 9.45).

The systems incorporated in inner vest for men can also be integrated in inner garment for women such as a brassier. The various sensors listed in the previous section can be incorporated in the e-bra and the signals from the sensors brought to the eNanoflex [36] module through printed conductive traces or conductive threads. Figure 9.46 shows a picture of the e-bra, the eNanoflex module used for data acquisition and wireless transmission, and the simple signal display interface that plots the data received from the eNanoflex module.

The data acquired by the e-vest can be transmitted wirelessly to a PC. The data received by the PC was then filtered using an adaptive filter algorithm to minimize the effect of motion on the ECG signal baseline [117]. The data acquisition and adaptive filter was developed using MATLAB. However, the same can be achieved on JAVA platform and can be deployed on a smartphone. Figure 9.47(a) shows the original 3 ECG signals Lead I, Lead II, and Lead III. It also plots the pulse

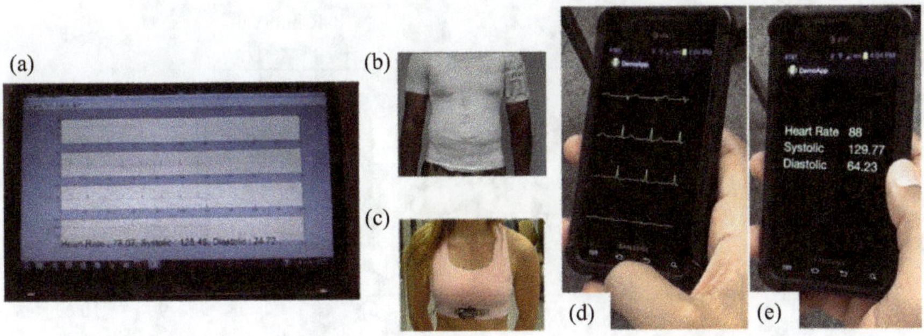

Figure 9.47 *(a) Real-time data plotted on the laptop; (b) the e-vest worn by the test subject; and (c) the e-bra worn by the test subject*

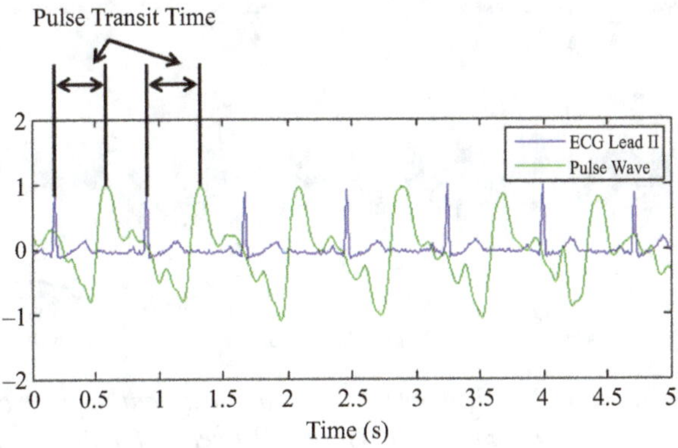

Figure 9.48 *Simultaneous plot of ECG Lead II and pulse wave depicting the derivation of PTT*

waveform, the heart rate, and the estimated systolic and diastolic blood pressure. Figure 9.47(b) shows the actual e-vest, arm band, and SEM module. The derived pulse transit time PTT values (Figure 9.48) are then used to estimate the systolic and diastolic blood pressure values based on the calibration equations previously obtained in [29]. Other sensor systems can be incorporated to develop wearable applications to monitor respiration, temperature, and blood oxygen level.

Post processing of ECG can also calculate heart rate variability (HRV), which is a prognostic and diagnostic tool. HRV is described as the sequence formed by concatenating the difference in heart rate between consecutive beats (Figure 9.49). It is calculated as the inverse of the difference in the intervals between consecutive R-peaks. The R-peak detection algorithm used for calculation of RR interval (RRI) was as given in [118]. Autoregressive (AR) power spectrum estimation technique was used to obtain the PSD plot of the RRI sequence. The characteristic LF and HF peaks were observed [119]. Figure 9.50 shows a plot of the RR interval series plotted against beats and the AR-PSD computed from the RRI series for the standing-up case. Figure 9.51 shows the same for supine ECG. AR-PSDs in both figure show a

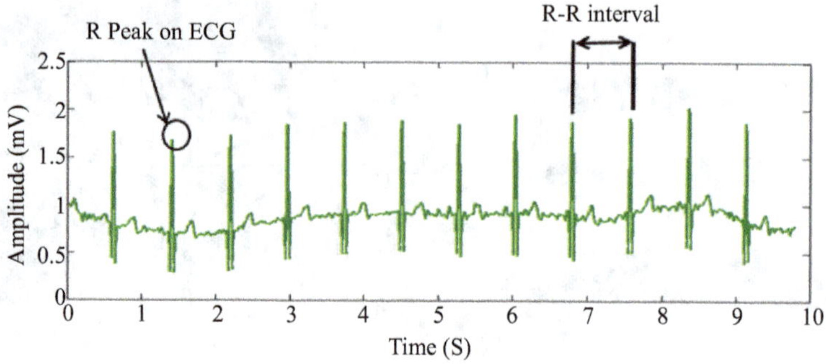

Figure 9.49 R-R interval calculations from ECG

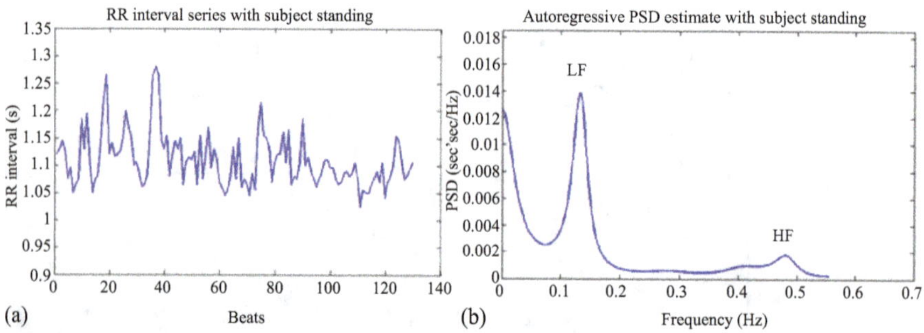

Figure 9.50 (a) Plot of the RR interval series against beat number and (b) plot of the AR-PSD computed from the RRI series for the standing case

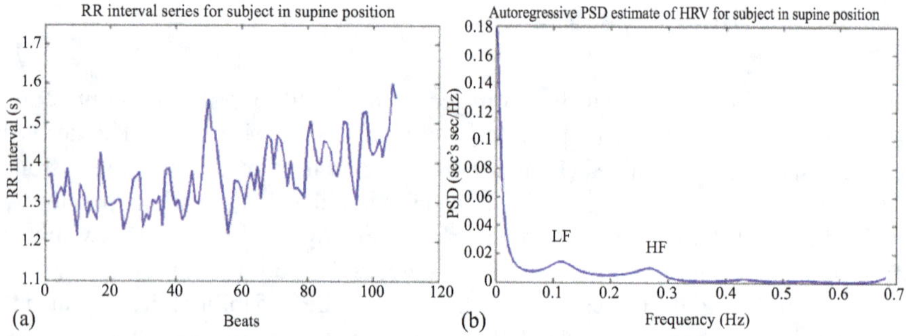

Figure 9.51 (a) Plot of the RR interval series against beat number and (b) plot of the AR-PSD computed from the RRI series for supine position

classic shift in the power distribution between LF and HF components with respect to total power. Thus, these implementations of e-vest and e-bra systems can be used for tracking of chronic conditions related to autonomous nervous regulation of cardiac activity.

Continuous multiple lead ECG monitoring can be used for detection of T-wave inversion, which is indicative of change in ventricular repolarization as mentioned previously in Section 9.6. Automated post processing of ECG by algorithm for detection of T-wave inversion can serve as an alarm system that will trigger a subroutine to initiate ECG signal relay through a remote server to doctor's office for diagnosis.

9.10 Biofeedback system for therapeutics

Since its introduction, smart wearable systems have been associated more with health monitoring than therapeutics. These systems are by nature dependent on biometric inputs for analysis and response, primarily to inform of audio-visual cues and/or suggestions. So while the systems described previously could only suggest medication or initiate medical intervention by doctors in remote locations, there have been proof-of-concept studies done to show that these systems can be of help in initiating audio-visual stimuli upon getting certain biosignals recorded by nanotextile biosensors. Such an automated system with decision algorithm can be used to help control and/or cure the underlying medical condition. In this section, such systems have been described that can be used therapeutically based on the biofeedback received by the sensors on board.

Music-based therapy has gained a lot of attention due to recent findings in the field of neuromusicology and music cognition. Music can evoke emotions, through autonomic correlates that have been shown to cause significant modulation of parameters like heart rate and blood pressure. Consequently, HRV analysis can be a powerful tool to explore evidence-based therapeutic functions of music and conduct empirical studies on effect of musical emotion on heart function [120,121].The HRV can be used to determine the autonomic nervous system activity [122]. Autonomic responses are known to vary according to the reports of valence (positive or negative) and arousal/intensity, which are considered as two dimensions of emotion [123], including musical emotions [124]. These emotions in turn have been shown to have strong autonomic correlates that regulate blood pressure and heart rate in the listener [125]. Although the initial results are promising, much research is needed to understand the autonomic nervous system activity involved in musically induced emotions, the variations in the brain processes with variations in intensity, and valence of the emotions experienced.

Few studies have been performed to examine brain process of emotions induced via auditory domain stimulus like music. Figure 9.52 shows some preliminary EEG measurements after application of auditory stimulus in the form of happy and sad music. While listening to both happy and sad music, there was an increase in alpha wave power in the right hemisphere of the brain. There was also an increase in the delta, theta, and gamma waves, but the extent of increase was greater during the happy music condition as compared to the sad music condition.

The textile-based nano-biosensor system for ECG monitoring, such as e-vest and e-bra, can be used for recording ECG while listening to music. The plots in Figure 9.53 show the periodogram of the HRV signals for each type of music that

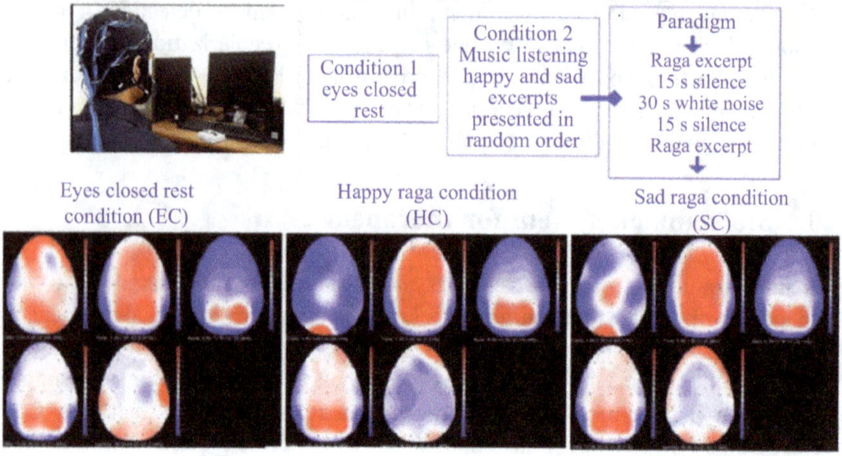

Figure 9.52 EEG measurements after the application of auditory stimulus in the form of happy and sad music (Raga), with eyes closed rest condition as reference

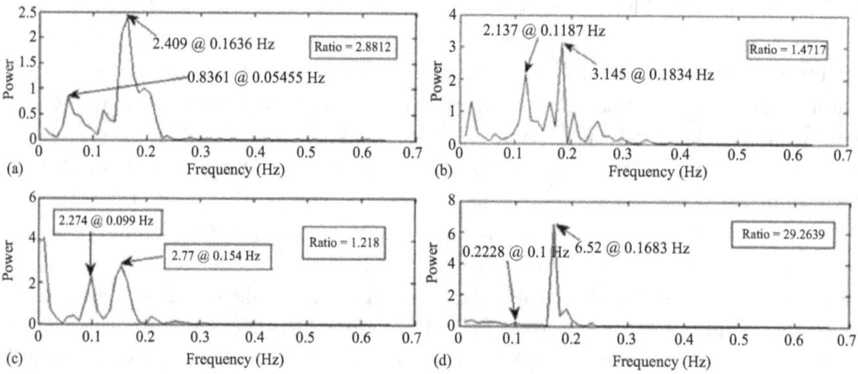

Figure 9.53 Periodogram of the HRV and high-/low-frequency power ratio for ECG data for (a) no music; (b) peppy; (c) happy; and (d) sad music

was played during recording. ECG data for 2 min, 1 min after the application of the stimulus music, were used for HRV analysis. The HRV analysis performed on this data shows that the emotion elicited by the type of music playing can modulate the autonomic regulation of the heart rate. A similar approach can be taken for EEG monitoring nanotextile biosensors. Music-based biofeedback therapy can be a potential treatment method to improve sensorimotor, language, and cognitive domains of functioning via music. It can be especially helpful in invoking neuroplasticity for rehabilitation of traumatic brain injury (TBI) patients [126].

Autism spectrum disorder (ASD) covers neurological disorders involving impeded development of social behavior, acquisition of skills, and language development. All these activities require the ability to mimic, by which children imitate and learn the appropriate forms of the aforementioned attributes. The ability to mimic is associated with a class of neurons called the mirror neurons found in insula and anterior cingulate cortex of the brain, which are responsible for interpretation of complex human intentions. The functioning of these neurons can be observed by monitoring Mu waves in those regions using EEG. Mu waves are synchronized patterns of electrical activity involving large numbers of neurons in the part of the brain that controls voluntary functions. In normal subjects, the Mu waves suppress when the voluntary function has been executed, whereas, in ASD subjects, the suppression does not happen because no function was executed.

Conventional EEG measurement systems use gel-based gold cup electrodes attached to the scalp with adhesive. It is obtrusive and wires sticking out of the electrodes to signal acquisition system make them impractical for use in sensitive subjects like infants and children with ASD. Alternatively, nanotextile biosensors can be incorporated with skull and baseball caps that are commonly used for infants and children. Textile-based multi-electrode EEG, EOG, and EMG monitoring system with embedded electronics for data acquisition and wireless transmission can be seamlessly integrated into these items for continuous detection of Mu waves (Figure 9.54). Textile electrodes at positions C3, CZ, and C4, according to 10–20 international system,

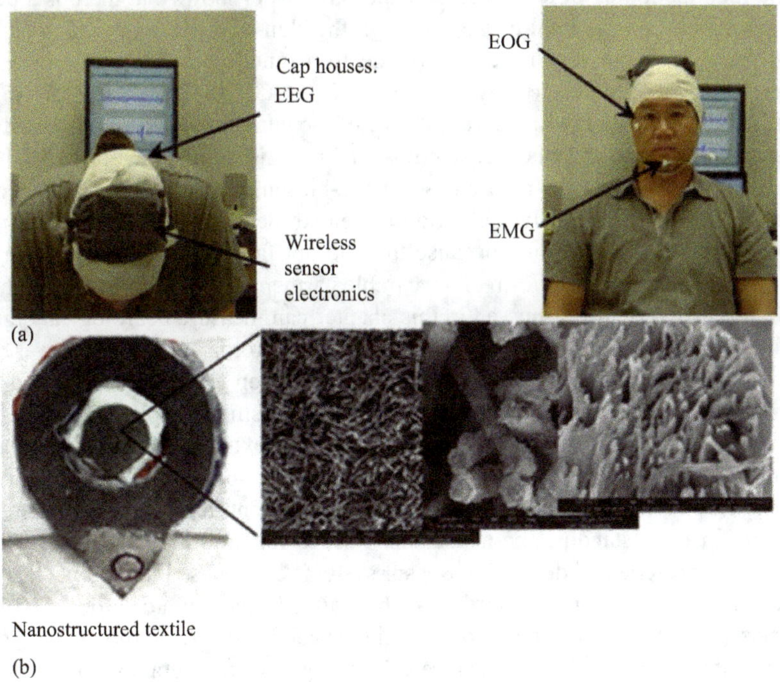

Figure 9.54 EEG, EOG, and EMG are acquired on the skin through conductive textile electrodes. The acquisition of EEG through the hair on the head was made possible by using nano-structured conductive electrode with long pillars of conductive textile to serve this purpose

can detect Mu waves. The textile-based system is ergonomic and can enable early diagnosis in infants and planning therapy for ASD patients [127].

Biofeedback can be used for therapeutic purposes for ASD by operant conditioning. By giving stimuli inform of a repetitive audio/visual cues to emulate a voluntary function (waving hand, saying hello, open/close fist) and monitoring the change in Mu wave intensity, the system can condition the subject. The audio-visual loop can be controlled by the embedded logic that breaks the loop only upon detecting Mu wave occurrence and suppression, after which, it will choose another cue from the audio-visual library on the system. This system can help in ASD diagnostics by monitoring Mu waves for detecting anomalies in brain wave patterns of mirror neurons [128]. The automated system can be used at home under parental guidance, while keeping the therapist updated at remote location through internet

9.11 Conclusion

Wearable textile-based nano-biosensor systems with mobile platform are a new class of unobtrusive continuous health monitoring with significant benefits for neurological and cardiovascular patients or high-risk patients. The basic science behind the functioning of textile-based nano-biosensors shows that there is a better flow of signal, i.e. electrical current through the skin-electrode interface, than silver-silver chloride gel electrode. The bioelectromagnetic principles of origination and propagation of bioelectric signals, i.e., EEG, EOG, EMG, and ECG, show that measured electric potential is representative of the cumulative electrical activity of the sources of these signals, i.e. neurons of brain tissue and myocytes of heart tissue. This means that the textile-based nano-biosensors can measure bioelectric signals better than silver-silver chloride gel electrode. The dry electrodes can be used for long-term monitoring because they do not face the problem of drying of gel and they are reusable in form of a wearable garment. The textile-based wearable nano-biosensor systems discussed in this chapter can measure neurological signals and identify anomalies for diagnosis of targeted neurological and cardiovascular disorders. These disorders range from chronic condition to safety to rehabilitation and improved quality of life. The bioelectromagnetism principles of neural and cardiac bioelectric signals and the performance of textile-based nano-biosensors provides a unique perspective to the potential for development of novel wearable systems that harness the potential of textile-based nano-biosensors and wireless platform for understanding the neural and cardiac function in and out of hospital setting in unprecedented detail. The sensor systems can be used for diagnostics and therapeutics of neurological disorders such as autistic spectrum disorder, TBI, and neuroprosthetics. They can be used for highly specialized cardiac monitoring, such as the vector cardiography, impedance cardiography or tomography, tumor detection, and prevention from sudden cardiac death by detection of T-wave alternans.

This chapter presents textile-based nanosensors combined with wearable electronics and wireless communications systems. These systems address some challenges in engineering fibers or filaments to introduce intelligent functionality, integration of computation on textile platforms and reliable communication with wireless devices for remote data access. For the patients and medical healthcare professionals to adapt to wearable medical device technology, the devices and systems haven meet challenges in form factor, device longevity, and wireless communication requirements that are specific to each application.

Patterns of design for wearable systems architecture are a burgeoning body of knowledge. With continuously evolving communication technology, miniaturization of electronics, advances in flexible electronics, garment technology, and battery technology, wearable technology is becoming a viable alternative to conventional medical health monitoring technology. Consensus standards for data communication, data security, and textile-based wearable devices and systems are key to acceptance of these devices and systems in all aspects healthcare including therapeutics.

References

[1] de la Maisonneuve, C., and Martins, J.O., "Public Spending on Health and Long-Term Care: A New Set of Projections," OECD Economic Policy Papers, no. 6. http://www.oecd.org/economy/public-spending-on-health-and-long-term-care.htm (2013).

[2] National Center for Chronic Disease Prevention and Health Promotion, "Chronic Disease Prevention and Health Promotion," http://www.cdc.gov/chronicdisease/index.htm (2013).

[3] Jorg, H., "My Heart," European Community 6th Framework Programme, www.cordis.europa.eu (2003–2008).

[4] Marculescu, D., Marculescu, R., Zamora, N.H., *et al.*, "Electronic Textile: A Platform for Pervasive Computing," Proceedings of IEEE, Vol. 91(12), pp. 1995–2018. (2003).

[5] Rai, P., Kumar, P.S., Oh, S., *et al.*, "Smart Healthcare Textile Sensor System for Unhindered-Pervasive Health Monitoring," SPIE Smart Structures + NDE, Nano., Bio. & Info-Tech. Sensors and Systems 2012, Vol. 8344, 83440E. (2012).

[6] Bergey, G.E., Squires, R.D., and Sipple, W.C., "Electrocardiogram Recording with Pasteless Electrodes," IEEE Transactions on Bio-medical Engineering, Vol. BME-18(3), pp. 206–211. (1971).

[7] David, R.M., and Portnoy, W.M., "Insulated Electrocardiogram Electrodes," Medical & Biological Engineering, Vol. 10(6), pp. 742–751. (1972).

[8] Khosla, A., "Smart Garments in Chronic Disease Management: Progress and Challenges," Proceedings of SPIE Nanosystems in Engineering and Medicine, Vol. 8548, 85482O. (2012).

[9] Tao, X., "Smart Fibers, Fabrics and Clothing," Cambridge, Woodhead. (2001).

[10] Zhang, X., and Tao, X., "Smart Textiles: Passive Smart," June 2001, pp. 45–49; "Smart Textile: Active Smart," July 2001, pp. 49–52; "Smart Textiles: Very Smart," pp. 35–37, Textile Asia (2001).

[11] Pejis, T., Chairman Smart Textiles Symposium (in call for papers), Plastic Electronics Conference, September 2005, Frankfurt, www.plastictronics.org. (2005).
[12] Park, S., Gopalsamy, C., Rajamanickam, R., and Jayaraman, S., "The Wearable Motherboard^{TM}: An Information Infrastructure or Sensate Liner for Medical Applications," Studies in Health Technology and Informatics, IOS Press, Vol. 62, pp. 252–258. (1999).
[13] Park, S., and Jayaraman, S., "Wearable Biomedical Systems: Research to Reality," Proceedings of IEEE International Conference Portable Information Devices PORTABLE07, pp. 1–7. (2007).
[14] Coosemans, J., Hermans, B., and Peurs, R., "Integrating Wireless ECG in Textiles," Sensors and Actuators A, Vol. 130–131, pp. 48–53. (2006).
[15] Lee, Y.D., and Chung, W.Y., "Wireless Sensor Network Based Wearable Smart Shirt for Ubiquitous Health and Activity Monitoring," Sensors and Actuators B, Vol. 140(2), pp. 390–395. (2009).
[16] Alzadi, A., Zhang, L., and Bajwa, H., "Smart Textile Based Wireless ECG System," Proceedings of IEEE Systems, Applications and Technology Conference (LISAT) 2012, pp. 1–5. (2012).
[17] Smart T-shirt by GOW Trainer. www.gowtrainer.com
[18] Wearable Wellness System from Smartex s.r.l. Pisa Italy. www.smartex.it
[19] Rattfalt, L., Chedid, M., Hult, P., Linden, M., and Ask, P., "Electrical Properties of Textile Electrodes," Engineering in Medicine and Biology Society, 2007. EMBS 2007. 29th Annual International Conference of the IEEE, pp. 5735–5738, 22–26 Aug. 2007.
[20] Aarts, E., and Marzano, S., "The New Everyday-Visions on Ambient Intelligence," 101 Publishers, Rotterdam, The Netherlands. (2003).
[21] Mazzoldi, A., De Rossi, D., Lorussi, F., Scilingo, E.P., and Paradiso, R., "Smart Textiles for Washable Motion Capture Systems," AUTEX Research Journal, Vol. 2(4), pp. 199–203. (2002).
[22] De Rossi, D., Carpi, F., Lorussi, F., *et al.*, "Electroactive Fabrics and Washable Biomonitoring Devices," AUTEX Research Journal, Vol. 3(4), pp. 180–185. (2003).
[23] Clemens, F., Wegmann, M., Graule, T., *et al.*, "Computing Fibers: A Novel Fiber for Intelligent Fabrics?" Advanced Engineering Materials, Vol. 5(9), pp. 692–687. (2003).
[24] Gimpel, S., Mohring, U., Muller, H., Neudeck, A., and Scheibner, W., "Textile-Based Electronics Substrate Technology," Journal of Industrial Textiles, Vol. 33(3), pp. 179–189. (2004).
[25] Numetrex® Heart Monitoring Apparel. www.numetrex.com
[26] Wearable Wellness System from Smartex s.r.l. Pisa Italy. www.smartex.it
[27] Post, E.R, and Orth, M., "Smart Fabric, or Washable Computing," Proceedings of IEEE International Symposium on Wearable Computers, Cambridge (MA), USA, Oct. 1997, pp. 167–168. (1997).
[28] Catrysse, M., "Wireless Power and Data Transmission for Implantable and Wearable Monitoring Systems," PhD Thesis, K. U. Leuven. (2004).

[29] Rai, P., Kumar, P.S., Oh, S., *et al.*, "Smart Healthcare Textile Sensor System for Unhindered-Pervasive Health Monitoring," SPIE Smart Structures + NDE, Nano., Bio. & Info-Tech. Sensors and Systems 2012, Vol. 8344, 83440E. (2012).

[30] Pantelopoulos, A., and Bourbakis, N., "A Survey on Wearable Biosensor Systems for Health Monitoring," 30th Annual International Conference of the IEEE, EMBS 2008, pp. 4887–4890. (2008).

[31] Pacelli, M., Caldani, L., and Paradiso, R., "Textile Piezoresistive Sensors for Biomechanical Variables Monitoring," Proceedings of 28th Annual International Conference IEEE EMBS 2006, pp. 5358–5361. (2006).

[32] Lim, Z.H., Chia, Z.X., Kevin, M., Wong, A.S.W., and Ho, G.W., "A Facile Approach Towards ZnO Nanorods Conductive Textile for Room Temperature Multifunctional Sensors," Sensors and Actuators B: Chemical, Vol. 151(1), pp. 121–126. (2010).

[33] Vigneshwaran, N., Kumar, S., Kathe, A.A., Varadarajan, P.V., and Prasad, V., "Functional Finishing of Cotton Fabric Using Zinc Oxide-Soluble Starch Nanocomposites," Nanotechnology, Vol. 17(20), pp. 5087–5095. (2006).

[34] Dhawan, A., Muth, J.F., Kekas, D.J., and Ghosh, T.K., "Optical Nanotextile Sensors Based on the Incorporation of Semiconducting and Metallic Nanoparticles into Optical Fibers," Smart Nanotextiles Symposium (Materials Research Society Symposium Proceedings), Vol. 920, pp. 121–126. (2006).

[35] Varadan, V.K., Oh, S., Kwon, H., and Hankins, P., "Wireless Point-of-Care Diagnosis for Sleep Disorder With Dry Nanowire Electrodes," Journal of Nanotechnology in Engineering and Medicine, Vol. 1(3), 031012 (11 pages). (2010).

[36] Varadan, V.K., Kumar, P.S., Oh, S., *et al.*, "e-Nanoflex Sensor Systems: Smartphone-Based Roaming Health Monitor," Journal of Nanotechnology in Engineering and Medicine, 2 (1), 011016 (11 pages). (2011).

[37] Maharani, D.K., Kartini, I., and Aprilita, N.H., "Nanosilica-Chitosan Composite Coating on Cotton Fabrics," Third Nanoscience and Nanotechnology Symposium 2010, Vol. 1284, pp. 87–91. (2010).

[38] Locher, I., Klemm, M., Kirstein, T., and Troster, G., "Design and Characterization of Purely Textile Patch Antennas," IEEE Transactions on Advanced Packaging, Vol. 29(4), pp. 777–788. (2006).

[39] Dhawan, A., Ghosh, T.K., Seyam, A.M., and Muth, J., "Woven Fabric-based Electrical Circuits, Part II: Yarn and Fabric Structures to Reduce Crosstalk Noise in Woven Fabric-based Circuits," Textile Research Journal, Vol. 74(11), pp. 955–960. (2004).

[40] Xue, P., and Tao, X.M., "Morphological and Electromechanical Studies of Fibers Coated with Electrically Conductive Polymer," Vol. 98(4), pp. 1844–1854. (2005).

[41] Jiang, S.Q., Newton, E., Yuen, C.W.M., and Kan, C.W., "Chemical Silver Plating on Polyester/Cotton Blended Fabric," Journal of Applied Polymer Science, Vol. 100(6), pp. 4383–4388. (2006).

[42] Jiang, S.Q., Newton, E., Yuen, C.W.M., and Kan, C.W., "Chemical Silver Plating on Cotton and Polyester Fabrics and its Application on Fabric Design," Textile Research Journal, Vol. 76 (1), pp. 57–65. (2006).

[43] Brevnov, D.A., "Electrodeposition of Porous Silver Films on Blanket and Patterned Aluminum-Copper Films," Journal of the Electrochemical Society, Vol. 153(4), pp. C249–C253. (2006).

[44] Wang, H., Wang, J, Hong, J., Wei, Q., Gao, W, and Zhu, Z., "Preparation and Characterization of Silver Nanocomposite Textile," Journal of Coatings Technology and Research, Vol. 4(1), pp. 101–106. (2007).

[45] Yeon, S.J., "Study on Texture Evolution and Properties of Silver Thin Film Prepared by Sputtering Deposition," Applied Surface Science, Vol. 221(1–4), pp. 281–287. (2004).

[46] Holme, I., "Nanotechnologies for Textiles, Clothing, and Footwear," Textile Magazine, Vol. 32(1), pp. 7–11. (2005).

[47] Tonezzer, M., and Lacerda, R.G., "Zinc Oxide on Carbon Microfiber as Flexible Gas Sensor," Physica E: Low-Dimensional Systems and Nanostructures, Vol. 44(6), pp. 1098–1102. (2012).

[48] Naik, A., "Future Impact of Nanotechnology in Textile Research," Proceedings of II International Symposium in Textile Engineering Natal Brazil, pp. 7–11. (2004).

[49] Munirathinam, V., Anandhan, G., and Paramasivam, M., "Enhanced Sensing of Anthraquinone Dye Using Multiwalled Carbon Nanotubes Modified Electrode," International Journal of Environmental Analytical Chemistry, Vol. 93(3), pp. 349–363. (2013).

[50] Sahoo, S., Ouyang, H., Goh, J.C.H., Tay, T.E., and Toh, S.L., "Characterization of a Novel Polymeric Scaffold Application in Tendon/Ligament Tissue Engineering," Tissue Engineering, Vol. 12(1), pp. 91–99. (2006).

[51] Barhate, R.S., and Ramakrishna, S., "Nanofibrous Filtering Media: Filtration Problems and Solutions from Tiny Materials," Journal of Membrane Science, Vol. 296(1–2), pp. 1–8. (2007).

[52] Oh, T.I., Yoon, S., Kim, T.E., *et al.*, "Nanofiber Web Textile Dry Electrodes for Long-Term Biopotential Recording," IEEE Transactions on Biomedical, Circuits and Systems, Vol. 7(2), pp. 204–211. (2013).

[53] Rai, P., "Hybrid Nanostructured Textile Bioelectrode for unobtrusive health monitoring," Doctoral Dissertation, University of Arkansas, ProQuest, UMI Dissertations Publishing, ISBN 9781303276231. (2013).

[54] Reneker, D.H., and Chun, I., "Nanometer Diameter Fibers of Polymer, Produced by Electrospinning," Nanotechnology, Vol. 7(3), pp. 216–223. (1996).

[55] Baker, W.R., "Microdenier Spinning,"International Fiber Journal, April 1992, pp. 7–15. (1992).

[56] Kandel, E.R., Schwartz, J.H., and Jessel, T.M., "Principles of Neural Science," McGraw-Hill Professional, pp. 324. (2000).

[57] Gray, H., "Anatomy of the Human Body," 40th Ed., Churchill Livingstone, Elsevier. ISBN 978-0-443-06684-9 (2008).

[58] Geselowitz, D.B. "Dipole Theory in Electrocardiography", American Journal of Cardiology, Vol. 14(3), pp. 301–306. (1964).

[59] Malmivuo, J., and Plonsey, R., "Bioelectromagnetism: Principles and Applications of Bioelectric and Biomagnetic Fields," Chapter 2, Oxford University Press, New York. ISBN 0-19-505823-2 (1995).

[60] Fine, E.J., Ionita, C.C., and Lohr, L., "The History of the Development of the Cerebellar Examination," Seminars in Neurology, Vol. 22(4), pp. 375–384. (2002).

[61] Reisner, A.T., Clifford, G.D., and Mark, R.G. "Course Materials for HST.582J/6.555J/16.456J," Biomedical Signal and Image Processing, Spring 2007. MIT Open Course Ware (http://ocw.mit.edu), Massachusetts Institute of Technology (2007).

[62] Niedermeyer, E. and da Silva F.L., "Electroencephalography: Basic Principles, Clinical Applications, and Related Fields," Lippincott Williams & Wilkins. ISBN 0-7817-5126-8 (2004).

[63] Wolpaw, J., Birbaumer, N., McFarland, D., Pfurtscheller, G., and Vaughan, T., "Brain-Computer Interfaces for Communication and Control," Clinical Neurophysiology, Vol. 113(6), pp. 767–791. (2002).

[64] Rita Levi-Montalcini, "Neurological disorders – Public health challenges," World Health Organization, Geneva (2006).

[65] "10 Leading Causes of Death by Age Group, United States 2005–2010," National Vital Statistics System, National Centre for Health Statistics, Centres for Disease Control (CDC). (2005–2010).

[66] Garcia, T.B., and Holtz, N.E., "12-lead ECG: The Art of Interpretation," Jones and Bartlett, Sudbury, M.A. (2001).

[67] Brosche, T.A.M., "EKG Handbook," Jones and Bartlett, Sudbury, M.A., Section 10, pp. 87–98. (2010).

[68] Jahrsdeorfer, M., Guiliano, K., and Stephens, D., "Clinical Usefulness of the EASI 12-Lead Continuous Electrocardiographic Monitoring System," Critical Care Nurse, Vol. 25(5), pp. 28–37. (2005).

[69] Khan. G. M, "Rapid ECG Interpretation", Humana Press; 3rd edition, pp. 25–297. (2007).

[70] "Electrical Impedance Tomography: Methods, History and Applications," Ed. Holder D.S., Institute of Physics, Series in medical Physics and Biomedical Engineering, London. (2005).

[71] Gehi, A.K., Stein, R.H., Metz, L.D., and Gomes, J., "Microvolt T-Wave Alternans for the Risk Stratification of Ventricular Tachyarrhythmic Events: A Meta-Analysis," Journal of the American College of Cardiology, Vol. 46(1), pp. 75–82. (2005).

[72] Maeda, S., Nishizaki, M., Yamawake, N., *et al.*, "Ambulatory ECG-based T-wave Alternans and Heart Rate Turbulence Predict High Risk of Arrhythmic Events in Patients with Old Myocardial Infarction," Circulation Journal , Vol. 73(12), pp. 2223–2228. (2009).

[73] Merchant, F.M., Ikeda, T., Pedretti, R.F., *et al.*, "Clinical Utility of Microvolt T-wave Alternans Testing in Identifying Patients at High or Low Risk of Sudden Cardiac Death." Heart Rhythm, Vol. 9(8), pp. 1256–64.e2. (2012).

[74] Klingenheben, T., Zabel, M., D'Agostino, R.B., Cohen, R.J., and Hohnloser, S.H., "Predictive Value of T-wave Alternans for Arrhythmic Events in

Patients with Congestive Heart Failure", The Lancet, Vol. 356(9230), pp. 651–652. (2000).
[75] Chow, T., Kereiakes, D.J., Bartone, C., *et al.*, "Prognostic Utility of Microvolt T-wave Alternans in Risk Stratification of Patients with Ischemic Cardiomyopathy", Journal of the American College of Cardiology, Vol. 47(9), pp. 1820–1827. (2006).
[76] Ribeiro, A.L., Rocha, M.O., Terranova, P., Cesarano, M., Nunes, M.D., and Lombardi, F., "T-wave Amplitude Variability and the Risk of Death in Chagas Disease," Journal of Cardiovascular Electrophysiology, Vol. 22(7), pp. 799–805. (2011).
[77] Cox, V., Patel, M., Kim, J., Liu, T., Sivaraman, G., and Narayan, S.M., "Predicting Arrhythmia-Free Survival Using Spectral and Modified-Moving Average Analyses of T-wave Alternans," Pacing and Clinical Electrophysiology, Vol. 30(3), pp. 352–358. (2007).
[78] Catanzaro, J.N., Meraj, P.M., Zheng, S., Bloom, G., Roethel, M., and Makaryus, A.N., "Electrocardiographic T-wave Changes Underlying Acute Cardiac and Cerebral Events", American Journal of Emergency Medicine, Vol. 26(6), pp. 716–720. (2008).
[79] Herzog, C.A., Mangrum, J.M., and Passman, R., "Sudden Cardiac Death and Dialysis Patients", Seminars in Dialysis, Vol. 21(4), pp. 300–307. (2008).
[80] Pun, P.H., Horton, J.R., and Middleton, J.P., "Dialysate Calcium Concentration and the Risk of Sudden Cardiac Arrest in Hemodialysis Patients", Clinical Journal of the American Society of Nephrology, Vol. 8(5), pp. 797–803. (2013).
[81] Wu, V.C., Lin, L.Y., and Wu, K.D., "QT interval Dispersion in Dialysis Patients," Nephrology (Carlton), Vol. 10(2), pp. 109–112. (2005).
[82] Wilson, M.G., Sharma, S., Carre, F., *et al.*, "Significance of Deep T-wave Inversions in Asymptomatic Athletes with Normal Cardiovascular Examinations: Practical Solutions for Managing the Diagnostic Conundrum," British Journal of Sports Medicine, Vol. 46(Suppl. I), pp. i51–i58. (2012).
[83] Eckart, R.E., Scoville S.L., Campbell, C.L., *et al.*, "Sudden Death in Young Adults: A 25-year Review of Autopsies in Military Recruits", Annals of Internal Medicine, Vol. 141(11), pp. 829–834. (2004).
[84] Go, A.S., Mozaffarian, D., Roger, V.L., *et al.*, American Heart Association Statistics Committee and Stroke Statistics Subcommittee. "Heart Disease and Stroke Statistics–2013 Update: A Report from the American Heart Association", Circulation, 127(1), e6-e245. (2013).
[85] Rawles, J., "Magnitude of Benefit from Earlier Thrombolytic Treatment in Acute Myocardial Infarction: New Evidence from Grampian Region Early Anistreplase Trial (GREAT)," British Medical Journal, Vol. 312(7025), pp. 212–215. (1996).
[86] Yu, S-N, and Cheng, J-C., "A Wireless Physiological Signal Monitoring System with Integrated Bluetooth and Wi-Fi Technologies," Engineering in Medicine and Biology Society, 2005. IEEE-EMBS 2005. 27th Annual International Conference of the, pp. 2203–2206, 17–18 Jan. 2006.

[87] Huang, J, Wang, L., Chen, L-K., and Zhang, Y-T., "A Low Power Wearable Transceiver for Human Body Communication," Engineering in Medicine and Biology Society, 2009. EMBC 2009. Annual International Conference of the IEEE, pp. 3802–3805, 3–6 Sept. 2009.

[88] Lee, R-G., Chen, K-C., Hsiao, C-C., and Tseng, C-L., "A Mobile Care System With Alert Mechanism," Information Technology in Biomedicine, IEEE Transactions on, Vol. 11(5), pp. 507–517. (2007).

[89] Pantelopoulos, A., and Bourbakis, N.G., "A Survey on Wearable Sensor-Based Systems for Health Monitoring and Prognosis," IEEE Transactions on Systems Man and Cybernetics—Part C: Applications and Reviews, Vol. 40(1), pp. 1–12. (2010).

[90] Shen, J., Barbera, J., and Shapiro, C.M., "Distinguishing Sleepiness and Fatigue: Focus on Definition and Measurement," Sleep Medicine Reviews, Vol. 10(1), pp. 63–76. (2006).

[91] Young, T., Palta, M., Dempsey, J., Skatrud, J., Weber, S., and Badr, S., "The Occurrence of Sleep-Disordered Breathing Among Middle-aged Adults," New England Journal of Medicine, Vol. 328(17), pp. 1230–1235. (1993).

[92] Flemons, W., Douglas, J., Kuna, T., Rodenstein, O., and Wheatley, J., "Access to Diagnosis and Treatment of Patients with Suspected Sleep Apnea," American Journal of Respiratory Care Medicine, Vol. 169(2), pp. 668–672. (2004).

[93] Iber, C., Ancoli-Israel, S., Chesson, A.L., and Quan, S., "The AASM Manual for the Scoring of Sleep and Associated Events: Rules, Terminology, and Technical Specifications," American Academy of Sleep Medicine, Westchester, IL, (2007).

[94] Lal, S.K.L., and Craig, A., "A Critical Review of the Psychophysiology of Driver Fatigue", Biological Psychology, Vol. 55(3), pp. 173–194. (2001).

[95] Ranney, T., Mazzae, E., and Goodman, M., "NHTSA Driver Distraction Research: Past, Present and Future," Driver Distraction Internet Forum, (2000).

[96] European Transport Safety Council, "The Role of Driver Fatigue in Commercial Road Transport Crashes," (2001).

[97] Vicki, N., Thomas, D., Shiela, K., Jeremy, S., and Michael, G., "An Overview of the 100-Car Naturalistic Study and Findings," Proceedings of the International Technical Conference on the Enhanced Safety of Vehicles, Vol. 19. (2005).

[98] Kamen, G., and Robertson, DGE, "Research Methods in Biomechanics," Chapter 8, pp. 171–181, Champaign, IL: Human Kinetics Publishers, (2004).

[99] Nilson, T., Nelson, T.M., and Carlson, D., "Development of Fatigue Symptoms During Simulated Driving", Accidents Analysis & Prevention, Vol. 29(4), pp. 479–488. (1997).

[100] Ramasamy, M., Oh, S., Varadan, V. K., and Harbaugh, R., "Textile Based Nanosensors and Wireless Mobile Communication Platform for Real Time

Driver Alertness Monitoring," Smart Nanosystems in Engineering and Medicine, Vol. 2(2), pp. 23–33. (2013).

[101] Lin, C-T., and Wu, R-C., "EEG-based Drowsiness Estimation for Safety Driving Using Independent Component Analysis", Circuits and Systems I: Regular Papers, Vol. 52(12), pp. 2726–2738.(2005).

[102] Jap, B.T., Lal, S., Fischer, P., and Bekiaris, E., "Using EEG Spectral Components to Assess Algorithms for Detecting Fatigue", Expert Systems with Applications, Vol. 36(2), pp. 2352–2359. (2009).

[103] Wolpaw, J., Birbaumer, N., McFarland, D., Pfurtscheller, G., and Vaughan, T., "Brain-Computer Interfaces for Communication and Control," Clinical Neurophysiology, Vol. 113(6), pp. 767–791. (2002).

[104] Chapin, J., Moxon, K., Markowitz, R., and Nicolelis, M., "Real-time Control of a Robot Arm Using Simultaneously Recorded Neurons in the Motor Cortex," Nature Neuroscience, Vol. 2(7), pp. 664–670. (1999).

[105] Wessberg, J., Stambaugh, C., Kralik, J., *et al.*, "Real-time Prediction of Hand Trajectory by Ensembles of Cortical Neurons in Primates," Nature, Vol. 408(6810), pp. 361–365. (2000).

[106] Serruya, M., Hatsopoulos, N., Paninski, L., Fellows, M., and Donoghue, J., "Instant Neural Control of a Movement Signal," Nature, Vol. 416(6877), pp. 141–142. (2002).

[107] Taylor, D., Tillery, S., and Schwartz, A., "Direct Cortical Control of 3D Neuroprosthetic Devices," Science, Vol. 296(5574), pp. 1829–1832. (2002).

[108] Nicolelis, M., "Brain-Machine Interfaces to Restore Motor Function and Probe Neural Circuits," Nature Reviews Neuroscience, Vol. 4(5), pp. 417–422. (2003).

[109] Bashashati, A., Fatourechi, M., Ward, R.K., and Birch, G.E., "A Survey of Signal Processing Algorithms in Brain-Computer Interfaces Based on Electrical Brain Signals," Journal of Neural Engineering, Vol. 4(2), pp. 32–57. (2007).

[110] Oh, S., Kumar, P.S., Kwon, H., and Varadan, V.K., "Wireless Brain-Machine Interface Using EEG and EOG: Brain Wave Classification and Robot Control," SPIE Proceedings, Vol. 8344, p. 83440U-1. (2012).

[111] Shyamkumar, P., Oh, S., Banerjee, N., and Varadan, V.K., "A Wearable Remote Brain Machine Interface using Smartphones and the Mobile Network," Proceedings of Advances in Science and Technology, Vol. 85, pp. 11–16. (2012).

[112] Yoon, H., Deshpande, D.C., Ramachandran, V., and Varadan V.K., "Aligned Nanowires Growth Using Lithography-Assisted Bonding of a Polycarbonate Template for Neural Probe Electrodes," Nanotechnology, Vol. 19, 025304 (8 pp.). (2008).

[113] Led, S., Martinez-Espronceda, M., Redondo, J., *et al.*, "Wearable Electrocardiogram (ECG) Recorder for a Mobile Point-of-Care Based on Recent Interoperability Standards," Point-of-Care Healthcare Technologies (PHT), 2013 IEEE, pp. 287–290. (2013).

[114] Noueihed, J., Diemer, R., Chakraborty, S., and Biala, S., "Comparing Bluetooth HDP and SPP for Mobile Health Devices," Body Sensor Networks (BSN), 2010 International Conference on, pp. 222–227, 7–9 June 2010.

[115] Laukhina, E., Pfattner, R., Mas-Torrent, M., Rovira, C., Veciana, J., and Laukhin, V., "Flexible Film-Based Sensors Structured with a High Piezoresistive Organic Molecular Conductor as an Active Component," Sensor Device Technologies and Applications (SENSORDEVICES), 2010 First International Conference on, pp. 4–9, 18–25 July 2010.

[116] Castellanos, J., Navas-Gonzalez, R., Ochoteco, E., and Vidal-Verdu, F., "Evaluation of a Low Cost Piezoresistive Material for High Resolution Tactile Sensors," Integration Issues of Miniaturized Systems – MOMS, MOEMS, ICS and Electronic Components (SSI), 2008 2nd European Conference & Exhibition on, pp. 1–3, 9–10 April 2008.

[117] Kwon, H., Oh, S., and Varadan, V.K., "Motion Artifact Removal Algorithm by ICA for E-bra: A Women ECG Measurement System," Proceedings of SPIE 8691, Nano-Bio- and Info_tech Sensors and Systems 2013, 86910A. (2013).

[118] Pan, J., and Tompkins, W.J., "A Real-Time QRS Detection Algorithm," Biomedical Engineering, IEEE Transactions on, Vol. 32(3), pp. 230–236. (1985).

[119] Task Force of the European Society of Cardiology the North American Society of Pacing Electrophysiology, "Heart Rate Variability: Standards of Measurement, Physiological Interpretation, and Clinical Use," Circulation, Vol. 93(5), pp. 1043–1065. (1996).

[120] Koelsch, S., and Siebel, W.A., "Towards a Neural Basis of Music Perception," Trends in Cognitive Sciences, Vol. 9(12), pp. 578–584. (2005).

[121] Khalfa, S., Isabelle, P., Jean-Pierre, B., and Manon, R. "Event-Related Skin Conductance Responses to Musical Emotions in Humans," Neuroscience Letters, Vol. 328(2), pp. 145–149. (2002).

[122] Zhou, P., Sui, F., Zhang, A., Wang, F., and Li, G., "Music Therapy on Heart Rate Variability," Proceedings of BMEI 3rd International Conference on Biomedical Engineering and Informatics, Vol. 3, pp. 965–968. (2010).

[123] Lang, P.J., Bradley, M.M., and Cuthbert, B.N., "Emotion and Motivation: Measuring Affective Perception," Journal of Clinical Neurophysiology, Vol. 15(5), pp. 397–408. (1998).

[124] North, A.C., and Hargreaves, D.J., "Liking, Arousal Potential, and the Emotions Expressed by Music," Scandinavian Journal of Psychology, Vol. 38(1), pp. 45–53. (1997).

[125] Bernardi, L., Porta, C., Casucci, G., *et al.*, "Dynamic Interactions Between Musical, Cardiovascular, and Cerebral Rhythms in Humans", Circulation, Vol. 119(25), pp. 3171–3180. (2009).

[126] Hegde, S., "Music Based Cognitive Remediation Therapy for Patients with Traumatic Brain Injury," Frontiers in Neurology, Vol. 5(34), pp. 1–7. (2014).

[127] Sahi, A., Rai, P., Oh, S., Ramasamy, M., Harbaugh, R.E., and Varadan, V. K., "Neural Activity Based Biofeedback Therapy for Autism Spectrum Disorder through Wearable Wireless Textile EEG Monitoring System," Proceedings of SPIE 8691, Nano-Bio- and Info_tech Sensors and Systems 2013, 86910A (2013).

[128] Ramachandran, V.S., and Oberman, L.M., "Broken Mirrors A Theory of Autism," Scientific American, Vol. 295(5), pp. 63–69. (2006).

Chapter 10

Meta-atoms and artificially engineered materials for antenna applications

Ravi Kumar Arya[1]*, Shiyu Zhang*[2]*, Shaileshachandra Pandey*[1]*, Ashwani Kumar*[3]*, Yiannis Vardaxoglou*[2]*, William Whittow*[2]*, and Raj Mittra*[4]

10.1 Introduction

In the "metamaterial" (MTM) world, there has been an enormous research effort into novel and unusual material properties that can find suitable uses in many microwave and antenna applications. In an attempt to produce a formalism that classifies these properties, in this chapter we introduce the concept of meta-atoms (MTAs). These "meso" scale particles have typical sizes much smaller than the wavelength at which they will be operating. They could therefore be assembled to form synthetic materials of some predefined properties required for a particular application. A research project [1] entitled "Synthesizing 3D Metamaterials for RF, Microwave and THz Applications (SYMETA)" has been established to study and classify these meta-atoms and produce structures that can be manufactured with additive manufacturing (AM) techniques, such as 3D printing. The first goal of this project is to synthesize a palette of 3D MTAs using suitable materials and construct designer-specified 3D arrangements of meta-atoms using process efficient AM to create MTMs. In light of this background on MTAs, this chapter introduces their implementation will focus on the topic of artificial synthesis of materials.

Composite structures that have a desired shape and desired set of properties over a wide range of frequencies. These structures could be viewed as artificial materials and provide a means of altering the electromagnetic (EM) properties of naturally occurring materials. In order to create these artificial materials, dielectric, magnetic,

[1]EMC Lab, The Pennsylvania State University, USA
[2]Wolfson School of Mechanical, Electrical and Manufacturing Engineering, Loughborough University, United Kingdom
[3]Department of Electronics, Delhi University, India
[4]EMC Lab, University of Central Florida, USA

or metallic inclusions in a host medium are arranged in defined or random lattices to alter the EM properties of a hybrid structure. The size of the inclusions is typically less than $\lambda/10$ [2,3], which distinguishes them from MTMs that utilize resonant inclusions and, hence, are dispersive narrowband and lossy. These synthetic materials typically have an effective permittivity, $\varepsilon_{r_{\text{eff}}}$ and effective permeability, $\mu_{r_{\text{eff}}}$ [4].

A meta-atom (defined here as a meso-scale particle, or inclusion) with constitutive parameters that can have a wide variation, for example, dielectric, ferrite and hybrids. These will build into a specific formation or a plurality of an array in 3D to invoke the required shape, then to achieve the required overall bulk change in $\varepsilon_{r_{\text{eff}}}$ and $\mu_{r_{\text{eff}}}$ where a wide range of values could be obtained.

Figure 10.1 shows the analogy between the model of natural materials and MTA-based materials. The MTAs in our case have typically meso-scale dimensions. These can also be conveniently represented by equivalent circuits since they are weakly coupled and assuming they are packed closely they can be nearly homogeneous. Figure 10.2 shows an exemplar of two primary MTA classifications of disconnected and connected pluralities. The former behaves primarily capacitive having an inductance as the frequency increases. The latter (connected meta-atoms) is predominant inductive with a soft capacitance appearing at the high frequencies.

In general, the $\varepsilon_{r_{\text{eff}}}$ and $\mu_{r_{\text{eff}}}$ depend on the volume fraction of the inclusion and host materials. Artificial heterogeneous materials can be constructed by adding the inclusions arranged in different lattices and this can control the effective EM properties including permittivity, permeability and losses [5–9]. Hence, engineers can design materials with bespoke EM properties and exterior shapes by changing the MTAs and their volume fraction. The MTAs could be arranged on a regular lattice in 3D by either stacking them or displacing them in their own plane, which can realize radially varied relative permittivity, see Figure 10.3.

With the advent of 3D printing, the currently available fused filament fabrication (FFF) 3D-printing is able to fabricate the composite materials. The FFF is an AM technique whereby layers of thermoplastic polymers are physically deposited onto each other, to create structures layers by layers and with shapes that may not otherwise be achievable using conventional manufacturing techniques such as

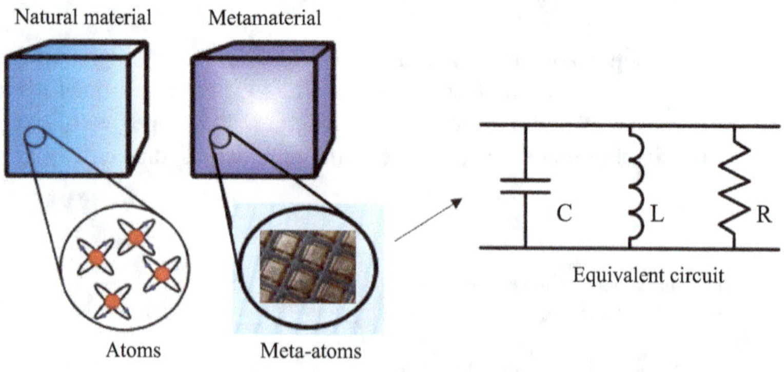

Figure 10.1 Spheres with connected meta-atom composition in a pentagonal lattice

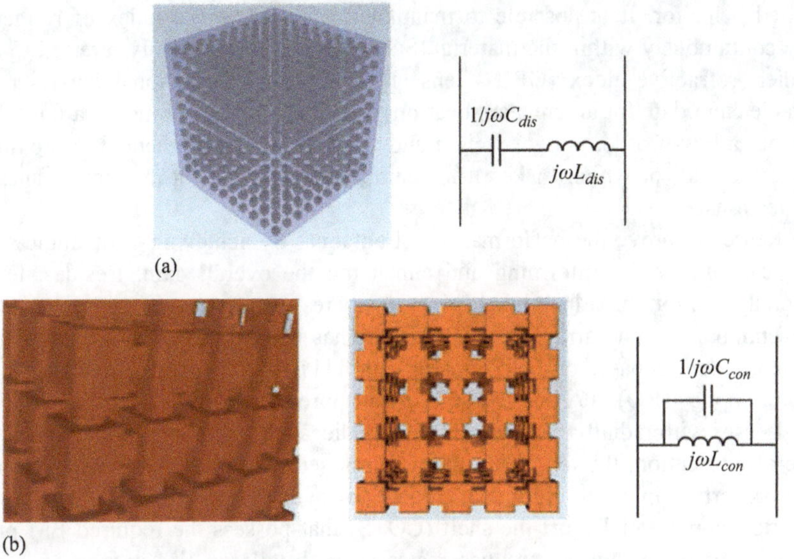

Figure 10.2 (a) Disconnected and (b) connected meta-atoms and their equivalent circuit

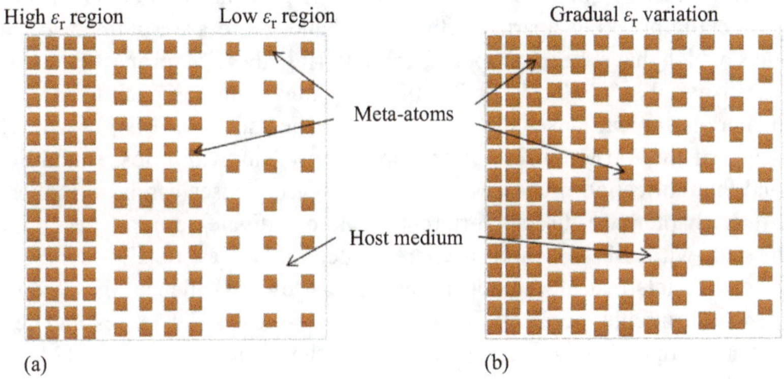

Figure 10.3 Varying the meta-atoms distribution to achieve (a) locally constant ε_r and (b) gradually varied ε_r

injection molding or computer (numerical)-controlled milling. The FFF is available for both single and multi-material printing whereby conductors can also be deposited, either from another filament extruder or as an ink.

10.2 Lens designs using MTMs

A lens is a commonly used antenna component that transforms a plane wavefront into a spherical one, or vice versa. In a conventional lens, the index of refraction is homogeneous within the material. The lens has a variable curvature that gives it its desired behavior. It is possible to manufacture lenses whose index of refraction vary continuously within the material. Such a design is commonly referred to as a gradient-refractive-index (GRIN) lens [10]. While a conventional lens is not a suitable candidate for antenna applications due to its curved geometry, a flat GRIN lens has a low-profile, is light-weight and can be easily implemented in proximity to the feed, all of which makes it a desirable candidate for a number of antenna applications.

Lenses improve the performances of antennas by achieving wide angle scanning capabilities, beamforming and enhancing the overall gain. Besides, lenses generally exhibit broadband behaviors that are not commonly associated with conventional antenna arrays. Recently, there has been significant development in flat lens antennas based on field transformation [11–13], transformation optics [14–16], ray optics (RO) [17] and transmit-array approaches [18].

Lenses with radially varying refractive indices are challenging to fabricate. For ease of fabrication, the lenses are often comprised of several zones with different EM properties in each individual zone. However, it is difficult to source the materials commercially off-the-shelf (COTS) that possess the required EM properties. Another method to produce materials with tailored EM properties is perforating a homogeneous slab with holes of variable diameters or separations with variable distances [19]. Precise machining and tight tolerance control are required for this approach, particularly for high-frequency applications. Sometimes the maximum number of holes needs to be limited to prevent the material from physically cracking/breaking apart. On the other hand, there are a number of methodologies which have been proposed for tailoring the EM properties by using synthetic materials [5,20,21,7,22]. Traditionally, the main approach to fabricating these materials for flat lenses has been based on the use of MTMs [23–26]. However, some of the MTM lenses suffer from narrow bandwidth, losses and dispersion, and the fabrication processes of the MTM-based lenses can be complicated.

In this chapter, we discuss different ways to mitigate some of the problems encountered with MTMs, and present strategies for artificially synthesizing dielectric materials that are broadband as well as low-loss; hence, they are useful for real-world antenna applications involving low-profile flat lenses and reflect-arrays, for example. The key to circumventing the difficulties with MTM, which we have identified above, is to steer clear of the common practice of using resonant inclusions or "particles" to achieve extreme material properties, such as $\gg 1$; $\ll 1$; negative index; and, zero index. Our strategy is to develop antenna designs that only call for material parameters that are realistic, so that they can either be acquired off-the-shelf, or by slightly tweaking the available materials by embedding small patches or apertures, often referred to as "particles" (see Section 10.5, for instance), whose dimensions are far removed from the resonance range. This obviates the problems of dispersion, narrow bandwidths and losses that plague the MTMs, at least those that fall in the "exotic" category, for example, the double-negative or DNG type. Although the RO approach leads to dielectric-only designs that do not need to use magnetic materials, these designs still typically require dielectric materials that may not be available off-the-shelf.

In the following sections, we present several examples to demonstrate the procedure for synthesizing artificial dielectrics for both single-layer and multilayer types—the latter to achieve better performance control including matching. After the procedure for synthesizing artificial dielectrics is presented, we design different lenses and fabricate them using a combination of conventional printed circuit board (PCB) and 3D-printing techniques. The first of these lenses was designed and fabricated using 3D-printing. The second one, a dial-a-dielectric (DaD) lens, which used state-of-the-art artificially engineered dielectric materials, was fabricated using 3D-printing and PCB laminate with patches on both sides. The third lens is designed and fabricated in a similar manner to the first lens but with 3D-printing material of higher permittivity. All three designs circumvented the difficulties in finding desired COTS materials for 3D-printed lenses. The lenses comprise several concentric dielectric rings with bespoke relative permittivities for transforming spherical waves into plane waves and vice versa. These lenses are low-cost and light-weight, but exhibit broadband and high gain performance. Measurement results show that all the lenses show comparable results to simulated designs over the designed frequency band of 12–18 GHz.

10.3 Lens design using RO

We begin by using the RO methodology to design the lens in the form of concentric rings, with variable permittivity values, as shown in Figure 10.4. The flat GRIN lens is in the form of a disk and its focal point is located at its axis of symmetry. The refractive index should be a maximum at the center of the lens and gradually decrease to the outermost region [27]. The required variation in the refractive index can be calculated by equating the phase delay of the different rays from the focal point until the point where they exit the lens. It is assumed that the angle between

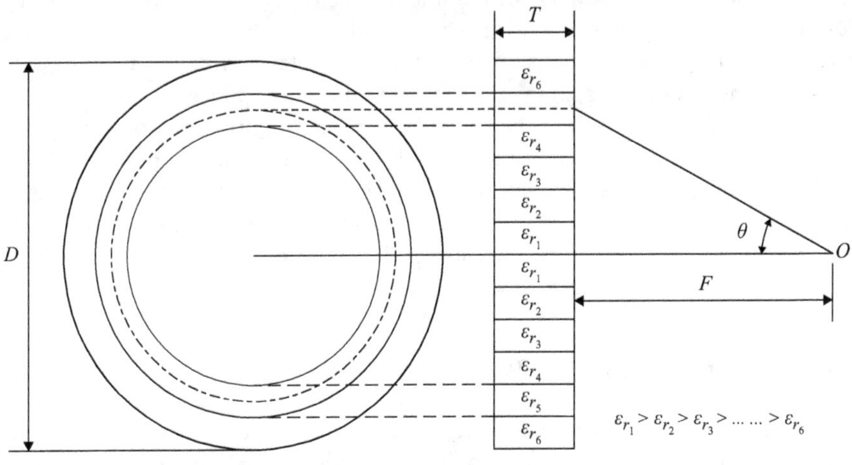

Figure 10.4 Sketch of lens design principle

Table 10.1 COTS dielectric materials

1.96	2.17	3	4.5	6	9.2	10.2
	2.2	3.02	4.7	6.15	9.8	
	2.33	3.2				
	2.5	3.27				
	2.75	3.55				
	2.94	3.6				
		3.66				

the path where the ray enters the lens and the lens axis is θ (see Figure 10.4). The relationship between the radially varied ε_r and θ is given in (10.1), and this becomes the fundamental design equation for the flat GRIN lens:

$$\frac{T}{F}\left(\varepsilon_r - \frac{2}{3}\sin^2\theta\right) = \sqrt{(\varepsilon_r - \sin^2\theta)}\left(\sqrt{\varepsilon_{r_{max}}}\frac{T}{F} - \sec\theta + 1\right) \tag{10.1}$$

where T is the thickness of the lens, F is the focal length, and $\varepsilon_{r_{max}}$ is the maximum permittivity at the center of the lens.

The design can be simplified by calculating ε_r for several given values of θ and then producing a smooth curve through the plotted points. For practical fabrication, it is considered reasonable to approximate the ideal smooth variation with a staircase version, which has a stepwise variation instead. Thus the lens can be fabricated by using a series of concentric dielectric cylindrical rings with different relative permittivities. The corresponding ε_r values for each ring are designed to ensure that they have the same focal point O in order to convert the spherical wave emanating from the focal point into a plane wave.

Before (10.1) can be applied to the lens design, it is necessary to know that the maximum value of θ_{max} is determined by the diameter of the lens (D) and the focal length (F). In addition, knowing the minimum value of $\varepsilon_{r_{min}}$ that can be produced by 3D-printing is equally essential. Structural considerations limit the minimum value of ε_r due to the resolution constraints of the 3D-printer and potential damage to structural integrity like cracking. Having determined the values of θ_{max} and $\varepsilon_{r_{min}}$, (10.1) was used to find the lens thickness T, hence the ε_r variation. Equation (10.1) can also be applied to find the focal length F for each cylindrical ring.

For completeness, we include the dielectric parameters of most commonly available COTS materials in Table 10.1. A quick check shows that many of the desired materials that are derived by using (10.1) are not commercially available from vendors such as Rogers Corporation.

10.4 3D-Printing technique

3D-printing is an AM technique which creates 3D-objects in successive layers. It provides a practical fabrication approach to produce highly customizable structures with the advantages of low-cost and fast, automated repeatable design and manufacturing. The 3D-printing process allows for the creation of embedded sub-millimeter internal structures, such as air voids within the 3D-object, in a single process without requiring any machining. Compared with perforating a solid material, the design can be easily modified and rapidly prototyped by using low-cost 3D-printing materials, and this is particularly useful for building laboratory prototypes such as lenses as in our case.

In the fabrication process described herein, a fused deposition modeling (FDM) Makerbot® Replicator™ 2X 3D-printer is used to fabricate the lenses utilizing the thermoplastic polylactic acid (PLA) as well as acrylonitrile butadiene styrene (ABS)-based 3D-printing materials. PLA material has $\varepsilon_r = 2.72$ with $\tan(\delta) = 0.008$, whereas ABS material (PREPERM® TP20280) has $\varepsilon_r = 4.4$ with $\tan(\delta) = 0.004$. The heated printer nozzle extrudes the printing material and creates the lens layer-by-layer from the bottom up. This process enables the fabrication of embedded micron-scale particles such as air voids in a single process without machining while keeping the wastage and cost at a low and acceptable level.

Although 3D-printing technology is becoming widespread and affordable day-by-day, current 3D-printers can only work with certain materials within a limited range of material parameters. Hence, this current work can be helpful for designing dielectric materials which cannot be directly handled in or realized/produced by 3D-printers. Typically, it is difficult to realize dielectric materials with permittivities that are higher than what is available from 3D-printing materials. In the next section, we explore some of the techniques that we utilize to design artificially engineered materials.

10.5 Design of artificially engineered materials

RO lenses are realized by using different dielectric materials as explained earlier in Section 10.3 where we have pointed out that not all the requisite materials are available commercially, consequently we need to design such artificially engineered materials.

Once we have determined the desired dielectric parameters by following the design strategy discussed in Section 10.3, we encounter three different possibilities:

(i) The desired permittivity value is the same as that of those commercially available. In this scenario, we move forward and use the available COTS material.

(ii) The desired dielectric constant is higher than that of the available COTS material. In this case, we use the methods presented in Sections 10.5.1 or 10.5.2.

(iii) The desired permittivity value is lower than that of the available COTS material/3D-printing material. In this case, we follow the strategies laid out in Sections 10.5.3 or Sections 10.5.4.

It is important to note that the above methods can be sometimes combined with each other to obtain the desired dielectric permittivity. We have also observed that stacking multiple layers provides considerable flexibility for achieving higher dielectric permittivity.

10.5.1 Designing higher-permittivity materials from low-permittivity COTS material: method-1

In this section, we present the technique for engineering the COTS materials to realize the dielectric parameters we desire by implementing the "DaD" scheme. In the DaD technique, to tweak the COTS materials, we use square patches (other shapes can be used as well), arranged in a circular/cartesian pattern as shown in Figure 10.5, and print them on top (bottom or both on top and bottom) of the dielectric rings to realize the desired ε_r values. Alternatively, we can print them on a mylar sheet and then place the sheet above the rings. To carry out the simulation, the concerned ring is discretized in the unit cells of appropriate periodicity (b). The periodicity is determined on the basis of phase value needed to compensate across the unit cell. In our tests, we found that the unit cell with periodicity around $\lambda/10$ provides satisfactory results while avoiding the issues of high loss, narrow

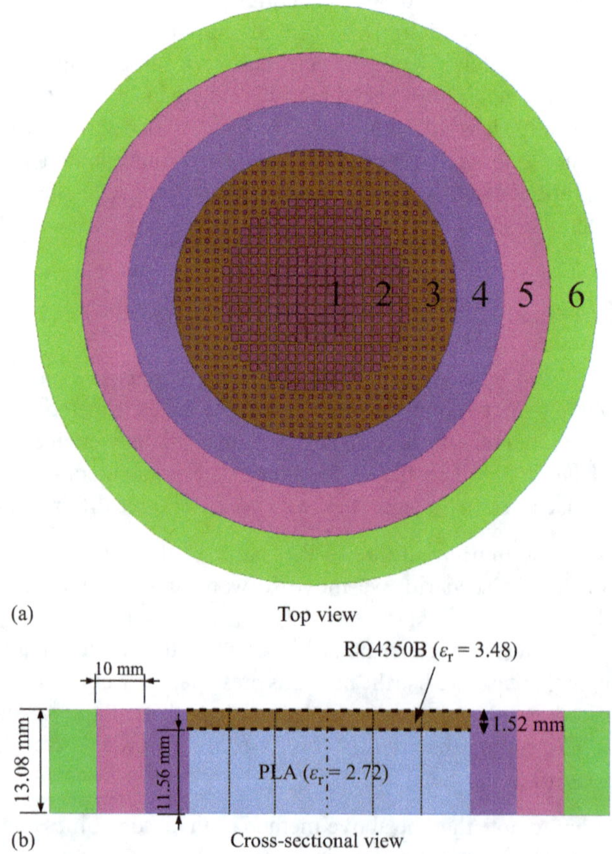

Figure 10.5 Proposed DaD lens

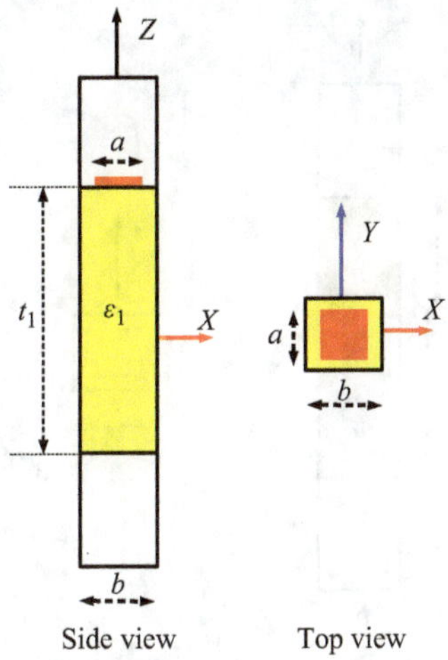

Figure 10.6 Unit cell for designing higher-permittivity materials from low-permittivity COTS materials

bandwidth and dispersion, typically associated with MTMs. We use unit cell of COTS material and patch combination (see Figure 10.6) to realize the artificial dielectric. We start this process by placing a patch with a very small side dimension on the COTS material layer. Initially, the phase of S_{21} for COTS material covered with small patch will be close to COTS-only material layer. After confirming this behavior, we increase the patch dimensions. The dimensions of the patches are chosen such that the phase of S_{21} of a dielectric-only layer, if available, would match the S_{21} of the COTS materials covered by the patch. Since the incremental change is relatively small, the patch-size needed to accomplish this phase shift behavior is such that it is far from its resonance, and this is the key to realizing a wideband low-loss design.

10.5.2 Designing higher-permittivity materials from low-permittivity COTS material: method-2

As mentioned in the previous section, there is a limit to the dielectric permittivity value we can achieve by following the approach presented therein. If we find that the required patch size becomes comparable to the local periodicity of the unit cell, insertion loss of the composite material becomes too high to be acceptable. In this scenario, we can modify the above approach as we will now explain. We use two dielectric blocks with one of the blocks of higher permittivity value and the other block of lower permittivity value and place the patch in the middle of the stack as shown in Figure 10.7. Since commercially available dielectric materials only come with pre-set thicknesses, we can use the patches to fine-tune the effective dielectric constant of the stack. The size of the patch introduces an additional

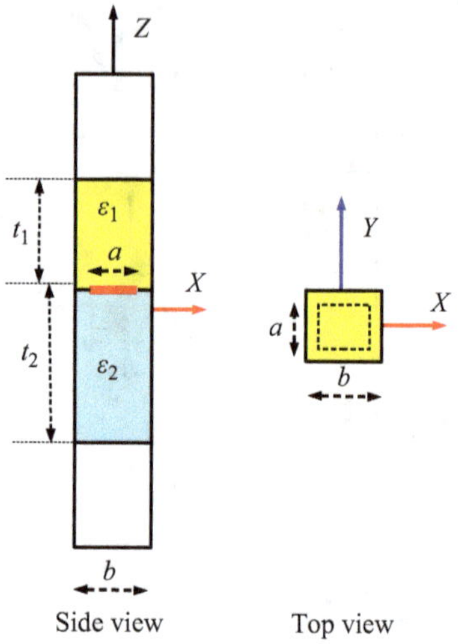

Figure 10.7 *Unit cell for designing higher-permittivity materials from low-permittivity COTS materials*

degree of freedom, and thus versatility, for tuning the permittivity of the unit cell as compared to using just two layers of dielectric blocks.

10.5.3 Designing lower-permittivity materials from high-permittivity COTS material

It is not uncommon to find that the desired value of permittivity is lower than that of the available COTS material and is closest to the desired one. In this event, we can use the following approach.

Take two COTS dielectrics, one with a lower and the other with a higher-permittivity value than the desired one. Stack them and adjust the height of both the dielectric materials (see Figure 10.8) until the desired phase of S_{21} is realized.

10.5.4 Designing lower-permittivity materials from high-permittivity 3D-printing material

The methods discussed in the previous sections, all have conductive patches as structural components and can be easily implemented by using traditional PCB techniques in which metal patches can be printed on the laminate. But most of these PCB laminates have pre-set thicknesses and permittivity values and thus provide limited flexibility to fine-tune the material permittivity. In such cases where the above-mentioned PCB technique does not work, we can use 3D printing materials alone to realize the desired value of permittivity. To accomplish this, we insert an air void in the host 3D-printing material which reduces the effective dielectric permittivity of the 3D-printing material. An example of a unit cell of such design is shown in Figure 10.9.

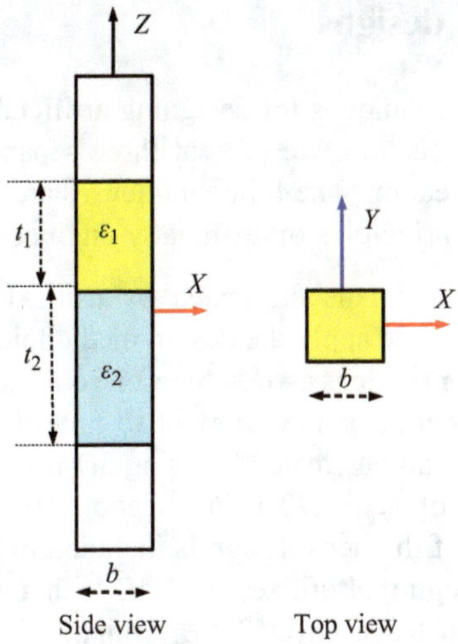

Figure 10.8 Unit cell for designing lower-permittivity materials from high-permittivity COTS materials

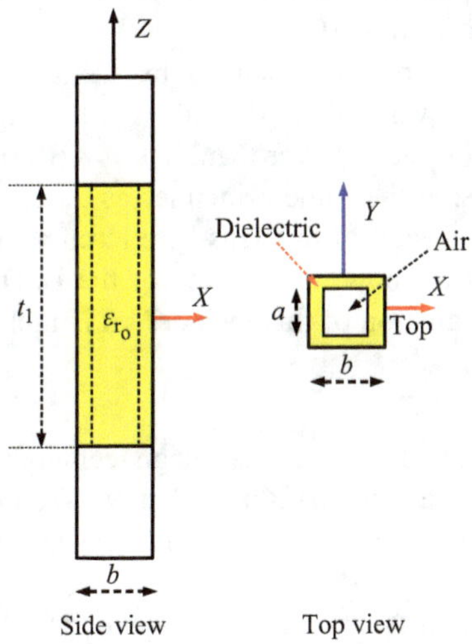

Figure 10.9 Unit cell for designing lower-permittivity materials from high-permittivity 3D printing material

The effective relative permittivity ($\varepsilon_{r_{\text{eff}}}$) of a 3D-printed unit cell with internal air void volume is given by:

$$\varepsilon_{r_{\text{eff}}} = \nu\varepsilon_{r_o} + (1 - \nu) \tag{10.2}$$

where ε_{r_o} is the relative permittivity of the 100% 3D-printing material and ν is the volume percentage indicated by the ratio of the volume of 3D-printing material to the volume of the entire unit cell.

10.6 Different lens designs

By using the appropriate techniques for designing artificially engineered materials discussed in the previous section, we present three separate designs of three flat graded-index lens in this section. The following lenses are designed to demonstrate the validity of the design principles of artificially engineered materials:

1. PLA Lens: a 3D-printed lens fabricated by using PLA as the 3D-printing material. In this design, we apply the design methodology where the maximum permittivity needed in the lens is less than or equal to the permittivity of the basic 3D-printing material. It is important to note that $\varepsilon_r = 2.72$ is the maximum value of ε_r that can be created by using 3D-printing technique with PLA 3D-printing material of $\varepsilon_r = 2.72$ using Section 10.5.4 exclusively. A more detailed explanation of this lens design is included in Section 10.6.1.
2. DaD Lens: a lens design that utilizes the DaD technique. We use PLA as 3D-printing materials and Rogers RO4350B laminate for printing patches on it. This design circumvents the design limitations of the PLA lens. We use the DaD technique in combination with 3D-printing technique to achieve the maximum permittivity of the lens, which is higher than that provided by the PLA 3D-printing material alone. A more detailed explanation of this lens design is included in Section 10.6.2.
3. ABS Lens: a 3D-printed lens designed by using ABS as the 3D-printing material. While there was no need for this lens design since the maximum permittivity value required is lower than that of ABS ($\varepsilon_r = 4.4$) and the design can be carried out using the same principles as those used to design the PLA lens. We include the design of this lens here to compare its performance with that of the DaD lens to show the efficacy of the DaD lens design technique. A more detailed explanation of this lens design is included in Section 10.6.3.

10.6.1 PLA Lens design

In this section, we explain in detail the lens design techniques that will be applied to designing the PLA as well as the ABS-based lenses. We introduce the 3D-printed lens design, its fabrication and measurement of its response to compare it with the simulated design.

The flat PLA GRIN lens is in the shape of a cylindrical disk and its focal point is located on the axis of symmetry (see Figure 10.4). The lens is divided into 6 discrete rings in the radial direction with the width of each ring to be 10 mm. The dielectric constants calculated for these rings satisfy the path length condition. The effective permittivity for the outermost ring is calculated to be 1.3, while the center ring has the highest effective permittivity of 2.72 as shown in Table 10.2 following the principles explained in Section 10.3. The derived values of dielectric constant for each ring are designed to ensure that the waves traveling through each ring collimate at the same focal point. Hence, the lens would render the plane waves entering from one side of the lens to focus at the focal point on the opposite side of the lens.

Table 10.2 Design parameters for the PLA lens

ε_{r_1}	ε_{r_2}	ε_{r_3}	ε_{r_4}	ε_{r_5}	ε_{r_6}
72	2.60	2.38	2.08	1.71	1.30

Table 10.3 Designed parameters of the 3D-printed lens

Parameter	**Value**
Diameter	$D = 120$ mm
Focal length	$F = 150$ mm
Thickness	$T = 18.5$ mm

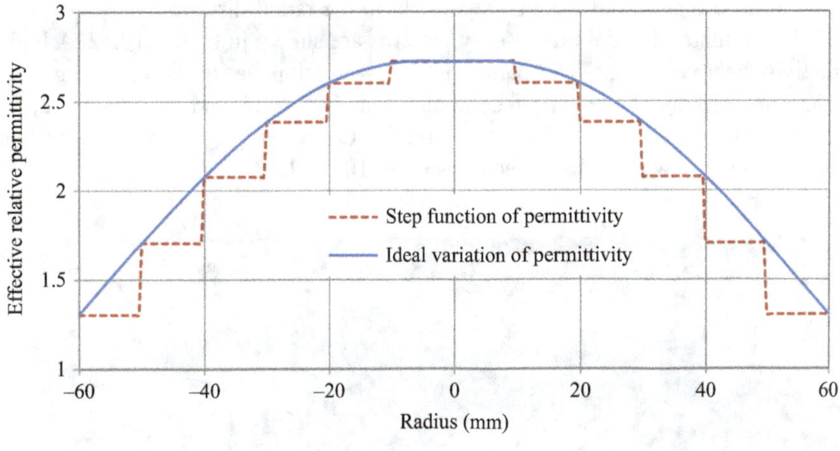

Figure 10.10 Effective relative permittivity for the 3D-printed lens as a function of its radius

The final design parameters are shown in Table 10.3. Figure 10.10 shows the step function approximation of the $\varepsilon_{r_{\text{eff}}}$ versus radial distance across the diameter of the lens. In order to obtain the bespoke $\varepsilon_{r_{\text{eff}}}$, the air volume fraction is gradually increased from the center to the outermost region by decreasing the PLA volume fraction. Increasing the number of rings for a smoother permittivity variation would improve the accuracy of the focal point and increase the gain of the lens. However, narrow rings with small variations of volume fractions are difficult to fabricate accurately due to the resolution of the printer that we are using.

A full-wave simulation using a commercial solver was carried out to verify the performance of the lens design. In order to simplify the simulation process and reduce the computational complexity, the PLA lens was modeled as six solid concentric cylindrical rings and the internal voids were not considered. Each ring had homogenous dielectric constant and the values of the ε_r follow exactly the step function from the center to the outermost ring as shown in Figure 10.10. The dielectric loss tangents for all the rings were set to be the loss tangent of 100% PLA material for simplicity (tan $\delta = 0.008$) in the numerical simulation. Generally, lower PLA volume percentages resulted in lower loss tangent values and the correlation of the two was approximately linear [28]. In reality the 3D-printed rings would have lower loss tangent values than the simulation, but since PLA is a low-loss material, the difference in performance is negligible.

A Ku-band conical feed horn was placed at the focal point which is 150 mm away from the lens. The simulated electric field of the lens antenna at 15 GHz is shown in Figure 10.11. It clearly shows that the spherical wavefronts generated from the horn are converted into a planar wavefront in the near-field region of the flat GRIN lens, and the lens in turn will have a highly directive radiation pattern in the far-field region. It is also evident from Figure 10.11 that not all the energy radiated from the feed antenna is captured by the lens. The middle of the lens has the highest field amplitude, which tapers out in the radial direction.

The simulated far-field directivity patterns are shown in Figure 10.12. A high-directivity beam in the far-field pattern in the z -direction was observed. The gain of the lens antenna and the feed horn composite was 18.0, 21.4, and 24.0 dBi at 12, 15, and 18 GHz, respectively. After achieving the desired simulated results, the lens was fabricated as will be discussed in Section 10.6.1.1.

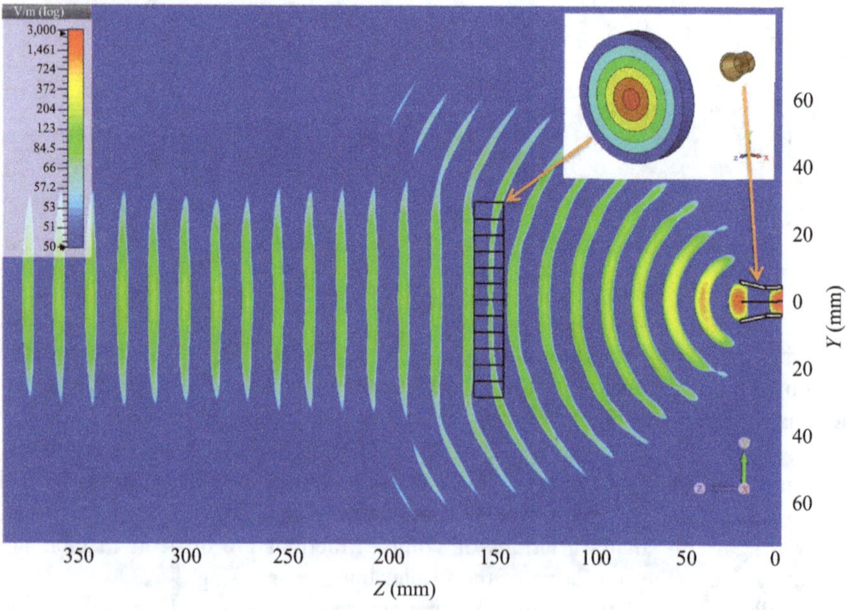

Figure 10.11 Simulated electric field of the GRIN lens at 15 GHz when fed by a circular Ku-band horn located at the focal point

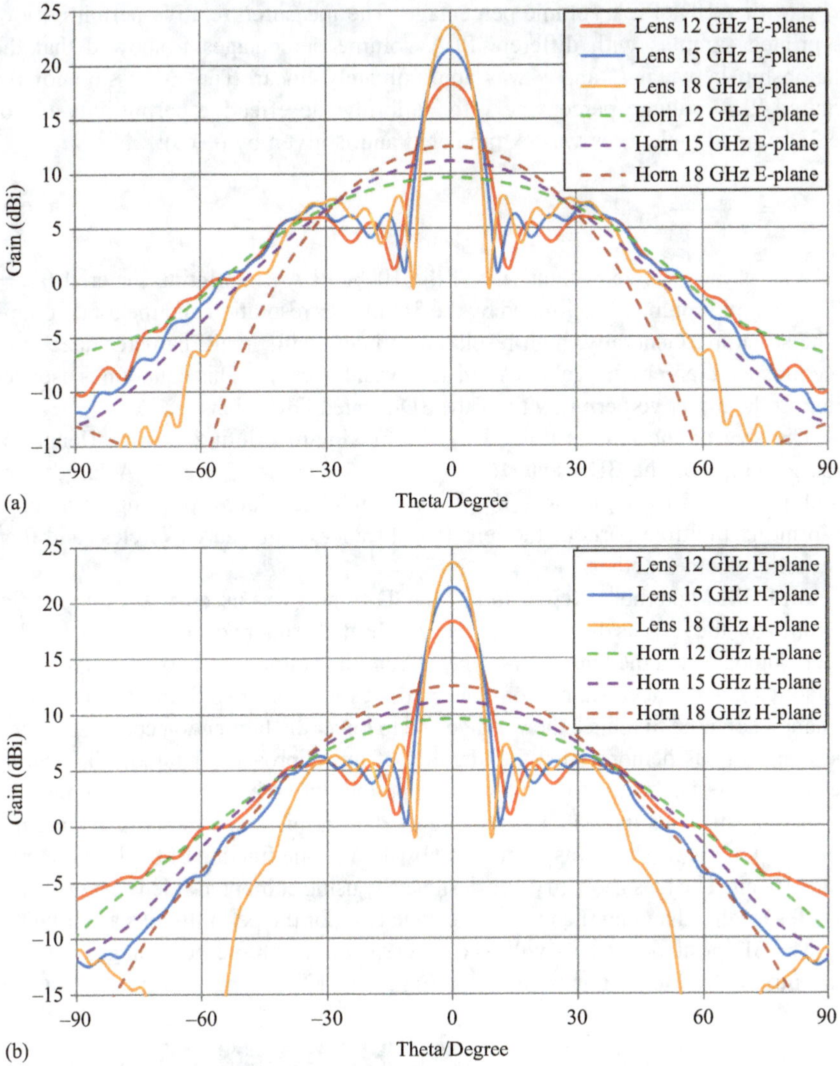

Figure 10.12 Simulated far-field gain pattern of the thick flat GRIN lens at 12, 15, and 18 GHz for: (a) E-plane and (b) H-plane

10.6.1.1 Lens fabrication

To fabricate the lens described in this section, thermoplastic polylactic acid (PLA) was used as the print material. The Nicolson-Ross and Weir (NRW) method [29] was used to measure the $\varepsilon_{r_{\text{eff}}}$ of the 3D-printed samples with different internal air void volumes. The PLA volume percentage (ν) indicated the ratio of the volume of PLA in the printed structure to the volume of the whole structure. Measurement results by using the NRW method indicated that the 100% solid/homogeneous PLA sample had a relative permittivity of 2.72 and the relative permittivity was reduced to 1.3 with a 18% PLA volume percentage. The measured relative permittivity of 3D-printed samples with different PLA volume percentages ν showed that the relationship between ε_r and ν was approximately linear. The expression for the required PLA volume percentage ν for tailoring the effective permittivity $\varepsilon_{r_{\text{eff}}}$ of the 3D-printed dielectrics was extrapolated and is given by (10.3):

$$\nu = \frac{\varepsilon_{r_{\text{eff}}} - 1}{\varepsilon_{r_o} - 1} \tag{10.3}$$

where ε_{r_o} is the relative permittivity of the 100% PLA 3D-printing material ($\varepsilon_{r_o} = 2.72$). The minimum $\varepsilon_{r_{\min}}$ is limited by the 3D-printer resolution and the total volume of the structure. Generally, high-resolution printers with small diameter nozzles or structures with larger total volumes can achieve a lower $\varepsilon_{r_{\text{eff}}}$. This equation is used for tailoring the effective permittivity of the 3D-printed GRIN lens.

The size of the lens is limited by the maximum printing volume (length × width×height) of the 3D-printer (24.6 cm × 15.2 cm × 15.5 cm). A larger lens could be realized by: (i) using a 3D-printer which has a larger printing volume; or (ii) printing multiple parts of the lens (for instance, four quarter circles) and then assembling them together.

An FDM Makerbot® Replicator™ 2X 3D-printer was used to fabricate the flat lens utilizing the PLA as the print material. The heated printer nozzle extruded the PLA material and created the lens layer-by-layer from the bottom up. This process enabled the fabrication of embedded micron-scale particles such as air voids in a single process without machining to reduce the wastage and to lower the fabrication cost. A previous research [28] has demonstrated the viability of this approach for rapidly fabricating dielectric materials with different relative permittivities by using 3D-printing. By introducing air voids into the host materials (PLA in this case), the effective permittivity ($\varepsilon_{r_{\text{eff}}}$) of the mixture was determined by the volume fraction of the host material relative to air. The lens geometry was designed by using computer-aided design (CAD) tools for locally changing the infill percentage to tailor the permittivities accordingly.

The 3D-printed flat lens with six different PLA volume percentages ν in the concentric cylindrical rings is shown in Figure 10.13. The matching layer was not

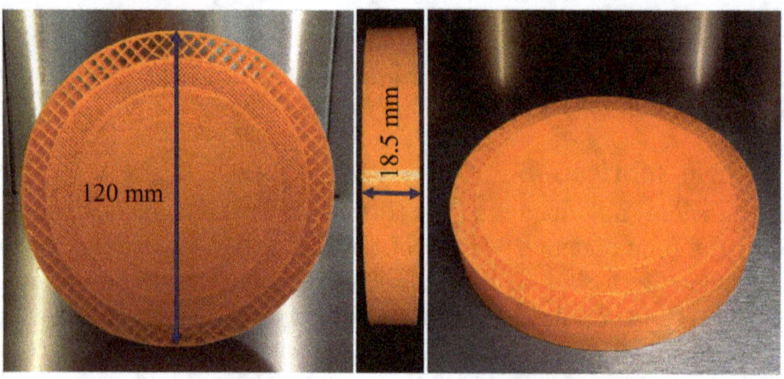

Figure 10.13 3D-printed flat PLA lens

Table 10.4 Parameters of 3D-printed concentric dielectric rings

Ring no.	$\varepsilon_{\mathbf{r}_{\text{eff}}}$	**ν**	**d**
	2.72	100%	100%
2	2.60	93%	90.8%
3	2.38	80.2%	77.2%
4	2.08	62.8%	58.0%
5	1.71	41.3%	35.1%
6	1.30	18%	10.1%

introduced in order to minimize the thickness of the lens. Equation (10.3) was used to determine the required ν for each bespoke ε_r. It is worth noting that conventional desktop FDM 3D-printers use "infill density (or infill percentage)" to describe the volume fraction of the thermoplastic to the total volume but excluding the exterior walls of the 3D-object. Generally the minimum wall thickness is equal to the nozzle diameter of the 3D-printer. Therefore, the "infill density" for the 3D-printing CAD software should always be smaller than the PLA volume percentages ν derived from (10.3). In this work, the infill density d for each cylindrical ring can be determined by using (10.4) and the results for each ring are shown in Table 10.4:

$$d = \frac{\nu\pi(R^2 - r^2) - 2\pi t(R + r)}{\pi(R^2 - r^2) - 2\pi t(R + r)} \tag{10.4}$$

where ν is the PLA volume percentage obtained from (10.3) and t is the exterior wall thickness. R and r are the exterior and interior radius of the cylindrical ring, respectively, and they include the wall thickness t. The Replicator 2X 3D-printer had a minimum wall thickness of 0.4 mm.

10.6.1.2 Lens measurement

A 180° azimuth-plane scan measurement was set up for measuring the performance of the 3D-printed lens. A Ku-band pyramidal horn antenna, which was placed at a distance of 1.5 m along the lens axis, served as the receiving antenna. This receiving horn was continuously moved from $\theta = -90°$ $\theta = +90°$ for azimuth-plane scanning. The sketch of the azimuth scanning system is shown in Figure 10.14. Note that the 1.5 m distance was slightly shorter than the far-field distance requirement ($2D^2/\lambda$) of the lens antenna below 15.6 GHz due to the limited indoor space available in the measuring chamber. A conical feed horn (open-end diameter of 22.75 mm) was used as the source for generating spherical wavefronts (see Figure 10.15(a)). The 3D-printed lens was held by a foam and was perpendicular to the azimuth-plane. The feed horn was mounted on a slider at the focal point of the lens axis, which was 150 mm away from the 3D-printed lens. The feed horn can be moved along both directions in the azimuth-plane by rotating the knobs for examining the performance off of focal point. The slider with the mounted feed horn is shown in Figure 10.15(b).

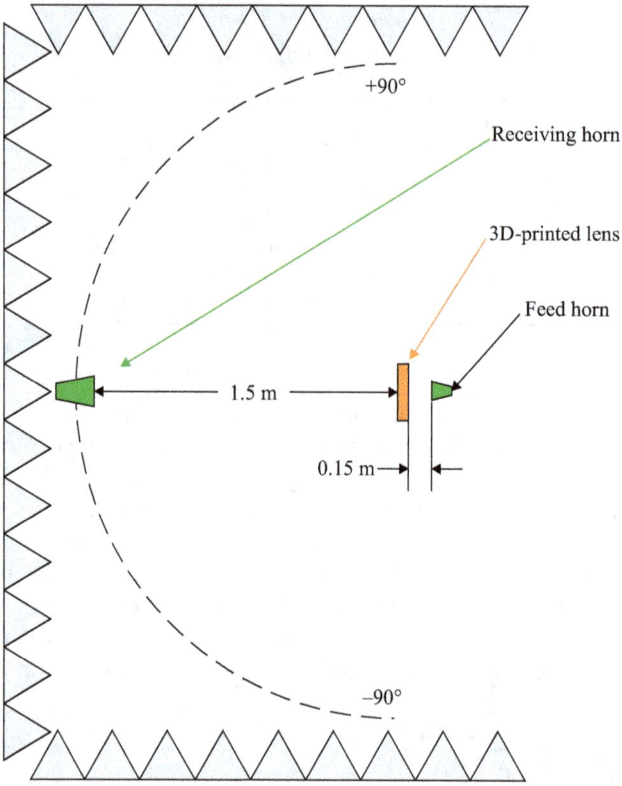

Figure 10.14 Sketch of the measurement setup for the 3D-printed lens antenna

10.6.1.3 Results

The first step was to place the feed horn on the lens axis with different feed distances to find the optimal focal length. The distance z indicated different distances between the feed horn and the lens surface and it was varied from 130 to 170 mm. Figure 10.16 shows the measured broadband gain results of the lens antenna at boresight ($\theta = 0°$), compared with the simulated gain with the feed horn placed 150 mm away. The gain of the 3D-printed lens antenna increased with frequency as was also found in the simulation results. The highest gain was observed when the feed horn was placed 150 mm away from the lens surface which matched the designed focal length. The lens antenna with far feeding (160 and 170 mm) had slightly higher gain compared to the close feeding cases (130 and 140 mm), particularly when the frequency was higher than 15 GHz. Moreover, 130 mm feeding had a lower gain compared with 140 mm feeding above 16.2 GHz. However, the measurement results indicated that the small amount of feed distance shifting had insignificant impact on the gain of this 3D-printed GRIN lens. The simulated gain varied from 18 to 24 dBi over the entire 12 to 18 GHz range, while the measured gain ranged from approximately 16 to 24 dBi.

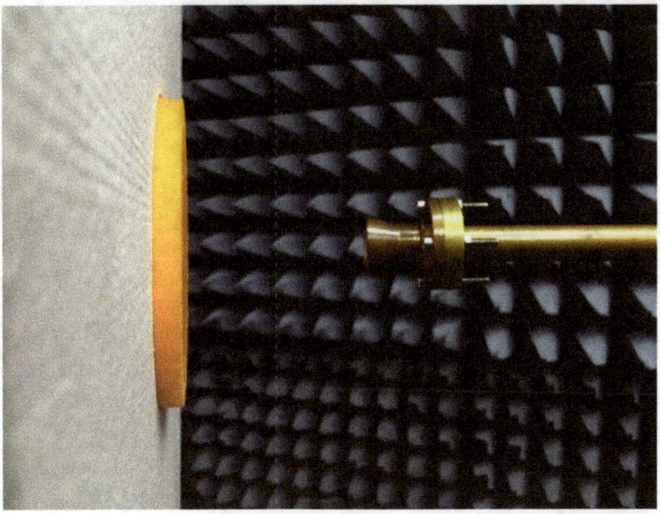

(a)

(b)

Figure 10.15 Measurement setup with the 3D-printed lens with a conical feed horn at the focal point of the lens. [Reprinted with permission from "3D-printed planar graded index lenses", IET Microwaves, Antennas & Propagation [30]]

The conical horn had a gain ranging from 7 to 13 dBi in the frequency range of 12–18 GHz; therefore, the lens provided a gain increase of 9 to 11 dB over the frequency of the Ku-band. The difference between simulated and measured gain values was due to the aberrations of the 3D-printed lens. The aberration was mainly due to fabrication

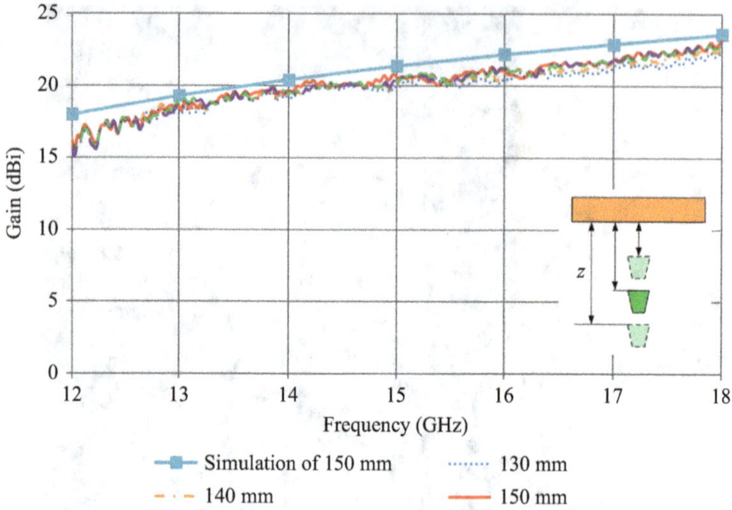

Figure 10.16 Measured broadband gain of 3D-printed lens with different feeding distances

tolerance of the infill density which was limited by the resolution of the printer. The interfaces between adjacent rings could also introduce differences between simulated and measured results.

The measured gain patterns in the E- and H-plane of the lens, at the frequencies of 12, 15, and 18 GHz, are shown in Figure 10.17, and good agreement can be observed with the simulated patterns. The radiation patterns had a higher directive main beam at higher frequencies. The secondary side-lobes were due to the feed source. At 15 GHz, the half-power beamwidth in the H-plane was approximately 11°, and 9.5° at 18 GHz. The peak of the main beam was approximately 13.8 dB higher than the first side lobe level at the center frequency of 15 GHz. The measured aperture efficiency of the fabricated lens was 18% at 12 GHz, 33% at 15 GHz, and 41% at 18 GHz.

Moving the feed horn off the lens axis resulted in a shift in the direction of the main beam and an increase of the sidelobe levels (SLLs). Figure 10.18 shows the measured results of the off-axis experiment at 15 GHz. The feed horn was moved by a distance x in the azimuth-plane. When the feed horn was moved one wavelength (20 mm at 15 GHz) away from the lens axis, an 8° shift of the main lobe was observed on the opposite side of the axis. The side-lobe level increased by approximately 1.5 dB at the same side of the feed horn and reduced by the same amount on the other side of the axis. The main beam was moved 11.5° off-axis after the feed horn was moved 3 cm away from the lens axis. The side-lobe level was increased by 0.8 dB at the feed horn side. Therefore, illuminating the lens with multiple feeds around the lens axis could achieve greater beam scanning coverage.

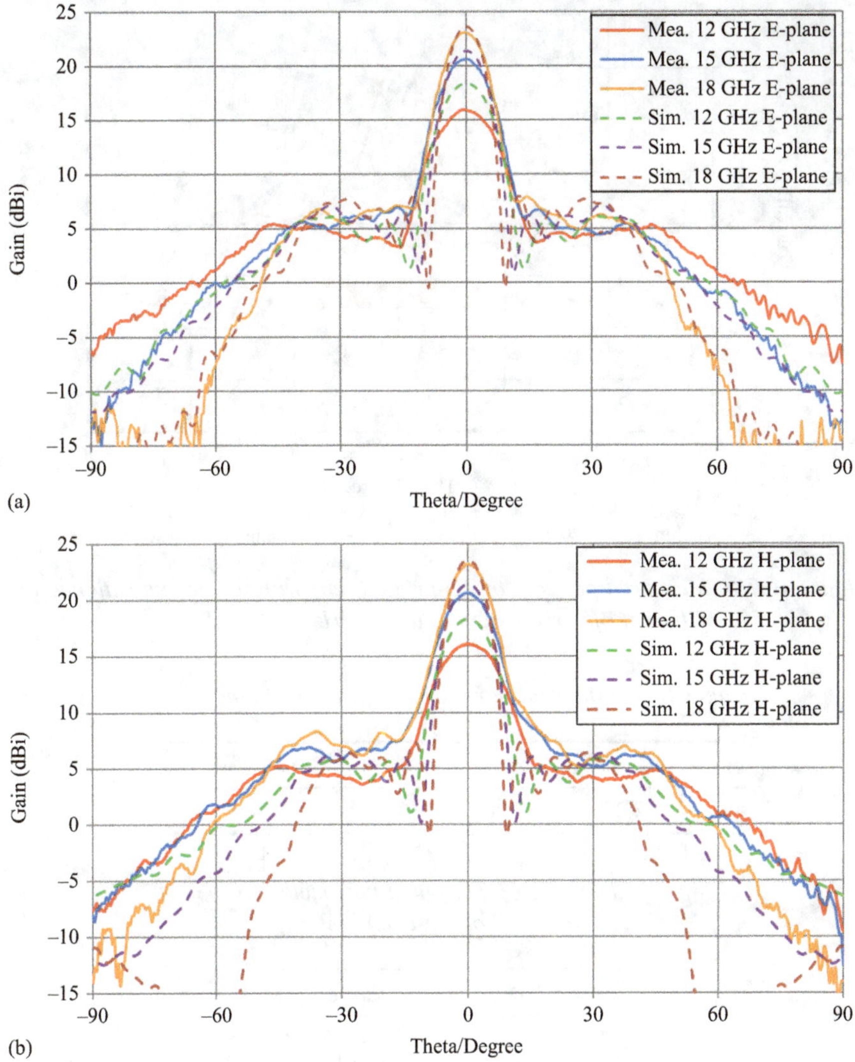

Figure 10.17 Measured and simulated far-field patterns of the GRIN lens at for: (a) E-plane and (b) H-plane

10.6.2 DaD lens design

In this section, we demonstrate the design process of the DaD lens. This new design will have ring dielectric values as shown in Table 10.5 and structural parameters as shown in Table 10.6. We see that the highest dielectric constant of this lens is higher than that provided by the PLA 3D-printing material, whose ε_r was 2.72.

A quick search of available materials reveals that not all of these materials are commercially available, off-the-shelf, for this lens. Design methodology discussed in

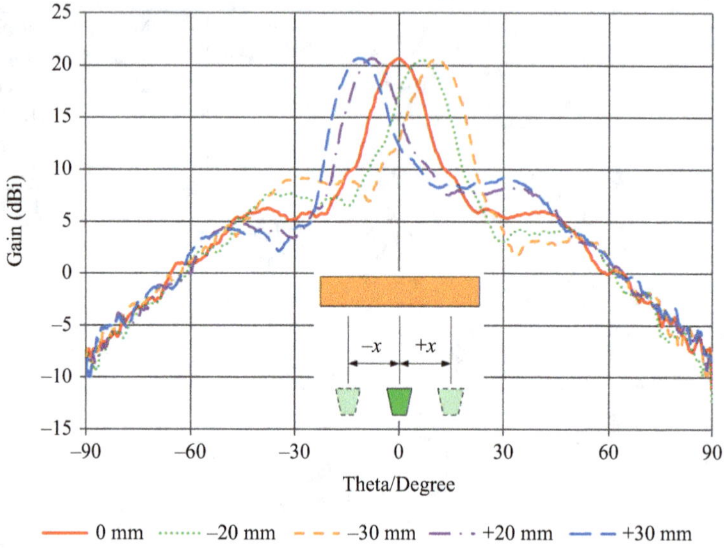

Figure 10.18 Measured patterns of 3D-printed lens with the horn positioned at different off-axis distances at 15 GHz

Table 10.5 Material parameters of the DaD lens

ε_1	ε_2	ε_3	ε_4	ε_5	ε_6
46	3.25	2.90	2.41	1.84	1.24

Table 10.6 Designed parameters of the DaD lens

Parameter	**Value**
Diameter	$D = 120$ mm
Focal length	$F = 150$ mm
Thickness	$T = 13.08$ mm

Section 10.6.1 for PLA lens also cannot be applied for the DaD lens as $\varepsilon_r > 2.72$ for the materials of some of the rings. Consequently, we use the DaD approach, in conjunction with the 3D-printing technique to realize these materials. For this design, we use the PLA 3D-material in combination with Rogers RO4350B ($\varepsilon_r = 3.48$ with thickness of 1.52 mm) laminate. If the permittivity needed for a particular ring exceeds 2.72, we design the material by using a combination of the DaD approach and the 3D-printing. We see in Table 10.5 that this lens requires higher permittivity values for rings 1–3 and lower values for rings 4–6 than those provided by PLA ($\varepsilon_r = 2.72$). Thus we use the DaD technique for rings 1–3 and pure dielectrics for rings 4–6, which can

be designed by using the PLA infill method as explained earlier in Section 10.5.4. In the DaD approach, we modify the permittivities of the COTS materials, by covering these materials with metallic patches while in the 3D-printing technique, we insert air voids in the 3D-printing material to realize the required permittivity as mentioned in Section 10.5. We have found that neither the DaD nor the 3D-printing approaches lead to material realizations that are lossy, dispersive and narrowband, as they would be if we had used resonant MTMs instead.

To tweak the material parameter of the COTS, we use square patches (other shapes can also be used) that are distributed in a Cartesian pattern as shown in Figure 10.5, and print them on both sides of the PCB laminate and place it on the PLA material for the inner 1–3 rings to realize the desired ε_r values. Unlike the previously published synthesis techniques, the presented approach does not rely on resonance properties of patches or apertures to realize the artificial dielectrics; hence, it circumvents the problem of losses and narrow bandwidths suffered by MTMs.

For the current DaD unit cell design, we choose the periodicity to be $2mm \times 2mm$ and use the patch-dielectrics composite unit cell shown in Figure 10.19. Design parameters for the unit cell are chosen to be $t_1 = 1.52$ mm, $t_2 = 11.56$ mm, $b = 2$ mm, $\varepsilon_{r_1} = 3.48$, $\varepsilon_{r_2} = 2.72$ and the patch size, a, is variable depending on the desired permittivity value. The dimensions of the patches are chosen such that the phase of S_{21} of a purely dielectric layer matches that of the COTS materials with patches. Since the incremental change is relatively small, the patch size required to accomplish this is such that it is far from the resonance range, and this is the key to realizing a wideband low-loss design. The full-wave simulations are conducted in HFSS for the DaD lens, as mentioned earlier. In Figures 10.20, 10.21, and 10.22, we show the S_{21} behavior of this composite unit cell when compared with pure

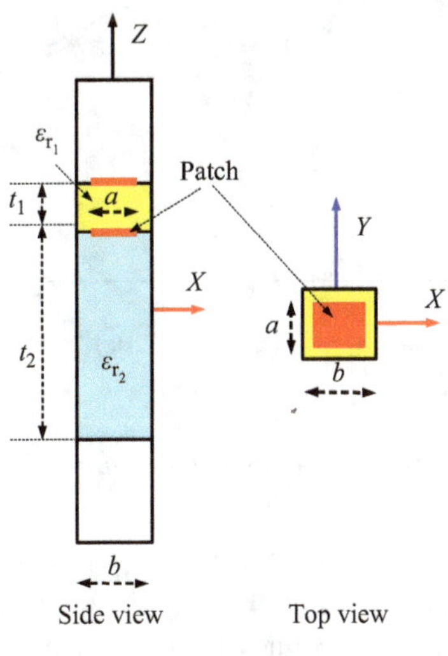

Figure 10.19 Unit cell used for the DaD lens design

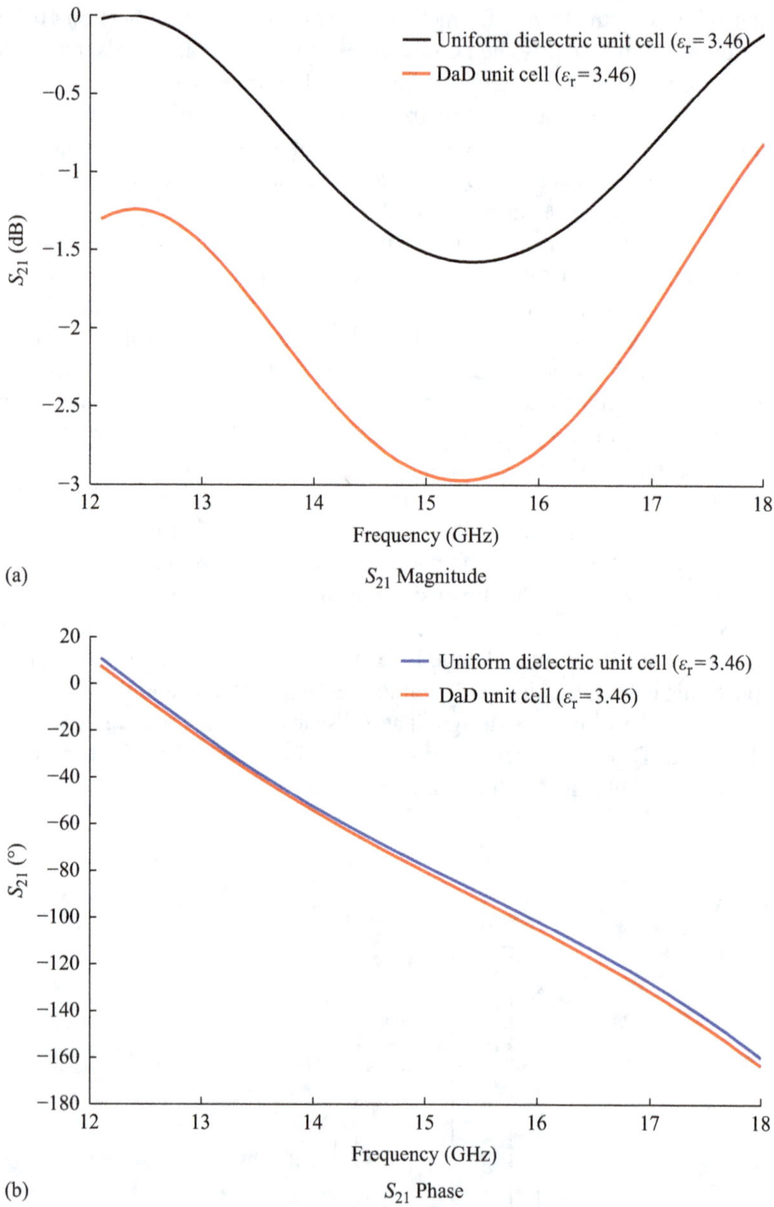

(a) S_{21} Magnitude

(b) S_{21} Phase

Figure 10.20 S_{21} parameter for unit cell of ring 1

dielectric unit cell for different rings. From these figures, we see that we are able to realize the comparable S_{21} phase values for rings 1–3, which means that we are able to achieve the required dielectric values for this patch-dielectrics composite unit cell. We also note that this composite unit cell shows lower transmission coefficient

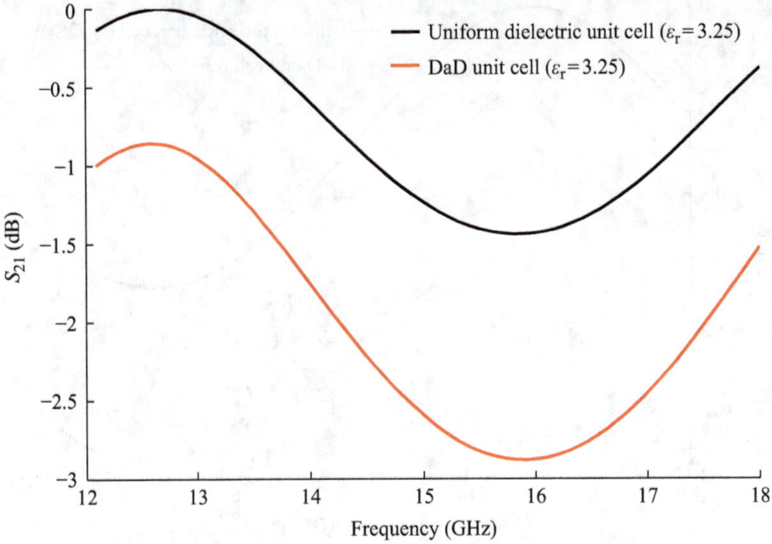

(a) S_{21} Magnitude

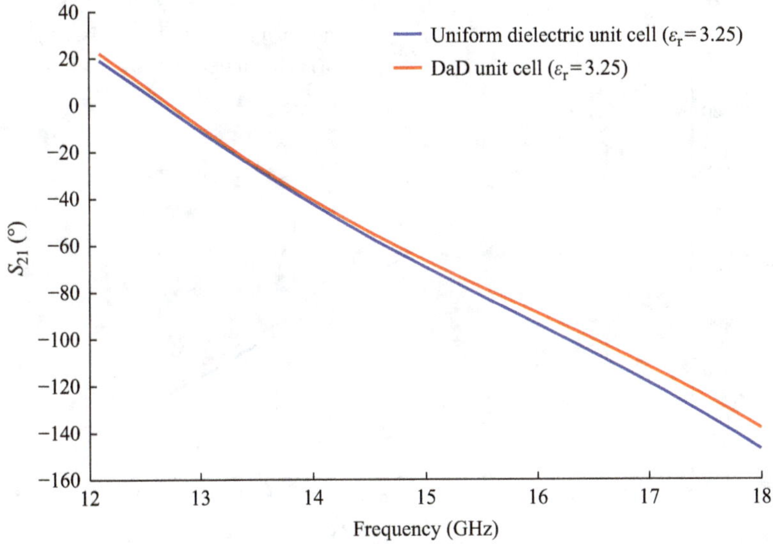

(b) S_{21} Phase

Figure 10.21 *S_{21} parameter for unit cell of ring 2*

as compared to purely dielectric unit cell. This is due to the large size of the patches that we used in these simulations. We can increase the transmission coefficient by decreasing the size of the patches but we ended up using the large-sized patches as we did not have the facility for printing smaller patches on the PCB laminates.

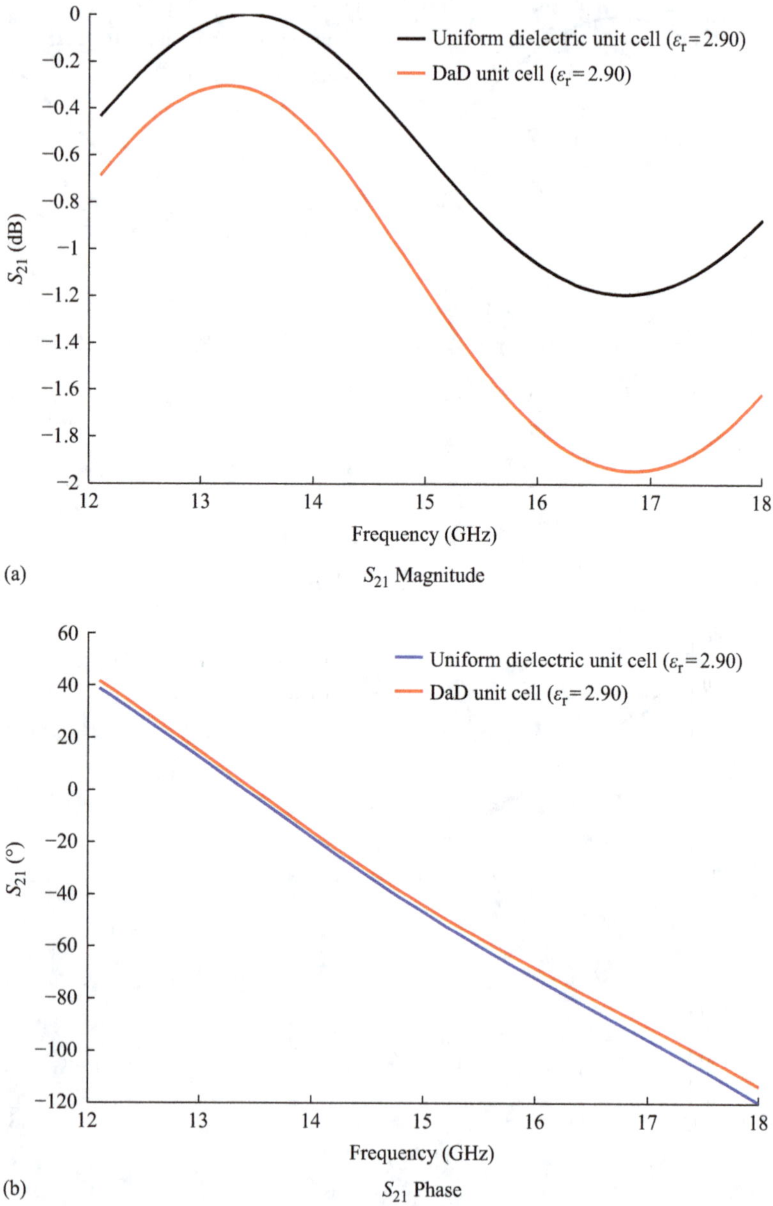

Figure 10.22 *S_{21} parameter for unit cell of ring 3*

Each dielectric ring is 10 mm wide. Each patches-dielectrics composite block is of size 2 mm × 2 mm so each dielectric ring can fit approximately 5 of such patch-dielectrics blocks in the radial direction. The height of dielectric ring is 13.08 mm. Table 10.7 shows the required size of the patches for rings 1–3 where the DaD technique

Table 10.7 Patch sizes and PLA volume density for different rings of the DaD lens

Ring No.	*a*	*d*
	1.72 mm × 1.72 mm	100%
	1.51mm × 1.51 mm	100%
	0.74 mm × 0.74 mm	100%
	N/A	84%
	N/A	52%
	N/A	17%

is used. We also show the infill density of 3D-printing material for different rings in Table 10.7.

10.6.2.1 Lens fabrication

We use the DaD technique only for rings 1–3 as explained earlier and Rogers RO4350B laminate with thickness of 1.52 mm is used. The laminate is cut into the same circular shape and patches are printed on both sides as shown in Figure 10.23. Then we 3D-print the lower part of the DaD lens and cover the patch-etched laminate on the 3D-printed material. 3D-printing technique used for DaD lens is similar to PLA lens and only the height and volume density are different than that of the PLA lens as listed in Table 10.7.

10.6.2.2 Results

We use similar measurement methodology for the DaD lens as was used for the PLA lens discussed in Section 10.6.1.2.

The simulated as well as measured far-field patterns in the H- and E-planes of DaD lens for the lowest operating frequency (12 GHz), the center frequency (15 GHz) and highest operating frequency (18 GHz) are shown in Figures 10.24, 10.25, and 10.26, respectively. Both simulated and measured results show comparable behavior.

In Figure 10.27, we show the gain behavior of the simulated and fabricated DaD lens. It can be observed that the DaD lens shows broadband behavior and its maximum gain increases with frequency as expected.

10.6.3 ABS lens design

For this lens design, we use the 3D-printing technique to design a lens with ABS as printing material. ABS material (PREPERM® TP20280) has $\varepsilon_r = 4.4$ with $\tan(\delta) = 0.004$. The design parameters of the ABS lens are the same as the DaD lens and are shown again in Tables 10.8 and 10.9 for completeness.

The main reason for using the DaD technique for the lens design in the previous section was due to the lack of dielectric material with $\varepsilon_r > 2.72$ for 3D-printing. But if we have 3D-printing material with $\varepsilon_r > 2.72$ as ABS material does provide, we can directly use 3D-printing for such a lens design. We follow this idea and 3D-printed the whole lens in a single step.

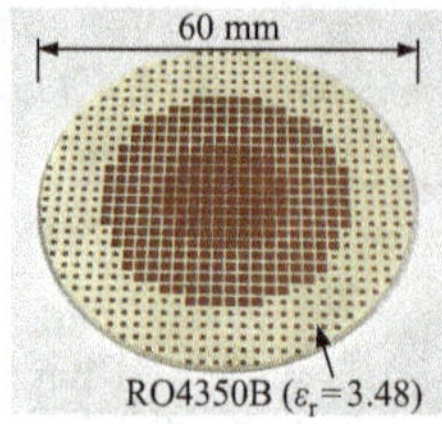

(a) PCB printed part of the DaD lens

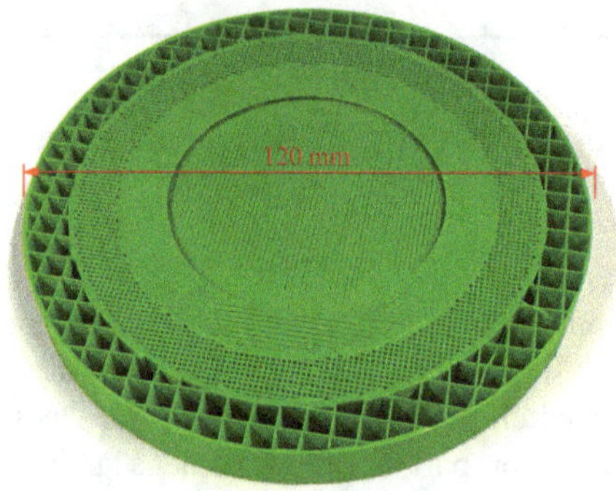

(b) 3D–printed part of the DaD lens

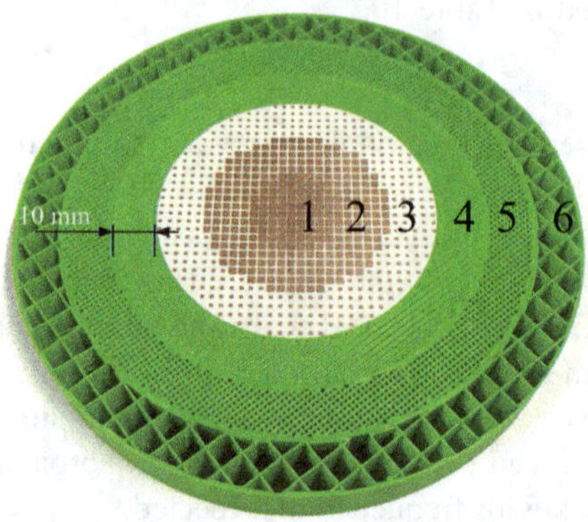

(c) Assembled DaD lens

Figure 10.23 Photographs of the DaD lens (see Figure 10.5 for all dimensions)

The ABS lens design methodology is similar to the PLA lens design and we use 3D-printing for both designs, readers can refer to Section 10.6.1.1 for more details. The infill density for ABS lens is shown in Table 10.10.

10.6.3.1 Results

A boresight measurement was set up for the measurements of the performance of the ABS lens. A Ku-band horn was used as the source to generate spherical wavefronts.

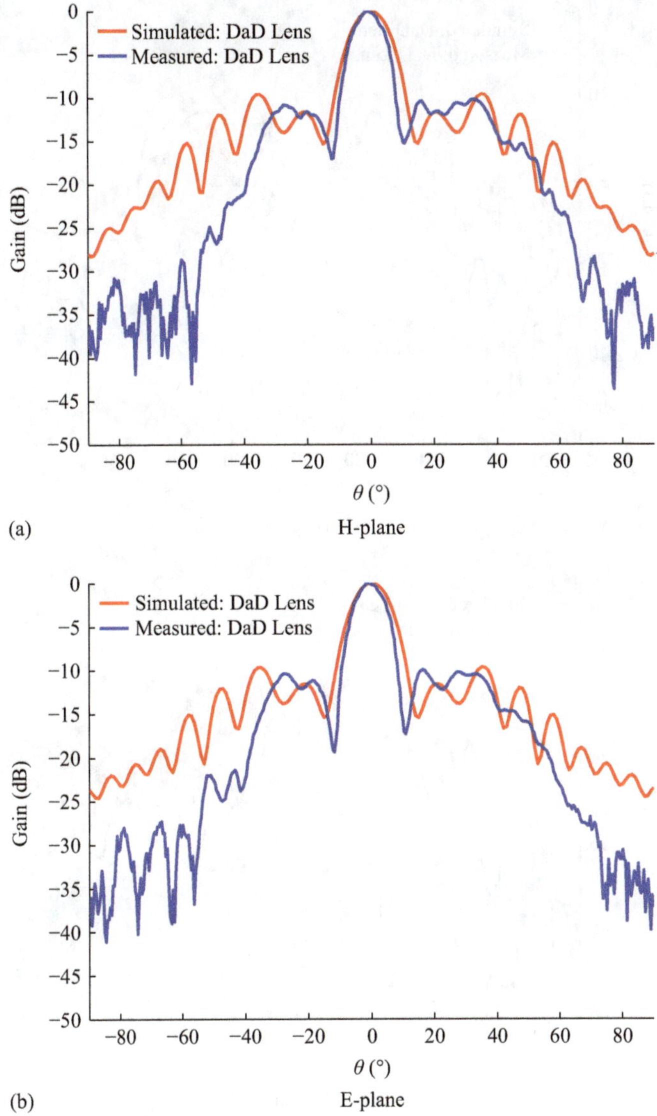

Figure 10.24 Far-field radiation patterns of DaD lens at 12 GHz

The source was placed at the focal point of the lens axis, which was 150 mm away from the ABS lens. The far-field patterns of the lens are measured similar to as mentioned in Section 10.6.1.2.

We use a conical horn (open-end diameter of 22.75 mm) as the feed source. This horn is placed at the focal point (O) and full wave simulation is conducted in

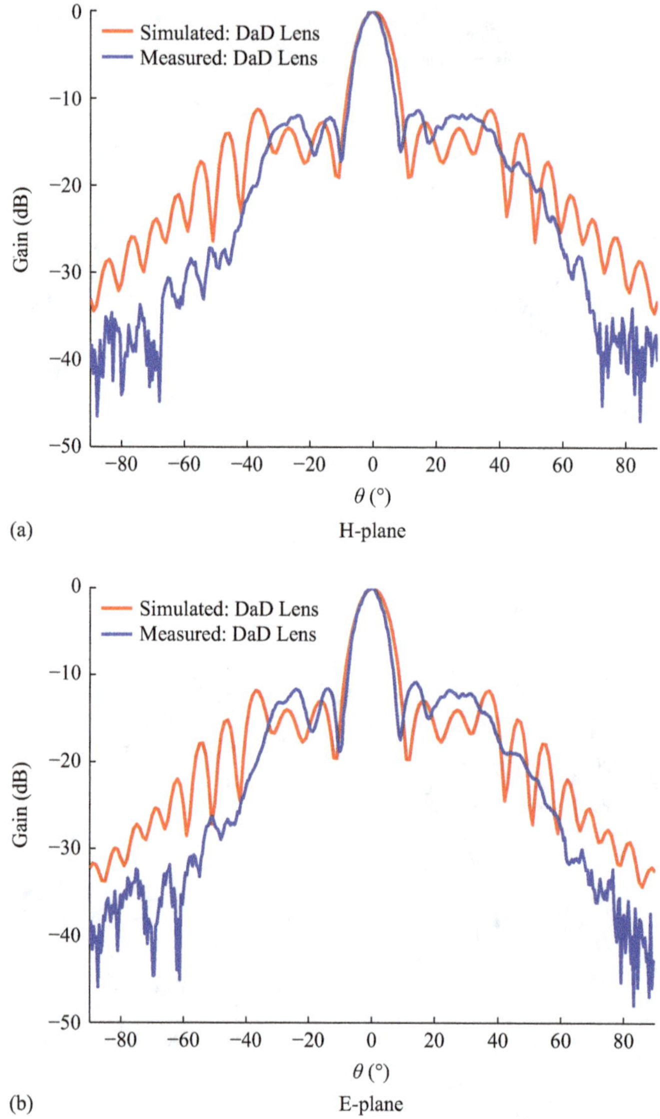

Figure 10.25 Far-field radiation patterns of DaD lens at 15 GHz

HFSS. The simulated and measured far-field pattern in the H- as well as E-plane for the ABS lens are shown in Figures 10.28, 10.29, 10.30, for 12, 15, and 18 GHz, respectively. Figure 10.31 shows the gain of ABS lens and we see that it shows broadband behavior. We also see that the gain increases with frequency.

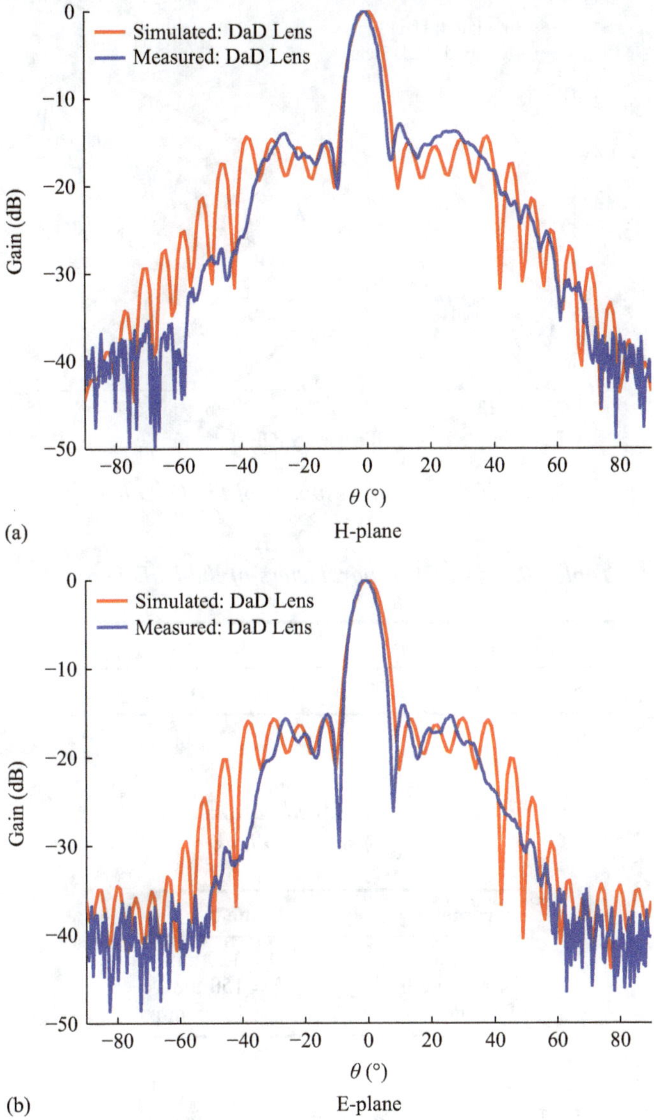

Figure 10.26 Far-field radiation patterns of DaD lens at 18 GHz

10.6.4 Comparison of DaD and ABS lenses

It can be observed in the previous sections that the design parameters are the same for the DaD and the ABS lens but the design methodology and material properties required to realize these lenses are different. One uses DaD technique by incorporating traditional PCB and 3D-printing while the other uses 3D-printing alone.

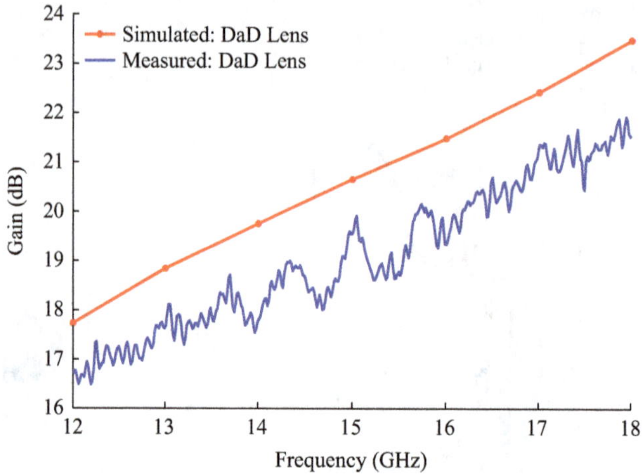

Figure 10.27 Gain response of the DaD lens

Table 10.8 Material parameters of the DaD lens

ε_1	ε_2	ε_3	ε_4	ε_5	ε_6
3.46	3.25	2.90	2.41	1.84	1.24

Table 10.9 Designed parameters of the DaD lens

Parameter	**Value**
Diameter	$D = 120$ mm
Focal length	$F = 150$ mm
Thickness	$T = 13.08$ mm

Table 10.10 Parameters of 3D-printed ABS lens dielectric rings

Ring no.	ε_r	d
1	3.46	72%
2	3.25	66%
3	2.90	56%
4	2.41	41%
5	1.84	25%
6	1.24	7%

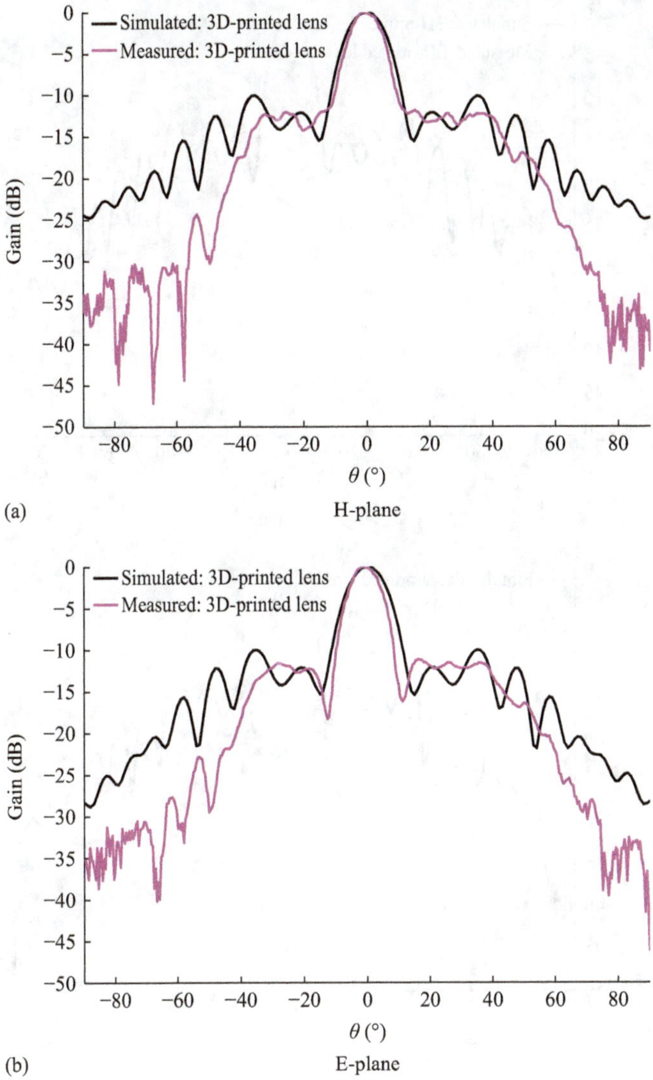

Figure 10.28 Far-field radiation patterns of 3D-printed ABS lens at 12 GHz

Results of both lenses are detailed in previous sections and are shown again in Figures 10.32, 10.33, and 10.34 for different frequencies for comparison. We also show comparison of the gain response for both lenses in Figure 10.35. From all these plots, we see that the results are comparable for both lenses. Due to the introduction of patches in the DaD approach, the DaD lens is more lossy as compared to ABS lens. We had also seen similar behavior when we studied the unit cell for the DaD lens as explained earlier.

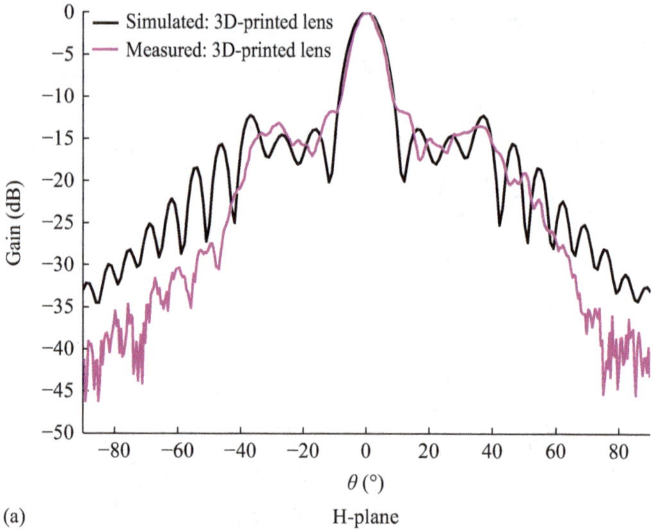

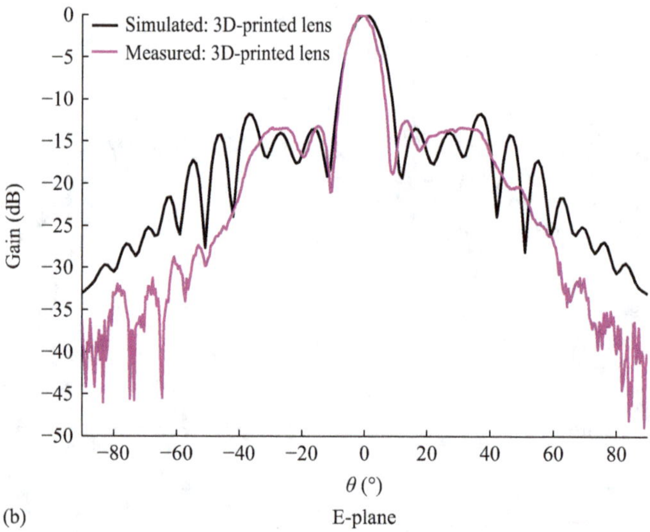

Figure 10.29 Far-field radiation patterns of 3D-printed ABS lens at 15 GHz

10.7 Summary

In this chapter we presented low-cost, light-weight and wideband 3D-printed/DaD flat lenses which can be rapidly prototyped and used for antenna applications (see Figure 10.36). Two of these are fabricated by using the direct 3D-printing technique while the remaining one is fabricated by using a combination of 3D-printing and PCB-printing techniques. For all three lenses, the dielectric materials are not

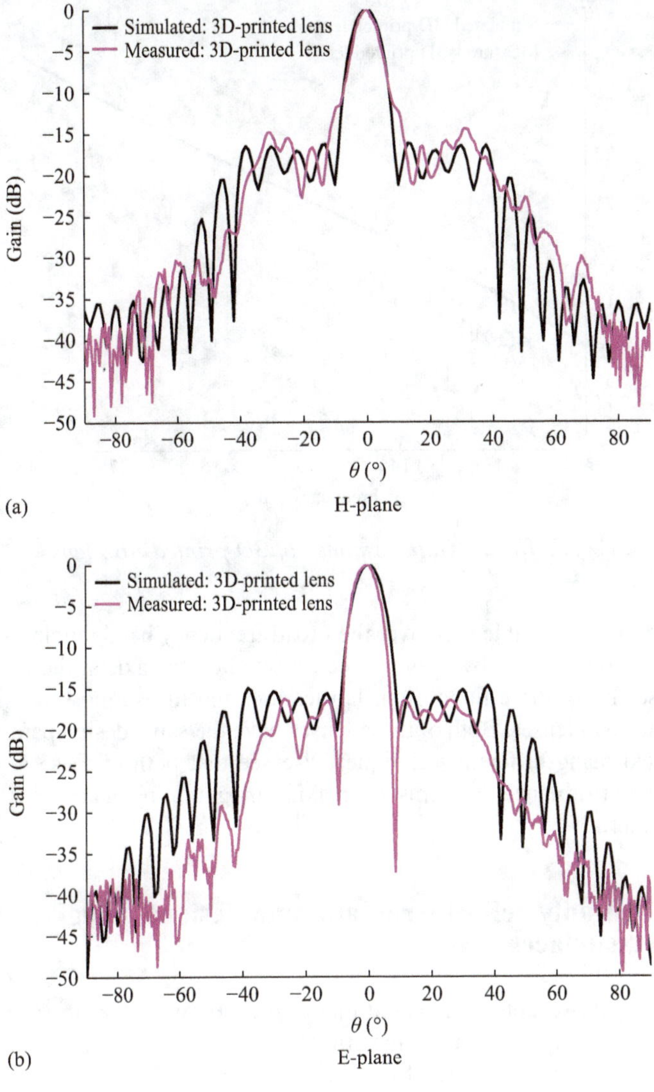

Figure 10.30 Far-field radiation patterns of 3D-printed ABS lens at 18 GHz

commercially available and they are realized by using the infill method as well as the DaD approach to achieve the desired permittivities for these lenses.

The entire PLA lens designed by using PLA as the 3D-printing material is light-weight and the material cost for this lens is less than $15. The 12 cm diameter lens is comprised of six concentric rings with different permittivity values. Air voids are created inside the unit cells of the lens during 3D-printing to reduce the permittivity of the 3D-printed material to bespoke values. The entire PLA lens is 3D-printed in a single process without the need for machining or assembling.

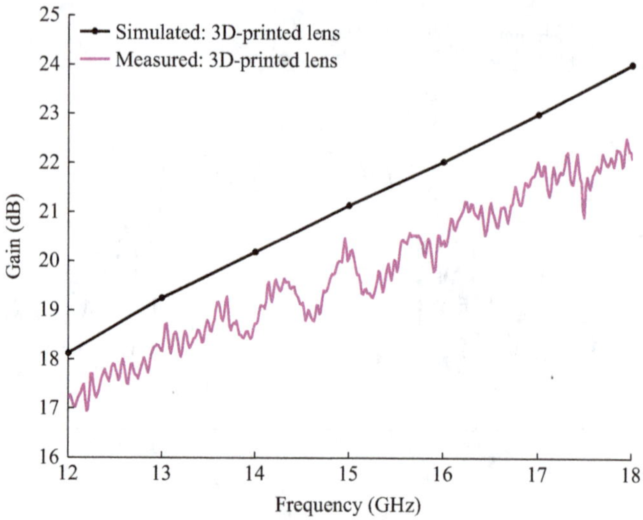

Figure 10.31 Gain response of 3D-printed ABS lens

The lens provides desirable gain over the broad frequency band ranging from 12 to 18 GHz when illuminated by a source feed located on the axis at the focal point.

We also demonstrate the design, fabrication, and measuring processes of the DaD and the ABS lenses. Both of these lenses have the same design parameters but are fabricated using different techniques. We see that both of these lenses show comparable performance in terms of maximum gain, frequency response, and radiation pattern.

10.8 Metal-only reflectarray antenna designs using metasurfaces

Before closing this chapter, we would like to mention two other interesting applications of artificially engineered materials, or MTMs. The first of these is the metal-only reflectarray antenna (MORA), described in this section, which finds applications as an alternative to reflectors, phased arrays and other high-gain aperture antennas. The second, which is discussed in the following section is a take-off on the Fabry–Perot type of resonator, which has found recent applications as a device for performance enhancement of antenna elements, or arrays, without the use of lenses of the type we have discussed in Section 10.6 in this chapter.

A reflectarray antenna [31–36], as its name implies, is a planar array of printed phasing elements illuminated by a prime-focus feed source. The reflectarray (RA) uses a reflecting surface, also referred to as an MSS, which controls the phase of the outgoing beam by placing phasing elements on its surface to transform an obliquely incident spherical wave into a planar wave front, mimicking the performance of a

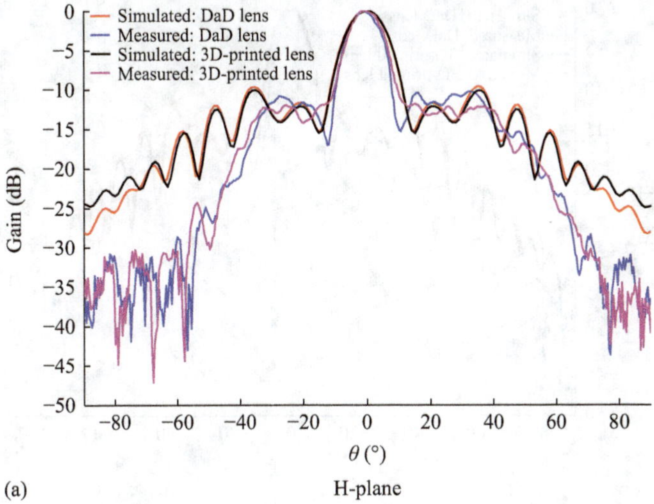

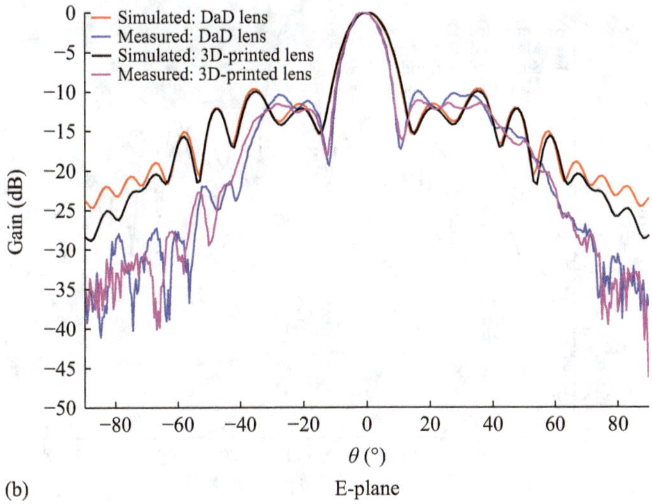

Figure 10.32 Far-field radiation patterns of DaD and 3D-printed ABS lens at 12 GHz

parabolic reflector, but with a flat surface. Mature and inexpensive manufacturing technique such as lithography can be used for fabricating the phasing elements in reflectarray antennas. Traditionally, printed dipoles, microstrip patches, rings or loops have been used as phase-shifting elements in reflectarrays [33,34]. These elements typically operate in the resonance range in order to provide the desired phase shift in the range of 0 to 360°; hence their bandwidth is typically narrow. Dielectric MSS have been proposed to mitigate this problem, broadband reflectarrays using such surface have been successfully designed [37].

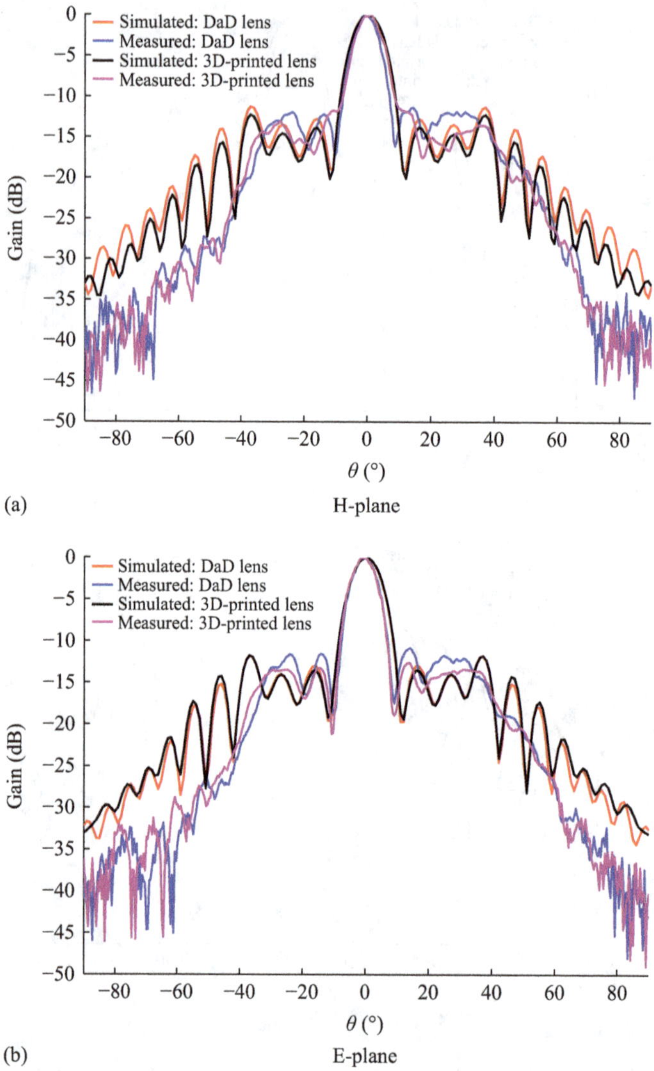

Figure 10.33 Far-field radiation patterns of DaD and 3D-printed ABS lens at 15 GHz

Recently, the MORAs have received attention because they neither require etching nor dielectric substrates; hence they are not only lowâ€"cost, but also less susceptible to environmental conditions, which makes them suitable for space applications. Several metal-only designs, which use slot-type phasing elements have been proposed in the literature [36,38,39–41]. The three-dimensional designs have also been described in [42,43].

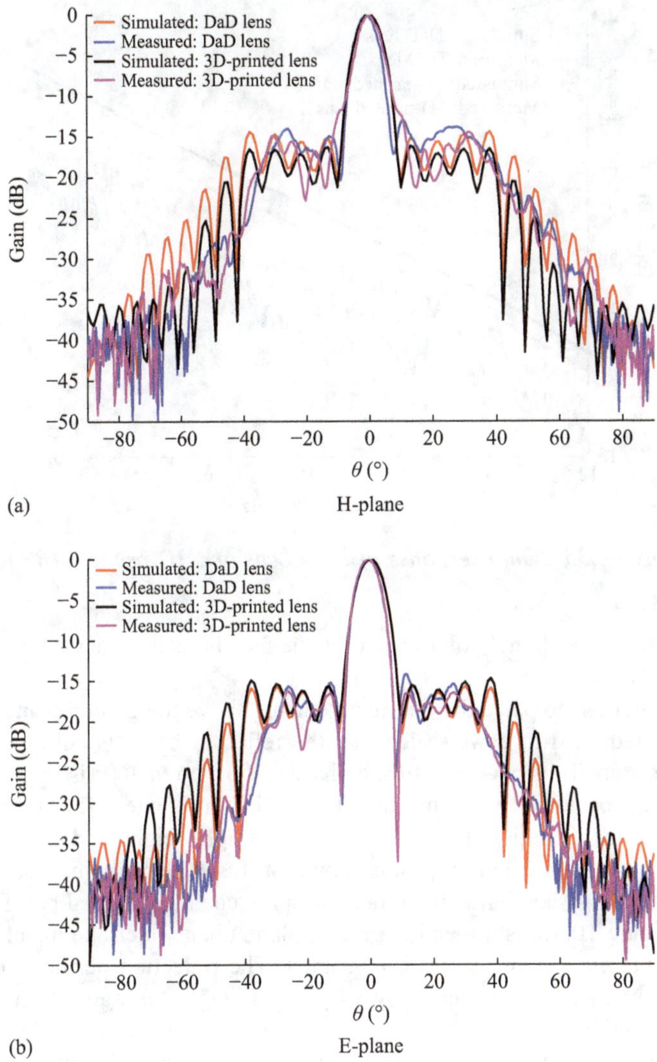

Figure 10.34 Far-field radiation patterns of DaD and 3D-printed ABS lens at 18 GHz

We will now present a novel 3D-cross type of phasing element, as shown in Figure 10.37(a), for an offset-fed MORA design. The proposed phasing element is designed for the Ka-band, and is then modified by introducing 3D quarter-section square loops (i.e., blue color) in the four quadrants for simultaneous operation in the W-band. The geometry of the phasing element for the dual-band metal-only reflectarray whose dimensions are 96.83 mm × 96.83 mm is shown in Figure 10.37(b).

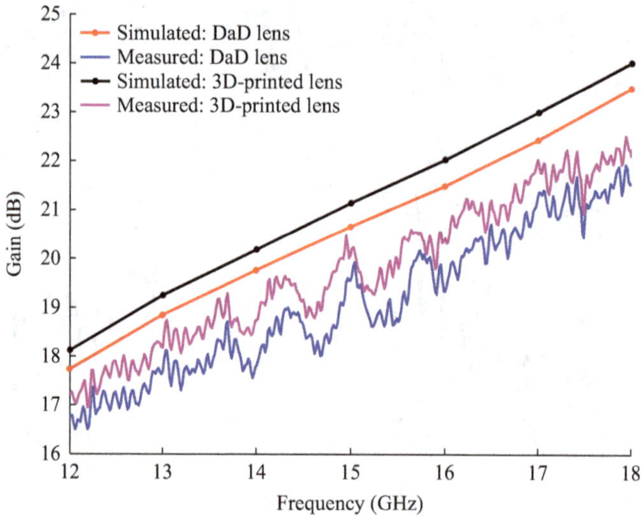

Figure 10.35 Gain response of DaD Lens and 3D-printed ABS lens

The reflectarray is obliquely illuminated by the feed horn at a skew angle of 30°, as depicted in Figure 10.38.

Our strategy is to place the phasing elements above the ground plane to reflect the illuminated field in a way such that the reflected beam is collimated in the desired direction. To accomplish this, a phase shift of up to 360° is imparted to the reflected field by linearly varying the element height of the reflectarray surface. It is worthwhile to mention that the reflectarray is lightweight, and a prototype can be conveniently fabricated by using low-cost fused deposition modeling 3D-printing technology combined with metallic spray coating. The red part in the unit cell (see Figure 10.37) is a metallic ground plane upon which the metal 3D-cross element, which is shown in green, is printed. The reflection performance of the phasing element can be evaluated by varying its height "*h*", and determining its phase contribution.

The dimensions for the phasing-element are: width W of 3D-cross = 0.5 mm; length A_L of 3D-cross = 4.21 mm; and, periodicity A of the element = 4.21 mm ($\lambda/2$). The reflection phase response of the unit cell is shown in Figure 10.39 for f (design frequency) = 35.6 GHz. Other relevant dimensions are: $A_x \times A_y$ (aperture size) where $A_x = A_y = 96.83$ mm (11.5λ); θ (tilt angle) = 30°; $\phi = 180$; H_f (height of feed) = 75 mm (8.9λ). We use a Ka-band pyramidal horn antenna as the feed antenna whose dimensions are 12.3 mm × 8.9 mm, and which is placed at a height H_f of 75 mm. The proposed phasing element is simulated by using the periodic boundary conditions with a plane wave incident on the unit cell from above. The incident field is the y-polarized field. The plot shows that we can realize the desired phase range of 0°–360°, by varying the 3D-cross phasing element height h from 0.15 to 4.9 mm, with the resolution of 0.05 mm.

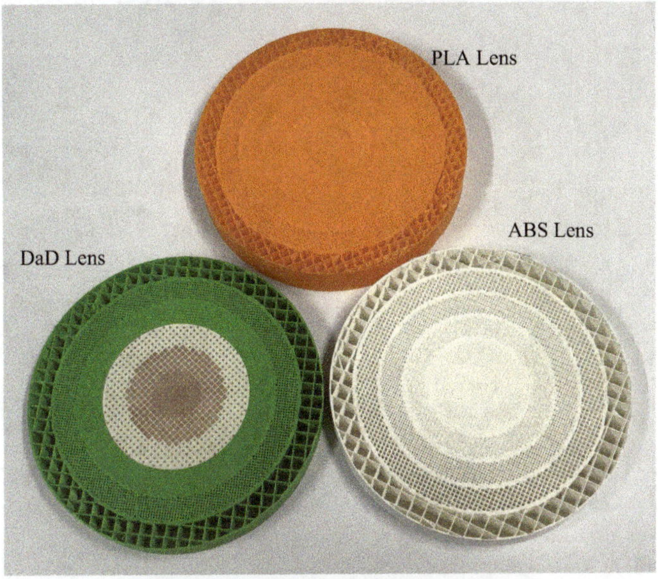

(a) Top view

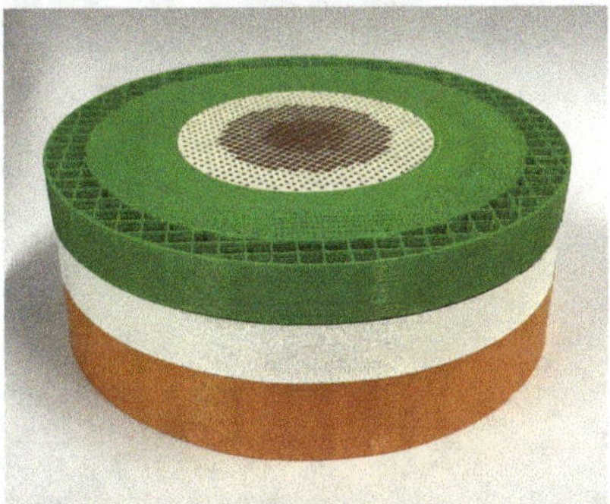

(b) Isometric view

Figure 10.36 Photograph of the different lenses

We choose the 3D-cross height "h" from Figure 10.39 based on the desired reflection phase [32] to radiate the beam in the direction normal to the reflectarray. Next, we place the 3D-cross elements along the x-axis for different values of y. The resulting offset-fed reflectarray antenna design is shown in Figure 10.38, with the exit beam designed to point along the normal of the reflectarray surface.

The simulated radiation pattern at 35.6 GHz is shown in Figure 10.40. The gain is found to be 28.9 dBi at 35.6 GHz, which is equivalent to aperture efficiency of 47%. The SLL is −19.7 dB below the beam maximum, which points to the normal direction.

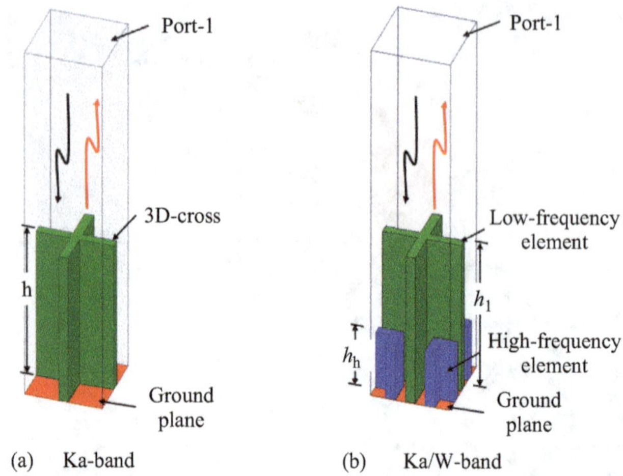

Figure 10.37 Proposed unit cell of metal-only phasing element: (a) Ka-band and (b) Ka/W-band

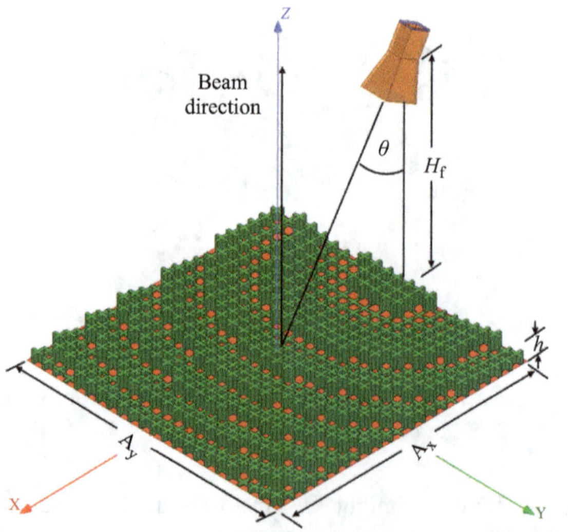

Figure 10.38 Proposed metal-only reflectarray

Figure 10.41 shows the 1-dB gain bandwidth of 10.9% (34.1–38 GHz). We note that the desired design frequency of 35.6 GHz, is satisfactorily covered by the design.

In summary, we have presented MSS designs in this section, which provide us the required phase variation needed for a dual-band reflectarry, operating at Ka and

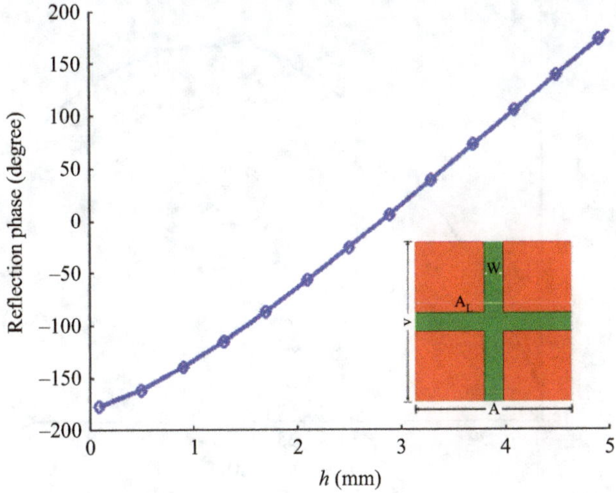

Figure 10.39 Simulated phase of proposed phasing element shown in Figure 10.37(a)

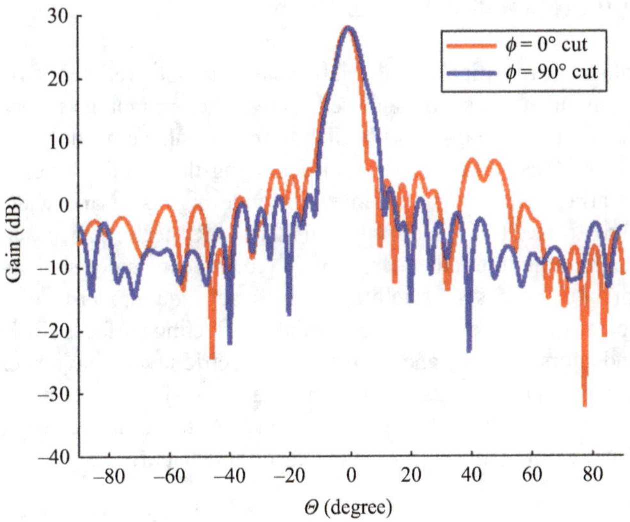

Figure 10.40 Simulated radiation pattern of MORA at 35.6 GHz

W bands. The design is suitable for space applications where it is difficult to find dielectric materials that are space-qualifiable. This is a relatively new area and is being actively researched [44] at present.

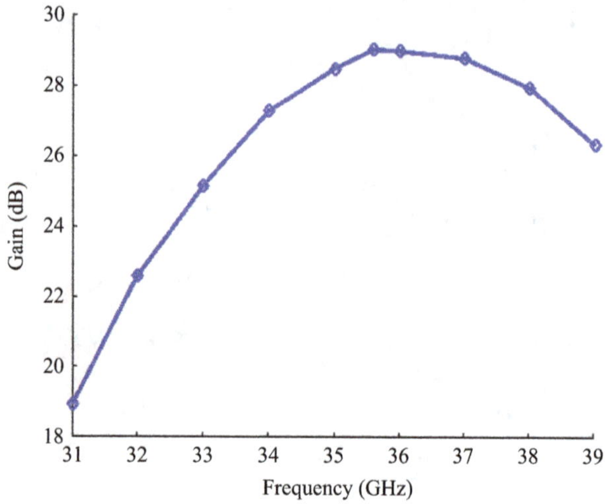

Figure 10.41 Simulated gain versus frequency of MORA

10.9 Performance enhancement of antenna and array antennas using metasurface superstrates

High-gain antennas are typically realized by using parabolic reflectors, reflect arrays, or other aperture antennas such as phased arrays when scanning is desired. Several techniques have been proposed in the literature for enhancing the various performance metrics of these antennas without degrading their radiation efficiency [45–49]. Antenna arrays which simultaneously provide high gain and wideband impedance matching, as well as wideband axial ratio (AR) performance, have attracted considerable attention of antenna designers in recent years, since such antennas find a plethora of applications in satellite, radar, space and modern communication. Frequency-selective surfaces, as well as partially reflecting surfaces (PRSs), artificial magnetic conductors (AMCs) and even plain dielectric layers, have been employed as superstrates to enhance the gain of the antenna [46–49]. In this section, we present two simple examples to show how we can use an MSS to simultaneously enhance the impedance bandwidth, gain, AR and the SLL of antenna arrays.

10.9.1 Example-1

We begin with a linearly polarized elliptical patch antenna as shown in Figure 10.42(a), referred to herein as antenna-1. The antenna is printed on an FR-4 substrate ($\varepsilon_r = 4.4$, $\tan\delta = 0.02$, $h = 1.6$ mm). This can produce a wideband and high gain circularly polarized antenna by adding an MSS. Following this, we use a PRS to further improve the gain of the antenna. The present example describes a wideband circularly polarized (CP) antenna with improved gain, which combines the MSS and the PRS to realize the desired performance.

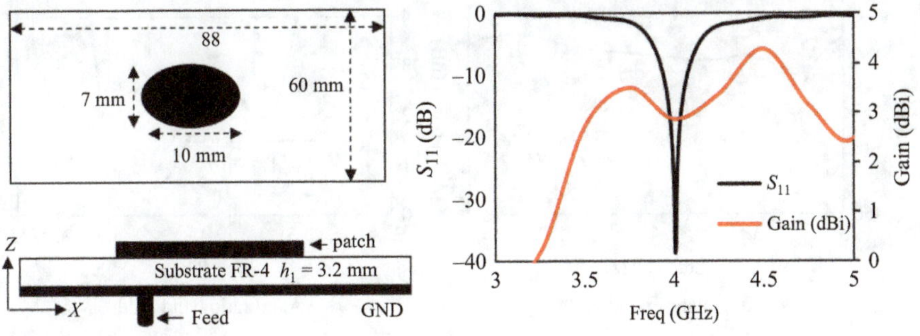

(a) Top view and front view (b) S_{11} and Gain of the antenna-1

Figure 10.42 (a) Top view of the antenna-1 with dimensions and (b) radiated gain and S_{11} characteristics of the antenna

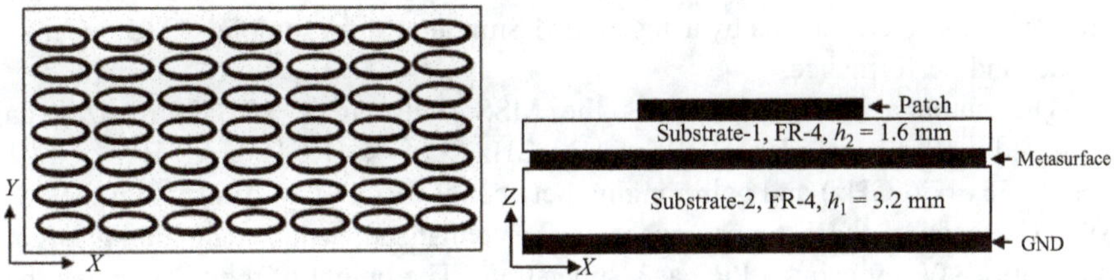

(a) Top view of array of 7 × 7metasurface (88 *mm* × 60 *mm*) (b) Front-view of antenna-2

Figure 10.43 Antenna-2 structure with MSS: (a) top view and (b) front view

Next, we turn to the design of the MSS, to be placed below antenna-1, to achieve the desired CP performance. For an anisotropic surface, however, the reflection phases are different for two orthogonal field polarizations of the incident wave. In the present work, the anisotropic MSS is realized with a 7 × 7 array of conducting elliptical rings on an FR-4 substrate ($\varepsilon_r = 4.4$, $\tan\delta = 0.02$, $h = 1.6$ mm) sandwiched between patch and ground. The radiating elliptical patch antenna is placed above the MSS. Figure 10.43(a) shows the top view of the MSS of size 88 mm×60 mm, while Figure 10.43(b) shows the front view patch antenna-2, together with the MSS.

The lengths of major and minor axes of outer elliptical ring are 4.4 and 2.8 mm, respectively, while the corresponding dimensions of the inner elliptical ring are 3.9 and 2.3 mm. The width of the elliptical ring is fixed at 0.5 mm. The proposed elliptical ring-shaped MSS behaves as a perfect magnetic conductor at 2.79 GHz, where its reflection phase is 0°. The reflection phase band for ±90° is in the range of 1.42–4.02 GHz. Within this band, the proposed MSS behaves as an AMC. The ellipticity of the ring elements of the MSS influences the resonance frequency of the antenna; its 3 dB AR BW; as well as its impedance BW. Figure 10.44(a) and (b)

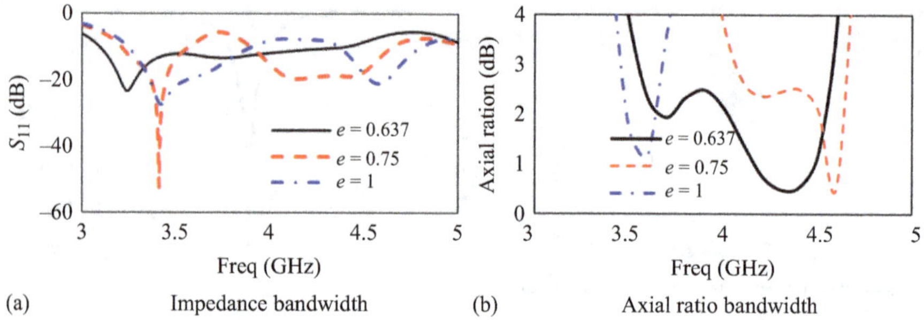

Figure 10.44 Effect of eccentricity (e) of elliptical ring MSS on (a) impedance BW and (b) axial ratio BW of antenna-2

demonstrates this aspect of ellipticity, for three different eccentricities, namely, $e = 0.637$, 0.75, and 1. We use this information to optimize the AR bandwidth of the circularly polarized antenna by using an EM Simulator, for example, CST, to realize a wideband performance.

The antenna-2 which includes the MSS, with an $e = 0.637$, realizes an improved −10 dB impedance BW of 1.33 GHz (3.08–4.41 GHz), AR BW of 1.01 GHz (3.55–4.56 GHz) and gain ranging between 7 dBi to 7.84 dBi, within the AR BW. The gain of the antenna can be further enhanced, while maintaining larger axial-ratio BW, by using a PRS as a superstrate. The height of the PRS, above the surface of the patch, influences both its gain and ARBW. The location of PRS is optimized with an EM Simulator, for example, CST.

Next, to further improve the antenna performances in term of gain, but without degrading the other performance characteristics, for example, return loss and AR bandwidths, we explore the use of a PRS as a superstrate to be placed at a height h above antenna-2. The PRS unit cell is chosen to be square-shaped, and its dimensions are 5 mm × 5 mm, and size of square metal patch is 4 mm × 4 mm. The PRS surface is designed with an array of 17 × 11 cells. The dimensions of the PRS superstrate are 88 mm × 60 mm; hence its size is the same as that of antenna-2 with the MSS. The PRS is designed on a Roger RT Duroid substrate, with ($\varepsilon_r = 2.2$, $h = 0.762$ mm). The gap width of 1 mm between the two-unit cells is kept constant in both the x- and y-directions; hence, unlike the MSS, it is an isotropic structure, which neither affects the circular polarization performance of antenna-2 nor its ARBW. Figure 10.45 shows the PRS superstrate and front view of the antenna-3, which includes both the MSS and the PRS.

Simulated and measured impedance bandwidth, gain and AR bandwidth of the proposed antenna-3 with PRS superstrate are shown in Figure 10.46. The optimized values for the height and thickness of the PRS are found to be 48 and 0.762 mm, respectively, and this choice the combination of height and thickness achieves a directive gain of 9.32 dBi with an ARBW of 1.0 GHz (24.69%). The antenna impedance BW of 1.32 GHz (35.29%) is not affected by the superstrate.

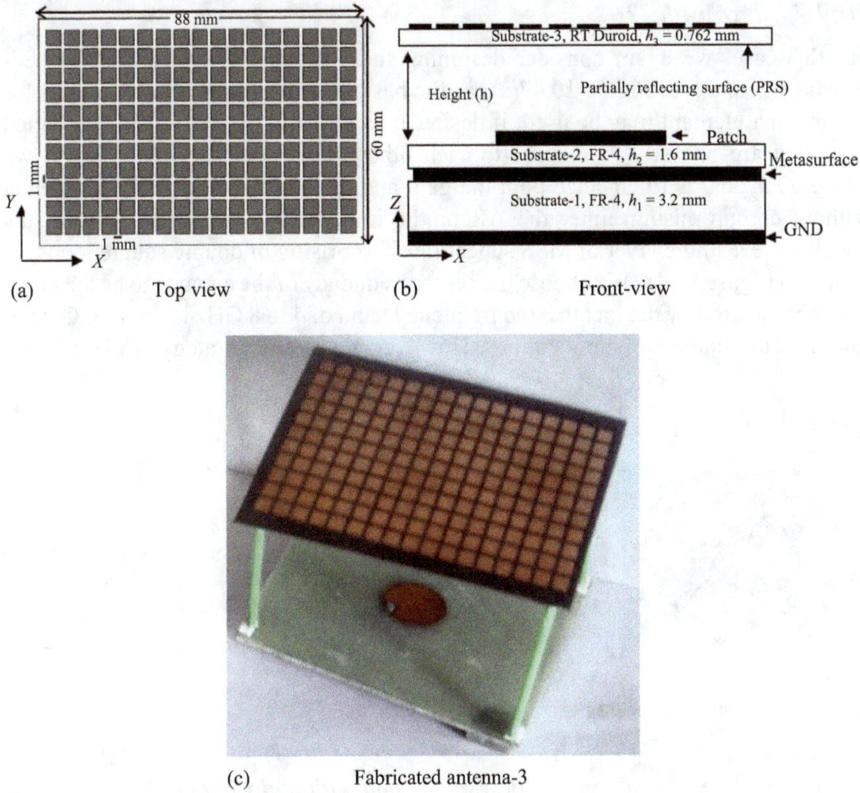

Figure 10.45 Antenna-3 with MSS and PRS superstrate: (a) top view; (b) front view; and (c) fabricated

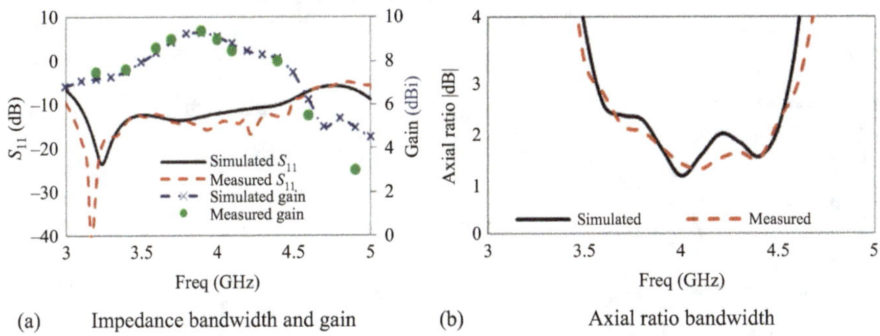

Figure 10.46 Simulated and measured impedance bandwidth, gain, and axial ratio bandwidth of the proposed antenna-3 with PRS superstrate

10.9.2 Example-2

For this case, we again consider designing the CP antenna by using a crossed-Bowtie antenna (see Figure 10.47(a)) as the basic radiator (other configurations for the antenna element may be used, if desired), comprising of two cloverleaf-shaped arms that are phased 90° relative to each other to generate the CP radiation we desire. Our goal is to enhance both the gain and SLL of the Bowtie array antenna without compromising either the AR or the impedance bandwidth. Toward this end, we use a single layer of MSS superstrate comprising of double square loops, as shown in Figure 10.48. We choose the center frequency of the antenna to be 5.2 GHz, which is dictated by the fact that the frequency band of 4 to 8 GHz is used for C-band satellite communication applications. However, the design strategy works equally

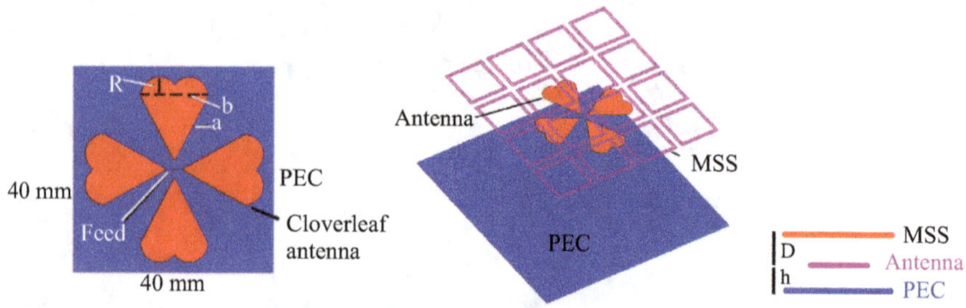

(a) Single antenna without MSS (Top side) (b) with MSS (c) Side view

Figure 10.47 Single cloverleaf Bowtie antenna with and without MSS: (a) single antenna without MSS (top side); (b) with MSS (single square loops are shown in the figure for visual clarity); and (c) with MSS (side view)

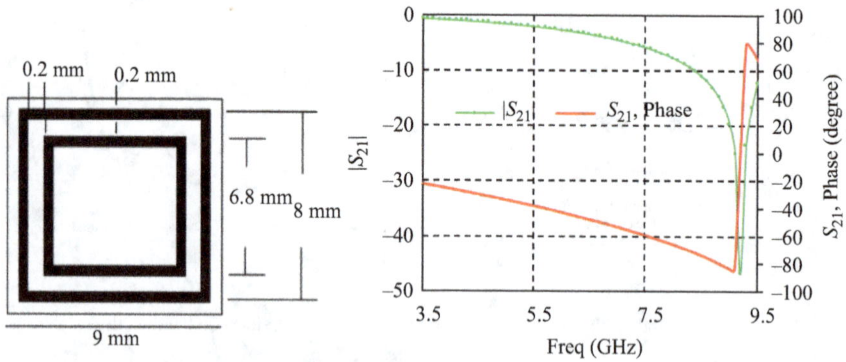

(a) Unit cell (b) S_{21} magnitude and phase

Figure 10.48 Unit cell of the MSS and its transmission magnitude and phase characteristics: (a) unit cell and (b) simulated S_{21} magnitude and phase well for higher frequencies, for example, at millimeter wavelengths. The dimensions of the antenna are: $a = 7.1$ mm, $b = 7$ mm, and $R = 1.95$ mm, where a, b, and R are defined in Figure 10.47(a).

Next, we place the antenna at a height of $h = 18.5$ mm ($\lambda/4$ at 5.2 GHz) above the PEC plane to realize a relatively wideband design. Following this, we insert a single layer of MSS at a height $D = \lambda/4$ above the antenna. We note that the transmission magnitude and phase in Figure 10.48 are −1.67 dB and 35°, respectively, at the frequency of 5.2 GHz. When D is varied from 0.5 to 10 mm, the associated return loss (S_{11}) bandwidth is found to be 10%. However, the return loss (S_{11}) bandwidth is found to improve to 38.5%, and the AR to 41.5%, by varying the height of the MSS, D, between 10 mm and 25 mm, as shown in Figure 10.49.

Figure 10.50(a) and (b) depicts the 3 × 3 and 5 × 5 cloverleaf Bowtie antenna arrays with and without the MSS superstrate. The dimensions of the antenna array are: $D_1 = 0.61\lambda$, $L_x = L_y = 90$ mm and $L_{x_1} = L_{y_1} = 200$ mm, where D_1, L_x, L_y, L_{x_1} and L_{x_1} are defined in Figure 10.50. The gain of the 3 × 3 array antenna without the

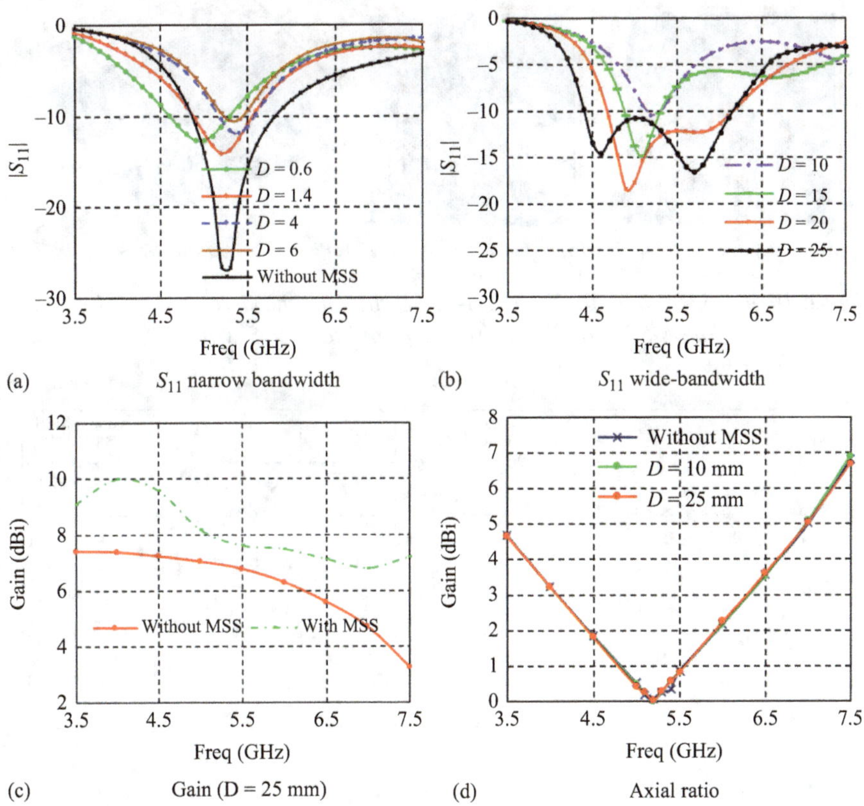

Figure 10.49 Single cloverleaf Bowtie antenna and simulated results with and without MSS: (a) S_{11} narrow bandwidth; (b) S_{11} wide-bandwidth; (c) gain ($D = 25$ mm); and (d) axial ratio

MSS are 13.89, 15.84, 17.36, and 14.8 dB, with SLLs of −13.7, −12.3, −9.3, and −0.6 dB for spacings of $D_1 = 0.47\lambda$, 0.60λ, 0.9λ and 1.3λ, respectively. The inter-element spacing is a crucial parameter that affects the gain and the SLL of the antenna array. The spacing D_1 between the two adjacent elements 0.6λ for all the designs, since this choice for D_1 provides not only an enhanced gain (see Figure 10.51) but also a desirable SLL.

Next, a single layer of MSS (12 × 12) is placed at a height D above the 3 × 3 antenna array. As we vary the height of the MSS from 10 mm to 15 mm to 20 mm, the gain changes to 17.6 dBi to 17.9 dBi and to 17.2 dBi, respectively, while the corresponding SLL changes to −16 dB to −16.1 dB and to −13.0 dB, respectively, as shown in Figure 10.51(a). By placing the MSS at $D = \lambda/4$, we take advantage of the fact that the *constructive interference* can be used to enhance the gain and SLL. By increasing the number of elements of the MSS from (12 × 12) to (16 × 16) to (20 × 20) the gain can be further improved to 17.9 dBi to 18.5 dBi and to 18.6 dBi,

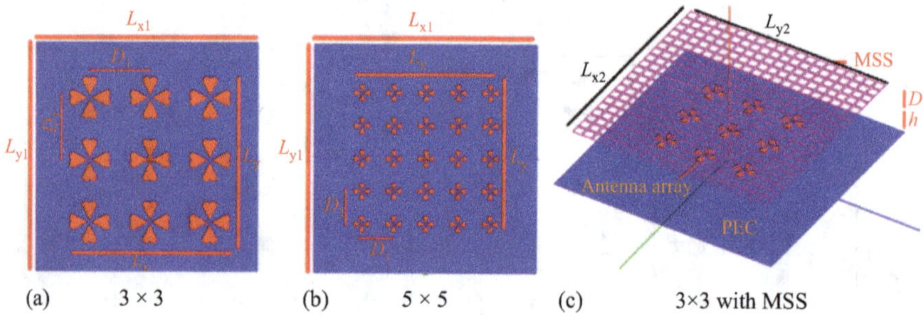

Figure 10.50 3 × 3 and 5 × 5 cloverleaf Bowtie antenna array with and without the MSS: (a) 3 × 3; (b) 5 × 5; and (c) 3 × 3 with MSS

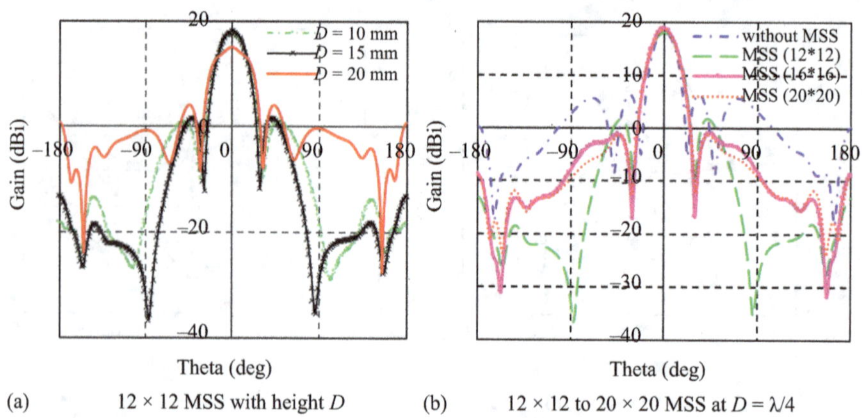

Figure 10.51 3 × 3 cloverleaf Bowtie antenna array with a metasurface superstrate: (a) 12 × 12 MSS with height D and (b) 12 × 12 to 20 × 20 MSS at D = λ/4

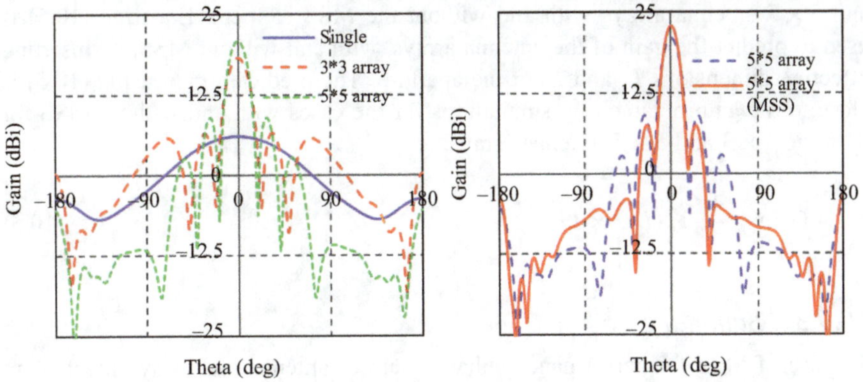

(a) 3 × 3 and 5 × 5 arrays without MSS (b) 5 × 5 array with and without MSS

Figure 10.52 3 × 3 and 5 × 5 antenna arrays with and without MSS superstrates: (a) 3 × 3 and 5 × 5 arrays without MSS and (b) 5 × 5 array with and without MSS

Table 10.11 Comparison of gain and SLL

For frequency = 5.2 GHz; different cases	Predicted gain (dBi)	Simulated gain (dBi)	Simulated SLL (dB)
3 × 3 antenna array $L_x = L_y =$ 90 mm× 90 mm	14.5	15.84	−12.3
3 × 3 antenna array with 12 × 12 MSS, $L_{x_2} = L_{y_2} =$ 107 mm × 107 mm	16.8	17.86	−16.0
3 × 3 antenna array with 16 × 16 MSS, $L_{x_2} = L_{y_2} =$ 143 mm × 143 mm	18.9	18.48	−19.3
3 × 3 antenna array with 20 × 20 MSS, $L_{x_2} = L_{y_2} =$ 179 mm × 179 mm	20.8	18.6	−20.9
5 × 5 antenna array $L_x = L_y =$ 163 mm × 163 mm	20.1	20.5	−11.8
5 × 5 antenna array with 28 × 28 MSS, $L_{x_2} = L_{y_2} =$ 51 mm × 51 mm	23.5	22.35	−14.72

respectively, while the SLL also improves to −16.0 dB to −19.3 dB and −20.9 dB, respectively, as shown in Figure 10.51(b).

Figure 10.50(b) shows the 5 × 5 cloverleaf Bowtie antenna array without the MSS. The 5 × 5 antenna array, has the dimensions: $L_x = L_y =$ 163 mm×163 mm and $D_1 = 0.6\lambda$, without the MSS. The gain of the 5 × 5 antenna without the MSS is 20.5 dBi, and its SLL is −11.8 dB. By placing the MSS above the 5 × 5 antenna at $D = \lambda/4$, the gain and SLL can further be improved to 22.5 dBi and −14.8 dB, respectively. Figure 10.52(a) compares the radiation pattern of the 3 × 3 and 5 × 5 antenna arrays, without the MSS, with that for the single antenna element, while Figure 10.52(b) compares the radiation patterns of the 5 × 5 antenna array, with and without the MSS. Table 10.11 compares the gain and SLL for both the 3 × 3 and 5 × 5 antenna arrays with and without the MSS element. Equation (10.5) is used to predict the gain of the antenna arrays, with and without MSS, by inserting effective dimensions L_x and L_y of the aperture. The predicted gain using (10.5) is close to the gain obtained by simulations for the cases with and without MSS for both the 3 × 3 and 5 × 5 antenna arrays:

$$\text{Gain} = \frac{4\pi}{\lambda^2} \times (Lx2 \times Ly2) \tag{10.5}$$

10.9.3 Summary

The use of MSS for performance enhancement of antenna and array antennas, in terms of improved impedance matching, higher gain and lower SLLs, as well as realization of wide-band AR performance of circularly polarized antennas, has been illustrated in this section. The proposed designs have the potential to replace conventional high-gain aperture antennas, such as reflectors and reflectarrays, with lower-profile alternatives, presented herein, which integrate the feed, which may be a specially designed array configuration with a low-loss feed network, together with the radiating aperture to realize the low-profile, high gain design.

References

[1] https://www.symeta.co.uk/

[2] L. Rayleigh, "On the influence of obstacles arranged in rectangular order upon the properties of a medium," *The London, Edinburgh, and Dublin Philosophical Magazine and Journal of Science*, vol. 34, no. 211, pp. 481–502, 1892.

[3] L. Lewin, "The electrical constants of a material loaded with spherical particles," *Journal of the Institution of Electrical Engineers – Part III: Radio and Communication Engineering*, vol. 94, no. 27, pp. 65–68(3), January 1947.

[4] A. H. Sihvola, *Electromagnetic Mixing Formulas and Applications*. No. 47, IET, 1999.

[5] Y. Zhang, R. Mittra, and W. Hong, "On the synthesis of a flat lens using a wideband low-reflection gradient-index metamaterial," *Journal of Electromagnetic Waves and Applications*, vol. 25, no. 16, pp. 2178–2187, 2011.

[6] D. R. Smith, D. Schurig, M. Rosenbluth, S. Schultz, S. A. Ramakrishna, and J. B. Pendry, "Limitations on subdiffraction imaging with a negative refractive index slab," *Applied Physics Letters*, vol. 82, no. 10, pp. 1506–1508, 2003.

[7] C. L. Holloway, E. F. Kuester, J. Baker-Jarvis, and P. Kabos, "A double negative (dng) composite medium composed of magnetodielectric spherical particles embedded in a matrix," *IEEE Transactions on Antennas and Propagation*, vol. 51, no. 10, pp. 2596–2603, 2003.

[8] C. C. Njoku, W. G. Whittow, and J. Vardaxoglou, "Effective permittivity of heterogeneous substrates with cubes in a 3-d lattice," *IEEE Antennas and Wireless Propagation Letters*, vol. 10, pp. 1480–1483, 2011.

[9] S. Zhang, W. Whittow, and J. Y. C. Vardaxoglou, "Additively manufactured artificial materials with metallic meta-atoms," *IET Microwaves, Antennas & Propagation*, vol. 11, no. 14, pp. 1955–1961, 2017.

[10] D. T. Moore, "Gradient-index optics: a review," *Applied Optics*, vol. 19, no. 7, pp. 1035–1038, 1980.

[11] G. Savini, P. A. Ade, and J. Zhang, "A new artificial material approach for flat THz frequency lenses," *Optics Express*, vol. 20, no. 23, pp. 25766–25773, 2012.

[12] M. K. T. Al-Nuaimi and W. Hong, "Compact size pyramidal horn lens antenna for e-band gigabit point-to-point communications," in *Antennas and Propagation (APCAP), 2014 3rd Asia-Pacific Conference on*, pp. 1183–1186, IEEE, 2014.

[13] E. Erfani, M. Niroo-Jazi, and S. Tatu, "A high-gain broadband gradient refractive index metasurface lens antenna," *IEEE Transactions on Antennas and Propagation*, vol. 64, no. 5, pp. 1968–1973, 2016.

[14] R. Yang, W. Tang, and Y. Hao, "A broadband zone plate lens from transformation optics," *Optics express*, vol. 19, no. 13, pp. 12348–12355, 2011.

[15] O. Quevedo-Teruel, W. Tang, R. C. Mitchell-Thomas, *et al.*, "Transformation optics for antennas: why limit the bandwidth with metamaterials?," *Scientific Reports*, vol. 3, p. 1903, 2013.

[16] C. Mateo-Segura, A. Dyke, H. Dyke, S. Haq, and Y. Hao, "Flat luneburg lens via transformation optics for directive antenna applications," *IEEE Transactions on Antennas and Propagation*, vol. 62, no. 4, pp. 1945–1953, 2014.

[17] Y. Zhang, R. Mittra, and W. Hong, "A zoned two-layer flat lens design," in *Antenna Technology (iWAT), 2011 International Workshop on*, pp. 412–415, IEEE, 2011.

[18] C. G. Ryan, M. R. Chaharmir, J. Shaker, J. R. Bray, Y. M. Antar, and A. Ittipiboon, "A wideband transmitarray using dual-resonant double square rings," *IEEE Transactions on Antennas and Propagation*, vol. 58, no. 5, pp. 1486–1493, 2010.

[19] A. Petosa and A. Ittipiboon, "Design and performance of a perforated dielectric fresnel lens," *IEE Proceedings-Microwaves, Antennas and Propagation*, vol. 150, no. 5, pp. 309–314, 2003.

[20] C. C. Njoku, W. G. Whittow, and J. Vardaxoglou, "Simulation methodology for synthesis of antenna substrates with microscale inclusions," *IEEE Transactions on Antennas and Propagation*, vol. 60, no. 5, pp. 2194–2202, 2012.

[21] X. Chen, H. Feng Ma, X. Ying Zou, W. Xiang Jiang, and T. Jun Cui, "Three-dimensional broadband and high-directivity lens antenna made of metamaterials," *Journal of Applied Physics*, vol. 110, no. 4, p. 044904, 2011.

[22] X. Cai, R. Zhu, and G. Hu, "Experimental study for metamaterials based on dielectric resonators and wire frame," *Metamaterials*, vol. 2, no. 4, pp. 220–226, 2008.

[23] D. Smith, J. Mock, A. Starr, and D. Schurig, "Gradient index metamaterials," *Physical Review E*, vol. 71, no. 3, p. 036609, 2005.

[24] T. Driscoll, D. Basov, A. Starr, *et al.*, "Free-space microwave focusing by a negative-index gradient lens," *Applied Physics Letters*, vol. 88, no. 8, p. 081101, 2006.

[25] H. F. Ma, B. G. Cai, T. X. Zhang, Y. Yang, W. X. Jiang, and T. J. Cui, "Three-dimensional gradient-index materials and their applications in microwave lens antennas," *IEEE Transactions on Antennas and Propagation*, vol. 61, no. 5, pp. 2561–2569, 2013.

[26] W. X. Jiang, T. J. Cui, H. F. Ma, X. M. Yang, and Q. Cheng, "Layered high-gain lens antennas via discrete optical transformation," *Applied Physics Letters*, vol. 93, no. 22, p. 221906, 2008.

[27] K. Kelleher and C. Goatley, "Dielectric lens for microwaves," *Electronics*, vol. 28, p. 142, 1955.

[28] S. Zhang, C. C. Njoku, W. G. Whittow, and J. C. Vardaxoglou, "Novel 3d printed synthetic dielectric substrates," *Microwave and Optical Technology Letters*, vol. 57, no. 10, pp. 2344–2346, 2015.

[29] W. B. Weir, "Automatic measurement of complex dielectric constant and permeability at microwave frequencies," *Proceedings of the IEEE*, vol. 62, no. 1, pp. 33–36, 1974.

[30] S. Zhang, R. K. Arya, S. Pandey, Y. Vardaxoglou, W. Whittow, and R. Mittra, "3d-printed planar graded index lenses," *IET Microwaves, Antennas & Propagation*, vol. 10, no. 13, pp. 1411–1419.

[31] J. Shaker, M. R. Chaharmir, and J. Ethier, *Reflectarray Antennas: Analysis, Design, Fabrication, and Measurement*. Boston, MA: Artech House, 2013.

[32] J. Huang and J. Encinar, *Reflectarray Antennas*. Hoboken, NJ: John Wiley & Sons, 2008.

[33] J. Budhu and Y. Rahmat-Samii, "Understanding the appearance of specular reflection in offset fed reflectarray antennas," in *Antennas and Propagation (APSURSI), 2011 IEEE International Symposium on*, pp. 97–100, IEEE, 2011.

[34] D. M. Pozar, S. D. Targonski, and H. Syrigos, "Design of millimeter wave microstrip reflectarrays," *IEEE Transactions on Antennas and Propagation*, vol. 45, no. 2, pp. 287–296, 1997.

[35] S. Pandey, R. K. Arya, and R. Mittra, "Low profile lens and reflectarray design for mm waves," in *Antennas and Propagation & USNC/URSI National Radio Science Meeting, 2015 IEEE International Symposium on*, pp. 11–12, IEEE, 2015.

[36] D. Berry, R. Malech, and W. Kennedy, "The reflectarray antenna," *IEEE Transactions on Antennas and Propagation*, vol. 11, no. 6, pp. 645–651, 1963.

[37] S. Pandey, "Techniques for designing microwave and millimeter wave antennas and components using artificially engineered materials and metasurfaces," PhD thesis, The Pennsylvania State University, 2017.

[38] G. Carluccio, A. Mazzinghi, and A. Freni, "Design of complementary reflectarray," *Electronics Letters*, vol. 50, no. 1, pp. 16–17, 2014.

[39] Y. H. Cho, W. J. Byun, and M. S. Song, "High gain metal-only reflectarray antenna composed of multiple rectangular grooves," *IEEE Transactions on Antennas and Propagation*, vol. 59, no. 12, pp. 4559–4568, 2011.

[40] R. Deng, F. Yang, S. Xu, and M. Li, "A low-cost metal-only reflectarray using modified slot-type phoenix element with 360° phase coverage," *IEEE Transactions on Antennas and Propagation*, vol. 64, no. 4, pp. 1556–1560, 2016.

[41] R. K. Arya and R. Mittra, "Offset-fed dielectric reflectarray antenna designs," in *Antennas and Propagation & USNC/URSI National Radio Science Meeting, 2017 IEEE International Symposium on*, pp. 93–94, IEEE, 2017.

[42] W. Lee, M. Yi, J. So, and Y. J. Yoon, "Non-resonant conductor reflectarray element for linear reflection phase," *Electronics Letters*, vol. 51, no. 9, pp. 669–671, 2015.

[43] H.-T. Chou, C.-Y. Lin, and M.-H. Wu, "A high efficient reflectarray antenna consisted of periodic all-metallic elements for the ku-band dtv applications," *IEEE Antennas and Wireless Propagation Letters*, vol. 14, pp. 1542–1545, 2015.

[44] M. H. Dahri, M. I. Abbasi, M. H. Jamaluddin, and M. R. Kamarudin, "A review of high gain and high efficiency reflectarrays for 5G communications," *IEEE Access*, vol. 6, pp. 5973–5985, 2018.

[45] Z.-G. Liu and W.-B. Lu, "Low-profile design of broadband high gain circularly polarized Fabry–Perot resonator antenna and its array with linearly polarized feed," *IEEE Access*, vol. 5, pp. 7164–7172, 2017.

[46] Z. Yu-wei, L. Shu, L. Ling, Y. Cai-tian, L. Sheng-chang, and L. Hao, "The simulation design of a low-side lobe level high gain and broadband microstrip patch antenna array," in *Antennas and Propagation (ISAP), 2016 International Symposium on*, pp. 742–743, IEEE, 2016.

[47] H. H. Tran and I. Park, "Compact wideband circularly polarized resonant cavity antenna using a single dielectric superstrate," *IET Microwaves, Antennas & Propagation*, vol. 10, no. 7, pp. 729–736, 2016.

[48] H. Malekpoor and S. Jam, "Improved radiation performance of low profile printed slot antenna using wideband planar AMC surface," *IEEE Transactions on Antennas and Propagation*, vol. 64, no. 11, pp. 4626–4638, 2016.

[49] N. Nasimuddin, Z. N. Chen, and X. Qing, "Bandwidth enhancement of a single-feed circularly polarized antenna using a metasurface: Metamaterial-based wideband CP rectangular microstrip antenna," *IEEE Antennas and Propagation Magazine*, vol. 58, no. 2, pp. 39–46, 2016.

Chapter 11

Microwave antennas based on metamaterials and metasurfaces

Wen Xuan Tang[1] and Tie Jun Cui[1]

Metamaterials (MTMs), first known as the left-handed materials (LHMs) or negative (refractive) index materials, have been a new frontier for both the scientific and the engineering communities in recent years. The LHMs possess both negative permittivity and negative permeability, and exhibit extraordinary physical properties such as negative refraction. The properties of LHMs were analyzed in 1968 in a pioneering theoretical work by Veselago [1]. It was over 50 years until Pendry *et al.* proposed several schemes to realize negative permittivity and negative permeability [2–4], along with some exciting potential applications of LHMs [5–7]. These contributions, accompanied with a successful experimental demonstration of negative refraction at microwave frequencies in 2001 [4], have brought about great expectation for this new kind of artificial material. However, most realized LHMs rely upon resonant structures and, therefore, have the distinct disadvantages of being high-loss and narrow-band [8]. Further development on LHMs emerged around 2005, when the concept of LHMs has been extended to MTMs. MTMs are no longer restricted to the materials with negative permittivity and/or permeability, but can also be referred to as other materials with designed electromagnetic (EM) parameters or extreme properties such as near-zero or extremely high refractive indices. Resonant or nonresonant units with designable EM response are arranged periodically or nonperiodically to compose such MTMs. In recent years, metasurfaces, plasmonic MTMs and coding MTMs have attracted a lot of attention and become new trends in the development of MTMs.

Thanks to the unique properties of MTMs, a lot of exciting, unusual, or novel physical phenomena have been discovered and experimentally verified, including negative refraction, perfect/super-resolution imaging, invisibility cloak, optical illusion, EM black hole, anomalous reflection, refraction, and radiation, etc. In the electrical and optical engineering communities, engineers seek to explore the applications of MTMs in various scenarios, for example, new kinds of microwave components, various types of antennas, polarization converters, antenna radomes, and thin metasurfaces for reducing the radar cross sections of targets.

[1]State Key Laboratory of Millimetre Waves, Southeast University, China

In this chapter, we will focus on introducing the microwave MTM antennas and their practical applications. We will demonstrate the design, realization, and measurement of several types of MTM antennas. Generally, they are categorized into two groups: the flat or curved gradient-refractive-index (GRIN) lens antennas composed of isotropic or anisotropic MTMs, and the transformation optics (TO)-based lens antennas which have distinct performances and novel functions compared with traditional antennas. In the end of this chapter, we will also give a brief introduction of metasurfaces, the two-dimensional (2D) version of MTMs, and introduce some important and latest applications of metasurfaces for antenna engineering.

11.1 GRIN MTM lens antennas

Antennas have played a significant role in modern wireless communication. In the early stage, MTMs were devoted to small antenna design because they have strong electric/magnetic responses to the outer fields. From 2005 when the concept of gradient index MTM was introduced [9], lens antennas using GRIN MTMs have soon become one of the most important applications for MTMs in engineering. In the first section, we will introduce a series of GRIN MTM lenses with flat or curved profiles.

11.1.1 MTM flat lens antenna

Lenses are commonly used in antenna systems to transform spherical waves into plane waves so as to create directive beams. Conventional dielectric lenses are usually made of homogeneous dielectrics with precisely curved surfaces decided in the basis of geometric optics. In contrast, MTM lenses can possess flat profiles because they make use of gradient refractive index to realize required phase changes instead of utilizing curved surfaces to achieve phase compensation.

11.1.1.1 High-gain flat lens

Figure 11.1(a) illustrates how to design a flat GRIN lens based on the geometric optics and Fermat principle. To generate plane waves beyond the lens, as is indicated in the figure, phases of EM fields on the upper aperture must be uniform.

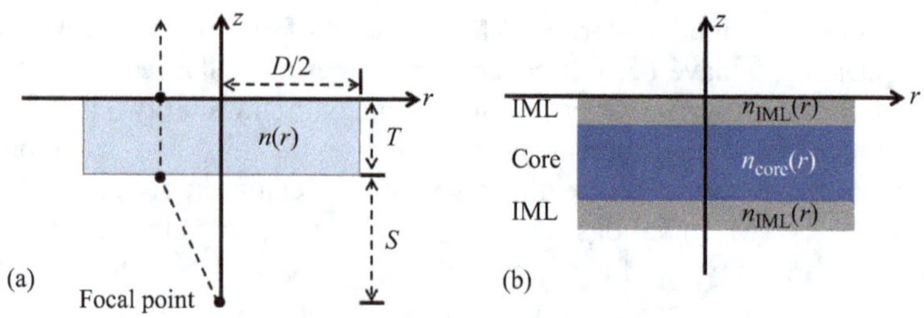

Figure 11.1 (a) The schematic diagram of a GRIN lens and (b) the whole flat lens composed of a core layer and two IMLs

In other words, every optical path from the focal point to the upper surface of the flat lens should have the same phase delay, with

$$S + n_0 T = \sqrt{r^2 + S^2} + n(r)T, \tag{11.1}$$

where $n(r)$ is the refractive index distributed in the radial direction. Note that for a flat lens with an electrically small thickness, the wave vector k is approximately in the $+z$-direction inside the lens. The refractive index distribution can be accordingly calculated as

$$n(r) = n_0 - \frac{\sqrt{r^2 + S^2} - S}{T}. \tag{11.2}$$

It should be pointed out that mismatch of wave impedance exists at the upper and lower surfaces between the lens and the air and, therefore, (11.2) cannot be directly applied in lens design. Instead, two impedance-matching layers (IMLs) are added to the flat lens in real application, as is illustrated in Figure 11.1(b). Note that usually the mismatching problem at the left and right surfaces is ignored as long as the lens is electrically thin. The IML can be considered as the one-stage quarter-wave transformer [10] and the optical path with the insertion of IMLs becomes

$$2T_{\mathrm{IML}} n_{\mathrm{IML}}(r) + T_{\mathrm{core}} n_{\mathrm{core}}(r) = Tn(r). \tag{11.3}$$

where T_{IML} and T_{core} are the thicknesses of the IML and the core, respectively, and n_{IML} and n_{core} are the radial functions of the index of refraction in the IML and the core, respectively. Since the wave impedance of the IML follows

$$\frac{\mu_0}{\varepsilon_{\mathrm{IML}}} = \sqrt{\frac{\mu_0}{\varepsilon_0}}\sqrt{\frac{\mu_0}{\varepsilon_{\mathrm{core}}}}, \tag{11.4}$$

there is a relation between n_{IML} and n_{core} as

$$n_{\mathrm{IML}} = \sqrt{n_{\mathrm{core}}}. \tag{11.5}$$

From (11.2), (11.3), and (11.5), one is able to straightforwardly decide the refractive index distribution for the flat lens.

GRIN MTMs are adopted to realize the designed refractive index distribution because they have Lorentz-like dispersive parameters, as is shown in Figure 11.2(a). According to the constructing units, MTMs are usually classified into two types: the resonant ones and the nonresonant ones. The resonant MTMs have strong EM responses and are able to realize large dynamic ranges of material parameters such as the refractive index. Since they are significantly dispersive, they operate in a narrow frequency band. In contrast, the nonresonant MTMs have weak EM responses and cannot achieve extreme values of material parameters. Since they are slightly dispersive, they operate in a wide frequency band. As long as the designed refractive index distribution has no extreme values, the nonresonant MTMs are preferable for wide-band GRIN lenses. Figure 11.2(b) and (c) presents two typical MTM units as constructing units—the metallic square ring and the drilling-hole dielectrics.

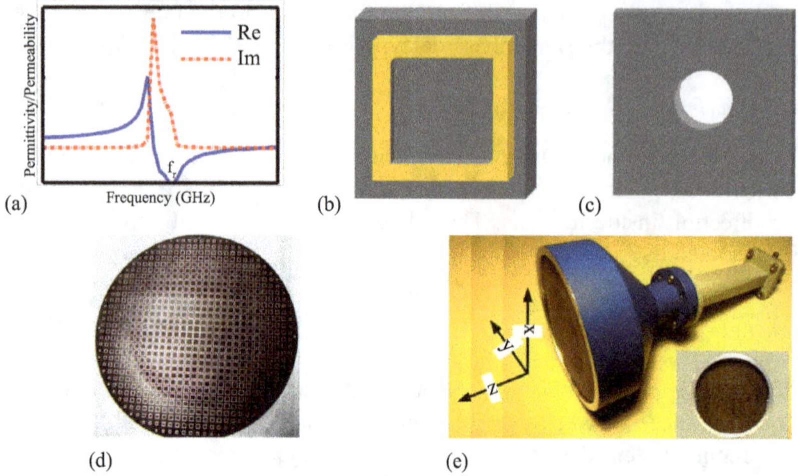

Figure 11.2 (a) Typical Lorentz dispersion curves for electrically or magnetically resonant metamaterials; (b) a metallic square ring unit cell; (c) a dielectric unit with drilled hole in dielectric substrate; (d) the top view of a sample MTM lens in [11]; (e) a horn antenna loaded with the MTM lens in [11]. ((d) and (e) reprinted with permission from X. Chen et al., three-dimensional broadband and high-directivity lens antenna made of metamaterials, J. Appl. Phys., 110: 044904. Copyright 2011 by the American Institute of Physics.)

The closed square-ring elements or air holes with variable sizes are distributed on planar substrates to satisfy the radial gradient index function as well as the axial IML configuration of the GRIN lens. Figure 11.2(d) is the top view of a sample lens. It is composed of a series of parallel layers and is located on the aperture of a horn antenna, as is shown in Figure 11.2(e).

We have presented theoretical modelling, design, and prototype demonstration for such MTM flat lens antennas in [11,12]. Measured results have verified that the prototypes maintain the low return loss and high directivity and gain in the full X band, from 8–12 GHz, for both horizontal and vertical polarizations. In particular, in [11], the measured gain of the lens antenna with the aperture size being 9.6 cm is nearly 6 dB higher than a traditional horn with the same aperture size at 12 GHz. In [12], an X-band lens with the aperture size being 1000 mm has been proved to have very high directivity as well as low sidelobes in both the E- and H-planes. Such kind of high-gain and light-weight MTM lens antennas have important applications in radar and communication systems.

11.1.1.2 High-gain and low-sidelobe flat lens

In principle, high gain is achieved by producing planar wave front on the antenna aperture. In many applications such as early warning plane radar, high gain and low sidelobes are required simultaneously. It is well known that in order to realize a

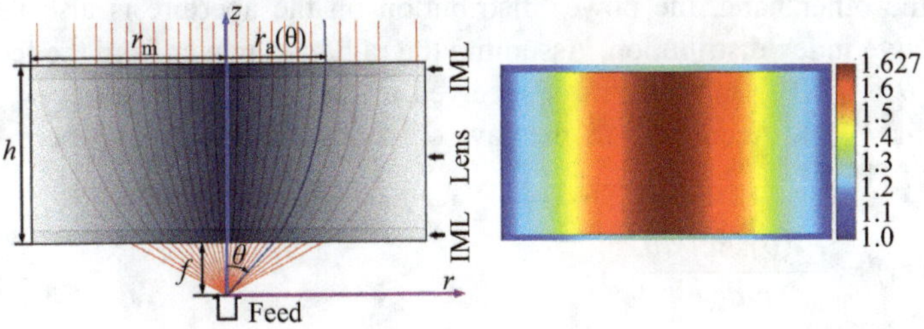

Figure 11.3 (a) Ray tracing in a flat GRIN lens and (b) the distribution of refractive index in the flat lens. ([2013] IEEE. Reprinted, with permission, from M. Q. Qi et al., tailoring radiation patterns in broadband with controllable aperture field using metamaterials, IEEE Trans. Ant. Propag., 61(11): 5792.)

high-gain lens antenna with low sidelobes, amplitude distribution on the lens aperture should approach Taylor circular distribution and the phase distribution should be uniform [13]. In this subsection, we introduce a method to tailor both the amplitude and phase of aperture field for a flat MTM lens, and so as to simultaneously achieve high gain and low sidelobes.

Figure 11.3(a) sketches such a GRIN lens with IMLs. In general, the distribution of refractive index determines the ray tracing in the lens, according to the law of refraction, and consequently determines the amplitude and the phase of aperture field. In other words, one is able to tailor the aperture field through changing the refractive index inside the lens. In [14], Qi *et al.* introduced a 10-order polynomial to express the refractive index distribution along the radial direction as

$$n(r) = n_5(r/r_m)^{10} + n_4(r/r_m)^8 + n_3(r/r_m)^6 + n_2(r/r_m)^4 + n_1(r/r_m)^2 + n_0. \tag{11.6}$$

where $r(m)$ is the diameter of the lens (see Figure 11.3(a)). Note that the coefficients n_0, n_1, n_2, n_3, n_4 and n_5 satisfy

$$n_0 + n_1 + n_2 + n_3 + n_4 + n_5 = 1 \tag{11.7}$$

so as to make sure there is no mismatch of impedance at the interface between the lens and the air.

Now let us look at the phase distribution on the aperture. Assuming a feeding horn antenna is located at the focal point of the lens, as is shown in Figure 11.3(a), EM wave enters the lens with an incident angle of θ will leave the lens on the other side at the point of $r_a(\theta)$. According to the Law of Refraction, one is able to know the phase at an arbitrary position on the aperture by calculating the optical path

$$\phi_a(r) = \int_0^{f_l+h} n(r,z)k_0\sqrt{1 + \left(\frac{dr}{dz}\right)^2}\, dz. \tag{11.8}$$

Note that at different aperture position r, the ray tracing is different, so is dr/dz.

On the other hand, the power distribution on the aperture is also related to the refractive index distribution. Assuming the radiation power from the feed in the angle of θ is $P_f(\theta)$, and the power received at the arbitrary position (r_a) on the aperture is $P_a(r)$. According to the law of energy conservation, there exists a relation of [15]:

$$P_a(r_a) = \frac{P_f(\theta)\sin\theta d\theta}{r_a dr_a}. \tag{11.9}$$

It is indicated from (11.9) that traces of the incident EM wave, which is controlled by the refractive index distribution inside the lens, finally decide how much power is received at each position on the aperture of the lens antennas.

Now, as is mentioned above, in order to achieve high gain lens antennas with low sidelobes, aperture phase distribution $\phi_a(r)$ should be uniform and meanwhile aperture power distribution $P_a(r_a)$ should approach Taylor circular array distribution (noted as $P_T(r_a)$). Global optimization algorithms, e.g., the particle swarm optimization [16], can be adopted to optimize the refractive index distribution. A fitness function is applied for optimization as

$$\text{Fitness} = A\cdot\max(|P_a(r_a) - P_T(r_a)|) + (1-A)\cdot(\max(\phi_a(r_a)) - \min(\phi_a(r_a))) \tag{11.10}$$

to balance the two aspects of amplitude and phase through the weights of A and $1 - A$, respectively. So, one is able to control the output of optimization through adjusting the value of A.

An optimized refractive index distribution was given in [14] in the form of (11.6) with $n_0 = 1.627$, $n_1 = -0.913604$, $n_2 = 0.492955$, $n_3 = -0.176123$, $n_4 = 0.00126724$, and $n_5 = 0.00193703$. Accordingly, the refractive index distribution in the IMLs is $\sqrt{n_r}$. The optimized distribution of refractive index in the lens is shown in color in Figure 11.3(b). Because the index ranges from 1.015 to 1.627 without extreme or infinite value, it can be realized through the method of drilling holes in dielectrics. The photograph of a fabricated prototype is given in Figure 11.4(a).

Additionally, since the GRIN lens is rotationally symmetric, in order to guarantee the power and phase distribution on the aperture, a proper feeding horn antenna with identical E- and H-plane radiation patterns is necessary. A waveguide antenna composed of a coaxial-waveguide transition and an open-ended rectangular waveguide loaded with a pair of horizontal umbrella-shaped metallic brims and stepped ladders has been proved to radiate nearly identical E- and H-plane radiation patterns [17]. The feeding horn antenna is also depicted in Figure 11.4(b). The configuration of the lens and the feed is shown in Figure 11.4(c).

Numerical simulation and experiment have been carried out in [14] for the demonstration of the lens antenna. Radiation patterns at three sample frequencies (12, 15 and 18 GHz) are depicted in Figure 11.5. The simulated results in (a–c) show that the gains are 26.3, 28.1 and 29.8 dBi at these frequencies, respectively, while the side-lobes are all below −28 dB in both the E- and H-planes.

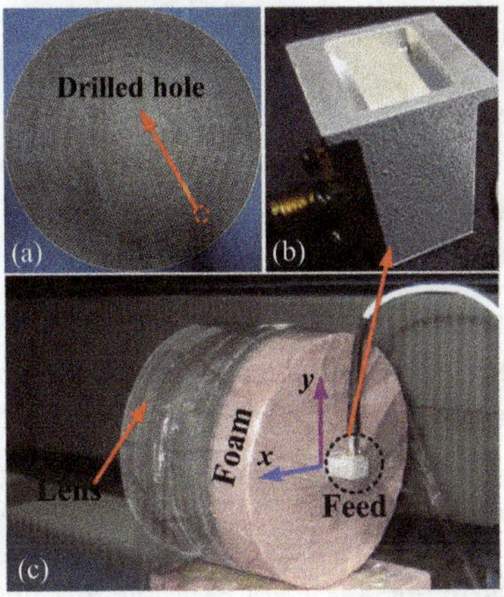

Figure 11.4 *(a) The drilled-hole dielectric plate; (b) the feed used in the lens antenna; and (c) configuration of the proposed lens antenna. ([2013] IEEE. Reprinted, with permission, from M. Q. Qi et al., tailoring radiation patterns in broadband with controllable aperture field using metamaterials, IEEE Trans. Ant. Propag., 61(11): 5792.)*

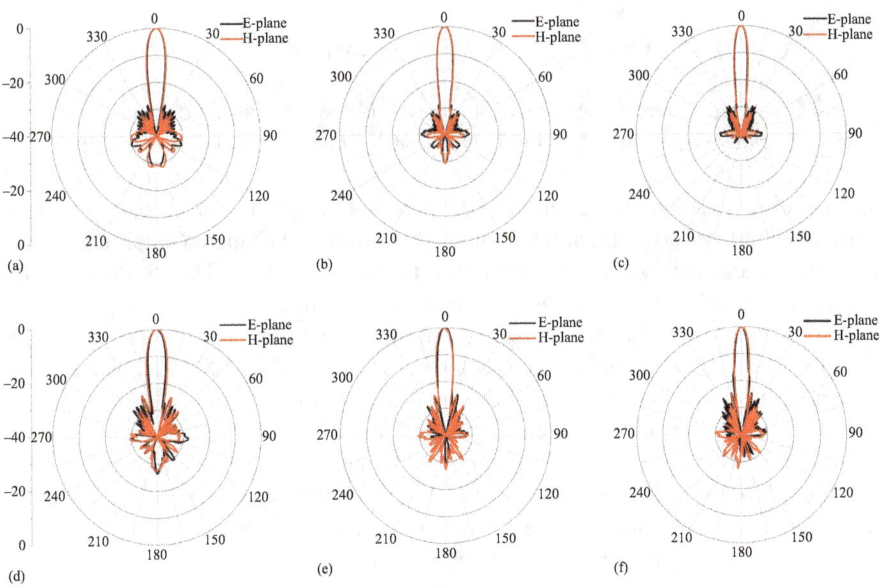

Figure 11.5 *Simulated (a)–(c) and measured (d)–(f) radiation patterns of the designed flat lens antenna at 12, 15, and 18 GHz, respectively. ([2013] IEEE. Reprinted, with permission, from M. Q. Qi et al., tailoring radiation patterns in broadband with controllable aperture field using metamaterials, IEEE Trans. Ant. Propag., 61(11): 5792.)*

The measured radiation patterns ((d)–(f)) have good agreements with the simulated ones with slight losses of gains and increases of sidelobes. Such degradation comes from the fact that constructing blocks with less-than-1.1 refractive index are neglected for the ease of fabrication. Finally, we clearly observe from measured results that high gain (higher than 25.5 dBi) and low sidelobes (lower than −24dB) are achieved simultaneously in the whole Ku-band from 12 to 18 GHz.

11.1.1.3 Loading-inside flat lens

The above subsection discussed an efficient method to manipulate the far-field radiation pattern of a lens antenna by controlling the aperture field distribution in terms of amplitude and phase. Such an MTM lens is placed away from its feed (a waveguide, a horn, etc.) as an added part of the antenna. Similar method can be applied to different existing antennas to improve directivity and depress side lobes from microwave to optical frequencies. For example, the horn antennas, which have been widely used in wireless communication systems, suffer from some inherent drawbacks such as huge discrepancy of radiation patterns in broadband and relatively high sidelobes in the E-plane. This is mainly due to the nonideal aperture field distribution. By applying an MTM flat lens inside the horn antenna, we will show in this subsection that the far-field radiation pattern can be improved significantly.

Based on the classic antenna theory [18], in a traditional pyramidal horn antenna, the amplitude on the aperture is approximately written as

$$|E_y(x,y)| = E_0 \cos\left(\frac{\pi}{C}x\right), |H_x(x,y)| = \frac{E_0}{Z} \cos\left(\frac{\pi}{C}x\right). \tag{11.11}$$

where C is the aperture dimension of the horn and the xyz coordinates are defined in Figure 11.6(a). Equation (11.11) indicates that the electric field is in the y-direction and its amplitude is nearly uniform on the aperture. In contrast, the magnetic field is in x-direction and its amplitude is tapered. The electric field distribution is visualized in Figure 11.6(a) and we can clearly observe the tapered amplitude in the x-direction and nearly uniform amplitude in the y-direction. This is the primary cause of high sidelobes in the radiation pattern. As is discussed in previous section, in order to reduce the sidelobe level, tapered distribution in the E-direction (y-direction) on the aperture is also required. In view of this, a MTM lens has been designed inside the horn to modify the aperture distribution [19].

The refractive index distribution along the y-direction for the MTMs lens is achieved through a procedure of calculation and optimization similar to that introduced in the previous subsection. We remark that since the E-field is in the y-direction, simple anisotropic electric resonators, the metallic strips shown in Figure 11.6(b), are chosen as the composing units. Units with longer strips are arranged at central areas to realize higher refractive index up to 1.7 while those with shorter strips are arranged at upper and lower areas to realize refractive index close to 1. Metallic strip arrays are printed on dielectric slices and in general the core layer includes four slices and each IML includes two slices, as depicted in Figure 11.6(c). These slices are assembled to form the flat lens and inserted inside

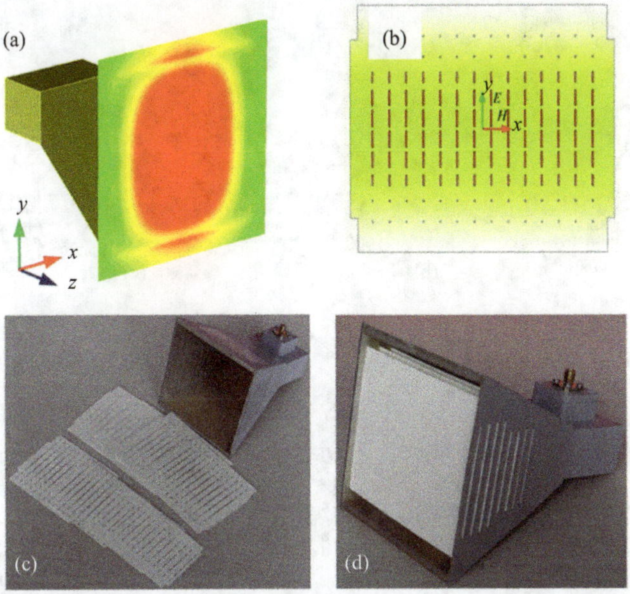

Figure 11.6 (a) Simulated amplitude of the electric field on the aperture of a horn antenna; (b) metallic strips are printed on dielectric substrates; (c) the air horn (an empty horn filled with the air) and the comprising MTMs; and (d) the meta-horn filled with the loading-inside MTMs lens. (Reprinted by permission from Nature Publishing Group. M. Q. Qi, et al., suppressing side-lobe radiations of horn antenna by loading metamaterial lens, Sci. Rep., 5: 9113. Copyright 2015.)

the pyramidal horn, as shown in Figure 11.66(d). In this way, the EM energy can be to some extent gathered to the central region of the aperture and a tapered field distribution is formed in both the x- and y-directions, guaranteeing low sidelobe levels in radiation pattern.

A prototype of the MTMs-loaded horn antenna has been designed and fabricated in [19]. The simulated and measured far-field radiation patterns of the empty horn (termed as the "airhorn" in the figure) and the MTMs-loaded horn (termed as the "metahorn" in the figure) at three sample frequencies (4, 5, and 6 GHz) are shown in Figure 11.7. Three features can be observed from the results. Firstly, the MTMs-loaded horn antenna has significant reduction of sidelobe and backlobe radiations in the operating frequency band compared with the empty horn, proving that one can effectively reduce the sidelobes of a horn antenna by loading MTMs lenses. Secondly, the beamwidths in the H-plane are almost the same for the empty horn and the MTMs-loaded horn, which is due to the fact that the gradient refractive index is only distributed in the E-direction. Thirdly, the gain of the MTMs-loaded horn is slightly lower than that of the empty horn. This is consistent with

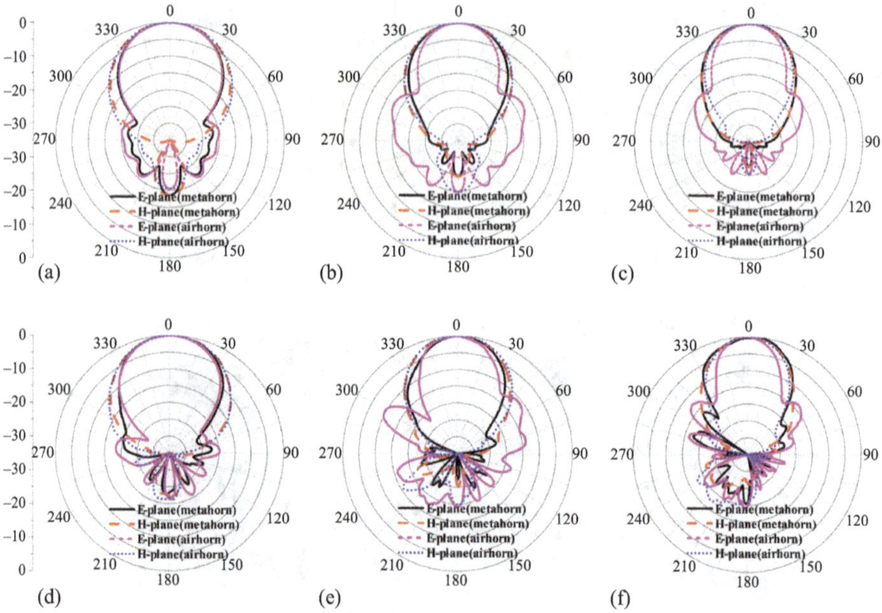

Figure 11.7 Simulated (a)–(c) and measured (d)–(f) far-field radiation patterns of the empty horn (airhorn) and the MTMs-loaded horn (metahorn) at 4, 5, and 6 GHz, respectively. (Reprinted by permission from Nature Publishing Group. M. Q. Qi, et al., suppressing side-lobe radiations of horn antenna by loading metamaterial lens, Sci. Rep., 5: 9113. Copyright 2015.)

the theory that the uniform distribution of aperture field, when compared with the tapered one, leads to higher gain.

The loading-inside flat lens is a handy component to tailor the radiation pattern of a horn antenna without increasing the antenna size. This is another good example to show the ability and flexibility of MTMs in manipulating trajectory of EM waves and modifying far-field performance of existing antennas.

11.1.1.4 Anisotropic flat lens

MTMs lenses discussed in Sections 11.1.1.1 and 11.1.1.2 are composed of nearly isotropic unit cells, which have almost the same responses to EM waves from all directions. In contrast, the lens discussed in Section 11.1.1.3 is composed of anisotropic units, the metallic strips, and has EM responds only if the incident electric field is parallel to the strips. This design indicates that anisotropic MTMs lens can provide independent control of differently polarized waves. An emblematic application on anisotropic MTMs is the polarization beam splitter (PBS), which serves to split orthogonally polarized beams/waves into different directions.

Figure 11.8 explains the scheme of a three-dimensional (3D) PBS. According to the geometrical optics (GO), a plane wave impinging on an isotropic GRIN

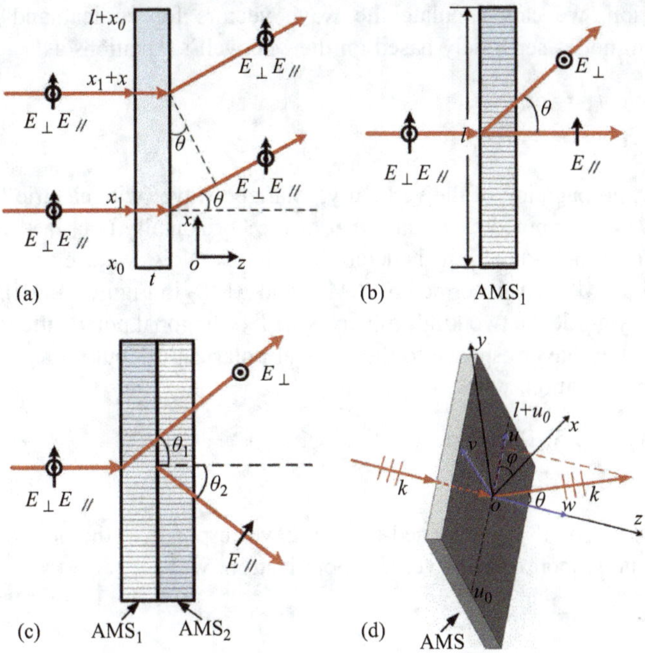

Figure 11.8 The schematic of the proposed 3D PBSs: (a) the waves deflected by passing through an isotropic GRIN planar lens, in which the vertical and horizontal polarizations cannot be separated; (b) the side view of a 3D PBS made of a single slab AMS_1, in which only the vertical polarization is deflected to the angle of θ; (c) the side view of a 3D PBS made of two slabs AMS_1 and AMS_2, in which both vertical and horizontal polarizations are deflected to angles of θ_1 and θ_2 independently; and (d) the 3D view of an arbitrary AMS, in which the vertical/horizontal polarization is deflected to a spatial direction of (θ, ϕ) by passing through the PBS. (Reprinted by permission from Nature Publishing Group. H. F. Ma, et al., independent control of differently polarized waves using anisotropic gradient-index metamaterials, Sci. Rep., 4: 6337. Copyright 2014.)

MTMs lens is bent with a deflection angle of θ related to the refractive index distribution n_0 inside the lens as

$$n(x) \approx n_0 + ((x - x_0)\sin\theta)/t, x_0 < x < l + x_0. \tag{11.12}$$

where θ, t, l and x are defined in Figure 11.8(a). As long as the lens is made of isotropic MTMs, both the vertically and horizontally polarized components, as are remarked by $E_\perp$ and $E_\parallel$, respectively, are deflected to the same direction. In order to independently control the two orthogonal polarizations, PBSs made of anisotropic MTMs are proposed in [20]. Supposing the incident wave propagates along the z-direction, we can calculate the wave vectors for vertical and horizontal transmission modes separately based on the Maxwell's equations as

$$\begin{aligned} k_{z\perp} &= \omega n_{\perp} = \omega\sqrt{\varepsilon_y \mu_x}, \\ k_{z\|} &= \omega n_{\|} = \omega\sqrt{\varepsilon_x \mu_y}. \end{aligned} \tag{11.13}$$

Clearly, propagation of the vertically polarized wave (with electric field in the x-direction) is determined by $n_{\perp}$ and that of the horizontally polarized wave (with electric field in the y-direction) is determined by $n_{\|}$. Now, we use two anisotropic MTMs slabs (AMSs), as is termed by AMS_1 and AMS_2 in Figure 11.8(b) and (c), to independently guide the two kinds of waves with orthogonal polarizations. AMS_1 is designed to only have response to the vertical polarization, but no response to the horizontal polarization, with

$$\begin{aligned} n_{1\perp}(x) &= \sqrt{\varepsilon_{1y}\mu_{1x}} = n_1(x), \\ n_{1\|}(x) &= \sqrt{\varepsilon_{1x}\mu_{1y}} = 1. \end{aligned} \tag{11.14}$$

Meanwhile, AMS_2 is designed to only have response to the horizontal polarization, but no response to the vertical polarization, with

$$\begin{aligned} n_{2\perp}(x) &= \sqrt{\varepsilon_{2y}\mu_{2x}} = 1, \\ n_{2\|}(x) &= \sqrt{\varepsilon_{2x}\mu_{2y}} = n_2(x). \end{aligned} \tag{11.15}$$

Note that $n_1(x)$ and $n_2(x)$ can be designed to determine the two deflection angles of θ_1 and θ_2, as is given in (11.12). In this way, the vertically polarized wave is deflected to one direction when passing through AMS_1 while the horizontally polarized one is deflected to another direction when passing through AMS_2. A 3D PBS is therefore formed by the two AMSs. Figure 11.8(d) illustrates a general design for an arbitrary AMS, in which the vertical/horizontal polarization is deflected to a spatial direction of (θ, ϕ). A series of AMSs can be combined to configurate a complicated PBS with desired function.

Next, we are looking for anisotropic MTM units which are competent to realize the designed PBS. In fact, anisotropic MTMs have been reported in literatures to provide novel solutions for manipulating the propagation of EM waves, for example, the well-known MTMs-based invisible cloaks and radar illusion devices [21,22]. Two key aspects should be taken into consideration. First, the anisotropic MTMs have independent responses to different polarizations. Second, the anisotropic MTMs have low loss and wide bandwidth, which means their impedances match to that of the air (or other surrounding media) in a broad frequency band. Accordingly, an anisotropic MTM unit has been carefully designed in [20], as is seen in Figure 11.9. The unit is essentially an electric resonator that responds only to EM waves with the electric field being parallel to the metallic fingers (along the y-axis in Figure 11.9). If the incident electric-field vector is along the y-axis and the magnetic-field vector along the x-axis, by varying the length of the fingers, one is able to change the refractive index from 0.9 to 2.2 and meanwhile maintain the effective impedance from 1.3 to 0.8 (which is matched to the wave impedance).

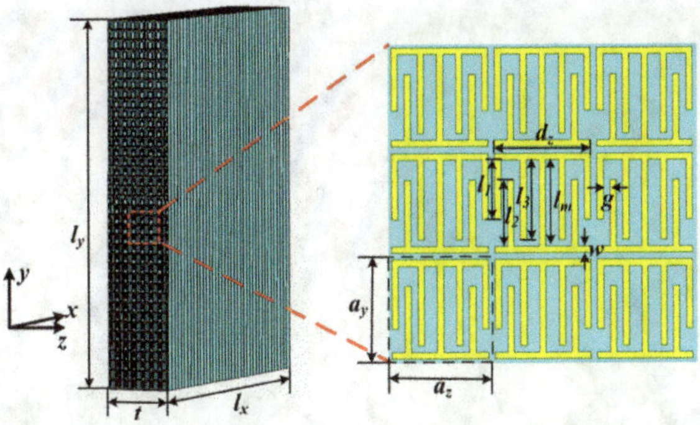

Figure 11.9 *An AMS with $\phi = 0$ (ϕ defined in Figure 11.8(d)). Details of the anisotropic metamaterial unit cells are zoomed in. (Reprinted by permission from Nature Publishing Group. H. F. Ma, et al., independent control of differently polarized waves using anisotropic gradient-index metamaterials, Sci. Rep., 4: 6337. Copyright 2014.)*

On the other hand, if the incident electric-field vector is along the x-axis and the magnetic-field vector along the y-axis, the refractive index remains 1.06 and the effective impedance remains 0.94 no matter how long the metallic fingers are. In other words, the unit cell has nearly no response to the second polarization. Therefore, a GRIN lens with desired θ and ϕ for vertical/horizontal polarization can be achieved using such anisotropic units.

A prototype PBS composed of two AMSs has been fabricated and measured. The PBS is designed to split two orthogonal polarizations and deflect each of them to different directions of ($\theta_1 = 30°$, $\phi_1 = 0°$) and ($\theta_2 = 30°$, $\phi_2 = 180°$), independently. Figure 11.10 shows the measured near-electric fields in the frequency range from 9 to 10.2 GHz, in which figures (a)–(c) give the vertically polarized waves deflecting to one direction and figures (d)–(f) give the horizontally polarized waves deflecting to another direction. Good polarization splitting performance has been observed in a wide frequency band with the absolute bandwidth reaching 1 GHz.

11.1.2 MTM curved lens antennas

In Section 11.1.1, we have introduced a series of flat GRIN lenses using isotropic or anisotropic MTMs. Aside from the flat lenses, some curved ones such as the Luneburg lens and the Maxwell's fisheye lens also have gradient-index distributions. In this section, we will show that the GRIN MTMs have provided us with a new means to realize curved lens antennas with easy design, good performance and low cost.

11.1.2.1 Luneburg lens

The Luneburg lens antenna has been proposed for over 70 years and played an important role for wireless communications on account of its outstanding ability of EM wave control [23]. A Luneburg lens is essentially a layered dielectric sphere

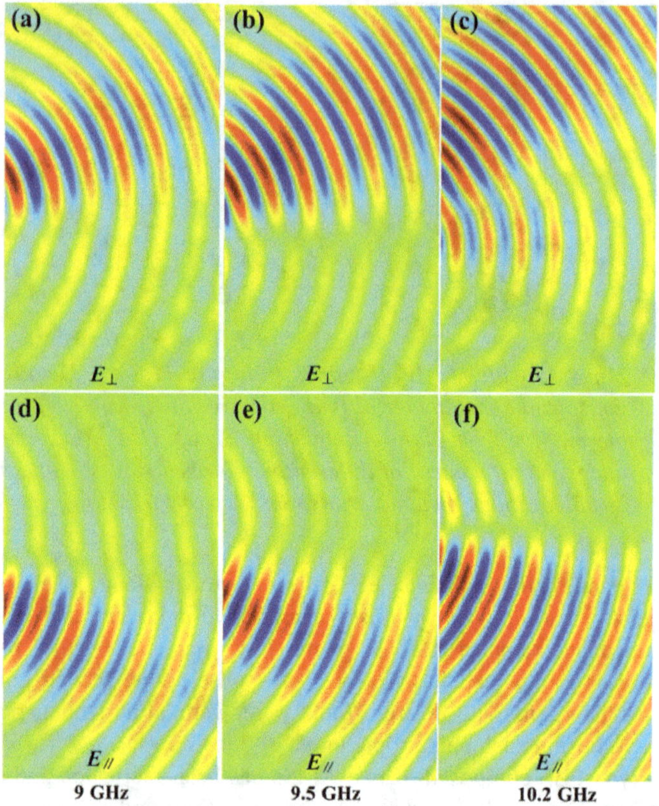

Figure 11.10 The measured near electric-field distributions at different frequencies of the PBS with AMS_1 and AMS_2: (a)–(c) the vertical polarizations at 9 GHz, 9.5 GHz, and 10.2 GHz; (d)–(f) the horizontal polarizations at 9 GHz, 9.5 GHz, and 10.2 GHz. (Reprinted by permission from Nature Publishing Group. H. F. Ma et al., independent control of differently polarized waves using anisotropic gradient-index metamaterials, Sci. Rep., 4: 6337. Copyright 2014.)

with radially gradient refractive index distribution. Theoretically, there are an infinite number of refractive solutions to Luneburg lenses to determine different focusing properties, and the simplest one is given by

$$n = \sqrt{2 - (r/R)^2}, \tag{11.16}$$

where R is the radius of the lens and $0 < r < R$. Due to the gradient distribution of refractive index, a Luneberg lens can transform the spherical waves from a point source on the lens surface into plane waves on the diametrically opposite side of the lens, and, reversely, guide an incoming plane waves from any direction to a focal point on the surface. The Luneburg lens possesses the merits of high gain, high degree of symmetry and flexible control of radiation beam, and hence has been applied in the mobile satellite communications in recent years. Current Luneburg lenses are mostly made of concentrically layered dielectrics with different permittivity values, and therefore are usually expensive and bulky in view of the existing materials and fabrication technology.

As a novel kind of artificial materials with designable EM parameters and profiles, MTMs are especially suitable for the realization of curved lens antennas. Firstly, GRIN MTMs have very small impedance change between two adjacent unit cells, and hence are able to generate high directivity with low return loss. Secondly, GRIN MTMs are not as expensive as precise compounds with assigned permittivities. Thirdly, GRIN MTMs have flexible profile and are easy to contain curved surfaces. In 2010, Ma *et al.* have presented a two-dimensional (2D) MTM Luneberg-like lens antenna [24]. Since the refractive index spans a limited range, nonresonant MTMs are preferred for broadband performance. The I-shaped metallic structures with broadband features [25] are chosen and aligned on substrates using the technique of printed circuit boards (PCBs). A foam with the relative permittivity close to 1 is applied to fix the PCB pieces, as is shown in Figure 11.11(a). In practice, the continuous distribution of refractive index for the Luneburg lens is discretized into sub-wavelength rectangular grids, as is illustrated in Figure 11.11(b). In each grid, one I-shaped structure is carefully designed to achieve the required refractive index. A sample has been fabricated and tested in a 2D near-field microwave scanning apparatus (2D mapper) through plotting the near electric field distributions [26]. Cylindrical waves from a line source located on the edge of the lens has been observed to be transformed to plane waves and propagate to the far-field region from 7 to 8.5 GHz.

Another configuration of the 2D Luneburg lens is to use circularly symmetric MTMs, and the refractive index distribution is discretized into concentrical layers, as is seen in Figure 11.11(c) and (d). I-shaped structures are also adopted as the composing units. The good beaming-forming performance is also proved from 7 to 8.5 GHz in experiment. Besides, a series of MTM-based 2D Luneburg lenses have also been reported in recent years [24,27–29].

MTMs have also been successfully used to construct 3D Luneburg lenses. Due to the symmetric property, the refractive index distribution of a 3D Luneburg lens can be achieved by rotating the 2D distribution (e.g., those in Figure 11.11(b) and (d)) on a diametral axis. A half spherical Luneburg lens combined with a perfectly electric conductor (PEC) plate has been realized by Ma *et al.* in 2013 [30]. Thanks to the symmetric property, spherical waves from a focal point on the lens surface can be reflected by the PEC plate and transformed to plane waves, as is illustrated in Figure 11.12(a).

In order to guarantee omnidirectional beam-forming performance, isotropic 3D MTMs are necessary to the realization of Luneburg lens. Strictly speaking, an ideal isotropic MTM unit should be a spherical dielectric shell (see Figure 11.12(b)). However, in the microwave frequency, it is difficult to fabricate 3D GRIN MTMs by using the spherical unit cells. An easy way to realize nearly isotropic GRIN

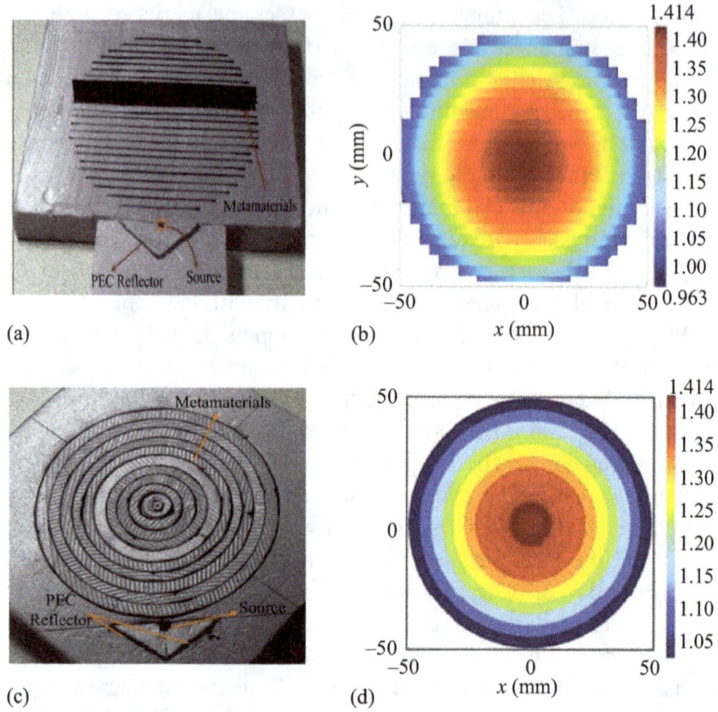

Figure 11.11 (a) The fabricated experimental sample of a 2D MTM Luneburg lens; (b) the relative refractive index distribution of the approximate Luneberg-like lens; (c) the sample of circularly symmetrical MTM lens; (d) the distributions of the discrete refractive index. (Reprinted by permission from Springer. H. F. Ma, et al., a broadband metamaterial cylindrical lens antenna, Chinese Sci. Bull., 83: 026601. Copyright 2011.)

MTMs is to use the drilled-hole dielectric unit cell, as is depicted in Figure 11.12(c), by using the PCB technology. Such a unit cell can be treated as approximately isotropic at the price of slightly deteriorated performance. The effective index of refraction of the unit cell can be controlled by changing the via-hole diameter, *D*. In this way, a prototype has been fabricated, as is given in Figure 11.12(d). We remark that two kinds of F4B substrates with different thicknesses have been employed in [30] to realize permittivity from 2 to 1 continuously from the spherical center to the surface.

Far fields of the Luneburg lens antenna has been measured in a fully anechoic microwave chamber. Two polarizations have been considered in measurements by adjusting the polarizations of feeding device, one with the magnetic field being parallel to the PEC plate (HPP) and the other with the electric field being parallel to the PEC plate (EPP). It has been observed that the sidelobes in the E- and H-planes

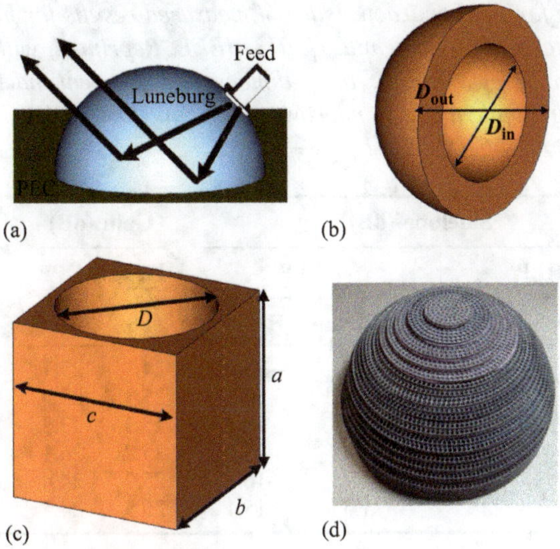

Figure 11.12 (a) A half-spherical Luneburg lens combined with a PEC plate; (b) a spherical dielectric shell unit cell; (c) a drilled-hole dielectric cuboid unit cell; and (d) the photograph of a fabricated half spherical Luneburg lens. ([2013] IEEE. Reprinted, with permission, from H. F. Ma et al., three-dimensional gradient-index materials and their applications in microwave lens antennas, IEEE Trans. Ant. Propag., 61(5): 2561.)

are lower than −10 dB and −17 dB, respectively, in the whole Ku-band. The gains range from 23 to 24.5 dB under the EPP polarization, and from 24.4 to 25.7 dB under the HPP polarization. The measured sidelobes, gain and efficiency have been listed in Table 11.1 for the whole Ku-band. Besides, the return loss of the MTM Luneburg lens has been proved to be lower than −14.2 dB by the Agilent Vector Network Analyzer (VNA) N5230c. All these results have demonstrated that 3D GRIN MTMs are competent to construct curved lens antennas with good performance.

11.1.2.2 Fish-eye lens

Another example of the MTM-based curved lens antenna is the Maxwell's fisheye lens, which was firstly conceived by Maxwell in 1854 [31]. A Maxwell's fisheye lens possesses the unique property that all light rays form circular trajectories and rays proceeding from any point in the lens will meet accurately in another point. In view of this, a 3D half-spherical fisheye lens serves to transform spherical wave from a point on the surface to plane wave on the other side of the lens, as is sketched in Figure 11.13(a). The refractive index distribution can be expressed by

$$n = 2(1 + r^2/R^2) - 1, \tag{11.17}$$

Table 11.1 The far-field characteristics of measured results for half-spherical Luneburg lens antenna. ([2013] IEEE. Reprinted, with permission, from H. F. Ma et al., *three-dimensional gradient-index materials and their applications in microwave lens antennas, IEEE Trans. Ant. Propag., 61(5): 2561.)*

f/GHz	Sidelobe (dB)				Gain (dB)		Efficiency (%)	
	HPP		EPP		HPP	EPP	HPP	EPP
	E-Plane	H-Plane	E-Plane	H-Plane				
	−13.3	−19.9	−10.6	−20.8	24.3	23.1	85.9	65.2
	−11.7	−17.2	−11.4	−20.8	25.3	23.7	93.3	65.3
	−9.2	−21.1	−10.5	−22.8	25.3	23.3	81.3	51.3
	−11.3	−18.9	−10.4	−25.0	25.7	24.1	78.3	54.2
	−10.5	−18.4	−11.9	−16.8	25.7	24.4	69.4	51.4
	−9.5	−22.1	−11.5	−15.1	25.3	24.5	56.4	46.9

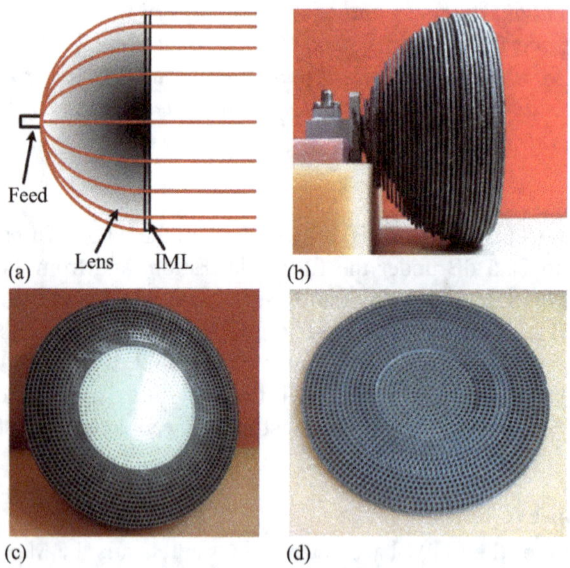

Figure 11.13 Sketch and fabrication of the half-spherical fisheye lens; (a) sketch of the lens; (b) photograph of the fabricated lens; (c) photograph of the fisheye lens body; and (d) photograph of the IML. ([2013] IEEE. Reprinted, with permission, from H. F. Ma et al., three-dimensional gradient-index materials and their applications in microwave lens antennas, IEEE Trans. Ant. Propag., 61(5): 2561.)

Table 11.2 The far-field characteristics of measured results for half-spherical fisheye lens antenna. ([2013] IEEE. Reprinted, with permission, from H. F. Ma et al., *three-dimensional gradient-index materials and their applications in microwave lens antennas, IEEE Trans. Ant. Propag., 61(5): 2561.)*

f/GHz	Sidelobe (dB)		Gain (dB)	Efficiency (%)
	E-Plane	H-Plane		
	−15.6	−33.6	19.8	42.6
	−16.9	−33.3	20.4	42.2
	−17.8	−35.6	20.3	36.7
	−15.9	−35.5	22.5	52.4
	−18.0	−36	23.3	55.6
	−23.2	−37.1	24.3	62.6

where R is the radius of the lens. (11.17) indicates that the refractive index at the surface of a Maxwell's fisheye lens is larger than 1, which is different from that of a Luneburg lens, and therefore IMLs are needed to reduce the reflection on the air-lens interface. Drilled-hole dielectric unit cells are again adopted to compose both the core lens and the IMLs. Configuration of the feeding source, lens, and the IMLs are depicted in Figure 11.13(b–d).

The 3D MTM-based fisheye lens has also been fabricated and tested in the anechoic microwave chamber, and the far-field performance has been recorded in Table 11.2 in terms of sidelobes, gain and efficiency. We notice that the sidelobes in the E- and H-plane patterns are lower than −15 dB and −30 dB over the whole Ku-band, respectively, with the gains ranging from 19.8 to 24.3 dB. Return loss has also been proved to be lower than −12 dB from 11.5 to 20 GHz using the Agilent VNA N5230c, indicating that reflection from the air-lens interface has been efficiently depressed by the IMLs.

Section 11.1.2 has introduced a handy and efficient method to realize curved lens antennas using GRIN MTMs. Aside from the Luneburg lens and the Maxwell's fisheye lens, other lenses with curved surfaces and inhomogeneous refractive index distributions, such as the Minano lens and the Eaton lens, can be designed and realized in a similar way.

11.2 MTM antennas using transformation optics

In Section 11.1, we have introduced a series of MTM lens antennas. Gradient refractive index distributions inside these lenses are mainly decided with GO method and ray-tracing method. In this section, we will discuss another category of MTM antennas using a recently developed strategy, the TO.

TO, also termed as "optical transformation" when referred to the behavior of rays, has been proposed in the last century when researchers studied Maxwell's equations in complex geometries [32–37]. This method is essentially based on the special

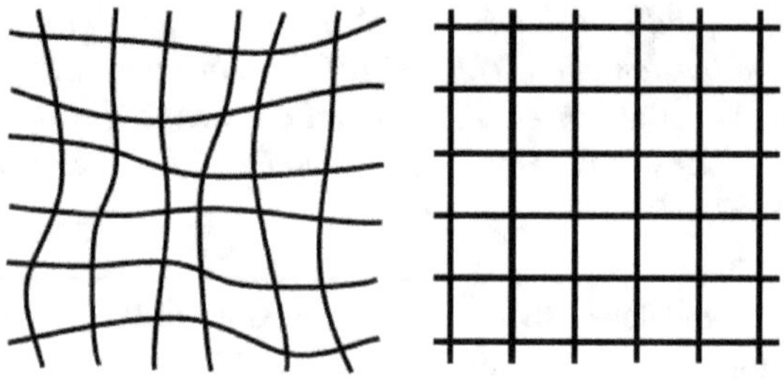

Figure 11.14 Left: the virtual space described in (x, y, z) coordinate system. Right: the physical space described in (x', y', z') coordinate system. ([2014] IEEE. Reprinted, with permission, from W. X. Tang et al., transformation electromagnetics for microwave antennas and radar illusion, IEEE Ant. Wireless Propag. Lett., 13: 1792.)

characteristics of Maxwell's equations in different coordinate systems. In general, Maxwell's equations have a form-invariant nature under TO, where the only change is a normalization of the EM parameters (refractive index n, or permittivity ε and permeability μ) of the background media of the space [35]. To clearly describe the TO method, two spaces characterized by two different coordinate systems are plotted in Figure 11.14. The left sub-figure presents a so-called "virtual space". It is a space where the coordinates can be "pulled" or "stretched" in line with ray's trajectories. The right sub-figure can be viewed as a "physical space", which is described in the Cartesian coordinates. Suppose the two coordinate systems are related as

$$x' = x'(x, y, z), y' = y'(x, y, z), z' = z'(x, y, z), \tag{11.18}$$

then the parameters in the physical space can be calculated by [38]:

$$\overline{\overline{\varepsilon}}' = \frac{\mathbf{J}\overline{\overline{\varepsilon}}\mathbf{J}^T}{det(\mathbf{J})}, \overline{\overline{\mu}}' = \frac{\mathbf{J}\overline{\overline{\mu}}\mathbf{J}^T}{det(\mathbf{J})}. \tag{11.19}$$

Solutions of the EM fields in the physical space are presented as

$$\mathbf{E}' = (\mathbf{J}^T)^{-1}\mathbf{E}, \mathbf{H}' = (\mathbf{J}^T)^{-1}\mathbf{H}. \tag{11.20}$$

In (11.19) and (11.20), J is the Jacobian transformation matrix between the two coordinate systems, defined as

$$J = \begin{pmatrix} \frac{\partial x'}{\partial x} & \frac{\partial x'}{\partial y} & \frac{\partial x'}{\partial z} \\ \frac{\partial y'}{\partial x} & \frac{\partial y'}{\partial y} & \frac{\partial y'}{\partial z} \\ \frac{\partial z'}{\partial x} & \frac{\partial z'}{\partial y} & \frac{\partial z'}{\partial z} \end{pmatrix}. \tag{11.21}$$

Obviously, the Jacobian matrix represents the distortion between the two coordinate systems. Equations (11.19) and (11.20) show that the distortion of EM fields can be maintained with the distortion of permittivity and permeability. In other words, if the physical space is filled with transformation media satisfying (11.19), an incident EM wave is "pulled" or "stretched" in the same way as in the virtual space. It should be pointed out that permittivity and permeability in (11.19) are spatially dependent in the physical space. Usually they are tensors if the transformation media are anisotropic. In view of this, MTMs are extremely competent as the realizing material.

The method of TO provides us with a method to connect the propagation of the wave to the property of the media. Equation (11.19) is the primary tool for controlling the propagation of EM waves and consequently creating novel functional devices. Applying the TO method, different devices with novel functionalities have been constructed, including invisible cloaks [21,25,39], EM rotators [40,41], EM concentrators [42–44], sensor cloaks [45], optical black holes [46,47], lens antennas [48,49], etc. In the following part of this section, we will introduce a series of TO-based MTM antennas with outstanding features or novel functions.

11.2.1 MTM flattened reflectors

In current antenna systems, many widely used devices have curved surfaces, such as parabolic reflectors and convex lenses. Using the TO method, we can design equivalent devices that operates in the same manner but have flexible profiles which are especially required in conformal antenna systems. A strategy called "quasi-conformal mapping" has been proposed for generating nearly isotropic transformation media, which are much easier to be realized than the anisotropic ones, as long as both the physical space and the virtual space have no strongly curved boundaries [50].

Figure 11.15 explains how to design a flattened reflector for instance [51]. In the virtual space perceived by the EM waves (see Figure 11.15(a)), the PEC reflector has a parabolic surfaces (with the focal length F being 108.6 mm and the aperture size being 180 mm), contains homogeneous and isotropic transformation media, and can be described with distorted coordinate systems. By using appropriate transformations, these distorted coordinate systems are mapped to the physical space (see Figure 11.15(b)), where the PEC reflector has a flat surface (with the same aperture size), contains spatially dispersive but isotropic materials, and are represented in new coordinate systems, the Cartesian coordinate system. Except for the PEC surfaces, the two spaces have the same boundaries to the air. We remark that in Figure 11.15(a) the focal point is outside of the transformation region, hence, in the physical space, a source at the focal point is not embedded in the transformation media.

Both spaces are discretized into small cells, and in each cell there is a set of local coordinates to be applied in the quasi-conformal mapping. Considering the Nyquist–Shannon sampling theorem [52] as well as the requirement of resolution, the size of each cell is set to be about 5 mm, which is approximately one-fifth

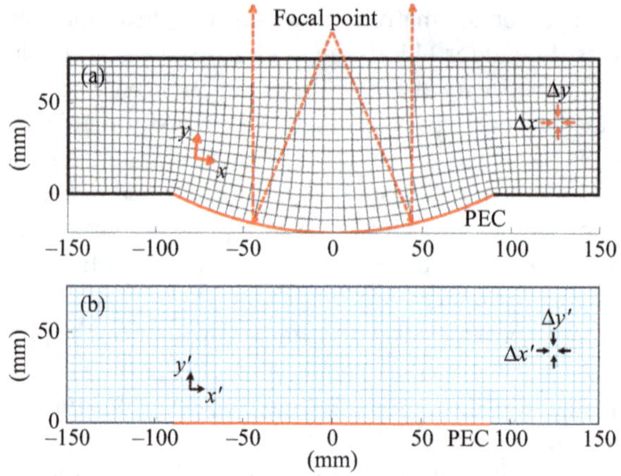

Figure 11.15 (Section view) (a) The virtual space with distorted coordinates. A parabolic reflector is placed in the free space, as illustrated by the red curve; (b) the physical space with orthogonal coordinates. The parabolic reflector is replaced with a flat PEC surface. ([2010] IEEE. Reprinted, with permission, from W. X. Tang et al., discrete coordinate transformation for designing all-dielectric flat antennas, IEEE Trans. Ant. Propag., 58(12): 3795.)

wavelength at 12 GHz [51]. To simplify the procedure, we can carry out the transformation in a 2D plane. The bold reduction from 3D to 2D is reliable because the parabolic reflector is axially symmetric and hence can be obtained by rotating the 2D models to its axis. In [51], it has been analyzed that if all cells in both the physical space and virtual space are near-orthogonal and approximately square-shaped with $\Delta x \simeq \Delta y$ and $\Delta x' \simeq \Delta y'$, the refractive index profile of the physical space can be directly retrieved from each cell within the grids, using (11.21):

$$n'^2 \simeq \frac{1}{\frac{\partial x'}{\partial x}\frac{\partial y'}{\partial y}} \simeq \frac{\Delta x \Delta y}{\Delta x' \Delta y'}. \tag{11.22}$$

For the E-polarization with electric field in the z-axis, the relative permeability of the device can be assumed to be isotropic and unity while the effective relative refractive index n' is only dependent on ε_z' as

$$\varepsilon_z'/\varepsilon_0 \simeq n'^2. \tag{11.23}$$

And for the H-polarization with magnetic field in the z-axis, the relative permittivity can be assumed to be isotropic and unity while the effective relative refractive index n' is only dependent on μ_z' as

$$\mu_z'/\mu_0 \simeq n'^2. \tag{11.24}$$

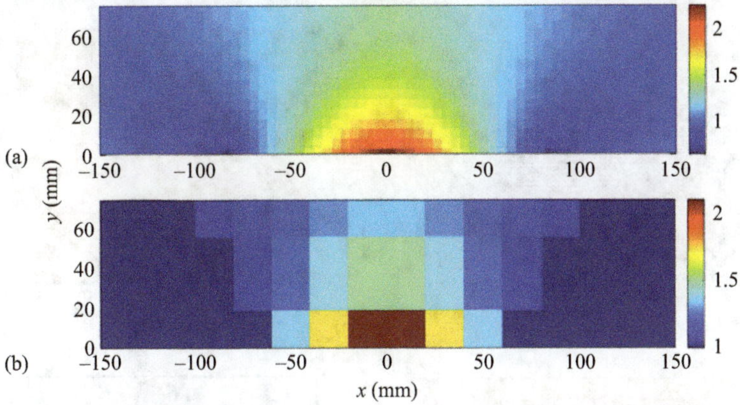

Figure 11.16 (a) Relative permittivity map consisting of 64 × 16 blocks and (b) relative permittivity map consisting of 16 × 3 blocks, without less-than-unity values. ([2010] IEEE. Reprinted, with permission, from W. X. Tang et al., discrete coordinate transformation for designing all-dielectric flat antennas, IEEE Trans. Ant. Propag., 58(12): 3795.)

Therefore, the designer can calculate the refractive index distribution, and choose to use isotropic MTMs with either ε or μ to compose the new reflector with flat surface.

Figure 11.16 gives the 2D permittivity distribution of a TO-based flat reflector delivered in [51], with the flat PEC surface being located at the bottom. We remark that a 3D map can be obtained by rotating the 2D map on its optical axis. The complete permittivity map is shown in Figure 11.16(a), consisting of 64 × 16 square blocks. In order to obtain a less-complicated permittivity map, two steps of simplifications can be adopted. Firstly, materials with relative permittivity of less-than-unity value are neglected and replaced with the background medium (e.g., the air in the free space). Secondly, according to the Nyquist–Shannon sampling theorem that an incident wave cannot resolve properties of media when their sizes are similar to or smaller than half a wavelength [52], resolution of the permittivity map in figure (a) can be reduced. Figure 11.16(b) shows the simplified map consisting of 16 × 3 big blocks. Since there are no extreme values existing in the map, non-resonant MTMs can be easily applied to construct this flattened reflector. These simplifications, however, may result in acceptable deteriorations of sidelobe level and gain, as discussed in [51,53]. Therefore, one has to make trade-offs between the antenna performance and its structure complexity in practical applications.

Another example of the flattened reflector antenna at the X-band has also been reported in [53] using the method of quasi-conformal TO. Nonresonant I-shaped MTMs are chosen in the demonstration of a 2D prototype shown in Figure 11.17(a). Metallic strips are fabricated on F4B substrate slices and supported by foam structure with permittivity close to 1. The 2D near-field scanning system is again

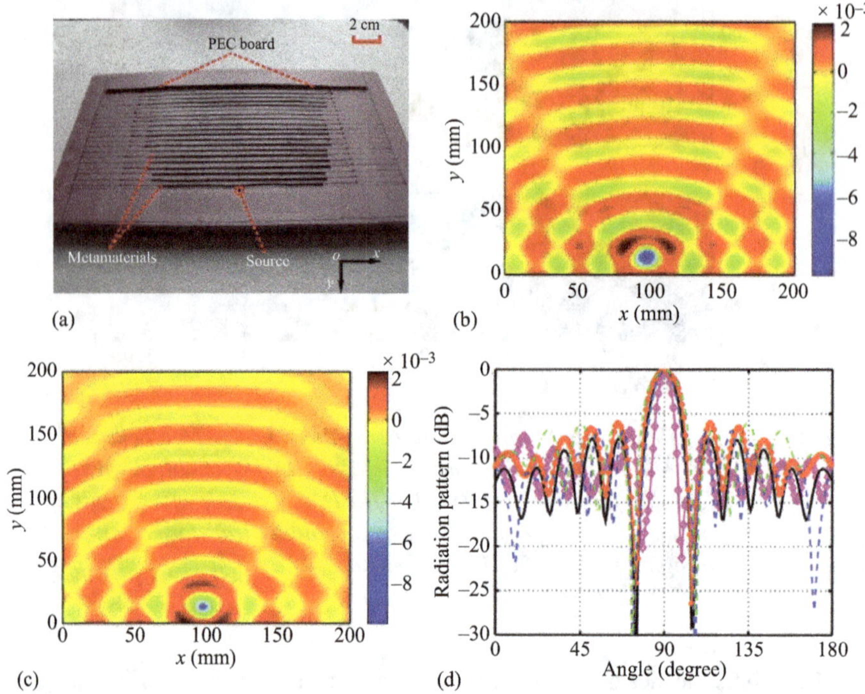

Figure 11.17 (a) Photograph of the fabricated 2D planar reflector antenna; (b) measured near-electric-field distributions of the original parabolic reflector at 10 GHz; (c) measured near-electric-field distributions of the flattened reflector at 10 GHz; and (d) normalized far-field radiation patterns for antennas at different frequencies. Magenta solid curve with diamond: the parabolic reflector; black solid curve: simulation data for the fabricated antenna at 10 GHz; green dash-dotted curve: measured result at 8.5 GHz; red solid curve with star: measured result at 10 GHz; blue dashed curve: measured result at 11.5 GHz. (The source of the material Z. L. Mei, J. Bai, and T. J. Cui, experimental verification of a broadband planar focusing antenna based on transformation optics, New J. Phys., 13: 063028, 2011, @IOP Publishing & Deutsche Physikalische Gesellschaft. Reproduced by permission of IOP Publishing. CC BY-NC-SA.)

implemented to plot the near-electric field distributions around the antenna. A probe is inserted as the line source for the antenna, as is seen in Figure 11.17(a). From Figure 11.17(b) and (c), it is observed that the original parabolic reflector and the TO-based flattened reflector have highly similar near-field distributions at 10 GHz. Good agreement between the two reflector antennas has also been given at other frequencies in the X-band. Furthermore, Huygens' principle is employed to

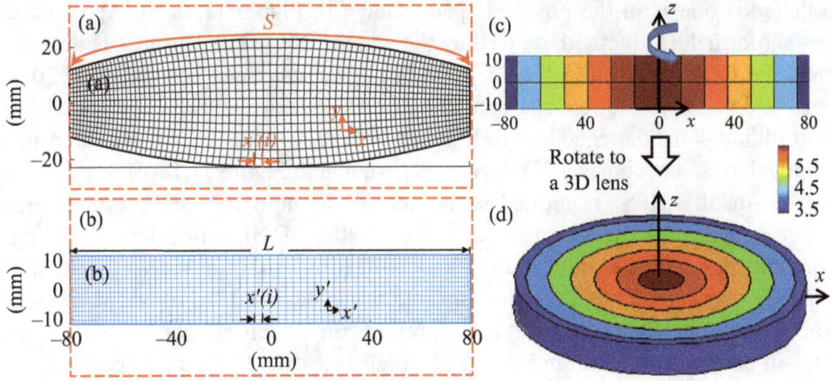

Figure 11.18 *(a) The virtual space with distorted coordinates where a convex lens made of homogeneous dielectric is embedded in the air; (b) the physical space with orthogonal coordinates where the flattened lens is inside the blue grid; (c) low-resolution permittivity map of the flattened lens consisting of* 14 × 1 *blocks. ([2010] IEEE. (a)–(c) Reprinted, with permission, from W. X. Tang et al., discrete coordinate transformation for designing all-dielectric flat antennas, IEEE Trans. Ant. Propag., 58(12): 3795.); (d) A 3D flattened convex lens is created by rotating the 2D permittivity map on its optical axis. The 3D lens is made of annular dielectric blocks. Color bar shows the relative permittivity values*

calculate far field radiation patterns using the measured near-field data. Directive beams have been proved at 8.5, 10, and 11.5 GHz (see Figure 11.17(d)).

In addition, flattened reflector antenna with a wide scanning angular has also been studied in [54]. Instead of moving or tilting the feed/reflector, Yang *et al.* employed an alternative way to manipulate the reflected emission by tuning the permittivity distribution derived from TO technique. This method has a merit of maintaining the profile of the feed-reflector combined system, therefore is potentially applicable for mounting beam steerable reflectors on the flat platform with low profiles.

11.2.2 MTM flattened convex and hyperbolic lenses

In this sub-section, we introduce another two kinds of MTM flattened lenses, the flattened convex lens and the flattened hyperbolic lens, both of which are designed using the method of quasi-conformal TO.

A convex lens is a common device in antenna systems, due to its function of transforming a spherical wave into a plane wave so as to increase the directivity. In [51], we have compressed a convex lens into flat profile at the C-band and the X-band through a procedure similar to that implemented in the flattened reflector design. A original convex lens is located in the virtual space shown in Figure 11.18(a) and the flattened one is in the physical space shown in Figure 11.18(b). Applying the quasi-conformal TO method, as well as the simplification strategy that is discussed above, we finally achieve a high-resolution permittivity map consisting of 20×4 blocks and a low-resolution one consisting of 14×1 blocks (see Figure 11.18(c)). Again, although the flattened lens is designed in 2D circumstance, when rotated about the optical axis, it becomes a 3D device, as is shown in Figure 11.18(d).

The Ansoft's HFSS is applied as another tool for numerical simulation to prove the performance of the original convex lens, the 3D flattened lens with high-resolution map and the 3D flattened lens with low-resolution map. Note that the flattened lenses are composed of ideal dielectric blocks instead of MTM unit cells in the simulation. Figure 11.19(a) to (c) presents the directivity patterns of the three lenses in both the E-plane and the H-plane at 4, 8, and 12 GHz, respectively. At 4 GHz, when the resolution of the 14×1-block lens is similar to half the wavelength, the 14×1-block lens has almost the same directivity as the 20×4-block one. When the frequency increases to 8 GHz, the two lenses still have very similar directivity patterns. The lower-resolution one has slightly lower peak directivity (the maximum directivity over all the directions, as defined by HFSS), and a bit increased sidelobes. As the operating frequency further increases to 12 GHz, the difference between the two lenses enlarges. The peak directivity of the low-resolution lens decreases further and the sidelobes increase as well. However, the low-resolution lens still holds an acceptable directivity pattern. Besides, the convex lens always has the best directivity in terms of the highest peak directivity and the lowest side lobes.

Figure 11.19(d) plots the peak directivity of the three lenses from 2 GHz to 16 GHz, so as to compare their operating bandwidth. The convex lens has a -3 dB bandwidth from about 8.5 GHz to about 16.7 GHz (8.2 GHz span), the 20×4-block lens has a bandwidth from about 7.7 GHz to about 16 GHz (8.3 GHz span), and the 14×1-block lens has a bandwidth from about 6.6 GHz to about 15.5 GHz (8.9 GHz span). Overall, in the entire frequency band, the peak directivity decreases as the convex lens, the 20×4-block lens and the 14×1-block lens is applied respectively. Differences between the three lenses increases as the frequency goes higher. It is also noticed that at lower frequencies, all three lenses have low directivity. The underlying physics is that when the aperture size of a lens is comparable to the wavelength, it cannot efficiently focus energy to the focal point, or reform the phase front of the wave. There results have demonstrated the performance of the flattened lenses and revealed their broadband properties.

In order to achieve directive beams, different lenses aside from the convex lens have been employed in antenna systems, including the Fresnel lenses, the hyperbolic lenses and the elliptical lenses. Among them, the hyperbolic lens works in broad frequency band but possesses bulky profile with curved surface, as is seen in Figure 11.20(a). A recent paper has reported a 3D flattened hyperbolic lens using the method of TO and the relative permittivity distribution of the lens is given in Figure 11.20(b) [55]. The flattened lens was manufactured making use of nano- and micro-sized titanates dispersed into a polymeric matrix material. This manufacturing technique is inexpensive and highly reproducible, and has the main

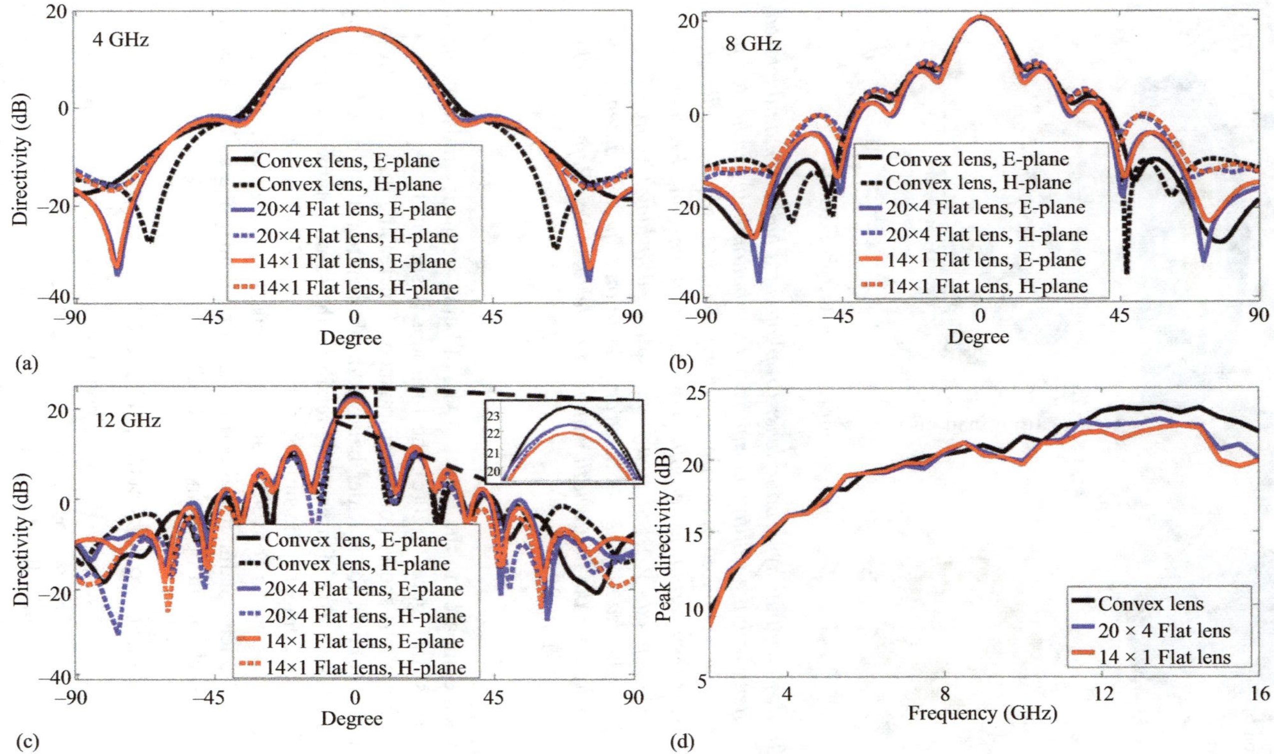

Figure 11.19 Comparison of the directivity of the convex lens, the 20×4-block flattened lens and the 14×1-block flattened lens. The directivity patterns are drawn in the E-plane and the H-plane at (a) 4 GHz; (b) 8 GHz; and (c) 12 GHz. In (d), the peak directivity of the three lenses is plotted from 2 to 16 GHz

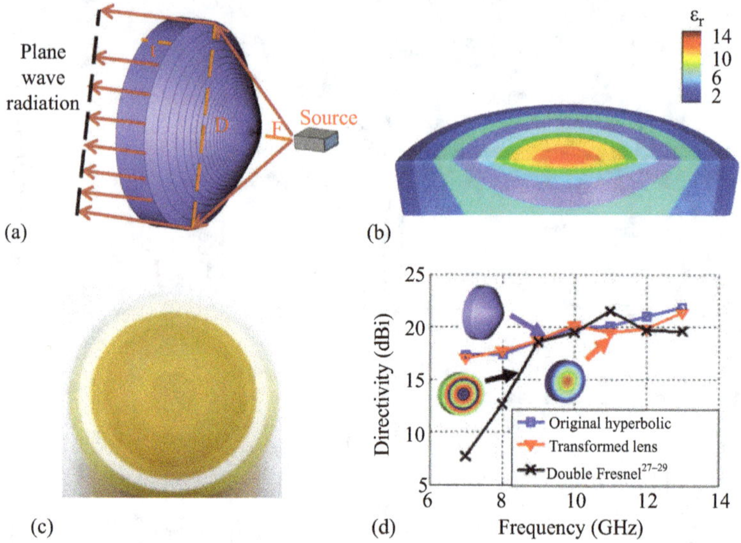

Figure 11.20 (a) Original hyperbolic lens; (b) photo of a half-cut manufactured lens with the relative permittivity distribution being indicated in color; (c) photo of the final prototype; and (d) simulated directivity for the original hyperbolic lens, flattened lens, and Fresnel lens. (Reprinted by permission from Nature Publishing Group. O. Quevedo-Teruel, et al., transformation optics for antennas: why limit the bandwidth with metamaterials? Sci. Rep., 3: 1903. Copyright 2013.)

advantage, when compared to the PCB MTMs, of being frequency independent of the EM properties. In Figure 11.20(c), the final lens is shown. In experiment, the flattened hyperbolic lens shows good performance of increasing directivity from 2 to 7 GHz. Also, it allows for a wider bandwidth of operation than other existing planar lenses such as Fresnel lenses while retaining the same volume and aperture size. Figure 11.20(d) compares the simulated directivities of the original hyperbolic lens, the flattened one, and a conventional Fresnel lens. Clearly we can see that the flattened hyperbolic lens maintains directivity of the original one and meanwhile surpasses the Fresnel lens in operating bandwidth.

11.2.3 MTM Luneburg lens with flattened focal surface

In Subsection 11.1.2, we discussed how to construct a Luneburg lens using GRIN MTMs. The MTM Luneburg lens obtains the advantages of high directivity and flexible beam control, but still has a spherical focal surface which is not welcomed by low-profile antenna systems and conformal antenna systems. In view of this, a great achievement has been made by generating a novel MTM Luneburg lens using the method of quasi-conformal TO [56,57]. The focal plane of the transformed lens

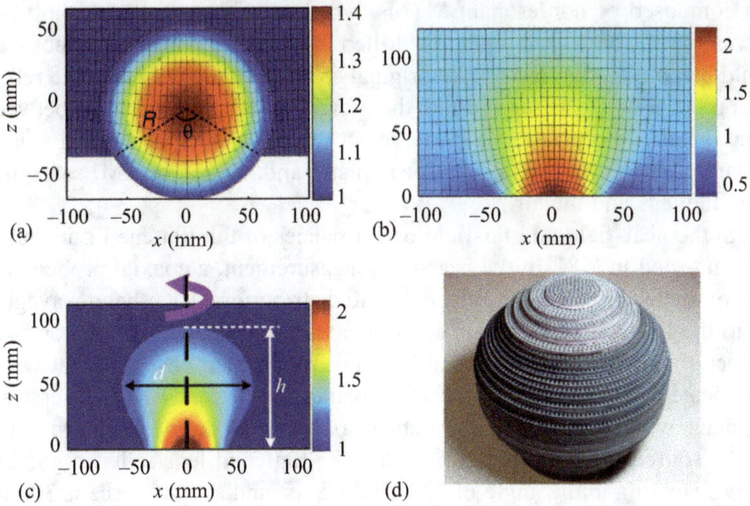

Figure 11.21 *(side view) (a) The original Luneburg lens in the virtual space; (b) the flattened Luneburg lens in the physical space and its refractive index distribution; (c) the simplified refractive index distribution; and (d) photograph of the fabricated 3D lens. (Reprinted by permission from Nature Publishing Group. H. F. Ma, et al., three-dimensional broadband and broad-angle transformation-optics lens, Nat. Commun., 1: 124. Copyright 2010.)*

becomes a planar surface instead of the spherical one, and incoming EM waves from extremely large angle can all be focused to the planar focal plane.

Figure 11.21 sketches how to create a 3D flattened Luneburg lens. The original spherical Luneburg lens is embedded in the air in the virtual space described with distorted coordinates, as is illustrated in 2D view in Figure 11.21(a). According to the theory of TO, the physical space described with Cartesian coordinates (see Figure 11.21(b)) has identical EM properties as the virtual space. Therefore, the physical space with specific distribution of refractive index can mimic a spherical Luneburg lens. We notice from Figure 11.21(b) that the refractive index in the physical space ranges from 0.5 to 2.2, with only a small portion being less than 1. A trade-off has been made between slight depression of radiation performance and significant increase of bandwidth, and finally the small area with $n < 1$ is ignored and replaced with the air. After such approximation, the simplified 2D refractive index profile is illustrated in Figure 11.21(c). After that, one is able to generate the 3D refractive index distribution by rotating the 2D profile about the optical axis, as is remarked in Figure 11.21(c), and finally achieve a 3D Luneburg lens in Figure 11.21(d) with the planar surface at the bottom being the new focal plane. It is observed from the color bar that the refractive index inside the Luneburg lens varies in a limited range from 1.1 to 2.1. This property indicates that nonresonant and isotropic MTMs are good candidates as composing materials. In [58], the 3D lens is composed of nonresonant MTMs, which are fabricated with multilayered dielectric plates, and each layer is drilled by different inhomogeneous holes. It should be pointed out that for the original spherical Luneburg lens the refractive index approaches 1 at the surface and, therefore, impedance is matched between the lens and the air. In contrast, for the flattened Luneburg lens the refractive index is much larger than 1 at the planar focal plane and, therefore, IMLs are needed between the lens and the air.

Both the near-field and far-field performances of the flattened Luneburg lens have been tested in [58]. In the near-field measurement, a coaxial probe is placed in front of the lens to detect the electric field distribution. A Ku-band waveguide is added to the lens as feeding source and located at different positions on the flattened focal plane to generate steerable beam. It has been observed that when the feed is located at the center of the focal plane, the flattened lens generates very good planar wavefronts along the optical axis (z-axis defined in Figure 11.21). When the source is 10 mm off the center, the flattened lens radiates good plane waves propagating in the angle of 20° off the axis; and as the source is 30 mm off the center, the plane waves propagate in the direction of 50° off the axis. In general, the radiation direction can be controlled in a large range by moving the feeding source on the flattened focal surface in the whole Ku-band. The far-field measurement is carried out in the anechoic chamber and recorded in Figure 11.22. Clearly, for both the horizontal (E-plane) and vertical (H-plane) polarizations, the flattened Luneburg lens has kept the ability to generate highly directive beams in a wide range from −50° to 50° with respect to the normal direction (along the z-axis). The broadband property has also been verified at 12.5, 15, and 18 GHz.

The 3D flattened Luneburg lens is a very important application of MTMs in antenna design. Its good overall performance, together with its low cost and easy fabrication, has made this MTM lens superior than the traditional Luneburg lens antennas. This demonstration has shown great advantages and potentials of MTMs for antenna engineering.

11.3 Metasurface antennas

MTM is a fast-developing research area. In recent years, a new version of MTMs, the metasurface which can be considered as the 2D planar MTM, has been developed. Metasurfaces are the period or the quasi-period surface structures which are composed of sub-wavelength scaled unit cells arranged under special regularity. Compared to the bulky 3D MTMs, metasurfaces possess the advantages such as low profile, easy fabrication, and planar or conformal geometry, and hence are highly expected as an efficient way to control EM waves from microwave to optical frequencies. In the limited space in this chapter, however, it is difficult to introduce in detail the metasurface antennas. So, instead, we only provide a brief discussion on the metasurface, and introduce some example applications that are conducted most recently in the State Key Laboratory of Millimeter Waves, Southeast University, China.

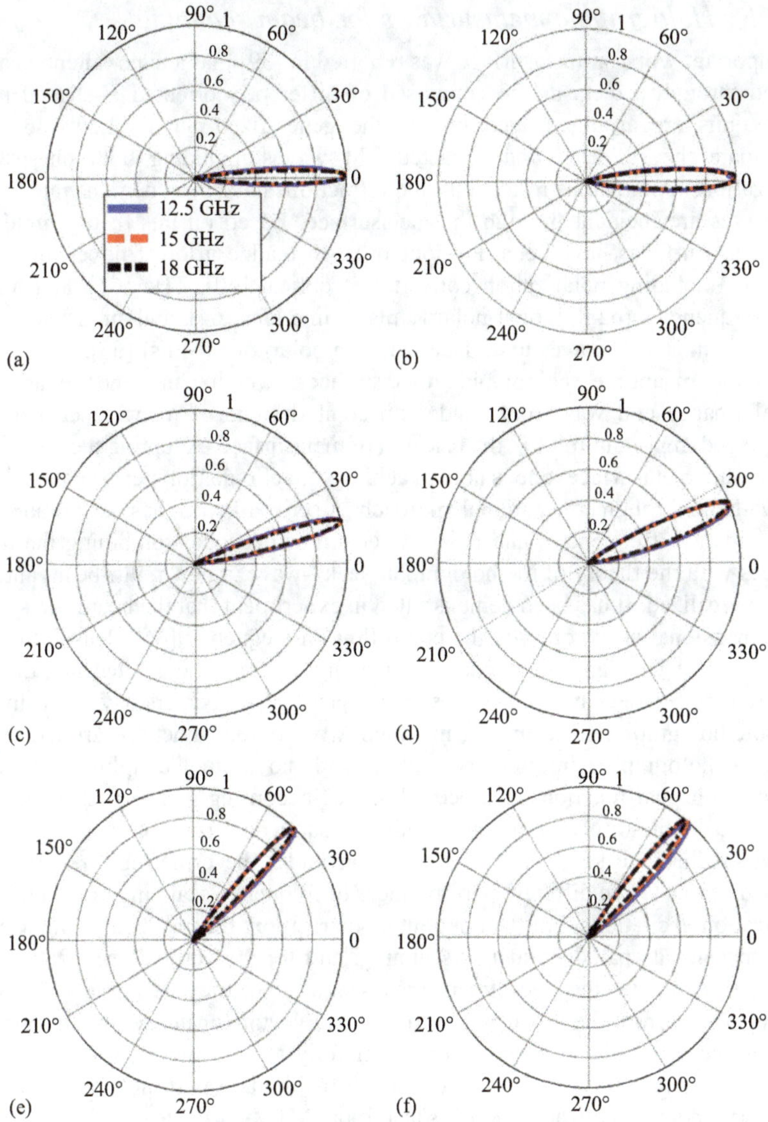

Figure 11.22 *Measured far-field results of the 3D flattened lens. (a, c, e) Show horizontal polarizations, whereas (b, d, f) present vertical polarizations at 12.5, 15, and 18 GHz. (a–b) The feeding source is located at the center of the focal plane (x = 0 and y = 0). (c–d) The feeding source is 10 mm off the center of the focal plane (x = –10 mm and y = 0). (e–f) The feeding source is 30 mm off the center of the focal plane. (Reprinted by permission from Nature Publishing Group. H. F. Ma, et al., three-dimensional broadband and broad-angle transformation-optics lens, Nat. Commun., 1: 124. Copyright 2010.)*

11.3.1 Holographic metasurfaces for beam scanning

An important work on metasurface was reported in [59] that a nano-antenna can be created through a metasurface composed of differently oriented V-shaped units. These units are arranged according to the generalized Snell's Laws so as to manipulate the reflected and refracted EM waves including both phases and polarizations. Anomalous reflections and refractions have been demonstrated when EM waves are incident through the metasurface. Based on this theory, gradient-phase metasurfaces have been developed fast to realize a lot of novel functional antennas, including polarization convertor at optical [60], THz [61], and microwave frequencies [62,63], dual-polarity plasmonic lens [64], and an efficient conversion from spatial waves to surface plasmon polaritons (SPPs) [65].

In the meantime, holographic metasurfaces, which can generate arbitrary optical images and wavefronts under different designs of interference patterns, have played important roles in the research of metasurface. Adopting the concept of holographic metasurface, one is able to achieve novel radiation performance that is not available through conventional approach. A series of work has been reported in the microwave frequencies and radar systems. For instance, combining the metasurfaces with the theory of the holographic leaky-wave, a new multi-beam antenna has been realized. It has been demonstrated in experiment that the beam can scan in two-dimensional space only by the controlling of frequency [66]. This is the first time to realize 2D scans of antennas by changing the frequency. After that, Li *et al.* have reported a new method to design planar dual-functional devices using an isotropic holographic metasurface, in which two different functions are written on the same holographic interference pattern with no mutual coupling. Two new devices with dual-function have been designed accordingly. The first device can realize the double-direction radiation under the same polarization, while the second one can realize the single-direction radiation under the orthotropic polarization. The two devices have different performances of frequency scanning under different polarization excitations. For the horizontal polarization, the beam can complete the one-dimensional spatial frequency scanning; and for the vertical polarization, the beam can complete the two-dimensional spatial frequency scanning [67]. Most recently, aside from the scalar (or isotropic) and vector (or uniaxially anisotropic) metasurfaces, the inhomogeneous tensor (or fully anisotropic) metasurfaces have been investigated by Wan *et al.*, in order to manipulate two independent beams with independent polarizations [68]. Anisotropic MTM particles have been adopted as composing units on the metasurface to achieve tensor surface impedance. This work has shown the great ability and flexibility of metasurface for beaming controlling. In view of these work, we expect metasurface to play more important roles in low-profile antenna systems.

11.3.2 Spoof SPP radiations

SPPs are highly localized surface wave that propagates on the interface between two media with opposite permittivities. In nature, SPPs only exist in the optical frequency because metals behave as PEC from THz downward. In 2004, Sir John

Pendry *et al.* firstly discussed the way of producing spoof SPPs in microwave and THz frequencies through making artificial structures on the metal surface [69]. Thereafter, 3D bulky plasmonic MTMs have been explored and investigated. In 2013, Cui *et al.* proposed that SPPs can be supported at microwave frequencies by an ultrathin corrugated metallic strip, which can be considered as a new kind of one-dimensional metasurface [70]. The plasmonic metasurface has been proved reliable when it is bent, folded and even twisted and, therefore, possesses great advantages and potentials in conformal antenna systems.

SPPs are in essence nonradiative EM modes with the wave number k_{SPP} being larger than k_0 (that in the air). In fact, a gradient-index metasurface has been proposed to convert spatial propagating waves to surface waves with nearly 100% efficiency [65]. Reversely, an SPP emitter has been created to convert surface-plasmon-like modes on the metasurface to spatial radiated modes [71]. Corrugated metallic strips are carefully designed with different geometry and, in this way, when spoof SPPs propagate along the gradient metasurface with periodic configuration, the wave number along the propagating direction is changed from k_{SPP} to k_0, realizing the perfect momentum matching from the slow wave to the radiation wave. In addition, compact feeding network has also been designed for array radiations of SPPs [72]. Efficient transition of SPP modes from a single corrugated metallic strip to two adjacently parallel corrugated metallic strips have been achieved to feed two branches of SPP emitters. Based on the theory of antenna array, the SPP waves can be effectively converted to far fields in the broadside direction. Due to the fact that the SPP waves can be easily controlled in terms of propagation constant, the far-field distribution of the antenna array is tunable. Attributing to this feature, the SPP antennas and antenna arrays are expected to have potential applications in highly integrated photonic circuits and phase-array antennas.

11.3.3 Coding metasurfaces

Last but not the least, we will give a short introduction of coding metasurfaces. In this chapter, we have discussed about MTMs, which can be described by the macroscopic effective medium parameters and regarded as analog artificial materials. As a counterpart of analog MTMs, digitally coding MTMs have been proposed most recently by Cui *et al.* [73]. The simplest coding MTMs are composed of only two kinds of unit cells with 0 and π phase responses, which are named as 0 and 1 elements. By coding 0 and 1 elements with controlled sequences (i.e., 1-bit coding), one is able to manipulate EM waves and realize different functionalities. The concept of coding MTMs can be extended from 1-bit coding to 2-bit or more. In 2-bit coding, four kinds of unit cells with phase responses 0, $\pi/2$, π, and $3\pi/2$ are required to mimic 00, 01, 10, and 11 elements, respectively, which have larger freedom to control EM waves. A unique MTM particle which has either 0 or 1 response controlled by a biased diode has been created in [73]. Based on the particle, digital MTMs with unit cells having either 0 or 1 state have also been conceived. Using the field-programmable gate array, one is able to control the digital MTMs digitally. By programming different coding sequences, a single digital MTM obtains distinct abilities in manipulating EM waves as the “programmable MTMs”.

The above concepts and physical phenomena have been confirmed experimentally in [73] through metasurfaces. Furthermore, abnormal reflections and diffusions have been generated through coding metasurfaces [74,75]. The well-designed coding metasurface reflects the incident plane waves in an abnormal way which is different from that of normal specular reflection. The phenomenon of abnormal reflection for coding metasurfaces include multibeam reflection, broadened beam reflection, angular scanning based on dual-beam reflection. More diversity of reflection can be achieved by combination of different codes along the two edges of coding metasurface. The design of coding metasurface has also been extended to THz region by realizing abnormal diffusion of the THz waves using a 2D coding metasurface. The coding MTMs, digital MTMS and programmable MTMs are very attractive for controlling the radiation beams of antennas. They have opened a new venue to control EM waves instantly, and have become an important trend of MTMs.

In this chapter, we have introduced a series of MTM-based antennas at microwave frequencies. As a novel kind of artificial materials, MTMs possess unique merits such as wide range of EM parameters, flexible properties and profiles, easy fabrications, and low lost. In light of these merits, MTMs are able to provide us with new solutions and breakthroughs in antenna engineering. Currently, most MTM antennas are related to GRIN MTM lenses and TO-based MTM lenses. With the fast development of metasurfaces, spoof SPPs waveguides and coding MTMs, MTMs-based antennas are expected to be more vitally applied in new-concept antenna systems and communication systems.

References

[1] V. G. Veselago. The electrodynamics of substances with simultaneously negative values of ε and μ. *Sov. Phys. Usp.*, 10: 509, 1968.

[2] J. B. Pendry, A. J. Holden, W. J. Stewart and I. Youngs. Extremely low frequency plasmons in metallic mesostructures. *Phys. Rev. Lett.*, 76: 4773, 1996.

[3] J. B. Pendry, A. J. Holden, D. J. Robbins and W. J. Stewart. Magnetism from conductors and enhanced nonlinear phenomena. *IEEE Trans. Micr. Theory Techn.*, 47: 2075, 1999.

[4] R. Shelby, D. R. Smith and S. Schultz. Experimental verification of a negative index of refraction. *Science*, 292: 77, 2001.

[5] J. Pendry. Negative refraction makes a perfect lens. *Phys. Rev. Lett.*, 85: 3966–3969, 2000.

[6] R. W. Ziolkowski. Superluminal transmission of information through an electro-magnetic metamaterial. *Phys. Rev. E*, 63: 046604, 2001.

[7] D. R. Smith, J. B. Pendry and M. C. K. Wiltshire. Metamaterials and negative refractive index. *Science*, 305: 788–792, 2004.

[8] R. Greegor, C. Parazzoli, K. Li and M. Tanielian. Origin of dissipative losses in negative index of refraction materials. *Appl. Phys. Lett.*, 82: 2356, 2003.

[9] D. R. Smith, J. J. Mock, A. F. Starr and D. Schurig. Gradient index metamaterials. *Phys. Rev. E*, 71: 036609, 2005.

[10] D. Pozar. Microwave Engineering, *3rd*. John Wiley & Sons, Hoboken, NJ, USA, 2005.

[11] X. Chen, H. F. Ma, X. Y. Zou, W. X. Jiang and T. J. Cui. Three-dimensional broadband and high-directivity lens antenna made of metamaterials. *J. Appl. Phys.*, 110: 044904, 2011.

[12] X. Y. Zhou, X. Y. Zou, Y. Yang, H. F. Ma and T. J. Cui. Three-dimensional large-aperture lens antennas with gradient refractive index. *Science China-Information Science*, 56: 120410, 2013.

[13] J. Kraus. Antennas. McGraw-Hill Education, New York, NY, USA, 1988.

[14] M. Q. Qi, W. X. Tang, H. X. Xu, H. F. Ma and T. J. Cui. Tailoring radiation patterns in broadband with controllable aperture field using metamaterials. *IEEE Trans. Ant. Propag.*, 61: 5792–5798, 2013.

[15] U. Sangawa. Gradient index lens antennas with controllable aperture field distributions. *IEICE Trans. Commun.*, E95b: 2051–2058, 2012.

[16] M. Clerc and J. Kennedy. The particle swarmłExplosion, stability, and convergence in a multidimensional complex space. *IEEE Trans. Evol. Comput.*, 6: 58–73, 2002.

[17] M. Q. Qi, H. X. Xu, H. F. Ma, M. P. Jin, W. Wang and T. J. Cui. A wideband waveguide antenna with nearly equal E- and H-plane radiation patterns. *Int. J. Antennas Propag.*, 2013: 608393, 2013.

[18] C. A. Balanis. Antenna Theory: Analysis and Design, 3rd Edition. Hoboken, NJ: Wiley-Interscience, 2005.

[19] M. Q. Qi, W. X. Tang, H. F. Ma, *et al.*, Suppressing side-lobe radiations of horn antenna by loading metamaterial lens. *Sci. Rep.*, 5: 9113, 2015.

[20] H. F. Ma, G. Z. Wang, W. X. Jiang and T. J. Cui. Independent control of differently polarized waves using anisotropic gradient-index metamaterials. *Sci. Rep.*, 4: 6337, 2014.

[21] D. Schurig, J. J. Mock, B. J. Justice, *et al.* Metamaterial electromagnetic cloak at microwave frequencies. *Science*, 314: 977–980, 2006.

[22] W. X. Jiang and T. J. Cui. Radar illusion via metamaterials. *Phys. Rev. E*, 83: 026601, 2011.

[23] R. K. Luneberg. Mathematical Theory of Optics. *Rhode Island: Brown University Press*, 1944.

[24] H. F. Ma, X. Chen, X. M. Yang, H. S. Xu, Q. Cheng and T. J. Cui. A broadband metamaterial cylindrical lens antenna. *Chinese Sci. Bull.*, 83: 026601, 2011.

[25] R. P. Liu, C. Ji, J. J. Mock, J. Y. Chin, T. J. Cui and D. R. Smith. Broadband ground-plane cloak. *Science*, 323: 366–368, 2009.

[26] B. Justice, J. Mock, L. Guo, A. Degiron, D. Schurig and D. Smith. Spatial mapping of the internal and external electromagnetic fields of negative index metamaterials. *Opt. Express*, 14: 8694–8705, 2006.

[27] Q. Cheng, H. F. Ma and T. J. Cui. Broadband planar Luneburg lens based on complementary metamaterials. *Appl. Phys. Lett.*, 95: 181901, 2009.

[28] O. Quevedo-Teruel, W. Tang and Y. Hao. Isotropic and nondispersive planar fed Luneburg lens from Hamiltonian transformation optics. *Opt. Lett.*, 37: 4850–4852, 2012.

[29] X. Wan, W. X. Jiang, H. F. Ma and T. J. Cui. A broadband transformation-optics metasurface lens. *Appl. Phys. Lett.*, 104: 151601, 2014.

[30] H. F. Ma, B. G. Cai, T. X. Zhang, Y. Yang, W. X. Jiang and T. J. Cui. Three-dimensional gradient-index materials and their applications in microwave lens antennas. *IEEE Trans. Ant. Propag.*, 61: 2561–2569, 2013.

[31] J. C. Maxwell. Solutions of problems. *Cambridge Dublin Math. J.*, 8: 188, 1854.

[32] L. S. Dolin. On a possibility of comparing three-dimensional electromagnetic systems with inhomogeneous filling. *Izv. Vyssh. Uchebn. Zaved. Radiofiz*, 4: 964–967, 1961.

[33] E. G. Post. Formal Structure of Electromagnetics; General Covariance and Electro-magnetics. New York: Interscience, 1962.

[34] M. Lax and D. F. Nelson. Maxwell equations in material form. *Phys. Rev. B*, 13: 1777, 1976.

[35] A. J. Ward and J. B. Pendry. Refraction and geometry in Maxwell's equations. *J. Mod. Opt.*, 43: 773–793, 1996.

[36] U. Leonhardt. Optical conformal mapping. *Science*, 312: 1777–1780, 2006.

[37] J. B. Pendry, D. Schurig and D. R. Smith. Controlling electromagnetic fields. *Science*, 312: 1780–1782, 2006.

[38] W. X. Jiang and T. J. Cui. Optical Transformation Theory in Metamaterials: Theory, Design, and Applications. Eds. Boston, MA: Springer, 2010.

[39] H. F. Ma, W. X. Jiang, X. M. Yang, X. Y. Zhou and T. J. Cui. Compact-sized and broadband carpet cloak and free-space cloak. *Opt. Express*, 17: 19947–19959, 2009.

[40] Y. Luo, H. Chen, J. Zhang, L. Ran and J. Kong. Design and analytical full-wave validation of the invisibility cloaks, concentrators, and field rotators created with a general class of transformations. *Phys. Rev. B*, 77: 125127, 2008.

[41] H. Chen, B. Hou, S. Chen, X. Ao, W. Wen and C. Chan. Design and experimental realization of a broadband transformation media field rotator at microwave frequencies. *Phys. Rev. Lett.*, 102: 183903, 2009.

[42] M. Rahm, D. Schurig, D. Roberts, S. Cummer, D. Smith and J. Pendry. Design of electromagnetic cloaks and concentrators using form-invariant coordinate transformations of Maxwell's equations. *Photon. Nanostruct. Fundam. Appl.*, 6: 87–95, 2008.

[43] A. Yaghjian and S. Maci. Alternative derivation of electromagnetic cloaks and concentrators. *New J. of Phys.*, 10: 115022, 2008.

[44] W. X. Jiang, T. J. Cui, Q. Cheng, *et al.* Design of arbitrarily shaped concentrators based on conformally optical transformation of nonuniform rational B-spline surfaces. *Appl. Phys. Lett.*, 92: 264101, 2008.

[45] A. Alù and N. Engheta. Cloaking a sensor. *Phys. Rev. Lett.*, 102: 233901, 2009.

[46] E. Narimanov and A. Kildishev. Optical black hole: Broadband omnidirectional light absorber. *Appl. Phys. Lett.*, 95: 041106, 2009.

[47] Q. Cheng, T. J. Cui, W. X. Jiang and B. G. Cai. An omnidirectional electromagnetic absorber made of metamaterials. *New J. of Phys.*, 12: 063006, 2010.
[48] F. Kong, B. Wu, J. Kong, J. Huangfu, S. Xi and H. Chen. Planar focusing antenna design by using coordinate transformation technology. *Appl. Phys. Lett.*, 91: 253509, 2007.
[49] N. Kundtz and D. Smith. Extreme-angle broadband metamaterial lens. *Nat. Mater.*, 9: 129–132, 2010.
[50] J. Li and J. Pendry. Hiding under the carpet: A new strategy for cloaking. *Phys. Rev. Lett.*, 101: 203901, 2008.
[51] W. Tang, C. Argyropoulos, E. Kallos, W. Song and Y. Hao. Discrete coordinate transformation for designing all-dielectric flat antennas. *IEEE Trans. Ant. Propag.*, 58: 3795, 2010.
[52] C. E. Shannon. Communication in the presence of noise. *Proc. IRE,*, 37: 10–21, 1949.
[53] Z. L. Mei, J. Bai and T. J. Cui. Experimental verification of a broadband planar focusing antenna based on transformation optics. *New J. of Phys.*, 13: 063028, 2011.
[54] R. Yang, W. Tang and Y. Hao. Wideband beam-steerable flat reflectors via transformation optics. *IEEE Ant. Wireless Propag. Lett.*, 10: 1290–1294, 2011.
[55] O. Quevedo-Teruel, W. Tang, R. C. Mitchell-Thomas, *et al.* Transformation optics for antennas: Why limit the bandwidth with metamaterials? *Sci. Rep.*, 3: 1903, 2013.
[56] D. Schurig. An aberration-free lens with zero f-number. *New J. Phys.*, 19: 115034, 2008.
[57] N. Kundtz, and D. R. Smith. Extreme-angle broadband metamaterial lens. *Nat. Mater.*, 9: 129, 2010.
[58] H. F. Ma and T. J. Cui. Three-dimensional broadband and broad-angle transformation-optics lens. *Nat. commun.*, 1: 124, 2010.
[59] N. Yu, P. Genevet, M. A. Kats, *et al.* Light propagation with phase discontinuities: Generalized laws of reflection and refraction. *Science*, 334: 333–337, 2011.
[60] Y. Yang, W. Wang, P. Moitra, I. I. Kravchenko, D. P. Briggs and J. Valentine. Dielectric meta-reflectarray for broadband linear polarization conversion and optical vortex generation. *Nano Lett.*, 14: 1394–1399, 2014.
[61] N. K. Grady, J. E. Heyes, D. R. Chowdhury, *et al.* Terahertz metamaterials for linear polarization conversion and anomalous refraction. *Science*, 340: 1304–1307, 2013.
[62] H. F. Ma, G. Z. Wang, G. S. Kong and T. J. Cui. Broadband circular and linear polarization conversions realized by thin birefringent reflective metasurfaces. *Opt. Mater. Express*, 4: 1717–1724, 2014.
[63] X. Gao, X. Han, W. P. Cao, H. O. Li, H. F. Ma and T. J. Cui. Ultrawideband and high-efficiency linear polarization converter based on double V-shaped metasurface. *IEEE Trans. Ant. Propag.*, 63: 3522–3530, 2015.
[64] X. Chen, L. Huang, H. Muhlenbernd, *et al.* Dual-polarity plasmonic metalens for visible light. *Nat. Commun.*, 3: 1198, 2012.

[65] S. Sun, Q. He, S. Xiao, Q. Xu, X. Li and L. Zhou. Gradient-index metasurface as a bridge linking propagating waves and surface waves. *Nat. Mater.*, 11: 426, 2012.
[66] Y. B. Li, X. Wan, B. G. Cai, Q. Cheng and T. J. Cui. Frequency-controls of electromagnetic multi-beam scanning by metasurfaces. *Sci. Rep.*, 4: 6921, 2014.
[67] Y. B. Li, B. G. Cai, Q. Cheng and T. J. Cui. Isotropic holographic metasurfaces for dual-functional radiations without mutual interferences. *Adv. Funct. Mater.*, 26: 29–35, 2016.
[68] X. Wan, T. Y. Chen, Q. Zhang, *et al.* Manipulations of dual beams with dual polarizations by full-tensor metasurfaces. *Adv. Opt. Mater.*, 4: 1567–1572, 2016.
[69] J. B. Pendry, L. Martin-Moreno and F. J. Garcia-Vidal. Mimicking surface plasmons with structured surfaces. *Science*, 305: 847–848, 2004.
[70] X. P. Shen, T. J. Cui, D. Martin-Cano and F. J. Garcia-Vidal. Conformal surface plasmons propagating on ultrathin and flexible films. *Proc. Natl. Acad. Sci. USA/PNAS*, 110: 40–45, 2013.
[71] J. J. Xu, H. C. Zhang, Q. Zhang and T. J. Cui. Efficient conversion of surface-plasmon-like modes to spatial radiated modes. *Appl. Phys. Lett.*, 106: 021106, 2015.
[72] J. J. Xu, J. Y. Yin, H. C. Zhang and T. J. Cui. Compact feeding network for array radiations of spoof surface plasmon polaritons. *Sci. Rep.*, 6: 22692, 2016.
[73] T. J. Cui, M. Q. Qi, X. Wan, J. Zhao and Q. Cheng. Coding metamaterials, digital metamaterials and programmable metamaterials. *Light: Science & Applications, 3*: e218, 2014.
[74] L. H. Gao, Q. Cheng, J. Yang, *et al.* Broadband diffusion of terahertz waves by multi-bit coding metasurfaces. *Light: Science & Applications, 4*: e324, 2015.
[75] L. Liang, M. Q. Qi, J. Yang, *et al.* Anomalous terahertz reflection and scattering by flexible and conformal coding metamaterials. *Adv. Optical Mater.*, 3: 1374–1380, 2015.

Chapter 12

Metamaterial-based zero-phase-shift-line loop antennas

Zhi Ning Chen[1], Xianming Qing[2], Jin Shi[3], and Yunjia Zeng[2]

12.1 Introduction

Metamaterials are artificially engineered structures to achieve unique electromagnetic properties which have never been found in their constituent components and any materials in nature so far. Therefore, the metamaterials definitely offer great opportunities for engineers to create a novel structure with unconventional properties incorporated in antenna systems to enhance their performance [1].

Among many metamaterial structures, two types of metamaterial-based transmission lines, namely, the composite right/left-handed transmission line (CRLH-TL) and zero-phase-shift-line (ZPSL) have been widely investigated and applied in engineering designs. The CRLH transmission lines are characterized by the existence of three different spectral regions: the left-handed region at lower frequencies supporting a backward wave wherein a negative propagation constant β is observed ($\beta < 0$), the right-handed region at higher frequencies supporting a forward wave wherein the propagation constant β is positive ($\beta > 0$), the frequency in between the right-handed and left-handed regions is the transition point with zero propagation constant ($\beta = 0$). The CRLH lines have been widely used for improving the performance of microwave devices [2,3] and leaky-wave antenna [4].

Different from the CRLH transmission lines, the ZPSL is another type of metamaterial-based transmission line that is a one-dimensional periodic structure. It can be configured by metallic strips/lines with lumped capacitors, metallic strips/lines with printed capacitors, or coupled lines [5] as exemplified in Figure 12.1. The key difference between the composite CRLH transmission line and the ZPSL lies in the fact that the CRLH transmission line is associated with a ground, while the ZPSL structure is essentially a ground-less single-line structure. The major feature of ZPSL is the very small phase lag associated with the current flowing along the line.

[1]Department of Electrical and Computer Engineering, National University of Singapore, Singapore
[2]Satellite Department, Institute for Infocomm Research (I^{2}R), Singapore
[3]School of Electronics and Information, Nantong University, PR China

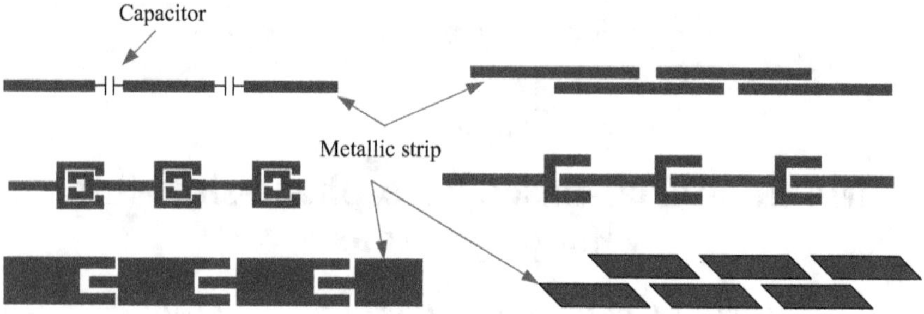

Figure 12.1 Zero-phase-shift-line configurations

12.2 State-of-the-art ZPSL loop antennas

ZPSL loop antennas were first investigated for ultra-high frequency (UHF) radio frequency identification (RFID) applications, where the reader antenna is required to generate strong and uniform magnetic field over an electrically large interrogation zone while the conventional solid-line loop antennas cannot make it. Dobkin *et al.* first presented the ZPSL loop antenna, or segmented loop antenna, consisting of a number of ZPSL sections in which each section is composed of a metal line and a series lumped capacitor [6]. In this structure, segmenting and combining the parasitic inductance of each section with a lumped capacitor make the electric current flowing along the electrically large loop uniform and in an unchanged direction. As a result, the loop generates even and strong magnetic field in the area enclosed by the loop. Later, Oliver presented three ZPSL lines using single, double, and triple broken lines, to avoid the use of lumped circuit components [7–9]. After that, the ZPSLs with various types of distributed capacitors have been proposed [10–12]. The most important characteristic of the ZPSL is a very small phase lag associated with the current flowing along the line, which makes it desirable for the design of electrically large loop antennas, for instance, for UHF near-field RFID readers. Compared with a conventional solid-line loop antenna, a single ZPSL loop antenna is capable of achieving a much larger interrogation zone with a perimeter up to several wavelengths, which is greater than that of a solid-line loop antenna (about half wavelength) [10,11,13–15].

In addition, the ZPSL dual-loop [16] and ZPSL grid-loop array [17] have been developed to further enlarge the interrogation zone with the perimeter up to four operating wavelengths. An artificial magnetic conductor (AMC)-based ZPSL loop has been proposed to achieve directional magnetic field distribution with enhanced magnetic field intensity [18].

Furthermore, the ZPSL has been applied to design antennas for generating omnidirectional radiation. A number of ZPSL loop-based horizontally polarized omnidirectional antennas and circularly polarized (CP) omnidirectional antenna have been reported for the access points of wireless local area networks (WLANs) and UHF far-field RFID readers [12,19,20].

12.3 Modeling of zero-phase-shift-line structure

12.3.1 Dispersion analysis of zero-phase-shift-line structure

The ZPSL is essentially a ground-less single-line structure, which can be considered as a planar Sommerfeld or Goubau line [21] periodically perturbed with discontinuities. The well-developed transmission-line approaches [22] in the metamaterial field can be applied to model and analyze these structures. The dispersion characteristic of a ZPSL reflects the phase information and therefore is the most important property. Without the dispersion relation, the design process of a ZPSL loop antenna becomes extremely time-consuming, in particular, for electrically large loop antennas. Moreover, it is difficult to optimize a design because of the unknown upper size limit of such antennas.

Two commonly used methods have been applied to characterize the dispersion relation of a periodic structure, namely, the eigen-mode approach based on eigenvalue solutions and the driven-mode approach based on the S-parameters [23].

12.3.1.1 Eigen-mode analysis

The Ansoft high-frequency structure simulator can be employed to perform the eigen-mode analysis. The simulation configuration is shown in Figure 12.2, in which the periodic boundary conditions truncate the unit cell in the longitudinal direction, taking into account the mutual coupling between the neighboring cells. Since the ZPSL is an open structure that exhibits strong radiation, perfectly matched layers (PMLs) are imposed in all the transversal directions to mimic the radiation space that extends to the infinity [24].

Figure 12.3 shows two types of commonly used ZPSL unit cells [25], where the solid-line unit cell is also shown for comparison. In simulations, a real phase

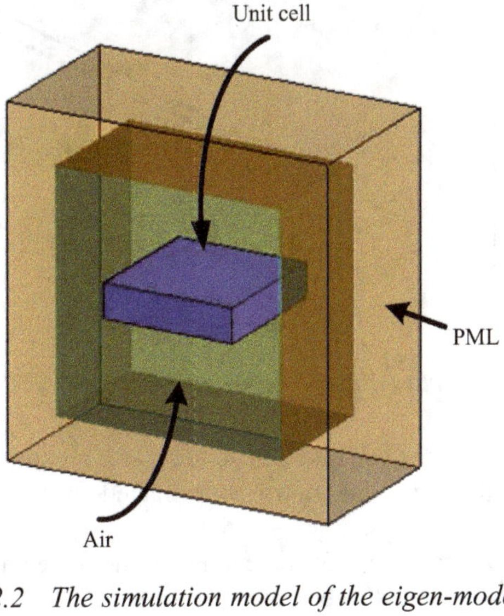

Figure 12.2 The simulation model of the eigen-mode analysis

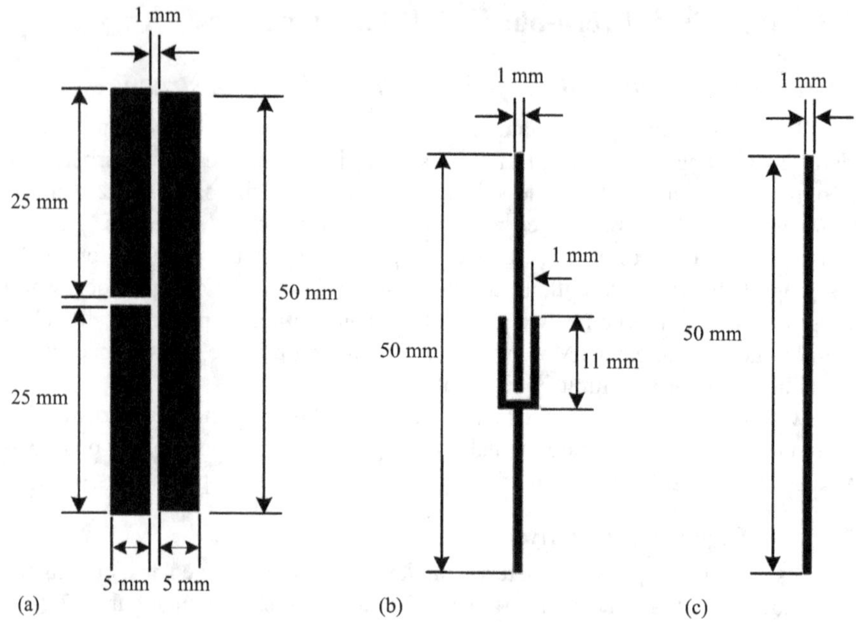

Figure 12.3 Unit cells for eigen-mode analysis: (a) the unit cell with dashed lines; (b) the unit cell with a fork-shaped distributed capacitor; and (c) the unit cell of the solid line

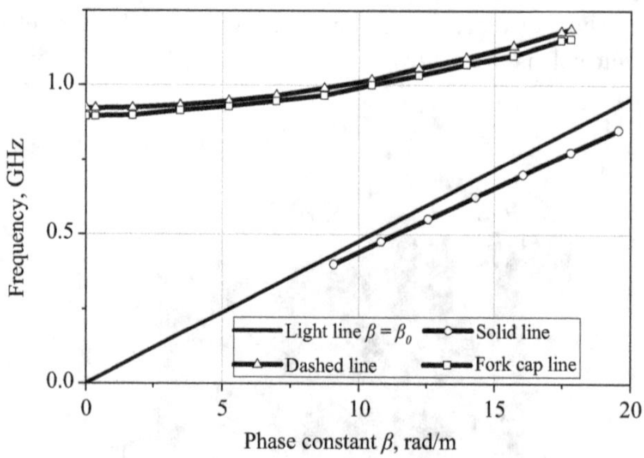

Figure 12.4 Dispersion curves of the two commonly used ZPSLs and the solid line

difference is assumed across the unit cell to solve for a complex frequency. The dispersion curve of the phase constant β is extracted from the real parts of the calculated complex frequencies.

Figure 12.4 illustrates the dispersion curves for the three unit cells as shown in Figure 12.3. Clear stop bands can be observed for both ZPSL unit cells, while the solid-line unit cell exhibits a linear dispersion relation. The light line represents the dispersion curve of the electromagnetic wave in the free space, wherein the phase constant is denoted as β_0. The light line delimits a small phase constant region to its upper left side, in which the phase constants at any single frequency are all smaller than the corresponding values on the light line. It is interesting to note that the two ZPSL structures have their dispersion curves entirely located in this small phase constant region, but the dispersion curve of the solid-line structure is positioned entirely in the lower right region to the light line.

Since the actual electromagnetic problem is associated with complex propagation constants, the complex frequencies are essentially the mathematical solutions. It is important to note that the complex frequencies obtained by using the eigen-mode solver might not necessarily represent the real physical solution that exists for the ZPSL [26]. As pointed out in [27], the interaction between the neighboring unit cells and the energy leakage are decoupled in the complex frequency approach, leading to the clear stopbands observed. On the other hand, when the dispersion curves reach the vertical axis, the zero-β points are produced at nonzero frequencies, resulting in the zeroth-order resonance. Nonetheless, the boundary condition for such zeroth-order resonance is a short circuit [28,29], which is difficult to implement with the ground-less ZPSL structure.

12.3.1.2 Driven-mode analysis with loop configuration

The driven-mode approach is applicable to the analysis of a ground-less ZPSL structure directly. An alternative way is to analysis the dispersion behaviors of the ZPSL structures under the commonly applied boundary conditions. Since the ZPSL structures have mainly been used in the loop configurations, the dispersion relations of the ZPSL structures can be examined by studying the loops that consist of the ZPSL unit cells. The CST full-wave driven-mode solver can be employed to analyze the loops. Figure 12.5(a) and (b) shows the loops that are configured using the unit cells exhibited in Figure 12.3(a) and (b), respectively. For a loop with a fixed number of unit cells, the phase reversal of the current flowing along the loop can be observed at a particular frequency. The guided wavelength at this frequency

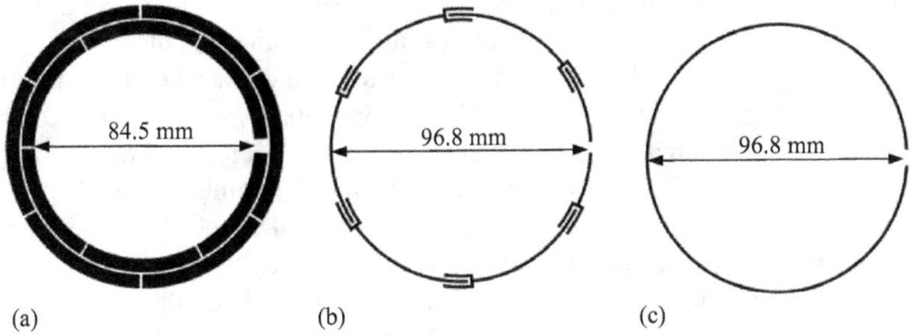

Figure 12.5 Configurations of the loops: (a) the loop with dashed lines; (b) the loop with fork-shaped distributed capacitors; and (c) the loop with the solid line

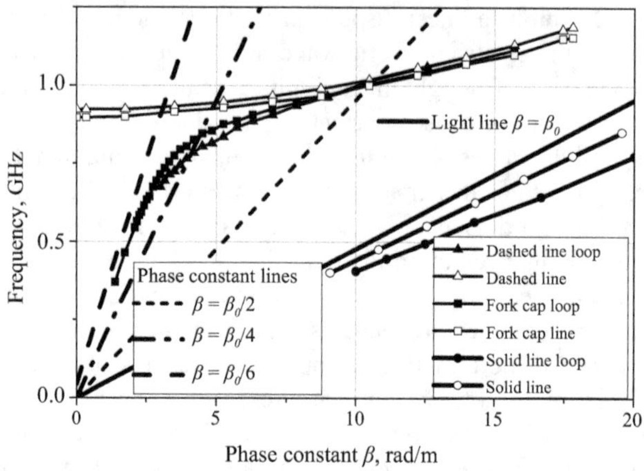

Figure 12.6 Dispersion curves of the two commonly used ZPSL structures and the solid-line structure

can be readily calculated as twice the perimeter of the loop. In this sense, the dispersion of the phase constant β of the loop can be obtained by varying the number of the unit cells in a loop while keeping the unit cell structure unchanged.

The extracted phase constants of the loops are plotted in Figure 12.6. The dispersion curves for the unit cells obtained using Eigen-mode approach are also plotted for comparison. Three phase constant lines are also plotted in Figure 12.6. These phase constant lines delimit the regions that have different guided wavelengths, and therefore can help to determine the largest ZPSL loop which is able to maintain an in-phase flowing current along the loop. For instance, for a dispersion point that is located to the right of the phase line $\beta = \beta_0/4$, the largest perimeter that the ZPSL loop could achieve is less than $2\lambda_0$, where λ_0 is the free-space wavelength at the corresponding frequency. It can be observed that the dispersion curves obtained from the driven-mode analysis deviate from the ones based on the eigen-mode analysis as decreases. The large deviation close to the points of the zeroth-order resonance could be attributed to the different boundary conditions in the loop structure. Nevertheless, it is interesting to note that the dispersion curves calculated from the two methods become close to each other when they are away from the clear edges of the stop bands. The reason is that when the phase constant of the ZPSL loop increases, the guided wavelength decreases and the largest perimeter of the ZPSL loop also decreases, reducing the number of the ZPSL unit cells that are used to form the loop. It essentially alleviates the effects of the interaction between neighboring unit cells and the overall energy leakage, and therefore the dispersion curve of the ZPSL loop approaches the corresponding dispersion curve from the Eigen-mode analysis. In addition, the fact that the two different ZPSL structures exhibit similar dispersion behaviors further verifies that both of the structures operate in a similar single-line fashion despite their different physical topologies.

12.3.1.3 Equivalent circuit

The physical insights of the ZPSL structure can be further gained by constructing an equivalent circuit. As shown in Figure 12.7, a unit cell of the ZPSL with a fork-shaped distributed capacitor can be modeled with lumped circuit components, where two circuit models are presented to analyze the dispersion characteristics that match the results in the previous sections, respectively.

The shunt capacitor C_R is given by

$$C_R = \frac{8\varepsilon}{\sqrt{\pi}}\sqrt{A} \tag{12.1}$$

where A is the area of the metal strips in the unit cell. The series circuit elements L_R, R_1, and C_L are obtained through an energy-based circuit parameter extraction method using Eigen-mode analysis [30]. The values of the circuit components are summarized in Table 12.1. The circuit elements R_2 and G can be tuned to produce a dispersion curve that fits the results of the Eigen-mode analysis. Moreover, the two resistances R_1 and R_2 represent the radiation of the structures into the free space.

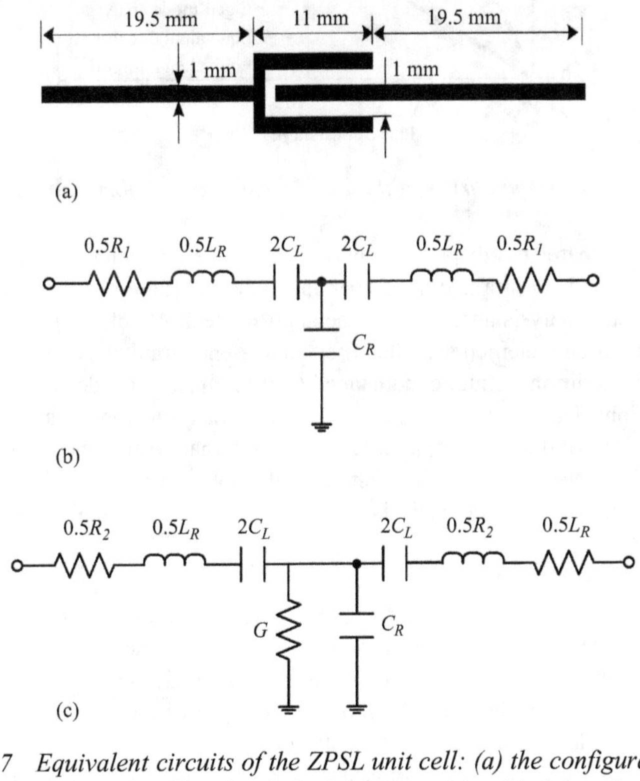

Figure 12.7 Equivalent circuits of the ZPSL unit cell: (a) the configuration of the ZPSL unit cell with a fork-shape capacitor; (b) the equivalent circuit with the dispersion curve matched to the driven-mode analysis; and (c) the equivalent circuit with the dispersion curve matched to the eigen-mode analysis

Table 12.1 Circuit elements for the equivalent circuit model

C_L	C_R	L_R	R_1
473.7 fF	341.4 fF	66.6 nH	72.3 Ω

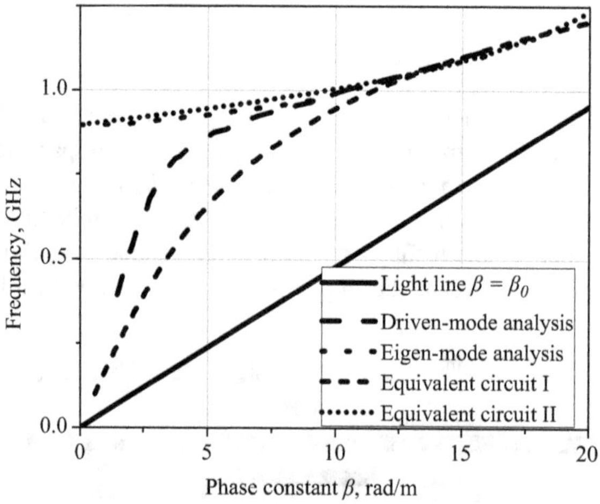

Figure 12.8 Dispersion curves of the ZPSL structure with fork-shaped capacitors

Figure 12.7 compares the dispersion curves, where the equivalent circuits shown in Figure 12.7(b) and (c) produce the dispersion curves that are close to the results of the driven-mode analysis and Eigen-mode analysis, respectively. It is remarked that while all the circuit elements in the first equivalent circuit have actual physical meanings, the additional shunt conductance G in Figure 12.7(c) does not correspond to the actual physical structure of the ZPSL cell. In fact, the conductance essentially blocks the effects of the shunt capacitor in the circuit, manifesting the series resonance of L_R and C_L. It also implies the deviation of the solution between the Eigen-mode analysis and the driven-mode analysis, which is closer to the actual setup.

12.3.2 Design guidelines

Based on the dispersion analysis on ZPSL, one can find the largest achievable perimeter of a ZPSL loop without any phase reversal of the flowing current. It is therefore useful to employ the dispersion analysis in the design process of the ZPSL loop antenna to explore the upper limit of the loop size with an in-phase current that can achieve the desired uniform field distribution.

Figure 12.9 exemplifies two fork-shaped ZPSL unit cells. Three dispersion points around 915 MHz for the ZPSL structures are calculated using driven-mode approach and exhibited in Figure 12.10. The second ZPSL structure is with a smaller F, which implies that it is able to configure larger loop without current phase reversion. The

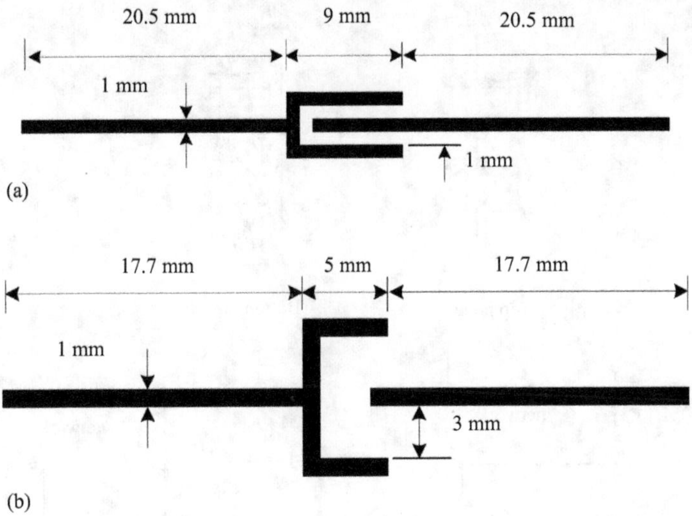

Figure 12.9 The unit cells with a fork-shaped distributed capacitor: (a) the first design and (b) the second design

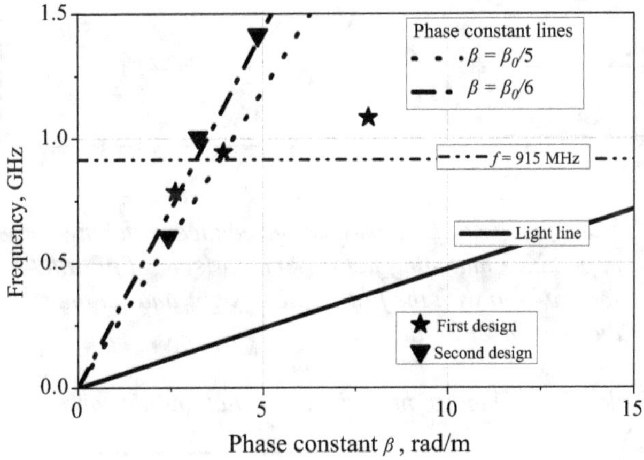

Figure 12.10 Dispersion plots of the proposed ZPSL structures

dispersion points are close to the phase constant lines of $\beta = \beta_0/5$ and $\beta = \beta_0/6$ at 915 MHz, it suggests that the perimeter of the corresponding ZPSL loop antenna can be up to $2.5\lambda - 3\lambda$ at the frequency of interest without current phase reversal.

Figure 12.11 exhibits two square ZPSL loop antennas that are constructed using the two ZPSL unit cells shown in Figure 12.9, respectively. A matching network consisting of parallel-line matching stubs is utilized to achieve desired impedance matching. The detailed dimensions of the input matching networks of the antennas are tabulated in Table 12.2. In both designs, the ZPSL unit cells are positioned along

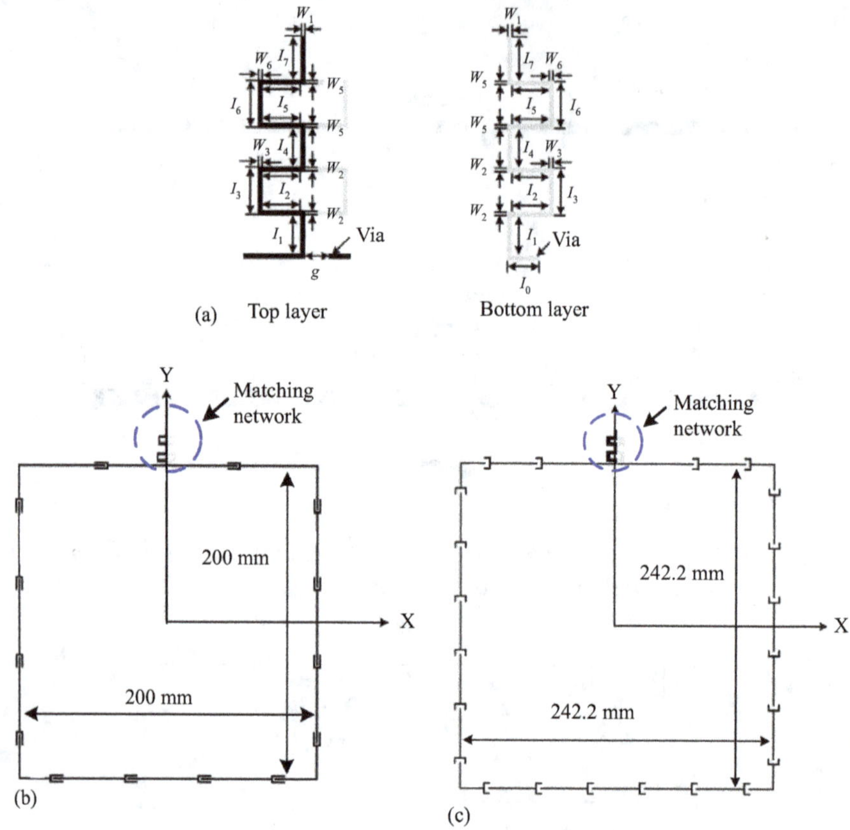

Figure 12.11 The square ZPSL loops with impedance matching networks: (a) the impedance matching network comprising of parallel-line stubs; (b) square loop with the first ZPSL design; and (c) the square loop with the second ZPSL design

Table 12.2 Dimensions for input-matching networks

Width (mm)	W_1	W_2	W_3	W_4	W_5	W_6	W_7
First design	1	1.5	0.5	1	1.5	0.5	1
Second design	2	2.5	3	2	2.5	3	2
Length (mm)	L_1	L_2	L_3	L_4	L_5	L_6	L_7
First design	2	5	6	5	4	6	4
Second design	1	3	8	3	3	7	5

the sides of the loop with identical separation except for the two unit cells close to the antenna input port, where the separation is different from the others for the purpose of impedance matching. The two loops are with a perimeter of 800 and 969.6 mm, which correspond to 2.44λ and 2.96λ at 915 MHz, respectively.

Figure 12.12 shows the reflection coefficients of the two ZPSL loop antennas. It can be seen that both the two antennas exhibit the reflection coefficients of less than −10 dB at 915 MHz. Figure 12.13 shows the current and the magnetic field distributions ($|H_z|$) of the antennas. In-phase flowing current is observed for both

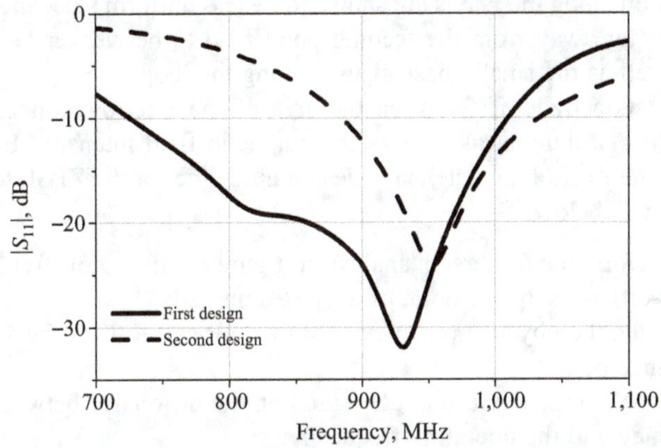

Figure 12.12 Simulated reflection coefficients of the ZPSL loop antennas

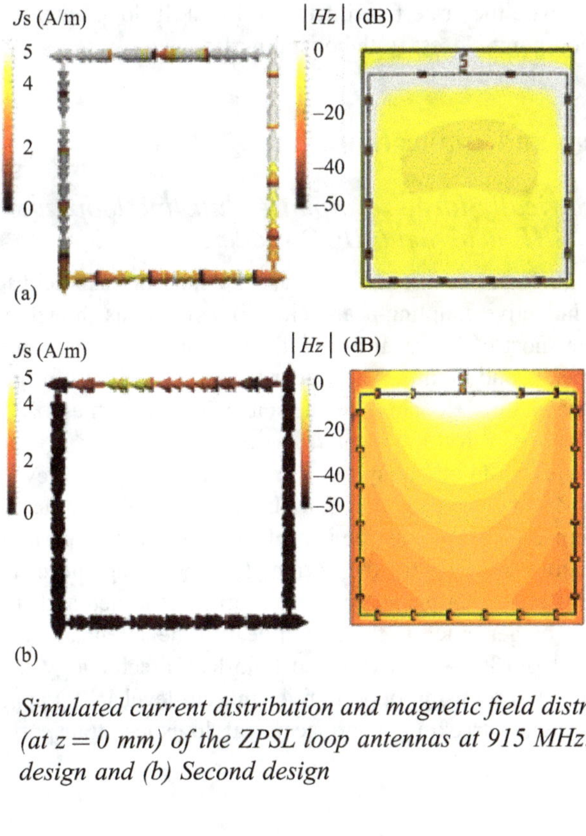

Figure 12.13 Simulated current distribution and magnetic field distribution ($|H_z|$) (at z = 0 mm) of the ZPSL loop antennas at 915 MHz: (a) First design and (b) Second design

designs. Note that the magnitudes of the currents along the two antennas are different. As shown in Figure 12.13(b), the larger ZPSL loop, namely, the second design, exhibits a smaller magnitude. The H_z distributions along the x-axis are relatively uniform for both ZPSL loops because of the unidirectional currents along the loops and the symmetry of the loop structures about the x-axis. The magnetic field distribution along the y-axis are shown to be less uniform since the currents on those portions far away from the feeding point tend to be weaker because of the radiation as well as the small phase change along the loop.

There exists a trade-off between the area of the interrogation zone and the magnetic field distribution, as well as the magnetic field intensity. Based on the case study of the ZPSL loop antenna, a design guideline for the ZPSL loop antenna is summarized as below.

Step 1. Perform the dispersion analysis to figure out the cutoff frequency or the zeroth-order-resonance point of a ZPSL unit cell;
Step 2. Tune the physical parameters of the unit cell to vary the zeroth-order-resonance point;
Step 3. Finalize the ZPSL unit cell based on the difference between the cutoff frequency and the operating frequency;
Step 4. Perform the driven-mode analysis on the ZPSL loop that consists of the designed ZPSL unit cells to find a few dispersion points near the operating frequency;
Step 5. Estimate the largest perimeter of the ZPSL loop antenna based on the dispersion point closest to the operating frequency.

12.4 Design and applications

12.4.1 Electrically large zero-phase-shift-line loop antennas for UHF near-field RFID readers

Loop antennas have been utilized as a reader antenna in near-field RFID systems for years. The inductive coupling near-field RFID system as shown in Figure 12.14 are preferred in most of practical applications because they can operate in close proximity to metals and liquids. The reader antenna is required to generate strong and uniform magnetic field close to the antenna so that the RFID tags can be detected at any position inside the interrogation zone.

The early near-field RFID systems operate at low frequency (LF, 125–134 kHz) and high-frequency (HF, 13.56 MHz) bands. The electrically small loop antennas implemented with solid line in such systems are with perimeter much less than half-operating wavelength, the current flowing along the electrically small solid-line loop is always in-phase, so that uniform distributed and strong magnetic field is easily to be generated in the region near to the antenna.

Currently, UHF (840–960 MHz) near-field RFID technology receives a lot of attention due to the promising opportunities in item-level RFID applications such as sensitive products tracking, pharmaceutical logistics, transport and medical

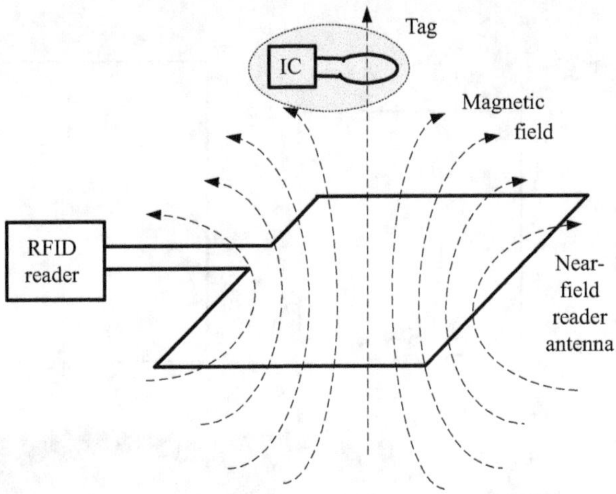

Figure 12.14 Coupling mechanism of inductive near-field RFID systems

products (blood, medicines, vaccines), biosensing applications (biohazard materials, security, etc.), and so on. With high reading rate, UHF RFID applications require a reader antenna with an adequate coverage (for example, 150 × 150 mm^2, the perimeter is about 2λ, where λ is the free space operating wavelength at 915 MHz) to detect a number of tags simultaneously. The loop antenna with such a large interrogation zone is no longer electrically small. It is known that the conventional solid-line loop antenna with a perimeter larger than 1λ cannot produce even magnetic field distribution in the near-field zone of the loop antenna because the current flowing along the loop experiences phase inversion and current nulls [25] as shown in Figure 12.15. The magnetic field is relatively weak in the central region of the loop antenna, which makes such an electrically large solid-line loop antenna not suitable for UHF near-field RFID applications.

The ZPSL enables flowing current keeping a very small phase lag and therefore an electrically large ZPSL loop antenna is able to generate strong and uniform distributed magnetic field over an enlarged interrogation zone, which makes the ZPSL loop antenna desired to UHF near-field RFID applications. So far there have been various designs reported [7–11,13–15].

12.4.1.1 Zero-phase-shift-line single-loop antenna

Figure 12.16 shows a ZPSL single-loop antenna with coupled line ZPSL cells for UHF near-field RFID applications. The antenna prototype printed onto a piece of FR4 substrate with an overall size of 175 × 180 × 0.5 mm, achieves a large interrogation zone of 154 × 154 mm with 10-dB return loss and uniform magnetic field distribution over an operating bandwidth of 820–1,050 MHz. In a setup shown in Figure 12.16(b), the antenna prototype is used as an RFID reader antenna and connected to an Impinj Speedway reader operating at 865–956 MHz with 30-dBm output to detect the Impinj button type tags (J21, 8 mm in diameter), where 25 tags

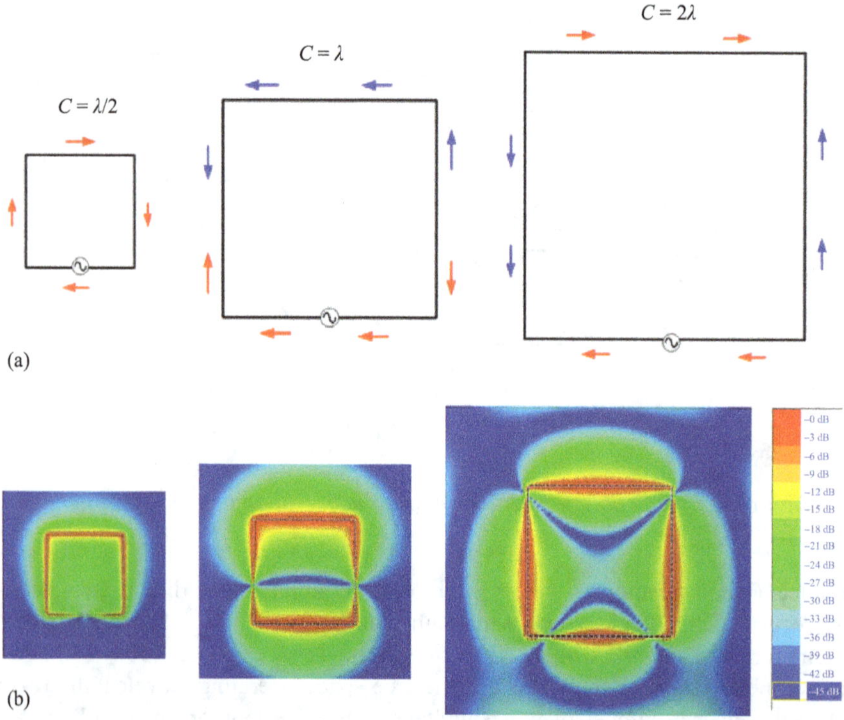

Figure 12.15 Simulated current and magnetic field distribution of solid-line loop antenna: (a) current distribution and (b) magnetic field distribution

are positioned symmetrically on a Styrofoam disc with a diameter of 160 mm as shown in Figure 12.17, the ZPSL loop antenna shows superior performance with a 100% bidirectional reading rate up to 24 mm, the solid-line single-loop antenna with the same size is unable to offer 100% detection even when the tags are positioned on the surface of the antenna.

12.4.1.2 Zero-phase-shift-line dual-loop antenna

For ZPSL single-loop antennas, there exists a trade-off between the area of the interrogation zone and the magnetic field distribution, as well as the magnetic field intensity. A smaller loop antenna provides smaller interrogation zone with stronger magnetic field intensity and more uniform field distribution. Increasing loop size enlarges the interrogation zone but weakens magnetic field intensity and degrades the magnetic field distribution. It is a challenge to design a ZPSL loop antenna with enlarged interrogation zone and stronger and uniformly distributed magnetic field simultaneously.

A number of ZPSL loop antennas with enlarged interrogation zone have been reported [16–18]. Figure 12.18 shows a design with ZPSL dual-loop antenna configuration. The antenna comprises a primary loop and a parasitic loop. The parasitic loop is positioned in the center of the primary loop to enhance the magnetic field intensity there and therefore generate strong and uniform magnetic field distribution

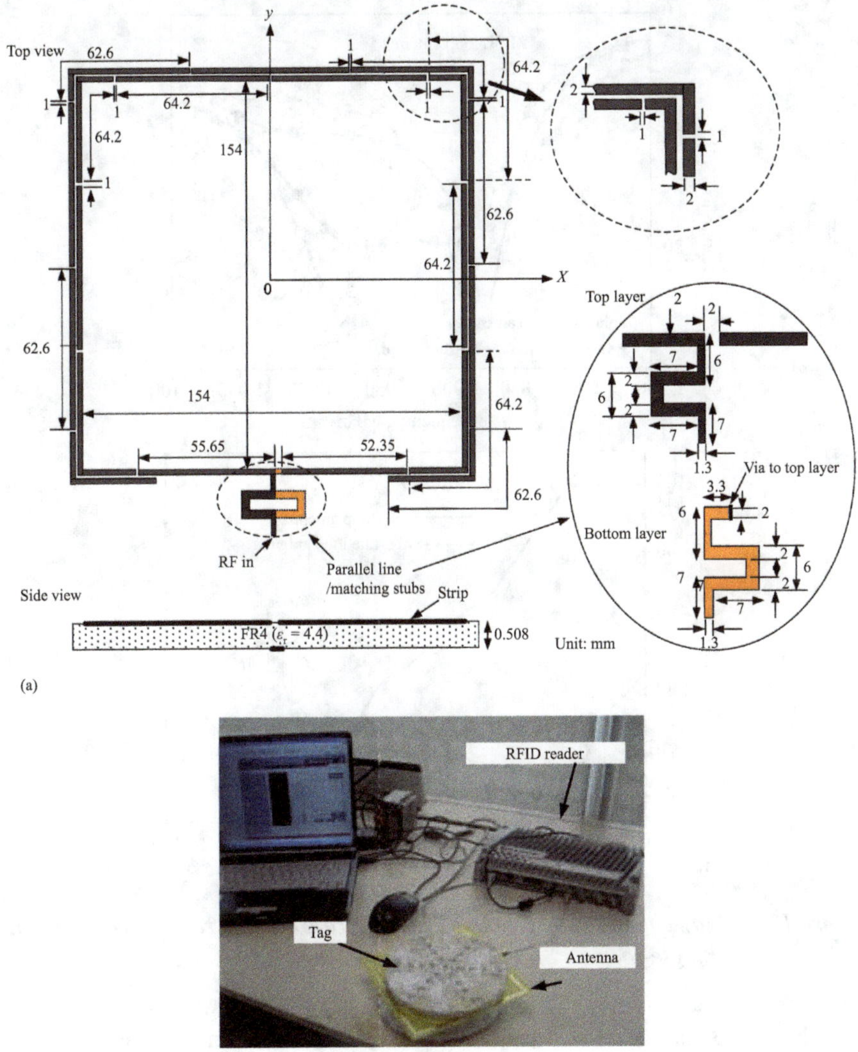

Figure 12.16 (a) The configuration of a single-loop ZPSL antenna with coupled line ZPSL unit cells and (b) near-field RFID measurement setup

over an interrogation zone with a perimeter of 3λ. Both loops are composed of a number of fork-shaped ZPSL unit cells. As shown in Figure 12.18(b), the currents flowing along each loop are no longer in-phase due to the coupling between the loops. The area enclosed by the main loop is divided into three regions by the parasitic loop. The upper and lower portions of the antenna can be considered as two loops with in-phase current. The magnetic field in the central portion enclosed by the parasitic loop is the superposition of the magnetic field generated by the primary loop as well as the parasitic loop.

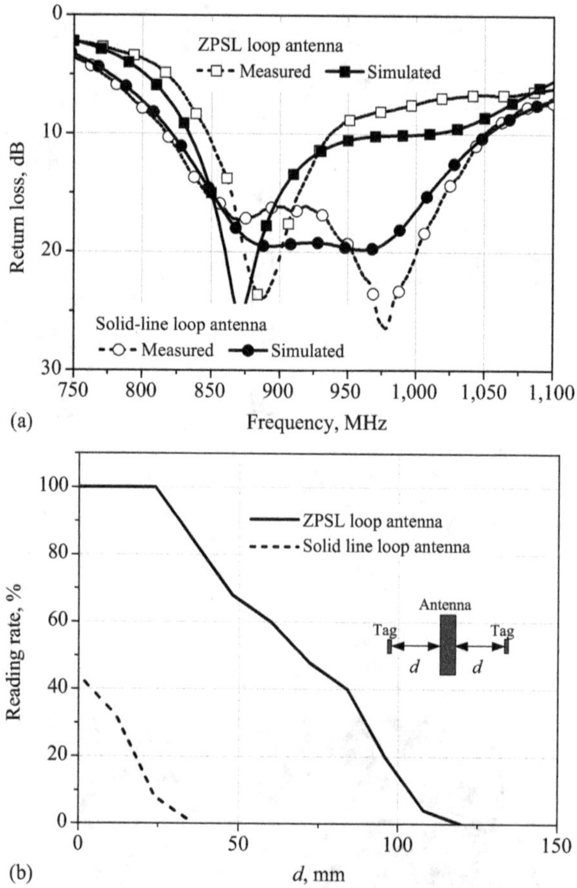

Figure 12.17 (a) The return loss and (b) reading rate against reading range of the single-loop ZPSL antenna and the single-loop solid-line antenna

The size of the parasitic loop shows significant effect on the magnetic field distribution of the dual-loop antenna. Figures 12.19 and 12.20 exhibit the effect of the length (L_1) and width (L_2) of the parasitic loop on the magnetic field distribution ($|H_z|$) along the x- and y-axis of the dual-loop antenna. The L_1 affects the magnetic field distribution significantly, especially along the x-axis. A larger L_1 offers stronger and more uniform magnetic field distribution. Therefore, the left and right sides of the parasitic loop should be arranged close to the sides of the main loop. The L_2 shows a slight effect on the magnetic field distribution while a smaller L_2 is more desirable for enhancing the magnetic field distribution.

Figure 12.21 exhibits the simulated two-dimensional (2-D) magnetic field distributions of the proposed ZPSL dual-loop antenna and a ZPSL single-loop with the same perimeter of 3λ at 915 MHz. It can be seen from Figure 12.24 that the magnetic field at the central portion of the ZPSL single-loop is greatly degraded

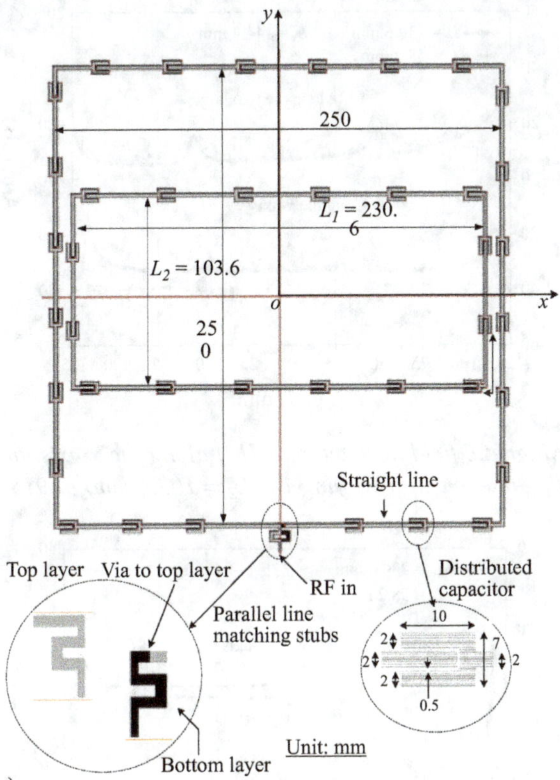

(a)

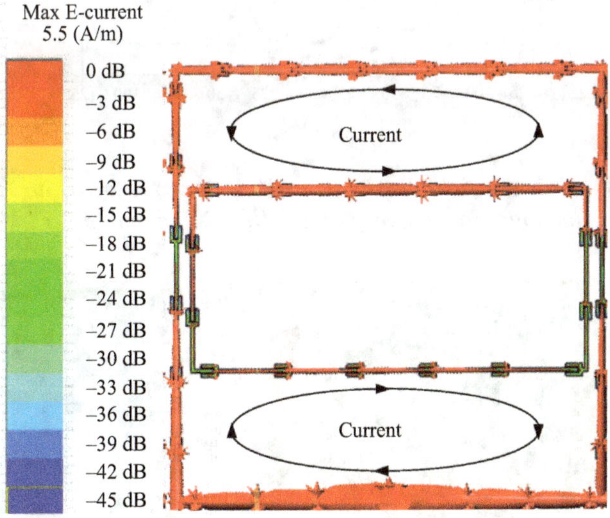

(b)

Figure 12.18 ZPSL dual-loop antenna: (a) configuration and (b) current distribution at 915 MHz

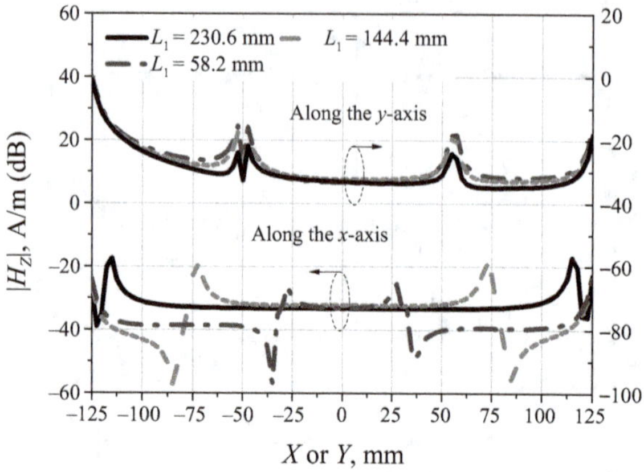

Figure 12.19 Magnetic field distribution ($|H_z|$) along the x-axis and the y-axis of the antenna with varying L_1 ($L_2 = 103.6$ mm) at 915 MHz

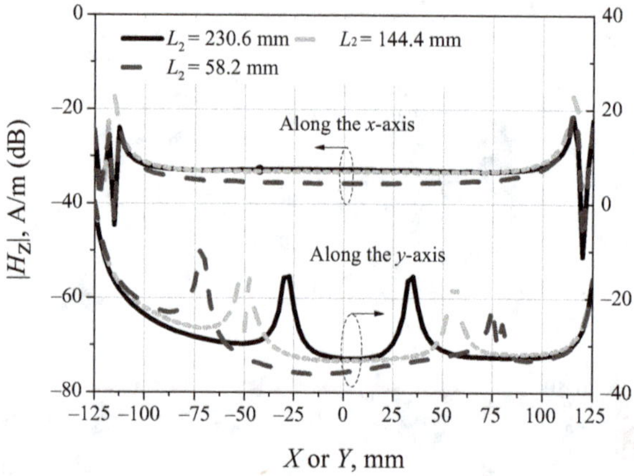

Figure 12.20 Magnetic field distribution ($|H_z|$) along the x-axis and the y-axis of the antenna with varying L_2 ($L_1 = 230.6$ mm) at 915 MHz

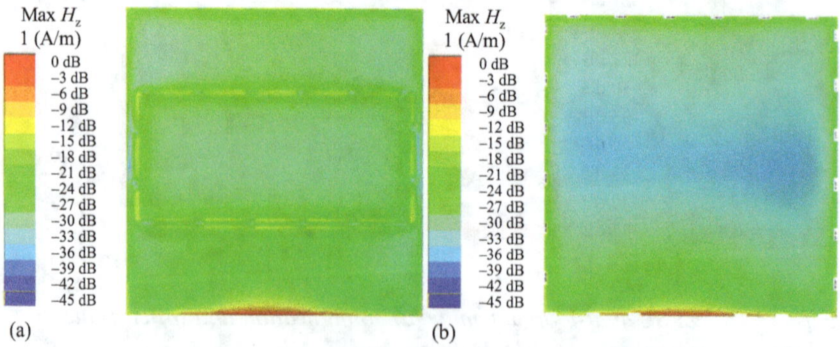

Figure 12.21 Simulated 2-D magnetic field distribution ($|H_z|$) at 915 MHz ($z = 0.5$ mm): (a) the ZPSL dual-loop antenna and (b) the ZPSL single-loop

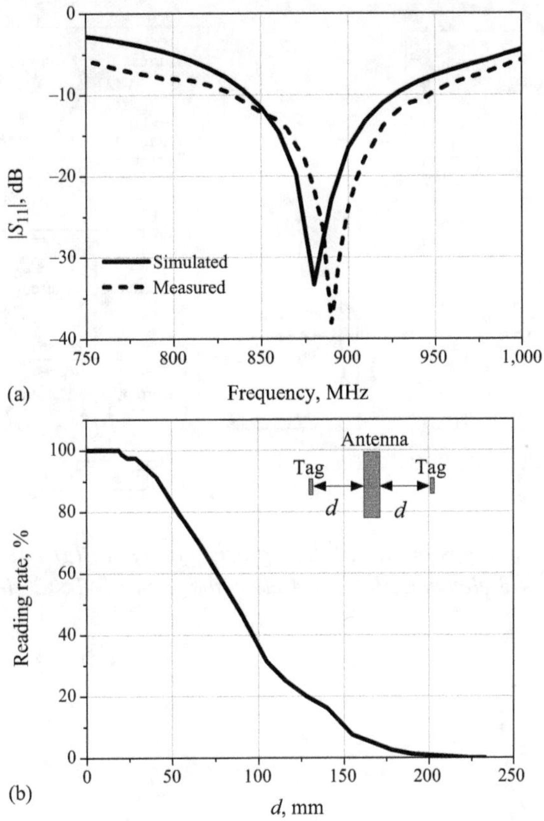

(a) (b)

Figure 12.22 Performance of the ZPSL loop antennas: (a) the reflection coefficient of the ZPSL dual-loop antenna and (b) the RFID tag reading rate/range

compared with the ZPSL dual-loop antenna, although the current along the ZPSL single-loop is in-phase.

Figure 12.22(a) compares the simulated and measured reflection coefficient of the ZPSL dual-loop antenna. –10-dB reflection coefficient is observed from 845 to 928 MHz. Acting as a UHF near-field RFID reader with same set-up shown in Figure 12.22(b) and 80 tags (J12, 15 × 8 mm), the ZPSL dual-loop antenna is able to detect all the tags with 100% up to a reading range of 19 mm (Impinj Speedway reader with 30-dBm output).

12.4.1.3 Zero-phase-shift-line grid-loop array antenna

ZPSL grid-loop array antenna is another solution to achieve an enlarged interrogation zone with enhanced magnetic field intensity. Figure 12.23 shows the topologies of the ZPSL grid-loop array antenna with 1 × 2, 1 × 3, 1 × 4, and 2 × 2 elements, where an example of the ZPSL loop cell is exhibited in Figure 12.24. To make the

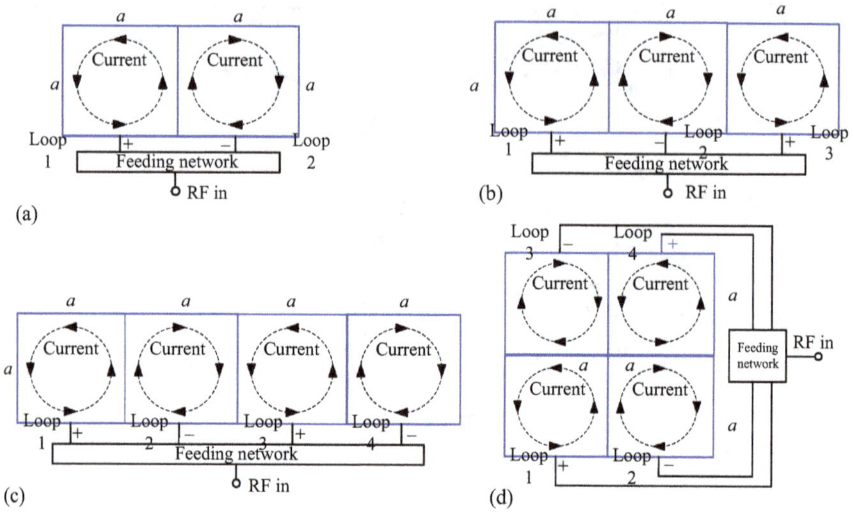

Figure 12.23 Topologies of the ZPSL grid-array antenna: (a) 1 × 2 elements; (b) 1 × 3 elements; (c) 1 × 4 elements; and (d) 2 × 2 elements

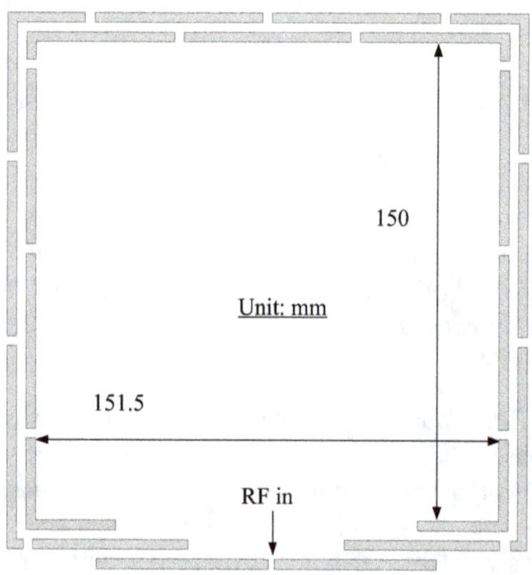

Figure 12.24 The configuration of the ZPSL loop element

magnetic field inside the interrogation zone as strong and uniform as possible, the following guidelines for configuring the grid-array antenna can be followed.

1. The currents flowing along adjacent loops should be in a reverse direction;
2. The input ports of the adjacent loops should be 180° out-of-phase;
3. The feeding network should be easy to control the in-phase or out-of-phase relationship between the output ports.

Square ZPSL loop as shown in Figure 12.24 is desired to be utilized as unit cell to configure the grid-loop array easily. The other ZPSL loops such as rectangle ZPSL loop, rhombic ZPSL loop, and so on can be used as unit cell to configure the grid-loop array as well.

In principle, more ZPSL loop cells are able to configure a larger grid-loop array antenna with an enlarged interrogation zone. However, the increased loss from the complicated feeding network may weaken the magnetic field intensity and degrade the detection range of the RFID systems, which should be taken into account in practical applications.

Figures 12.25 and 12.26 show a 1×2 ZPSL grid-loop array antenna design and the results, respectively. The antenna is composed of two ZPSL loops with total length of 0.94λ and width of 0.46λ (λ is the free space wavelength at 915 MHz). The antenna prototype is with an enlarged interrogation zone with perimeter up to

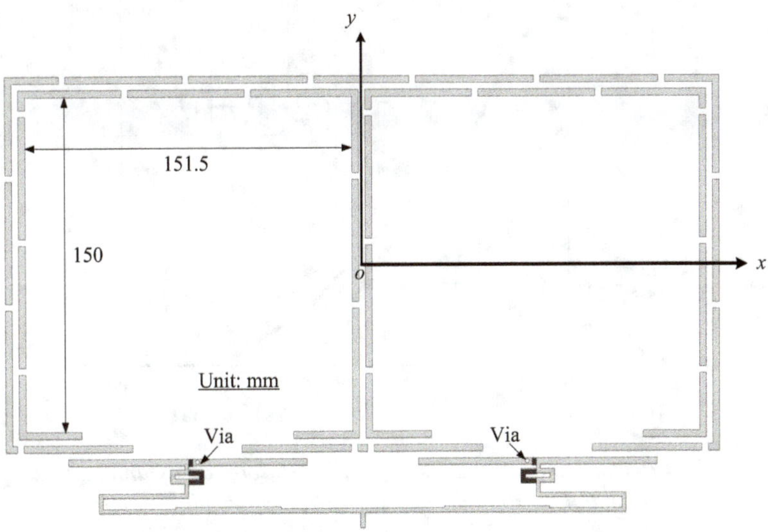

(a)

(b)

Figure 12.25 1× 2 grid-ZPSL loop array antenna: (a) configuration and (b) 2-D magnetic field distribution ($|H_z|$) at 915 MHz ($z = 0.5$ mm)

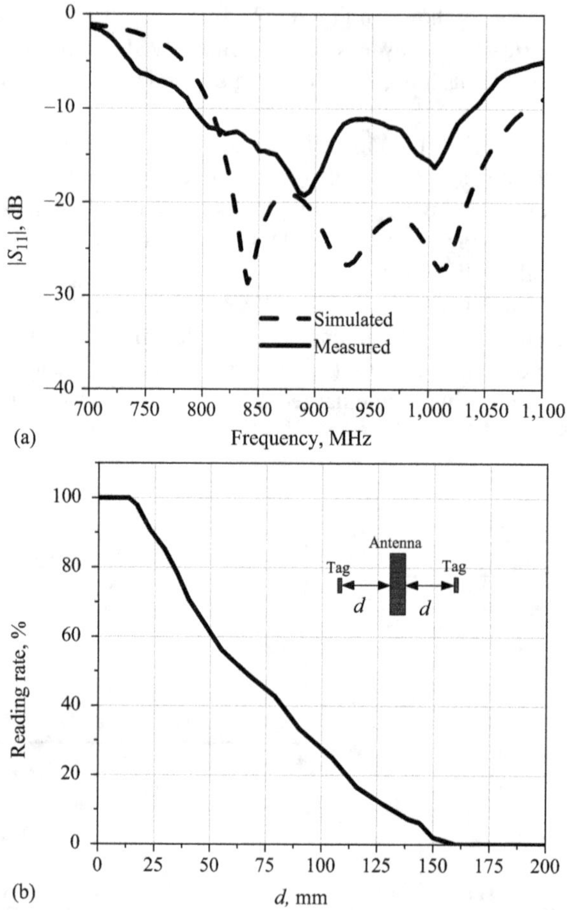

Figure 12.26 Performance of the 1× 2 grid-ZPSL loop array antenna: (a) reflection coefficient and (b) RFID tag reading rate/range

3λ uniform distributed and strong magnetic field is observed. The −10-dB reflection coefficient is achieved from 790 to 1040 MHz as shown in Figure 12.26(a). Moreover, it is able to provide 100% reading rate for tags in the interrogation zone up to 13.5 mm (Impinj Speedway reader with 30-dBm output and J12 tags), as shown in Figure 12.26(b).

12.4.1.4 AMC-backed directional ZPSL loop antenna

Conventional loop antennas are with bidirectional radiation, such a characteristic is not desired for near-field applications since the bidirectionally distributed field will be coupled with the tags at both sides of the antenna, which causes interference and degrades the system detection accuracy. A ZPSL loop antenna with a directional magnetic field distribution is preferred in near-field applications. In order to fulfill the requirement of a directional distribution, a large metal plate is usually used to back a bidirectional ZPSL loop antenna as a reflector [15]. However, the out-of-phase currents on such a metal plate tend to disturb the magnetic field distribution and weaken the magnetic field intensity in the near-field region especially when the plate is placed electrically close to the loop antenna.

AMC features the characteristic of zero reflection phase for normal incidence electromagnetic wave, which has been widely utilized as a reflector to design low-profile directional antennas. Figure 12.27 exhibits an AMC-backed ZPSL loop antenna with a perimeter of 2.44λ at 915 MHz. The 1 $\times$ 2 ZPSL grid-loop array is composed of two rectangular ZPSL loops with same length while different width. Such configuration is able to enhance the uniformity of the magnetic field distribution, especially in the region that is far away from the feeding point. The ZPSL unit cell is a symmetric structure that consists of three solid-line sections and two parallel-line sections, which are alternately distributed along the line as shown in Figure 12.27(a). Figure 12.27(b) shows the side view of the antenna, the ZPSL grid-loop and the AMC structure are printed onto a 0.8128 mm thick FR4

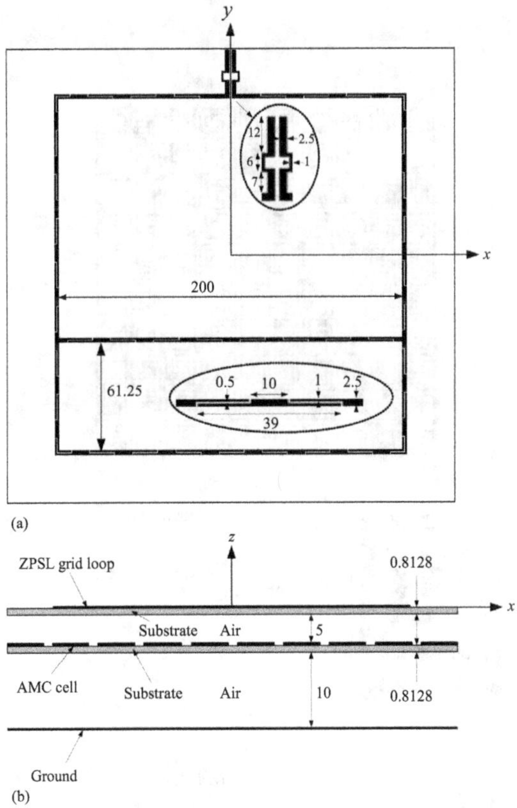

Figure 12.27 AMC-backed grid-ZPSL grid-loop array antenna: (a) top view and (b) side view (unit: mm)

PCB, respectively, and separated with a distance of 5 mm. A metallic plate is positioned 10 mm away from the AMC structure to form the low profile AMC-based reflector.

Figure 12.28 shows the square AMC reflector that consists of 100 AMC unit cells and a metallic ground plate. Each AMC unit cell comprises four spiral branches. The symmetric structure ensures the same reflection phase response in the x- and y-directions. Such intertwined spiral configuration exhibits wideband AMC properties with a small unit cell size of 25.5 × 25.5 mm [31]. The air layer between the AMC cells and the ground is to enhance the bandwidth of the AMC reflector.

Figure 12.29 illustrates the reflection phase of the AMC reflector. In simulation using CST full-wave driven-mode solver, the unit cell structure is illuminated by a plane wave at normal incidence with periodic boundary condition. The AMC reflector shows an operating frequency band of 850–979 MHz, wherein the reflection phase is within the range of ±−90°.

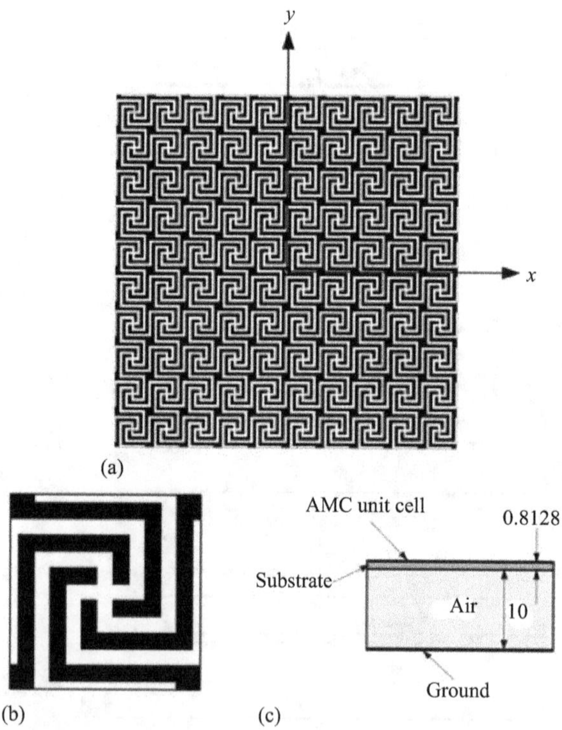

Figure 12.28 AMC reflector configuration: (a) square AMC with 10 × 10 unit cells; (b) top view of the AMC unit cell; and (c) side view of the AMC unit cell (unit: mm)

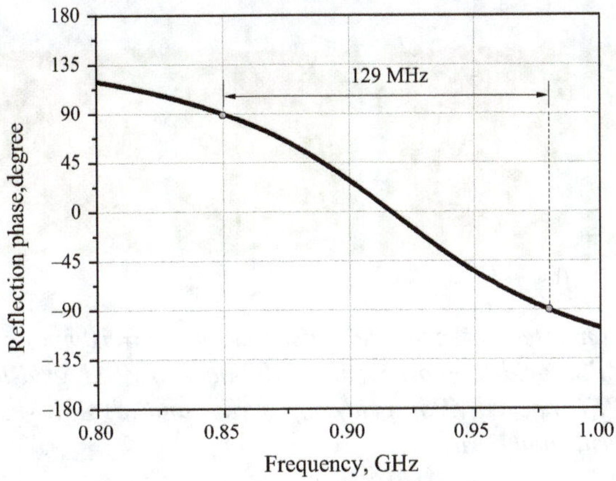

Figure 12.29 Simulated reflection phase response of the AMC unit cell

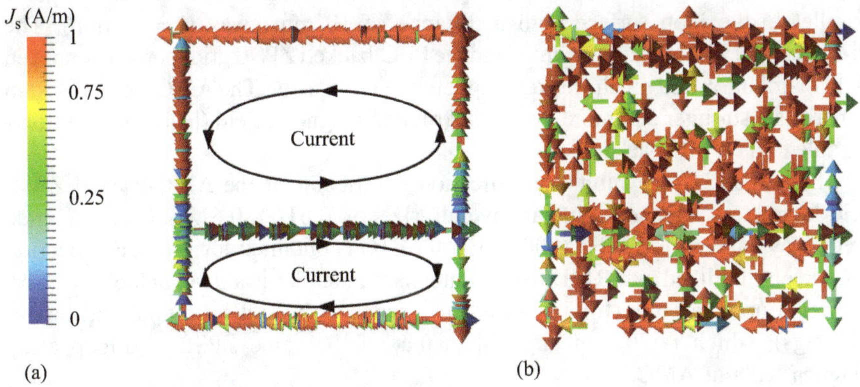

Figure 12.30 Simulated current distributions of the AMC-backed ZPSL grid-loop array at 915 MHz: (a) the ZPSL grid-loop array and (b) the AMC surface

Figure 12.30 shows the simulated current distribution of the AMC-backed ZPSL grid-loop array antenna, where an in-phase current distribution is observed for both loops. The current distribution on the AMC surface is shown in Figure 12.30(b), several closed-loop currents can be observed in the areas that are below the inner-loop regions of the two loops, and each of the closed-loop current flows in the same way as that on the corresponding loop.

Figure 12.31 compares the simulated magnetic field distributions of a ZPSL grid-loop array, a perfect magnetic conductor (PMC)-backed ZPSL grid-loop array, a perfect electric conductor (PEC)-backed ZPSL grid-loop array, and an AMC-backed

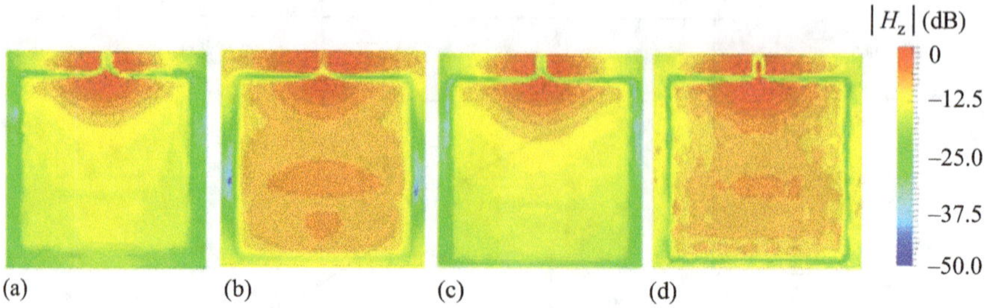

Figure 12.31 Simulated magnetic field distributions (z = 10 mm) at 915 MHz: (a) ZPSL grid-loop array; (b) PMC-backed ZPSL grid-loop array, (c) PEC-backed ZPSL grid-loop array; and (d) AMC-backed ZPSL grid-loop array

ZPSL grid-loop array, wherein the ZPSL grid-loop array design is kept unchanged; the PMC reflector is placed 5 mm below the antenna while the PEC reflector is positioned a quarter-wavelength away; and the magnetic field distribution is observed in a plane parallel to the loop surface with a distance of 10 mm away ($z = 10$ mm). The bi-directional ZPSL grid-loop array and the PEC-backed ZPSL grid loop show uneven field distribution and much weaker magnetic field intensity. The AMC-backed design generates a stronger and more evenly distributed magnetic field ($|H_z|$) which is comparable to the PMC-backed design.

Figure 12.32(a) exhibits the reflection coefficient of the AMC-backed ZPSL grid-loop array antenna with an overall size of $0.61\lambda \times 0.61\lambda \times 0.05\lambda$, a wide frequency band of 140 MHz (900 to 1,040 MHz) is obtained for −10-dB reflection coefficient. Utilized as a RFID reader antenna, the reader is able to achieve a 100% detection up to 90 mm (Impinj Speedway reader with 30-dBm output and Impinj J41 tags), which is 70 mm larger than that of the same ZPSL grid-loop array antenna without AMC.

12.4.2 Horizontally polarized omnidirectional antenna for WLAN access points

Generally, the antenna for WLAN access point and portable devices requires an omnidirectional radiation in order to cover a large service area. Most of the current WLAN systems are vertically polarized, wherein the dipole-like omnidirectional antennas are used to radiate radio wave equally in the plane aligning the *H*-field (*H*-plane). However, in the urban or indoor wireless environment, the polarization of the propagating radio wave may change significantly because of the complicated multiple reflections or scatterings. It has been reported that using horizontally polarized antennas at both the transmitter and receiver can achieve a 10-dB improvement in system gain as compared to vertically polarized antennas at both ends of the link [32,33]. Furthermore, polarization diversity can be used to improve the reliability of a communication link, where the orthogonal polarization allows for the frequency to be reused and isolation to be increased between the independent LANs.

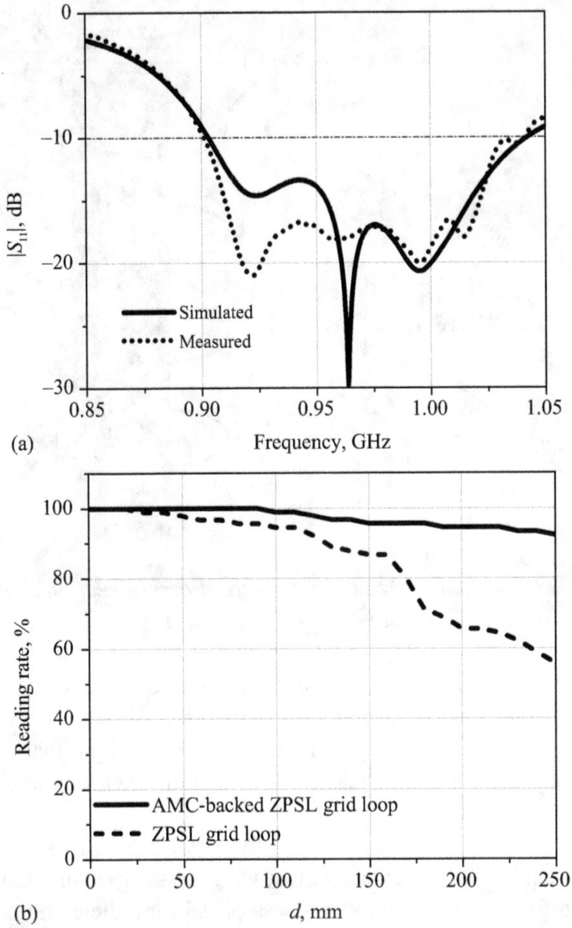

Figure 12.32 Performance of the AMC-backed ZPSL grid-loop array antenna: (a) reflection coefficient and (b) the RFID tag reading rate/range

Therefore, a horizontally polarized omnidirectional antenna seems more preferable for WLAN applications [34–37].

A horizontally polarized omnidirectional antenna is required to radiate equally in the plane aligning the *E*-field (*E*-plane), which is more challenging. An electrically small solid-line loop is able to generate horizontally polarized omnidirectional radiation because the existence of the in-phase and uniform current flowing along the loop. However, such an electrically small solid-line loop antenna has a very small radiation resistance and a large reactance, which makes the antenna very difficult to match. A solid-line loop antenna with the perimeter comparable to a wavelength has a reasonable radiation resistance and reactance for impedance matching while there is no longer horizontally polarized omnidirectional radiation because of the out-of-phase current distribution on the adjacent loop sections.

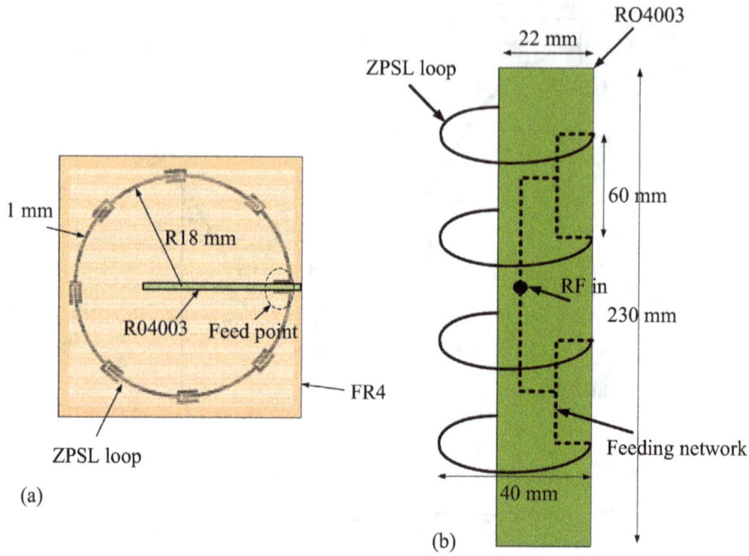

Figure 12.33 ZPSL-based horizontally polarized omnidirectional antenna array: (a) ZPSL loop and (b) 1×4 antenna array

The ZPSL loop antenna is able to generate horizontally polarized omnidirectional radiation because it can to keep the flowing in-phase current with uniform distribution even if the perimeter of loop antenna is larger than one operating wavelength. In addition, the antenna is easily matched with simple and low cost configuration.

Figure 12.33 shows a 1×4 horizontally polarized omnidirectional antenna array, where the ZPSL loop antenna element with fork-shaped distributed capacitors is printed onto a FR4 substrate with thickness of 0.5 mm, dielectric constant of 4.4, and dielectric loss tangent of 0.02. The four radiators are connected to the outputs of a parallel line feeding network, respectively. The parallel line feeding network is etched on the opposite sides of a 0.8 mm thick RO4003 substrate with dielectric constant of 3.38 and dielectric loss tangent of 0.0023 and positioned inside the radiators, which makes the antenna array compact. The radius of the ZPSL loop is 18 mm, and the perimeter of the loop is about 0.92λ at 2.45 GHz.

Figure 12.34 shows the photograph of the 1×4 horizontally polarized omnidirectional antenna array prototype as well as the measured results. The measured reflection coefficient of -10 dB is achieved from 2.35 to 2.55 GHz. The measured gain is greater than 6 dBi over 2.35–2.55 GHz. Good omnidirectional radiation in E-plane is achieved at 2.45 GHz with an omnidirectionality of ± 2.3 dB.

12.4.3 CP omnidirectional antenna for UHF far-field RFID readers

Omnidirectional CP antennas have caught much attention because of their omnidirectional radiation patterns and CP property. The omnidirectional radiation

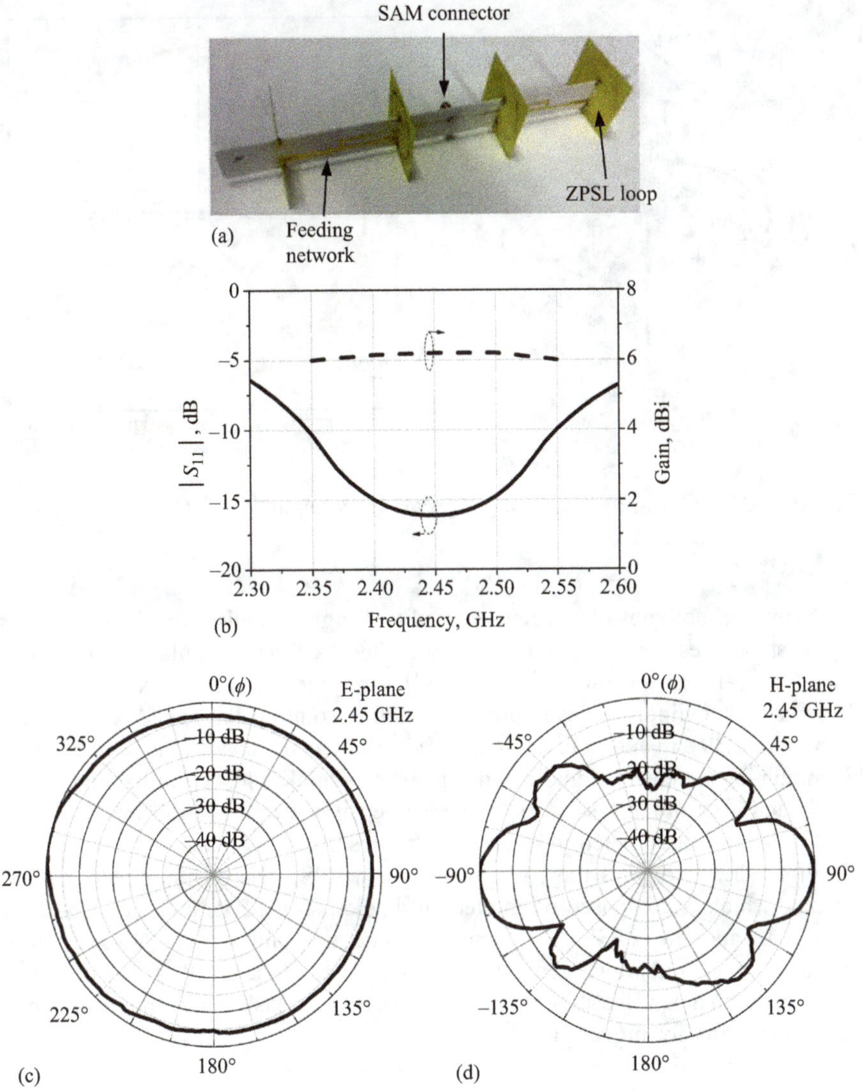

Figure 12.34 The horizontally polarized omnidirectional antenna array and measured results: (a) photograph of the antenna array prototype; (b) $|S_{11}|$ and gain; (c) E-plane radiation pattern at 2.45 GHz; and (d) H-plane radiation pattern at 2.45 GHz

enables one antenna covering a large area and thus simplify system configuration; the CP property can support a free alignment between the receiving and transmitting antennas. Therefore, the omnidirectional CP antenna is desirable for many wireless systems such as RFID, WLAN, global positioning systems, and so on.

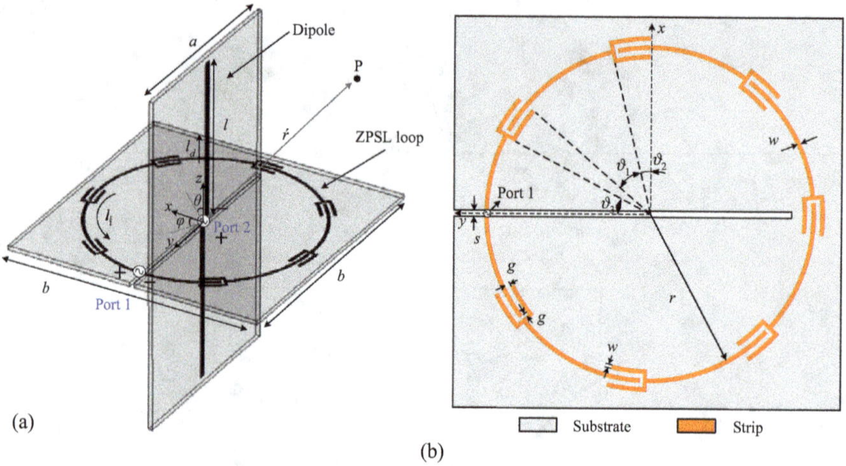

Figure 12.35 ZPSL-based CP omnidirectional antenna: (a) 3-D view and (b) ZPSL loop

Many technologies have been reported to design omnidirectional CP antennas over past decades. One way is to make the tilted half-wave dipole antennas to be wound helically in the same direction [38]. Arraying the directional CP antenna elements on a cylindrical plane provides another solution for omnidirectional CP radiation [38,39]. In addition, the omnidirectional CP antenna is able to be realized by combining an electric dipole and a magnetic dipole [40,41], utilization of the zeroth and first-order resonance modes of epsilon negative transmission lines [42,43], using slotted patch [44], or dielectric resonator [45], and so on.

Figure 12.35 depicts the configuration of a ZPSL-based omnidirectional CP antenna, which is composed of an electric dipole and a ZPSL loop. The in-phase fed dipole and the ZPSL loop generate vertically and horizontally polarized omnidirectional radiation, respectively, and a combined CP omnidirectional radiation. Such a configuration makes the antenna very easily to be implemented in an antenna array for higher gain.

Assume that the current distribution on the dipole is $I_d = I_{d_0}\sin[(2\pi/\lambda)\times(l\pm z)]$, and the ZPSL loop is ideal case wherein the current is uniform and in-phase i.e. at any position $I_l = I_{l_0}$, the far-field of the antenna is given by

$$\boldsymbol{E}_{\text{total}} = \boldsymbol{E}_d + \boldsymbol{E}_l$$

$$= \frac{j60[I_{d_0}]}{r'}\left[\frac{\cos[(2\beta l\cos\theta)/2] - \cos(\beta l)}{\sin\theta}\right]\hat{\boldsymbol{\theta}} + \frac{60\pi\beta r[I_{l_0}]}{r'}J_1(\beta r\sin\theta)\hat{\boldsymbol{\phi}} \quad (12.2)$$

where, $[I_{d_0}] = I_{d_0}e^{-j2\pi r'/\lambda}$, I_{d_0} is the peak value of the current on dipole, l is the length of arm of the dipole and $l = \lambda/4$, β is the propagation constant, r' is the distance between the origin and the observation point P, I_{l_0} is the amplitude of the current on

ZPSL loop, r is the radius of the ZPSL loop, and J_1 is the Bessel function of the first order and of argument ($\beta a \sin\theta$).

The far-fields of the dipole and the ZPSL loop are orthogonal with 90° phase difference when Ports 1 and 2 are fed identically, which meet the basic conditions of an omnidirectional CP radiation. The axial ratio (AR) of the ideal case can be given by

$$AR = \left| 20\log_{10}\left(\left| \frac{\{\cos[(2\beta l\cos\theta)/2] - \cos(\beta l)\}[I_{d_0}]}{\pi\beta r[I_{l_0}]J_1(\beta r\sin\theta)\sin\theta} \right| \right) \right| \tag{12.3}$$

Assume $I_{d_0} = I_{l_0}$, and considering $\theta = 90°$ to achieve good AR in the azimuth plane (xy-plane), (12.3) can be written as:

$$AR = \left| 20\log_{10}\left(\left| \frac{1}{\pi\beta r J_1(\beta r)} \right| \right) \right| = 0 \tag{12.4}$$

Then, βr can be worked out as about 0.8357, which suggests the initial radius r of the ZPSL loop is about 0.13λ or the perimeter of ZPSL loop is about 0.82λ.

The antenna shown in Figure 12.35 generates the right-hand CP (RHCP) radiation. The left-hand CP (LHCP) radiation can be generated by exchanging the positive/negative parts of the feeding point of the ZPSL loop to reverse the flowing direction of the current along the ZPSL loop.

Figure 12.36 shows a 1×4 CP omnidirectional antenna array at 915 MHz. The antenna array comprises four CP omnidirectional antenna elements as exhibited in Figure 12.35 and a parallel strip-line feeding network, which are all printed onto RO4003C PCB (dielectric constant of 3.38, dielectric loss tangent of 0.0023, thickness of 0.8128 mm). The upper two ZPSL loops are fed at the right side and the lower two are at the left side to reduce the effect of the feeding network on the omnidirectionality. Moreover, the excitations for upper two ZPSL loops and the lower two ZPSL loops should be out-of-phase to keep the currents flowing along all the ZPSL loops the same direction, which is achieved by two swaps on the 1:8 parallel strip-line power divider.

The dimensions of the CP omnidirectional antenna array at 915 MHz are $a =$ 70 mm, $b = 104$ mm, $l = 85.23$ mm, $r = 45$ mm, $w = 0.5$ mm, $g = 0.3$ mm, $s =$ 0.508 mm, $\theta_1 = 32.4°$, $\theta_2 = 13°$ and $\theta_3 = 37.3°$, $d = 0.5\lambda$. Figure 12.36(c) shows the -10-dB reflection coefficient is achieved from 900 to 935 MHz. Figure 12.37 shows that the axial ratio is less than 1.72 dB and the gain is between 3.38 dBic and 5.40 dBic in the azimuth plane at 915 MHz. The gain variation of the antenna is 2.02 dB, namely, the omnidirectionality is ± 1 dB. The simulated total efficiency of the antenna array is 78.2% at 915 MHz. As shown in Figure 12.38, the measured axial ratio is less than 3 dB over the frequency range from 893 MHz to 933 MHz, and the measured gain is between 5.14 and 5.4 dBic from 900 to 930 MHz. The radiation pattern in azimuth plane shown in Figure 12.39(a) exhibits small axial ratio; the maximum radiation is achieved at $\theta = 90$ in elevation plane with 3-dB axial ratio beamwidth of about 60°.

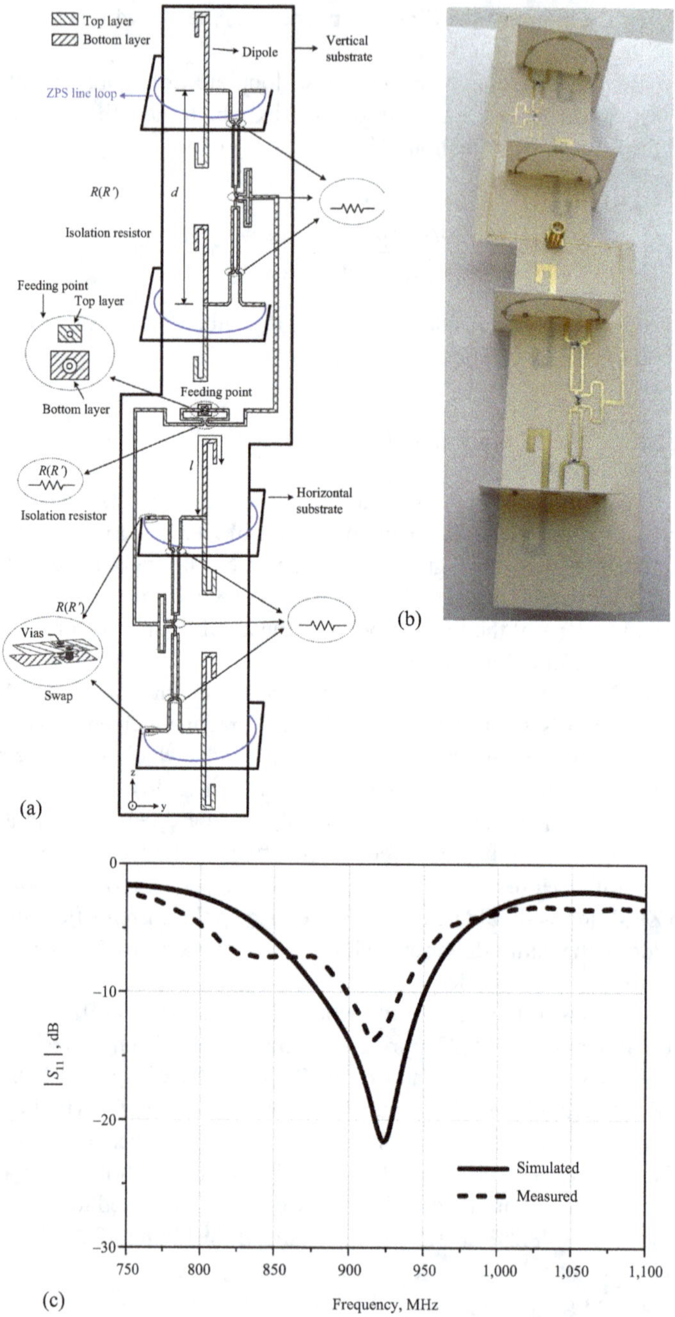

Figure 12.36 1 × 4 CP omnidirectional antenna array: (a) antenna array configuration; (b) photograph of the antenna array prototype; and (c) $|S_{11}|$ of the antenna prototype

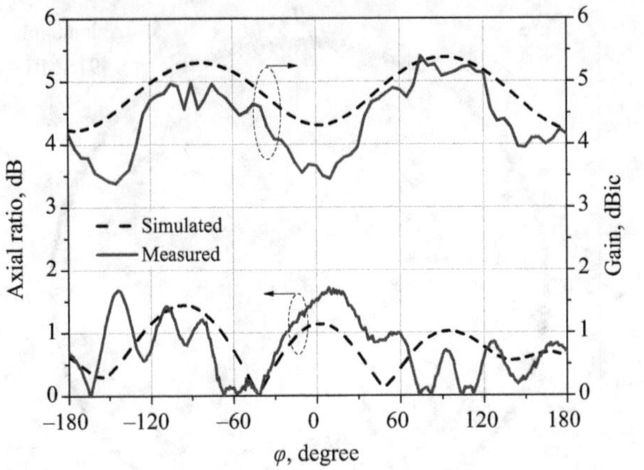

Figure 12.37 *Axial ratio and gain of the antenna array in the azimuth plane at 915 MHz*

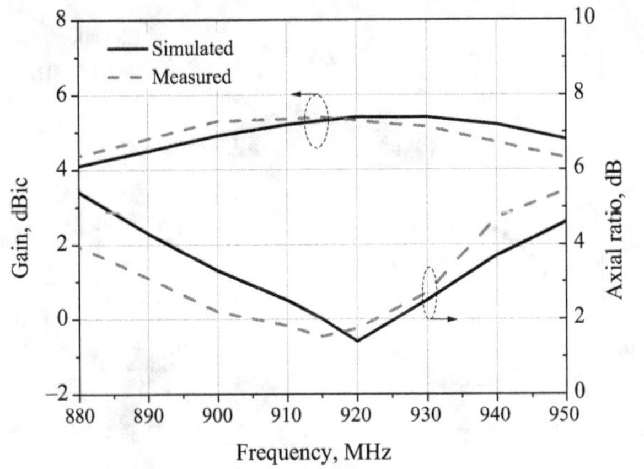

Figure 12.38 *The maximum axial ratio and gain of the antenna array in the azimuth plane*

12.5 Summary

The metamaterial-based ZPSL structures have exhibited unique dispersion characteristic which enables the flowing current along it featuring very small phase lag. This chapter has outlined the ZSPL configurations, the dispersion characteristic analysis approaches based on electromagnetic simulation and circuit model, and the design guidelines. A number of ZPSL loop antenna designs have been exemplified for different applications. The relevant discussions are believed to benefit the antenna researchers, engineers, and students to further understand of the ZPSL structure and development new ZPSL-based antennas.

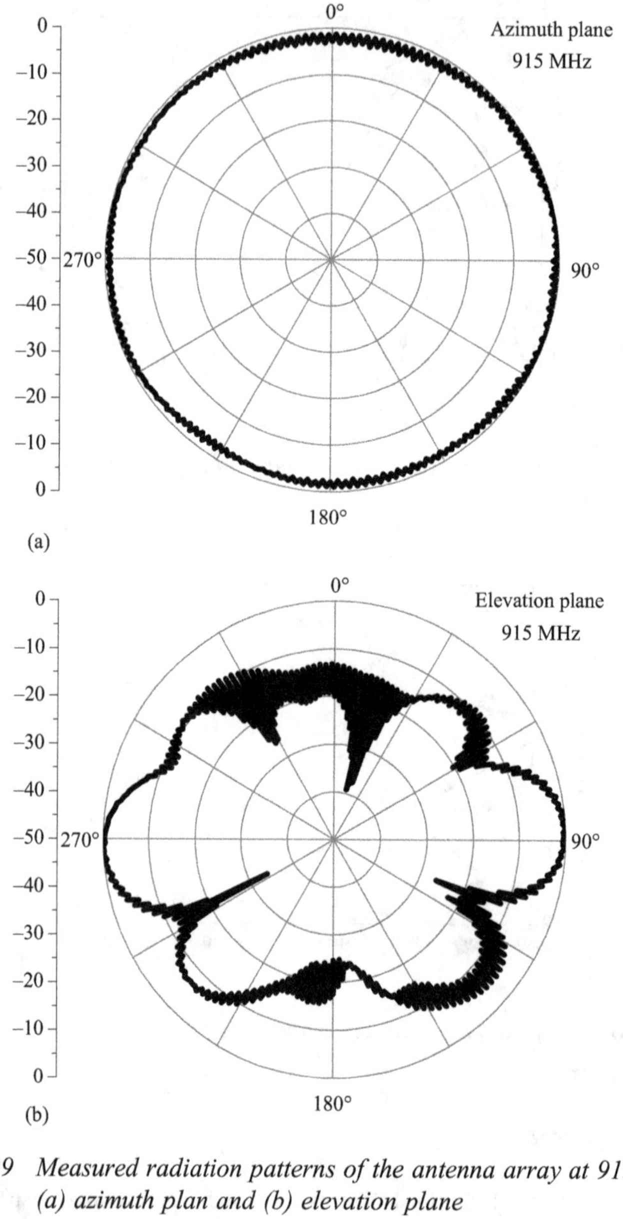

Figure 12.39 Measured radiation patterns of the antenna array at 915 MHz: (a) azimuth plan and (b) elevation plane

References

[1] V. Slyusar, "Metamaterials in antenna technology: a history and basic principles," *Electronics: Science, Technology, Business*, vol. 7, pp. 70–79, 2009.

[2] A. Sanada, C. Caloz, and T. Ito, "Novel zeroth-order resonance in composite right/left-handed transmission line resonators," *Asia-Pacific Microwave Conference*, pp. 1588–1591, Nov. 2003.

[3] C. Caloz, A. Sanada, and T. Itoh, "A novel composite right-/left-handed coupled-line directional coupler with arbitrary coupling level and broad bandwidth," *IEEE Trans. Microw. Theory Techn.*, vol. 52, no. 3, pp. 980–992, Mar. 2004.

[4] L. Liu, C. Caloz, and T. Itoh, "Dominant mode leaky-wave antenna with backfire-to-endfire scanning capability," *IET Electron. Lett.*, vol. 38, no. 2, pp. 1414–1416, Dec. 2002.

[5] X. Qing, Z. N. Chen, J. Shi, and C. K. Goh, "Zero-Phase-Shift Line Antennas," *IEEE International Workshop Antenna Technology*, pp. 179–182, Mar. 2013.

[6] D. M. Dobkin, S. M. Weigand and N. Iye, "Segmented magnetic antennas for near-field UHF RFID," *Microw. J.*, vol. 50, no. 6, Jun. 2007.

[7] R. A. Oliver, "Broken-loop RFID reader antenna for near field and far field UHF RFID tags," *U.S. Patent D574 369 S*, Aug. 2008.

[8] R. A. Oliver, "Broken-loop RFID reader antenna for near field and far field UHF RFID tags," *U.S. Patent D574 370 S*, Aug. 2008.

[9] R. A. Oliver, "Broken-loop RFID reader antenna for near field and far field UHF RFID tags," *U.S. Patent D570 337 S*, Jun. 2008.

[10] Y. S. Ong, X. Qing, C. K. Goh, and Z. N. Chen, "A segmented loop antenna for UHF near-field RFID," in *Proc. APSURSI*, Toronto, pp. 1–4, Jul. 2010.

[11] X. Qing, Z. N. Chen, and C. K. Goh, "UHF near-field RFID reader antenna with capacitive couplers," *IET Electron. Lett.*, vol. 46, no. 24, pp. 1591–1592, Dec. 2010.

[12] K. Wei, Z. Zhang, and Z. Feng, "Design of a wideband horizontally polarized omnidirectional printed loop antenna," *IEEE Antennas Wireless Propag. Lett.*, vol. 11, pp. 49–52, Jan. 2012.

[13] X. Qing, C. K. Goh, and Z. N. Chen, "Segmented loop antenna for UHF near-field RFID applications," *IET Electron. Lett.*, vol. 45, no. 17, pp. 872–873, Aug. 2009.

[14] X. Li, J. Liao, Y. Yuan, and D. Yu, "Segmented coupling eye-shape UHF band near field antenna design," in *Proc. IEEE Microw. Asia Pacific Conf.*, Singapore, pp. 2401–2404, Dec. 2009.

[15] X. Qing, C. K. Goh, and Z. N. Chen, "A broadband near-field UHFRFID antenna," *IEEE Trans. Antennas Propag.*, vol. 58, no. 12, pp. 3829–3838, Dec. 2010.

[16] J. Shi, X. Qing, Z. N. Chen, and C. K. Goh, "Electrically large dual loop antenna for UHF near-field RFID reader," *IEEE Trans. Antennas Propag.*, vol. 61, no. 3, pp. 1019–1025, Nov. 2013.

[17] J. Shi., X. Qing, and Z. N. Chen, "Electrically large zero-phase-shift line grid-array UHF near-field RFID reader antenna," *IEEE Trans. Antennas Propag.*, vol. 62, no. 4, pp. 2201–2208, Apr. 2014.

[18] Y. J. Zeng, Z. N. Chen, X. M. Qing, and J. M. Jin, "An artificial magnetic conductor backed zero-phase-shift line grid-loop antenna for near-field applications," *IEEE Trans. Antennas Propag.*, vol. 65, no. 4, pp. 1599–1606, Apr. 2017.

[19] X. Qing and Z. N. Chen, "A compact metamaterial-based horizontally polarized omnidirectional antenna array," *IEEE Intl. Symp. Antennas and Propagation*, (Orlando), pp. 1–4, Jul. 2013.

[20] J. Shi, X. Wu, X. M. Qing, and Z. N. Chen, "An ominidirectional circularly polarized antenna array," *IEEE Trans. Antennas Propag.*, vol. 64, no. 2, pp. 574–581, Feb. 2016.

[21] G. Goubau, "Surface waves and their application to transmission lines," *J. Appl. Phys.*, vol. 21, no. 11, pp. 1119–1128, Dec. 1950.

[22] G. V. Eleftheriades and K. G. Balmain, *Negative-Refraction Metamaterials: Fundamental Principles and Applications*, Hoboken, NJ: Wiley-IEEE Press, 2005.

[23] Y. J. Zeng, Z. N. Chen, X. Qing, and J. M. Jin, "Modeling and characterization of zero-phase-shift lines and optimization of electrically large ZPSL loop antennas for near-field systems," *IEEE Trans. Antennas Propag.*, vol. 64, no. 11, pp. 4587–4594, Nov. 2016.

[24] J.-P. Berenger, "A perfectly matched layer for the absorption of electromagnetic waves," *J. Comput. Phys.*, vol. 114, no. 2, pp. 185–200, Oct. 1994.

[25] X. Qing, and Z. N. Chen, "UHF near-field segmented loop antennas with enlarged interrogation zone," *IEEE International Workshop Antenna Technology*, pp. 132–135, Mar. 2012.

[26] S. Otto, A. Al-Bassam, A. Rennings, K. Solbach, and C. Caloz, "Transversal asymmetry in periodic leaky-wave antennas for Bloch impedance and radiation efficiency equalization through broadside," *IEEE Trans. Antennas Propag.*, vol. 62, no. 10, pp. 5037–5054, Oct. 2014.

[27] Y. Ding and R. Magnusson, "Band gaps and leaky-wave effects in resonant photonic-crystal waveguides," *Opt. Express*, vol. 15, no. 2, pp. 680–694, Feb. 2007.

[28] K. P. Wei, Z. J. Zhang, Z. H. Feng, and M. F. Iskander, "A MNG-TL loop antenna array with horizontally polarized omnidirectional patterns," *IEEE Trans. Antennas Propag.*, vol. 60, no. 6, pp. 2702–2710, Jun. 2012.

[29] C. Caloz and T. Itoh, *Electromagnetic Metamaterials: Transmission Line Theory and Microwave Applications*. Hoboken, NJ: Wiley-IEEE Press, 2005.

[30] S. Otto, A. Rennings, T. Liebig, C. Caloz, and K. Solbach, "An energy based circuit parameter extraction method for CRLH leaky wave antennas," in *Proc. 4th Eur. Conf. Antennas Propag.*, Barcelona, Spain, pp. 1–5, Apr. 2010.

[31] A. Bellion and M. Cable, "A new wideband and compact high impedance surface," in *Proc. Int. ANTEM. Symp.*, Toulouse, pp. 1–5, Apr. 2012.

[32] C. Soras, M. Karaboikis, G. Tsachtsiris, and V. Makios, "Analysis and design of an inverted-F antenna printed on a PCMCIA card for the 2.4 GHz ISM band," *IEEE Antennas Propag. Mag.*, vol. 44, no. 1, pp. 37–44, Feb. 2002.

[33] D. Chizhik, J. Ling, and R. A. Valenzuela, "The effect of electric field polarization on indoor propagation," *Proc. IEEE Int'l Conf. Universal Personal Comm.*, Florence, Italy, pp. 459–462, Oct. 1998.

[34] N. J. McEwan, R. A. Abd-alhameed, E. M. Ibrahim, P. S. Excell, and J. G. Gardiner, "A new design of horizontally polarized and dual-polarized uniplanar conical beam antennas for HIPERLAN," *IEEE Trans. Antennas Propag.*, vol. 51, no. 2, pp. 229–237, Feb. 2003.

[35] D. Wu, M. Zhao, Y. Fan, and Y. Zhang, "A wideband 8-element omnidirectional array for wireless system," *Microw. Opt. Tech. Lett.*, vol. 49, no. 12, pp. 2944–2946, Dec. 2007.

[36] H. Nakano, R. Satake, and J. Yamauchi, "Horizontally polarized, omnidirectional with a single feed," *IEEE Int'l Conf. Wireless Info Tech Systems*, Honolulu, HI, pp. 1–4, Aug. 2010.

[37] C. C. Lin, L. C. Kuo, and H. R. Chuang, "A horizontally polarized omnidirectional printed antenna for WLAN applications," *IEEE Trans. Antennas Propag.*, vol. 54, no. 11, pp. 3551–3556, Nov. 2006.

[38] V. Galindo and K. Green, "A near-isotropic circularly polarized antenna for space vehicles," *IEEE Trans. Antennas Propag.*, vol. 13, no. 6, pp. 872–877, Nov. 1965.

[39] K. Nakaym, T. Kawano, and H. Nakano, "A conformal spiral array antenna radiating an omnidirectional circularly-polarized wave," *IEEE Antennas Propag. Soc. Int. Sym.*, Orlando, FL, pp. 894–897, Jul. 1999.

[40] K. L. Wong, F. R. Hsiao, and C. L. Tang, "A low-profile omnidirectional circularly polarized antenna for WLAN access point," *IEEE Antennas Propag. Soc. Int. Sym.*, Monterey, CA, pp. 2580–2583, Jun. 2004.

[41] B. Li, S.-W. Liao, and Q. Xue, "Omnidirectional circularly polarized antenna combining monopole and loop radiators," *IEEE Antennas Wireless Propag. Lett.*, vol. 12, pp. 607–610, Apr. 2013.

[42] B. C. Park and J. H. Lee, "Omnidirectional circularly polarized antenna utilizing zeroth-order resonance of epsilon negative transmission line," *IEEE Trans. Antennas Propag.*, vol. 59, no. 7, pp. 2717–2721, Jul. 2011.

[43] B. C. Park and J. H. Lee, "Dual-band omnidirectional circularly polarized antenna using zeroth and first order modes," *IEEE Antennas Wireless Propag. Lett.*, vol. 11, pp. 407–410, Apr. 2012.

[44] A. Narbudowicz, X. L. Bao, and M. J. Ammann, "Omnidirectional circularly-polarized microstrip patch antenna," *IET Electron. Lett.*, vol. 48, no. 11, pp. 614–615, May 2012.

[45] W. W. Li and K. W. Leung, "Omnidirectional circularly polarized dielectric resonator antenna with top-loaded Alford loop for pattern diversity design," *IEEE Trans. Antennas Propag.*, vol. 61, no. 8, pp. 4246–4256, Aug. 2013.

Index

acrylonitrile butadiene styrene (ABS) lens design 357, 362, 377–81
DaD and ABS lenses, comparison of 381–4
results 378–81
acute myocardial infarction 304
additive manufacturing (AM) techniques 351
adenosine triphosphate (ATP) molecules 293
Alford loop 224
Altair FEKO 2, 5, 12, 17–18
characteristic mode simulations in 20
analog to digital converter (ADC) 315, 329
anisotropic flat lens 416–18
Ansoft's HFSS 432
apparent diversity gain (ADG) 114
array lenses: *see* transmitarrays
artificially engineered materials 357
designing higher-permittivity materials from low-permittivity COTS material 358–60
designing lower-permittivity materials
from high-permittivity 3D-printing material 360–1
from high-permittivity COTS material 360
artificial magnetic conductor (AMC) 244
-backed grid-ZPSL grid-loop array antenna 467–8
-based flexible wearable antennas 244, 247–9
-based ZPSL loop 446
artistic flexible antennas 260
atria 291
autism spectrum disorder (ASD) 338
autoregressive (AR) power spectrum estimation technique 335
axial ratio (AR) bandwidth 191, 196, 475
axon fiber 287

band-aid-type e-Nanoflex 329
bandwidth enhancement of platform-based antennas 15–31
barium strontium titanate (BSTO) 221
beam scanning 129, 138, 152–4, 163–5, 174, 179, 184, 188
beam-steerable liquid metal antennas 224–7
beam steering 174
benzocyclobutene (BCB) 204, 208–9
bicuspid valve 292
bioelectric events 287, 289
biopotential signals 282, 301, 312–15, 317, 320–1
Bluetooth 244–5, 309–10, 312, 324, 329
body area network (BAN) 282, 310–11
body sensor networks (BSN) 246–7
Bowtie antenna 398–401
brain 286–7
brain machine interface (BMI) 312, 323
branch power ratio (BPR) 112

cable model 294
cardiac muscle cells 292
cardiological signal measurements of diagnostic value 302–7
cardiovascular anatomy and electrophysiology 291
derivation of ECG from dipole vector 296–8
dipole theory for ECG 294–6
cardiovascular disorders 304
cardiovascular health monitoring 327
ECG signal acquisition 329–37
hardware system 328–9
cerebellum 286–7
cerebrum 286–7
channel capacity 110, 114–17
characteristic base pattern (CBP) 50–9

characteristic bases (CBs) 36, 64
-based procedure for placement of plurality of antennas 52
chassis excited by six dipoles 58–62
four microstrip patch antennas on an FR4 substrate 52–5
four PIFA antennas on FR4 substrate 57
topside of ship excited by monopoles 55–7
comparison between characteristic modes and, 65–72
generation of 64–5
pattern control using 50–2
characteristic basis function method (CBFM) 36–7
for locating antennas on mobile phone platforms 42–5
characteristic modes (CMs) 36–40
comparison between characteristic bases and, 65–72
generation of 64
characteristic mode theory 1–6
antenna design examples using 6
bandwidth enhancement of platform-based antennas 15–31
chassis-based MIMO antennas 12–14
circularly polarized antennas 7–11
wideband antennas 11–12
chassis-based MIMO antennas 12–14
Chu-McLean limit 16
circular beam-steering antenna 224
circularly polarized (CP) omnidirectional antenna for UHF far-field RFID readers 472–8
circularly polarized antennas 7–11, 31, 394, 402
circular polarization (CP) 97–8, 101, 189–90, 192, 244
classic antenna theory 414
cognitive radio front-ends 119–20
commercially off-the-shelf (COTS) 354, 359, 373
common mode rejection ratio (CMRR) 315, 328
composite right/left-handed (CRLH) based reconfigurable LWA 130, 136, 152–6
fixed-frequency beam scanning 152–5
with tunable beamwidth and radiation angle 155–6
composite right/left-handed transmission line (CRLH-TL) 135–7, 445
computer-aided design (CAD) tools 366
conductive fabrics 244, 284
conformal ground plane 99, 102
conformal wearable antennas 247, 250
constructive interference 400
conventional loop antennas 466
coplanar waveguide (CPW) antenna 217–18, 222
correlation coefficient (CC) 112
co-to-cross-pol ratio 90–3, 95
CST Microwave Studio 2, 161
cumulative distribution functions (CDF) 114

dendrites 287
dial-a-dielectric (DaD) lens design 355, 358, 362, 371
DaD and ABS lenses, comparison of 381–4
lens fabrication 377
results 377
dielectric liquid 205
-based reconfigurable antennas 222
reconfigurable antennas using 220–4
dielectric resonator antennas (DRAs) 122, 221, 251, 253–5
dipole theory for ECG 294–6
dipole vector, derivation of ECG from 296–8
diversity gain (DG) 113–14
double-layer AMC (D-AMC) 248
drilled-hole dielectric unit cells 422, 425
drowsiness 318
dry electrodes 314
dry textile-based nano-biosensor electrodes 304
dual band antennas 99, 102
dual-band wearable flexible antenna 246
dual polarization 98–9, 102

e-bra 334–6
effective DG (EDG) 114
effective relative permittivity 361, 363
E-field 12, 471
eigen-currents 2–4
eigen-values 2, 4
Einthoven's Triangle 302
electrical impedance tomography (EIT) technique 303–4, 327

electrical signals from the brain and heart 286–91
electrocardiography (ECG) 284, 293, 302, 327–8
  derivation from dipole vector 296–8
  dipole theory for 294–6
electroencephalogram (EEG) 287, 289–90, 298, 321, 323
  and its characteristics 300
  international 10–20 placement of electrodes for 299
electromagnetic bandgap (EBG) surface 244
electromyography (EMG) 301
electrooculogram (EOG) 299, 321
electrowetting 209
e-Nanoflex 328–9, 331, 334
enhancing field-isolation 116
enhancing port-isolation 115–16
envelope correlation coefficient (ECC) 112
epoxy 208, 214, 216
e-textile 245–6, 282
eutectic gallium – indium (EGaIn) 210, 212–14, 217–18
e-vest 332–6
evoked potentials 287
Expeditionary Fighting Vehicle (EFV) model 29–31

Fabry–Perot type of resonator 386
Fabry–Pérot (FP) antennas: *see* partially reflective surface (PRS) antennas
Faraday's constant 289
fiberonics technology 284
fifth-generation (5G) wireless systems 171
fish-eye lens 423–5
5G antenna systems 123
flexible and stretchable liquid metal antennas 210–13
flexible and wearable antennas 243
  AMC-based flexible wearable antennas 247–9
  inkjet-printed wearable antennas 249–50
  textile-based wearable antennas 250–71
  wearable antennas for biomedical applications 246–7
flexible liquid metal patch antenna 212
FlexPIFA antenna 243
fourth-generation (4G) wireless communication standard 109
fractal geometries 281
free-standing-aligned nanostructures 284
frequency and polarization reconfigurable cross-dipole antenna 219
frequency-reconfigurable dielectric resonator antenna (DRA) 221
frequency-reconfigurable liquid metal antennas 213–20
frequency-reconfigurable monopole antennas 214
frequency-reconfigurable PRS antenna 182–3
frequency-reconfigurable slot antenna 218, 220
frequency-scanning radars 129
frequency tunable CPW folded slot antenna 217
frequency-tunable monopole 215, 228
front-to-back ratio (FBR) 92, 95, 247
fused deposition modeling (FDM) 357, 366
fused filament fabrication (FFF) 3D-printing 352–3

Galinstan 210, 213, 217
geometrical optics (GO) 416
global optimization algorithms 412
global positioning system (GPS) antenna 35
Global System for Mobile Communication (GSM) 243
gold nanowire electrodes 314
gradient-refractive-index (GRIN) lens 354–5, 408
  MTM curved lens antennas 419–25
  MTM flat lens antenna 408–19
GSM 1900 244

half-loop antenna 22–3
half-width microstrip LWA (HW-MLWA) 137–8, 156–67
  multistate microcell 157–64
Harrington's approach 1
health monitoring, smart textile for 282–6
heart 291
heart rate variability (HRV) 335, 337–8
high-gain and low-sidelobe flat lens 410–14
high-gain beam-steering antenna arrays 225
high-gain flat lens 408–10

high-impedance surface (HIS) 132, 141, 146, 182
Holter monitoring system 307–8
Home Sleep Test (HST) 312
hypertrophic cardiomyopathy (HCM) 306

impedance-matching layers (IMLs) 409
Impinj button type tags 457
inductive near-field RFID systems, coupling mechanism of 457
inkjet-printed monopole antenna with AMC surface 246
inkjet-printed wearable antennas 249–50
intelligent textiles 283
intelligent wearable sensor systems 280
internet of things (IoT) 243
inverse problems 290

Jacobian matrix 427
Jacquard loop weaving 284

Ka-band 389
Kapton Polyimide 249
King Fahd University of Petroleum and Minerals (KFUPM) 109
Ku-band conical feed horn 364
Ku-band pyramidal horn antenna 367
Ku-band waveguide 436

Laplacian equation 297
large-scale MIMO 122
Law of Refraction 411
leaky-wave antennas (LWAs) 129
  classification of 131
    one-dimensional quasi-uniform LWA 132
    one-dimensional uniform LWA 132
    two-dimensional LWA 132
  experimental results 149–51
    CRLH-based reconfigurable LWA 152
    reconfigurable half-width microstrip LWA 156
  history of 130
    basic operating principle 130–1
  passive frequency-scanning LWA structures 132
    composite right/left-handed transmission line and LWA 135–7
    half-width microstrip LWA 137–8
    one-dimensional (1-D) Fabry–Pérot LWA 132–5
  reconfigurable LWAs
    1-D FP-reconfigurable LWAs 138–9
    two-dimensional (2-D) FP-reconfigurable LWA 146
left-hand circular polarization (LHCP) 192, 196
left-hand CP (LHCP) radiation 475
left-handed materials (LHMs): *see* metamaterials (MTMs)
lens designs
  using metamaterials (MTMs) 353–5
  using ray optics (RO) 355–6
linear polarization (LP) 189–90, 192
liquid crystal polymer (LCP) layer 214
liquid-metal free devices 205
liquid metal monopole array 216, 224
liquid metals 204, 210
loading-inside flat lens 414–16
loop antennas 456
Lorentz-like dispersive parameters 409
L-probe: *see* wideband L-probe patch antenna
Luneburg lens 419–23

macrocell, defined 160
magneto-electric dipole (MD-ME) concept 246
Makerbot Replicator 2X 3D-printer 357, 366
massive-MIMO (m-MIMO) 122
Maxwell-Garnett mixing formula 221
Maxwell's equations 270, 418
meandering probe (M-probe) 77–8
  feeding mechanism 83–91
meta-atoms (MTAs) 351–2
meta-atoms and artificially engineered materials 351
  3D-printing technique 356–7
  acrylonitrile butadiene styrene (ABS) lens design 362, 377
    results 378–81
  artificially engineered materials, design of 357
    designing higher-permittivity materials from low-permittivity COTS material 358–60

designing lower-permittivity materials from high-permittivity 3D-printing material 360–1
designing lower-permittivity materials from high-permittivity COTS material 360
DaD and ABS lenses, comparison of 381–4
dial-a-dielectric (DaD) lens design 358, 362, 371
lens fabrication 377
results 377
lens designs
using metamaterials (MTMs) 353–5
using ray optics (RO) 355–6
metal-only reflectarray antenna (MORA) designs using metasurfaces 386–94
metasurface superstrates
performance enhancement of antenna and array antennas using 394–402
polylactic acid (PLA) lens design 362
lens fabrication 365–7
lens measurement 367–8
results 368–71
metallic strip arrays 414
metallized plates 205
metal-only reflectarray antenna (MORA) designs using metasurfaces 386–94
metamaterial-based zero-phase-shift-line loop antennas 445
design and applications 456
CP omnidirectional antenna for UHF far-field RFID readers 472–8
electrically large zero-phase-shift-line loop antennas for UHF near-field RFID readers 456–70
horizontally polarized omnidirectional antenna for WLAN access points 470–2
state-of-the-art ZPSL loop antennas 446
design guidelines 452–6
dispersion analysis of zero-phase-shift-line structure 447–52
metamaterials (MTMs) 407
antennas, using transformation optics 425
MTM flattened convex and hyperbolic lenses 431–4
MTM flattened reflectors 427–31
MTM Luneburg lens with flattened focal surface 434–6
curved lens antennas 419
Fish-eye lens 423–5
Luneburg lens 419–23
flat lens antenna 408
anisotropic flat lens 416–18
high-gain and low-sidelobe flat lens 410–14
high-gain flat lens 408–10
loading-inside flat lens 414–16
lens designs using 353–5
metasurface antennas 436
coding metasurfaces 439–40
holographic metasurfaces for beam scanning 438
spoof SPP radiations 438–9
metasurface superstrates
performance enhancement of antenna and array antennas using 394–402
Metelics MSV34069-E28X varactors 155
microfluidically reconfigurable antennas 203
beam-steerable liquid metal antennas 224–7
fabrication and actuation techniques 205–9
flexible and stretchable liquid metal antennas 210–13
frequency-reconfigurable liquid metal antennas 213–20
reconfigurable antennas using dielectric liquids 220–4
reconfigurable antennas using microfluidically repositionable metallized plates 227–36
microfluidic beam-steering focal plane array 227, 230–3, 235
microfluidic channels 204, 217
microfluidics-based reconfiguration principles 204
microstrip antenna design 243
microstrip leaky-wave antennas 130
microwave antennas based on metamaterials and metasurfaces 407
GRIN MTM lens antennas 408
MTM curved lens antennas 419–25
MTM flat lens antenna 408–19
metasurface antennas 436

coding metasurfaces 439–40
holographic metasurfaces for beam scanning 438
spoof SPP radiations 438–9
MTM antennas using transformation optics 425
MTM flattened convex and hyperbolic lenses 431–4
MTM flattened reflectors 427–31
MTM Luneburg lens with flattened focal surface 434–6
mitral/bicuspid valve 292
mm-wave beam-steering array 225
mobile phones and handheld devices 116–19
modal excitation coefficient 4
modal significance (MS) 5, 7–8, 10
motorized gimbaled antenna systems 225
multi-band MIMO antenna systems 118
multiple antennas on a complex platform 45
CB-based approach 50–2
illustrative examples 52
chassis excited by six dipoles 58–62
four microstrip patch antennas on an FR4 substrate 52–5
four PIFA antennas on FR4 substrate 57
topside of ship excited by monopoles 55–7
TCM-based approach 45–9
multiple-input-multiple-output (MIMO) antenna systems 109
design challenges 115–16
examples 116
cognitive radio front-ends 119–20
mobile phones and handheld devices 116–19
USB dongle MIMO implementations 120
wireless access point MIMO implementations 121–2
for 5G-enabled systems 122
base station 5G solutions 122–3
mobile terminal 5G solutions 122
performance metrics 110–15
multiplexing techniques 109
music-based therapy 337–8
Mu waves 339–40
myocytes 292

nano-biosensor electrode system 325
nanostructured electrode surface 285
nanostructured textile-based electrodes 314
nanotextile-based biosensors 286
nanotextile-based wireless biosensor systems 281
near-field RFID systems 456
negative (refractive) index materials: *see* metamaterials (MTMs)
Nernst voltage 289
nerve cell 287–8
neurological disorder 301
neurological disorder monitoring by wearable wireless nano-bio-textile sensors 311–27
neurological signal measurements 298–302
Nicolson-Ross and Weir (NRW) method 365
nontoxic liquid metals 209
Nyquist–Shannon sampling theorem 427

omnidirectional CP antennas 472
one-dimensional (1-D) Fabry–Pérot LWA 132–5, 138–9
analysis on reconfigurable FP LWA 139–42
antenna prototype and measured results 142–6
one-dimensional quasi-uniform LWA 132
one-dimensional uniform LWA 132
open-circuit (O.C.) condition 226–7
optical transformation 425–6
orthogonal frequency division multiplexing 109
oxide skin 210
oxygen-depleted blood 291

parasympathetic nervous system 293
partially reflective surface (PRS) antennas 132, 134, 140, 172
reconfigurable PRS antennas 182–96
pattern-reconfigurable PRS antenna 183–9
perfect electric conductor (PEC) 135, 243
perfect magnetic conductor (PMC) 244
peristaltic pumps 209
phased array antennas 174
phased arrays
disadvantages of 178
using RDMS phase shifters 177

phase-sensitive demodulator 303–4
photoplethysmography arm band 332
piezoelectric pumps 209
PIN diodes 164, 185
planar-inverted F-antenna (PIFA) 243
planar square monopole antenna 11
platform-based antennas, bandwidth enhancement of 15–31
platform-mounted antennas 23, 29
  simulated normalized current distributions and radiation patterns 23
point-of-care (POC) 279, 327
Poisson's equation 290
polarization reconfigurable antennas 205
polarization reconfigurable crossed microstrip patch antenna 223
polarization-reconfigurable PRS antenna 189–96
polydimethylsiloxane (PDMS) 204, 206–8, 211, 222–3
polylactic acid (PLA) lens design 357, 362
  lens fabrication 365–7
  lens measurement 367–8
  results 368–71
polysomnography (PSG) 312
polytetrafluoroethylene (PTFE) tubes 216
port isolation 111
power spectral density (PSD) 325
printed circuit board (PCB) 100, 103, 132, 204, 209, 355, 375, 381, 421
proper characterization 116
propylene glycol monomethyl ether acetate (PGMEA)-based solution 207
proximity-coupled textile-based wearable antenna 246
Purkinje fiber cells 293–4
P-wave 294

QT interval dispersion 306
quality factor 232
quasi-conformal mapping 427

radiated energy 19
radio frequency (RF) devices 203
ray optics (RO), lens designs using 355–6
reconfigurable and uniform LWAs comparison 164–7
reconfigurable antennas (RAs) 171–2
  using dielectric liquids 220–4
  using microfluidically repositionable metallized plates 227–36
reconfigurable-defected microstrip structure (RDMS) 176–7, 186
reconfigurable half-width microstrip LWA 156
  analysis of 1-D reconfigurable structure using multistate macrocell 157–61
  analysis of the LWA using macrocell states 162–4
  antenna design using binary unit cells 161–2
  comparison of reconfigurable LWA with uniform LWA 164
  measured results 164–7
reconfigurable high-gain antennas for wireless communications 171
  reconfigurable array antennas 172–81
  reconfigurable PRS antennas 182
    frequency-reconfigurable PRS antenna 182–3
    pattern-reconfigurable PRS antenna 183–9
    polarization-reconfigurable PRS antenna 189–96
reconfigurable LWAs 138
  1-D FP-reconfigurable LWAs 138–9
  two-dimensional (2-D) FP-reconfigurable LWA 146
reflectarray (RA) 180, 386
refractive index distribution 409
remote patient monitoring (RPM) 279, 308
resonator length 232
reversibly frequency-reconfigurable crossed dipole antenna 218
right-hand circular polarization (RHCP) 192, 196
R-peak detection algorithm 335

sample under test (SUT) 256
screen printing 322
selectively metallized plate 207, 209, 234–6
semiconductor and ferroelectric varactors 203
sensor electronics module (SEM) 332
serial port profile (SPP) 329
Sheffield Mark 1 304
sidelobe levels (SLLs) 370
Silicone TC5005 elastomer 213

siloxane 206
single- and multi-layer multi-Bowtie conformal antennas 250–3
single-layer AMC (S-AMC) 248
single-loop ZPSL antenna 459
single-neuron bioelectric events 287
singular value decomposition (SVD) 50
sinoatrial node 293
skin–electrode interface 285–6
sleep disorders 312
smart garments 280
smart textile for health monitoring 282–6
soma 287
spontaneous activity 287
square patch antenna 8–9
standing wave ratio (SWR) 77–9
straight resonant feed network 231–2
stretchable frequency tunable liquid metal patch antenna 212
substrate integrated waveguide (SIW) 130
sudden cardiac arrest (SCA) 304
sudden cardiac death (SCD) 304
surface plasmon polaritons (SPPs) 438–9
Sylgard 184 kit (Dow Corning) 207
sympathetic nervous system 293
system integration 115

Tajima machine 265–6
textile-based nano-biosensor system 282, 308, 337
textile-based sensor systems 282
textile-based wearable antennas 250
  dielectric resonator antennas for wearable application 253–5
  single- and multi-layer multi-Bowtie conformal antennas 250–3
  wearable artistic antennas for WLAN-band 255
    applications and art gallery exhibition 270
    art gallery exhibition 270
    design challenges 260–2
    motivation 257–9
    simulated and measured results 266–9
    simulation models, prototyping, and measurement 262–5
    Tajima machine used for prototyping 265–6
textile-integrated sensors systems 311, 332
textronics 283–4
theory of characteristic modes (TCM) 35
  analysis of mobile phone antenna and antenna-plus-platform 73–4
  to designing antennas for mobile phone platforms 37–41
therapeutics, biofeedback system for 337–40
thermoplastic polylactic acid (PLA) 357
three-dimensional (3D) PBS 416
3D-printing technique 355–7, 381
T-network switching concept 234
total active reflection coefficient (TARC) 111
transcranial direct current stimulation (tDCS) 301
transcranial magnetic stimulation (TMS) 301
transmembrane potential (TP) difference 292
transmitarrays 180
transverse equivalent network (TEN) 132
transverse resonance equation (TRE) 134
transverse resonance method (TRM) 139
trapezoidal patch antenna 9–10
tricuspid valve 291
tunable reflectarrays 178
T-wave alternans (TWA) 305
T-wave inversion (TWI) 306
two-dimensional (2-D) FP reconfigurable LWA 146
  antenna configuration 146
  dispersion and structure 146–9
  experimental results 149–52
two-dimensional LWA 132

UHF near-field RFID technology 456–7
ultra-high frequency (UHF) 446
ultra-wide-band (UWB) 119–20, 211, 245
universal asynchronous receive/transmit (UART) interface 315
USB dongle MIMO implementations 120

Vagus nerve 293
varactor diodes 181
varactor-loaded devices 203
ventricles 291
ventricular arrhythmias 304
voltage standing wave ratio (VSWR) 16, 79–80, 85, 251–2

volume conductor 289
volume source 289

waveguide-based LWA structures 130, 132
wearable antennas for biomedical applications 246–7
wearable artistic antennas for WLAN-band 255
  applications and art gallery exhibition 270
  art gallery exhibition 270
  design challenges 260–2
  motivation 257–9
  simulated and measured results 266–9
  simulation models, prototyping, and measurement 262–5
  Tajima machine used for prototyping 265–6
wearable nano-biosensor systems 323
wearable technology and mobile platform 279
  biofeedback system for therapeutics 337–40
  cardiovascular anatomy and electrophysiology 291
    derivation of ECG from dipole vector 296–8
    dipole theory for ECG 294–6
  cardiovascular health monitoring 327
    ECG signal acquisition 329–37
    hardware system 328–9
  electrical signals from the brain and heart 286–91
  monitoring and diagnosis
    cardiological signal measurements of diagnostic value 302–7
    neurological signal measurements 298–302
  monitoring systems 307–11
  neurological disorder monitoring by wearable wireless nano-bio-textile sensors 311–27
  smart textile for health monitoring 282–6
wearable wireless brain-machine interface 324
wearable wireless nano-bio-textile sensors
  neurological disorder monitoring by 311–27
wideband antennas 11–12
wideband L-probe patch antenna 77
  characteristics 78
    L-probe feeding mechanism 78–82
    M-probe feeding mechanism 83–91
  L-probe and M-probe fed patch antenna, development of 97
    circular polarization 97–8, 101
    conformal ground plane 99, 102
    dual band 99, 102
    dual polarization 98–9, 102
    fusion of 100–3
    printed circuit board (PCB) 100, 103
  parametric studies 91
    performance with different aspect ratio 93–7
    performance with different P_h 91–3
Wi-Fi-based systems 309
Wi-Fi-Bluetooth integrated systems 309
wireless access point MIMO implementations 121–2
wireless body area network (WBAN) 248
wireless communication electronics 281
wireless local area network (WLAN) systems 172, 182, 309
wireless personnel area network (WPAN) 312
wireless telemedicine system 314, 317–18
wireless wearable textile-based nano-biosensor systems 318
WLAN access points
  horizontally polarized omnidirectional antenna for 470–2
WLAN-band, wearable artistic antennas for 255–71
WLAN frequency bands 244
worldwide Interoperability for Microwave Access (WiMAX) bands 248

Yagi–Uda antenna design 224
Yagi–Uda monopole arrays 216

zero-phase-shift-line (ZPSL) 445
  design guidelines 452–6
  dispersion analysis of 447
    driven-mode analysis with loop configuration 449–50

Eigen-mode analysis 447–9
equivalent circuit 451–2
electrically large ZPSL loop antennas for UHF near-field RFID readers 456–70
AMC-backed directional ZPSL loop antenna 466–70
dual-loop antenna 458–63
grid-loop array antenna 463–6
single-loop antenna 457–8
ZigBee 309, 316, 333

www.ingramcontent.com/pod-product-compliance
Lightning Source LLC
LaVergne TN
LVHW011551200926
848484LV00001B/302